T0141909

Advances in Intelligent Systems and Computing

Volume 1043

The series "Advances in Intelligent Systems and Computing" contains publications on theory, applications, and design methods of Intelligent Systems and Intelligent Computing. Virtually all disciplines such as engineering, natural sciences, computer and information science, ICT, economics, business, e-commerce, environment, healthcare, life science are covered. The list of topics spans all the areas of modern intelligent systems and computing such as: computational intelligence, soft computing including neural networks, fuzzy systems, evolutionary computing and the fusion of these paradigms, social intelligence, ambient intelligence, computational neuroscience, artificial life, virtual worlds and society, cognitive science and systems, Perception and Vision, DNA and immune based systems, self-organizing and adaptive systems, e-Learning and teaching, human-centered and human-centric computing, recommender systems, intelligent control, robotics and mechatronics including human-machine teaming, knowledge-based paradigms, learning paradigms, machine ethics, intelligent data analysis, knowledge management, intelligent agents, intelligent decision making and support, intelligent network security, trust management, interactive entertainment, Web intelligence and multimedia.

The publications within "Advances in Intelligent Systems and Computing" are primarily proceedings of important conferences, symposia and congresses. They cover significant recent developments in the field, both of a foundational and applicable character. An important characteristic feature of the series is the short publication time and world-wide distribution. This permits a rapid and broad dissemination of research results.

** Indexing: The books of this series are submitted to ISI Proceedings, EI-Compendex, DBLP, SCOPUS, Google Scholar and Springerlink **

More information about this series at http://www.springer.com/series/11156

Zhaojie Ju · Longzhi Yang ·
Chenguang Yang · Alexander Gegov ·
Dalin Zhou
Editors

Advances in Computational Intelligence Systems

Contributions Presented at the 19th UK
Workshop on Computational Intelligence,
September 4–6, 2019, Portsmouth, UK

 Springer

Editors
Zhaojie Ju
School of Computing
University of Portsmouth
Portsmouth, UK

Chenguang Yang
Bristol Robotics Laboratory
University of the West of England
Bristol, UK

Dalin Zhou
School of Computing
University of Portsmouth
Portsmouth, UK

Longzhi Yang
Department of Computer and Information
Sciences
Northumbria University
Newcastle upon Tyne, UK

Alexander Gegov
School of Computing
University of Portsmouth
Portsmouth, Hampshire, UK

ISSN 2194-5357   ISSN 2194-5365 (electronic)
Advances in Intelligent Systems and Computing
ISBN 978-3-030-29932-3   ISBN 978-3-030-29933-0 (eBook)
https://doi.org/10.1007/978-3-030-29933-0

This Springer imprint is published by the registered company Springer Nature Switzerland AG
The registered company address is: Gewerbestrasse 11, 6330 Cham, Switzerland

Preface

Welcome to UKCI 2019, the 19th Annual UK Workshop on Computational Intelligence held at the University of Portsmouth. UKCI series has been one of the most important academic meetings in the field of computational intelligence in the UK and continuously supported by Springer. This year, UKCI 2019 is held in conjunction with EDMA-2019, the 3rd International Engineering Data- and Model-Driven Applications Workshop.

UKCI 2019 covers both theory and applications in computational intelligence systems and has been most successful in attracting a total of 63 submissions addressing the state-of-the-art development and research covering topics related to fuzzy systems, intelligence in robotics, deep learning approaches, optimisation and classification, detection, inference and prediction, hybrid methods, emerging intelligence, intelligent healthcare and engineering data- and model-driven applications. Following the rigorous reviews of the submissions, a total of 45 papers (71% acceptance rate) were selected to be presented in the conference during 6–8 September 2019. We hope that the published papers of UKCI 2019 will prove to be technically constructive and helpful to the research community of computational intelligence.

We would like to express our sincere acknowledgement to the attending authors and the 5 distinguished plenary speakers. The acknowledgement is also given to the UKCI 2019 programme committee for their efforts in the rigorous reviewing process. Special thanks are extended to Dalin Zhou in appreciation of his contribution to the organisation throughout UKCI 2019. Last but not least, the help from Rini Christy and Thomas Ditzinger of Springer is appreciated for the publishing.

We hope that you enjoy the academic discussions and the waterfront city of Portsmouth.

Organisation

Programme Committee

Giovanni Acampora	University of Naples Federico II, Italy
Khulood Alyahya	University of Exeter, UK
Peter Andras	Keele University, UK
Plamen Angelov	Lancaster University, UK
Atta Badi	University of Reading, UK
Abdelhamid Bouchachia	Bournemouth University, UK
Fei Chao	Xiamen University, China
Tianhua Chen	University of Huddersfield, UK
Mihaela Cocea	University of Portsmouth, UK
George Coghill	University of Aberdeen, UK
Sonya Coleman	University of Ulster, UK
Simon Coupland	De Montfort University, UK
Alexandra Cristea	Durham University, UK
Keeley Crockett	Manchester Metropolitan University, UK
Jon Garibaldi	University of Nottingham, UK
Alexander Gegov	University of Portsmouth, UK
Hani Hagras	University of Essex, UK
Hongmei He	Cranfield University, UK
Chris Hide	Loughborough University, UK
Xia Hong	University of Reading, UK
Thomas Jansen	Aberystwyth University, UK
Robert John	University of Nottingham, UK
Zhaojie Ju	University of Portsmouth, UK
Dermot Kerr	University of Ulster, UK
Ahmed Kheiri	Lancaster University, UK
Naoyuki Kubota	Tokyo Metropolitan University, Japan
Caroline Langensiepen	Nottingham Trent University, UK
Miqing Li	University of Birmingham, UK

Alberto Cabri	University of Genoa, Italy
David Richardson	Jaguar Land Rover, UK
Unal Yildirim	University of Bradford, UK

Contents

Deep Learning Approaches

Optimisation and Classification

Fuzzy Systems

Fuzzy Control of Uncertain Nonlinear Systems with Numerical Techniques: A Survey

Raheleh Jafari[1]([✉]), Sina Razvarz[2], Alexander Gegov[3], and Wen Yu[2]

[1] Centre for Artificial Intelligence Research (CAIR), University of Agder, Grimstad, Norway
raheleh.jafari@uia.no
[2] Departamento de Control Automático, CINVESTAV-IPN (National Polytechnic Institute), Mexico City, Mexico
srazvarz@yahoo.com, yuw@ctrl.cinvestav.mx
[3] School of Computing, University of Portsmouth, Buckingham Building, Portsmouth PO1 3HE, UK
alexander.gegov@port.ac.uk

Abstract. This paper provides an overview of numerical methods in order to solve fuzzy equations (FEs). It focuses on different numerical methodologies to solve FEs, dual fuzzy equations (DFEs), fuzzy differential equations (FDEs) and partial fuzzy differential equations (PFDEs). The solutions which are produced by these equations are taken to be the controllers. This paper also analyzes the existence of the roots of FEs and some important implementation problems. Finally, several examples are reviewed with different methods.

Keywords: Fuzzy equations · Solutions · Numerical methods

1 Introduction

Fuzzy numbers have been used in many studies to deal with uncertainties in recent years [1–16]. Several approaches used the parametric form of fuzzy numbers for uncertainties in crisp dynamic systems [17]. In [18], the extension principle is implemented and suggests that the coefficients can be real or complex fuzzy numbers, where the system uncertainties are represented by fuzzy coefficients.

FEs are the equations whose parameters can be changed from the form of the fuzzy set [19]. The solutions of the FEs can be applied directly for modeling and nonlinear control. The results of the feedback control in reference to the wave equation is described in [20], while the open loop control corresponding to the wave equation is shown in [21]. However, the solutions are not easy to obtain. [22] introduced Newton's method. In [23], the fixed-point technique is used in order to obtain the solution of FEs. The numerical solutions of the FEs can be obtained applying the iterative method [24], the interpolation method [25]

© Springer Nature Switzerland AG 2020
Z. Ju et al. (Eds.): UKCI 2019, AISC 1043, pp. 3–14, 2020.
https://doi.org/10.1007/978-3-030-29933-0_1

and Runge-Kutta method [26]. It can also be implemented to the differential equations. [27] portrays Euler's numerical technique methodology to solve the FDE. The extension of classical fuzzy set theory in [28] results in obtaining a numerical solution of FDE. The wave solutions associated with two nonlinear PDE systems have been investigated in [29]. In [30] a static neural network is suggested to solve FDE. In [31] neural network is used in order to obtain the solution of the ordinary differential equation (ODE).

In this work, a survey is presented on the numerical solutions of the FEs, DFEs, FDE, and PFDE. The privileges of numerical techniques regarded to the precision are explained. The study of prior works demonstrates detail explanations in order to obtain numerical solutions for these equations. This paper is structured as follows. In Sect. 2, some basic definitions and notions used in the rest of the paper are given. In Sect. 3, some numerical methods in order to find the numerical solutions of the FEs and the DFEs are presented. Section 4 describes some numerical methods in order to find the numerical solutions of the FDEs and PFDEs. In Sect. 5, some examples are provided with different methods in order to compare the efficiency of the numerical methods to approximate the solution of DFEs and FDEs. Conclusions are included in Sect. 6.

2   Mathematical Preliminaries

The following definitions are used in this paper.

Definition 1. *If x is: (1) normal, there exists $\zeta_0 \in R$ in such a manner that $x(\zeta_0) = 1$; (2) convex, $x[\lambda\zeta + (1-\lambda)\xi] \geq \min\{x(\zeta), x(\xi)\}, \forall \zeta, \xi \in R, \forall \lambda \in [0,1]$; (3) upper semi-continuous on R, $x(\zeta) \leq x(\zeta_0) + \varepsilon, \forall \zeta \in N(\zeta_0), \forall \zeta_0 \in R, \forall \varepsilon > 0$, $N(\zeta_0)$ is a neighborhood; or 4) $x^+ = \{\zeta \in R, x(\zeta) > 0\}$ is compact, then x is a fuzzy variable, and the fuzzy set is defined as E, $x \in E : R \to [0,1]$.*

The fuzzy variable x can also be represented as

$$x = A(\underline{x}, \bar{x}) \tag{1}$$

where $\underline{x}$ is the lower-bound variable, $\bar{x}$ is the upper-bound variable, and A is a continuous function. The membership functions are utilized to implicate the fuzzy variable x. The best known membership function is the triangular function

$$x(\zeta) = F(a, b, c) = \begin{cases} \frac{\zeta - a}{b - a} & a \leq \zeta \leq b \\ \frac{c - \zeta}{c - b} & b \leq \zeta \leq c \\ 0 & otherwise \end{cases} \tag{2}$$

Definition 2. *A fuzzy number x associates with a real value with α-level as*

$$[x]^\alpha = \{a \in R : x(a) \geq \alpha\} \tag{3}$$

where $0 < \alpha \leq 1$, $x \in E$.

If $x, y \in E$, $\lambda \in R$, the fuzzy operations are as follows:
Sum,

$$[x \oplus y]^\alpha = [x]^\alpha + [y]^\alpha = [\underline{x}^\alpha + \underline{y}^\alpha, \bar{x}^\alpha + \bar{y}^\alpha] \tag{4}$$

subtract,

$$[x \ominus y]^\alpha = [x]^\alpha - [y]^\alpha = [\underline{x}^\alpha - \underline{y}^\alpha, \bar{x}^\alpha - \bar{y}^\alpha] \tag{5}$$

or multiply,

$$\underline{z}^\alpha \leq [x \odot y]^\alpha \leq \bar{z}^\alpha \text{ or } [x \odot y]^\alpha = A\left(\underline{z}^\alpha, \bar{z}^\alpha\right) \tag{6}$$

where $\underline{z}^\alpha = \underline{x}^\alpha y^1 + \underline{x}^1 y^\alpha - \underline{x}^1 y^1$, $\bar{z}^\alpha = \bar{x}^\alpha \bar{y}^1 + \bar{x}^1 \bar{y}^\alpha - \bar{x}^1 \bar{y}^1$, and $\alpha \in [0,1]$.

Therefore, $[x]^0 = x^+ = \{\zeta \in R, x(\zeta) > 0\}$. Since $\alpha \in [0,1]$, $[x]^\alpha$ is a bounded interval such that $\underline{x}^\alpha \leq [x]^\alpha \leq \bar{x}^\alpha$. The α-level of x between $\underline{x}^\alpha$ and $\bar{x}^\alpha$ is given as

$$[x]^\alpha = A\left(\underline{x}^\alpha, \bar{x}^\alpha\right) \tag{7}$$

3   Numerical Techniques for Solving Fuzzy Equations

In most cases, the analytical solution for FEs may not be found. Instead of using analytical methods that are not suitable for solving FEs and DFEs, numerical methods are proposed to solve these equations. In this section, we describe three numerical methods that are among the most important techniques.

3.1   Newton Method

In 1671, Isaac Newton introduced a novel algorithm [32] for solving a polynomial equation which was demonstrated on the basis of an example as $x^3 - 2x - 5 = 0$. To obtain a precise root of the mentioned equation, initially, a starting value should be assumed, here $x \approx 2$. By assuming $x = 2 + q$ and substituting it into the original equation, the result is obtained as $q^3 + 6q^2 + 10q - 1 = 0$. As q is presumed to be minute, $q^3 + 6q^2$ is neglected in comparison with $10q - 1$. The previous equation generates $q \approx 0.1$, so a superior approximation of the root is $x \approx 2.1$. The repetition of this process is feasible and $q = 0.1 + a$ is extracted. The substitution gives $a^3 + 6.3a^2 + 11.23a + 0.061 = 0$, henceforth $a \approx -0.061/11.23 = -0.0054...$, so a novel approximation of the root is $x \approx 2.0946$. It is the requirement to repeat the process till the expected number of digits is achieved. In his methodology, Newton did not explicitly utilize the concept of the derivative, he just applied it on polynomial equations.

In [33] Newton's methodology is proposed for fuzzy nonlinear equations in lieu of standard analytical methodologies, as they are not appropriate everywhere. The primary intention is to extract a solution for fuzzy nonlinear equation $G(x) = b$. Primarily the cited researchers have mentioned fuzzy nonlinear equation in parametric form as illustrated below

$$\begin{cases} \underline{G}(\underline{x}^\alpha, \bar{x}^\alpha) = \underline{b}^\alpha \\ \bar{G}(\underline{x}^\alpha, \bar{x}^\alpha) = \bar{b}^\alpha \end{cases}$$

so they resolved it by utilizing Newton's methodology. Also, the convergence of the method is proved.

Newton's method is relatively expensive, since the calculation of the Hessian on the first iteration is needed. Accordingly, the analytic expression for the second derivative is often complicated or intractable, requiring a lot of computation.

3.2   Genetic Algorithm Method

A genetic algorithm is shown to solve the FE, $R(x) = y$ in [34], where x and y are considered real fuzzy numbers sampled with k, it is also considered that R is a fuzzy function relying on x. The motivation is to obtain an adequate value of the fuzzy argument x in such a way that the calculated value of the polynomial, $R(x)$, is very adjacent to the provided objective value y. The genetic algorithm presented uses a different demonstration of the fuzzy numbers that allows the implementation of simple genetic operators. The algorithm is self-sufficient to find multiple solutions associated with FEs. Unfortunately, no method has been used for an identical problem involved in the area of neural networks that can be taken over. Due to the different discrete criteria of fuzzy arithmetic, the only realistic approach to solve this problem is to design a dedicated genetic algorithm [35].

In [36] genetic algorithm is used for resolving non-linear equations of the form $g(x) = 0$, where x and $g(x)$ may be real, complex or vector quantities. At first $g(x) = 0$ is transformed into a minimization problem, then genetic algorithm is applied for finding the minimum. The method is extended for systems of non-linear equations.

The genetic algorithm represents the most consistent results in terms of accuracy and convergence but it is expensive in computational costs.

3.3   Neural Network Method

Approximation methods such as fuzzy neural networks are also effective tools to overcome the limitations of the other numerical methods. The major advantage of using fuzzy neural networks is training a large amount of data sets, quick convergence and high accuracy.

In [37] an architecture related to the fuzzy neural network is suggested in order to obtain a real root at par with fuzzy polynomials which is illustrated in the form mentioned below

$$C_1 x + \ldots + C_n x^n = C_0 \tag{8}$$

where $x \in \Re$ as well as $C_0, C_1, \ldots, C_n$ are fuzzy numbers. A learning algorithm associated with the cost function in order to adjust the crisp weights has been suggested. The methodology mentioned in [37] has drawbacks. It is solely capable of extracting a crisp solution of fuzzy polynomials, and this neural network cannot extract a fuzzy solution.

In [38] an architecture of fuzzy neural networks is suggested for solving dual fuzzy polynomial equations. A learning algorithm of fuzzy weights of two-layer feedforward fuzzy neural networks is used whose input-output relations are defined by extension principle.

4 Numerical Techniques for Solving Fuzzy Differential Equations

Analytical methods can not be applied for solving FDEs. Therefore, numerical methods become essential, especially for PDEs. In this section three different important methods are described.

4.1 Taylor Method

In [39], an approach on the basis of the 2nd Taylor technique is illustrated in order to resolve linear as well as nonlinear FDEs. The convergence order of the Euler technique in [40] is $O(h)$, whereas the convergence order in [39] is $O(h^2)$. The better solutions are extracted by [39].

In [41] the Taylor method of order p is utilized for solving FDEs. The algorithm is explained by resolving some linear and nonlinear fuzzy Cauchy problems. The convergence order of the Taylor method is $O(h^p)$.

The drawback of Taylor series technique is the computation of higher derivatives, that by increasing the order the calculation process becomes increasingly complicated. However, Runge-Kutta method is generally considered to be the most effective one-step technique.

4.2 Runge-Kutta Method

In [42] an effective s-stage Runge-Kutta technique is employed for extracting the numerical solution of FDE. In that paper, Runge-Kutta method is applied for a more generalized category of problems and a convergence definition as well as error definitions are given at par with FDEs theory. Furthermore, convergence related to s-stage Runge-Kutta method is analyzed. This technique, when compared with developed Euler technique, performs superior. Although Euler technique is suitable, it is embedded with the disadvantage that, when analyzing the convergence of Euler technique [40], the authors generally investigate on the convergence of the ODEs system which happens while resolving numerically.

In [43] a numerical algorithm in order to solve linear as well as nonlinear fuzzy ODE on the basis of Seikkala derivative of fuzzy process is suggested. A numerical technique on the basis of the Runge-Kutta methodology of order five is elaborately investigated and this is carried on by going through an analysis of complete error. This technique with $O(h^5)$ outperforms than improved Euler's technique with $O(h^2)$.

In [44] a numerical algorithm in order to solve FDEs on the basis of Seikkala's derivative of a fuzzy process is suggested. A numerical technique based on a

Runge-Kutta Nystrom technique of order three is employed for solving the initial value problem, also it is illustrated that this methodology is superior in comparison with the Euler method by considering the convergence order of Euler methodology $(O(h))$ as well as Runge-Kutta Nystrom methodology $(O(h^3))$.

The major advantage of Runge-Kutta technique is that it is easy to apply. The main drawback of Runge-Kutta technique is that it needs more computer time when compared with multi-step techniques, also it does not easily yield desirable global approximations of the truncation error. The neural network is comparatively simple as well as computational rapid. Due to the superior estimation abilities of neural networks, the estimated solution for FDE is extremely near to the exact solution.

4.3   Neural Network Method

In [45] a modified technique is proposed in order to obtain the numerical solutions of fuzzy PDEs by utilizing fuzzy artificial neural networks. Utilizing modified fuzzy neural network ensures that the training points get selected over an open interval without training the network in the range of first and end points. This novel technique is on the basis of substituting each x in the training set (where $x \in [a, b]$) by the polynomial $Q(x) = \epsilon(x + 1)$ in such a manner that $Q(x) \in (a, b)$, by selecting an appropriate $\epsilon \in (0, 1)$. Also, it can be suggested that the proposed methodology can deal efficiently with all types of fuzzy PDEs as well as to generate precise estimated solution entirely for all domain and not only at the training set. Hence, one can utilize the interpolation methodologies (to be mentioned as curve fitting methodology) in order to obtain the estimated solution at points in the midst of the training points or at points outside the training set.

In [46] a novel technique on the basis of learning algorithm associated with the fuzzy neural network as well as Taylor series is laid down for extracting numerical solution of FDEs. A fuzzy neural network on the basis of the semi-Taylor series (in concerned to the function e^x) for the first (and second) order FDE is utilized. It is possible to use the same approach for solving high order FDE as well as fuzzy PDE. A fuzzy trial solution related to the fuzzy initial value problem is presented as an addition of two parts. The primary part suffices the fuzzy initial condition and it includes Taylor series, also contains no fuzzy adjustable parameters. The secondary part includes a feed-forward fuzzy neural network having fuzzy adjustable parameters (the fuzzy weights). Therefore by development, the fuzzy primary condition is sufficed and the training of the fuzzy network is carried out in order to suffice the FDE. The preciseness of this technique relies on the Taylor series that is selected for the trial solution. This selection is not distinct, hence, the preciseness is different from one problem to another problem. The suggested technique gives more precise estimations. Superior outcomes will be possible if more neurons or more training points are used. In addition, after resolving a FDE the solution is achievable at any arbitrary point in the training interval (even in the midst of training points).

5 Numerical Examples

In this section, two application examples are established to compare the performance of numerical techniques in order to estimate the solution of DFEs and FDEs.

Example 1. The water tank system contains two inlet valves V_1, V_2, as well as two outlet valves V_3, V_4, see Fig. 1. The areas of the valves are uncertain as the triangle function (2), $C_1 = G(0.021, 0.023, 0.024)$, $C_2 = G(0.008, 0.018, 0.038)$, $C_3 = G(0.012, 0.013, 0.015)$, $C_4 = G(0.038, 0.058, 0.068)$. The velocities of the flow (controlled by the valves) are $g_1 = (\frac{\vartheta}{10})e^\vartheta$, $g_2 = \vartheta cos(\Pi\vartheta)$, $g_3 = cos(\frac{\Pi\vartheta}{8})$, $g_4 = \frac{\vartheta}{2}$. If the outlet flow is aimed to be $z = (4.088, 6.336, 36.399)$, what is the quantity of the control variable ϑ? The mass balance of the tank is [47]:

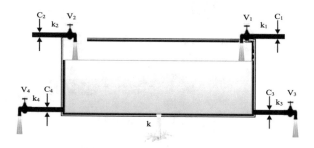

Fig. 1. Water tank system

$$\rho C_1 g_1 \oplus \rho C_2 g_2 = \rho C_3 g_3 \oplus \rho C_4 g_4 \oplus z$$

where ρ is considered to be the density of the water. The exact solution is taken to be $\vartheta_0 = 2$ [47]. To approximate the solution, we use three popular methods: Newton method, Genetic algorithm method, and Neural network method. The errors of these methods are shown in Table 1. It can be seen that all three methods can approximate the solutions of the dual fuzzy equations. Neural network method is more suitable for solving these kind of equations. In this table k is the number of iterations. The small approximation errors can be obtained by making the number of iteration larger. By increasing the number of iterations the estimated errors of the neural networks based algorithm are less than the other methods. Neural network method is more robust when compared with the other methods.

Example 2. The vibration mass system displayed in Fig. 2 is modeled by,

$$\frac{d}{dt}u(t) = \frac{c}{m}x(t), \quad u(t) = \frac{d}{dt}x(t) \tag{9}$$

Table 1. Approximation errors

k	Newton	Genetic algorithm	Neural network
1	0.18635	0.33463	0.43967
2	0.29602	0.24791	0.32375
3	0.36175	0.13012	0.21763
⋮	⋮	⋮	⋮
119	0.07982	0.04563	0.00316
120	0.07526	0.03952	0.00286

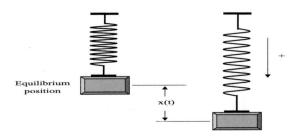

Fig. 2. Vibration mass

Table 2. Approximation errors

α	Taylor	Runge-Kutta	Neural network
0	[0.0604,0.1088]	[0.0407,0.0889]	[0.0209,0.0606]
0.2	[0.0701,0.1191]	[0.0609,0.1092]	[0.0308,0.0704]
0.4	[0.0511,0.0993]	[0.0211,0.0692]	[0.0102,0.0501]
0.6	[0.0403,0.0880]	[0.0211,0.0691]	[0.0011,0.0409]
0.8	[0.1009,0.1493]	[0.0712,0.1192]	[0.0512,0.0913]
1	[0.1104,0.1104]	[0.0806,0.0806]	[0.0602,0.0602]

where the spring constant is considered to be $c = 1$, as well as the mass is $m = (0.73, 1.123)$. If the initial position is taken to be $x(0) = (0.73 + 0.23\alpha, 1.123 - 0.123\alpha)$, $\alpha \in [0, 1]$, so the exact solutions of (9) are [48],

$$x(t, \alpha) = \left[(0.73 + 0.23\alpha)e^t, (1.123 - 0.123\alpha)e^t \right] \qquad (10)$$

where $t \in [0, 1]$. In order to estimate the solution (10), we utilize three popular techniques: Taylor technique, Runge-Kutta technique, and Neural network technique. The errors related to these techniques are demonstrated in Table 2. Corresponding solution plots are displayed in Fig. 3. All three methods are suitable for resolving the FDEs. The leaning procedure of the neural network method

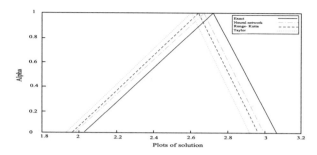

Fig. 3. Comparison plot of three popular methods and the exact solution

is more quickly than the other methods. Also, the robustness of neural network method is better when compared with the other methods.

6 Conclusion

In this review, recent numerical methods are considered to solve FEs. It addresses numerical methodologies to solve FEs, DFEs, FDEs, and PFDEs. Research in this field continues to grow with new types of numerical methods and new strategies. Emphasis is given to current developments in the solving strategies in the last two decades, which demonstrates their significant improvements.

References

1. Jafari, R., Yu, W.: Uncertainty nonlinear systems control with fuzzy equations. In: IEEE International Conference on Systems, Man, and Cybernetics, pp. 2885–2890 (2015). https://doi.org/10.1109/SMC.2015.502
2. Jafari, R., Razvarz, S.: Solution of fuzzy differential equations using fuzzy sumudu transforms. Math. Comput. Appl. **23**, 1–15 (2018)
3. Razvarz, S., Jafari, R., Yu, W.: Numerical solution of fuzzy differential equations with Z-numbers using fuzzy sumudu transforms. In: IEEE International Conference on Industrial Technology (ICIT), Toronto, ON, 2017, pp. 890–895 (2017). https://doi.org/10.1109/ICIT.2017.7915477
4. Jafari, R., Yu, W.: Uncertainty nonlinear systems control with fuzzy equations. In: IEEE International Conference on Industrial Technology (ICIT), Toronto, ON, 2017, pp. 890–895 (2015). https://doi.org/10.1109/ICIT.2017.7915477
5. Jafari, R., Yu. W., Li, X.: Solving fuzzy differential equation with Bernstein neural networks. In: IEEE International Conference on Systems, Man, and Cybernetics (SMC), Budapest, pp. 001245–001250 (2016). https://doi.org/10.1109/SMC.2016.7844412
6. Jafari. R., Yu, W.: Artificial neural network approach for solving strongly degenerate parabolic and burgers-fisher equations. In: 12th International Conference on Electrical Engineering, Computing Science and Automatic Control (CCE), Mexico City, pp. 1–6 (2015). https://doi.org/10.1109/ICEEE.2015.7357914

7. Jafarian, A., Measoomy nia, S., Jafari, R.: Solving fuzzy equations using neural nets with a new learning algorithm. J. Adv. Comput. Res. **3**(4), 33–45 (2012)
8. Jafari, R., Razvarz, S., Gegov, A., Paul, S.: Fuzzy modeling for uncertain nonlinear systems using fuzzy equations and Z-numbers. In: Advances in Computational Intelligence Systems, UKCI 2018. Advances in Intelligent Systems and Computing, vol. 840, pp. 96–107 (2019)
9. Jafari, R., Razvarz, S., Gegov, A., Paul, S., Keshtkar, S.: Fuzzy sumudu transform approach to solving fuzzy differential equations with Z-numbers. In: Advanced Fuzzy Logic Approaches in Engineering Science, pp. 1–31 (2019). https://doi.org/10.4018/978-1-5225-5709-8.ch002
10. Jafari, R., Razvarz, S., Gegov, A.: Solving differential equations with Z-numbers by utilizing fuzzy sumudu transform. In: IntelliSys 2018. Advances in Intelligent Systems and Computing, vol. 869, pp. 1125–1138. Springer, Cham (2019). https://doi.org/10.1007/978-3-030-01057-7.82
11. Jafari, R., Razvarz, S., Gegov, A., Paul, S.: Modeling and control of uncertain nonlinear systems. In: International Conference on Intelligent Systems (IS), Funchal - Madeira, Portugal, pp. 168–173 (2018). https://doi.org/10.1109/IS.2018.8710463
12. Jafari, R., Razvarz, S., Gegov, A.: Fuzzy differential equations for modeling and control of fuzzy systems. In: 13th International Conference on Theory and Application of Fuzzy Systems and Soft Computing, ICAFS 2018. Advances in Intelligent Systems and Computing, vol. 896, pp. 732–740. Springer, Cham (2019). https://doi.org/10.1007/978-3-030-04164-9-96
13. Jafari, R., Razvarz, S., Gegov, A.: A novel technique to solve fully fuzzy nonlinear matrix equations. In: 13th International Conference on Applications of Fuzzy Systems and Soft Computing. Advances in Intelligent Systems and Computing. Springer (2018)
14. Jafari, R., Razvarz, S., Gegov, A.: A new computational method for solving fully fuzzy nonlinear systems. In: Computational Collective Intelligence, ICCCI 2018. Lecture Notes in Computer Science, vol. 11055, pp. 503–512. Springer (2018). https://doi.org/10.1007/978-3-319-98443-8-46
15. Razvarz, S., Jafari, R., Gegov, A., Yu, W., Paul, S.: Neural network approach to solving fully fuzzy nonlinear systems. In: Fuzzy Modelling and Control Nova Science Publishers, pp. 46–68 (2018)
16. Yu, W., Jafari, R.: Fuzzy Modeling and Control of Uncertain Nonlinear Systems (IEEE Press Series on Systems Science and Engineering), 1st edn., 208 p. Wiley-IEEE Press (2019). ISBN: 978-1119491552
17. Friedman, N., Ming, M., Kandel, A.: Fuzzy linear systems. Fuzzy Sets Syst. **96**, 201–209 (1998). https://doi.org/10.1016/S0165-0114(96)00270-9
18. Buckley, J., Qu, Y.: Solving linear and quadratic fuzzy equations. Fuzzy Sets Syst. **35**, 43–59 (1990). https://doi.org/10.1016/0165-0114(90)90099-R
19. Buckley, J., Hayashi, Y.: Can fuzzy neural nets approximate continuous fuzzy functions. Fuzzy Sets Syst. **61**, 43–51 (1994). https://doi.org/10.1016/0165-0114(94)90283-6
20. Gibson, J.: An analysis of optimal modal regulation: convergence and stability. SIAM J. Control. Optim. **19**, 686–707 (1981). https://doi.org/10.1137/0319044
21. Kröner, A., Kunisch, K.: A minimum effort optimal control problem for the wave equation. Comput. Optim. Appl. **57**, 241–270 (2014). https://doi.org/10.1007/s10589-013-9587-y
22. Abbasbandy, S., Ezzati, R.: Newton's method for solving a system of fuzzy nonlinear equations. Appl. Math. Comput. **175**, 1189–1199 (2006). https://doi.org/10.1016/j.amc.2005.08.021

23. Allahviranloo, T., Otadi, M., Mosleh, M.: Iterative method for fuzzy equations. Soft Comput. **12**, 935–939 (2008). https://doi.org/10.1007/s00500-007-0263-y
24. Kajani, M., Asady, B., Vencheh, A.: An iterative method for solving dual fuzzy nonlinear equations. Appl. Math. Comput. **167**, 316–323 (2005). https://doi.org/10.1016/j.amc.2004.06.113
25. Waziri, M., Majid, Z.: A new approach for solving dual fuzzy nonlinear equations using broyden's and newton's methods. Adv. Fuzzy Syst., 1–5 (2012). https://doi.org/10.1155/2012/682087
26. Pederson, S., Sambandham, M.: The Runge-Kutta method forhybrid fuzzy differential equation. Nonlinear Anal. Hybrid Syst. **2**, 626–634 (2008). https://doi.org/10.1016/j.nahs.2006.10.013
27. Tapaswini, S., Chakraverty, S.: Euler-based new solution method for fuzzy initial value problems. Int. J. Artif. Intell. Soft. Comput. **4**, 58–79 (2014). https://doi.org/10.1504/IJAISC.2014.059288
28. Hüllermeier, E.: An approach to modeling and simulation of uncertain dynamical systems. Int. J. Uncertain. Fuzziness Knowl. Based Syst. **5**, 117–137 (1997). https://doi.org/10.1142/S0218488597000117
29. Guo, S., Mei, L., Zhou, Y.: The compound $\frac{G'}{G}$ expansion method and double nontraveling wave solutions of (2+1)-dimensional nonlinear partial differential equations. Comput. Math. Appl. **69**, 804–816 (2015). https://doi.org/10.1016/j.camwa.2015.02.016
30. Effati, S., Pakdaman, M.: Artificial neural network approach for solving fuzzy differential equations. Inform. Sci. **180**, 1434–1457 (2010). https://doi.org/10.1016/j.ins.2009.12.016
31. Agatonovic-Kustrin, S., Beresford, R.: Basic concepts of artificial neural network (ANN) modeling and its application in pharmaceutical research. J. Pharm. Biomed. Anal. **22**, 717–727 (2000). https://doi.org/10.1016/S0731-7085(99)00272-1
32. Newton, I.: The method of fluxions and infinite series. London: Henry Woodfall, **3**, pp. 43–47. https://archive.org/details/methodoffluxions00newt. ISBN: 1498167489, 9781498167482, (1671)
33. Abbasbandy, S., Asady, B.: Newton's method for solving fuzzy nonlinear equations. Appl. Math. Comput. **159**, 349–356 (2004). https://doi.org/10.1016/j.amc.2003.10.048
34. Brudaru, O., Leon, F., Buzatu, O.: Genetic algorithm for solving fuzzy equations. In: 8th International Symposium on Automatic Control and Computer Science, Iasi. ISBN973-621-086-3
35. Michalewicz, Z.: Genetic algorithms+data structures=evolution program. In: Artificial Intelligence, 2nd ed. Springer, Berlin. https://doi.org/10.1007/978-3-662-03315-9
36. Mastorakis, N.: Solving non-linear equations via genetic algorithms. In: 6th WSEAS International Conference on Evolutionary Computing, Lisbon, Portugal, pp. 24–28 (2005)
37. Abbasbandy, S., Otadi, M.: Numerical solution of fuzzy polynomials by fuzzy neural network. Appl. Math. Comput. **181**, 1084–1089 (2006). https://doi.org/10.1016/j.amc.2006.01.073
38. Mosleh, M., Otadi, M.: A new approach to the numerical solution of dual fully fuzzy polynomial equations. Int. J. Ind. Math. **2**, 129–142 (2010)
39. Abbasbandy, S., Allahvinloo, T.: Numerical solutions of fuzzy differential equation. Math. Comput. Appl. **7**, 41–52 (2002). https://doi.org/10.3390/mca7010041

40. Ma, M., Friedman, M., Kandel, A.: Numerical solutions of fuzzy differential equations. Fuzzy Sets Syst. **105**, 133–138 (1999)
41. Abbasbandy, S., Allahvinloo, T.: Numerical solutions of fuzzy differential equations by taylor method. Comput. Methods Appl. Math. **2**, 113–124 (2002). https://doi.org/10.2478/cmam-2002-0006
42. Palligkinis, S., Papageorgiou, G., Famelis, I.: Runge-kutta methods for fuzzy differential equations. Appl. Math. Comput. **209**, 97–105 (2009). https://doi.org/10.1016/j.amc.2008.06.017
43. Jayakumar, T., Maheskumar, D., Kanagarajan, K.: Numerical solution of fuzzy differential equations by Runge-Kutta method of order five. Appl. Math. Sci. **6**, 2989–3002 (2012)
44. Kanagarajan, K., Sambath, M.: Runge-Kutta nystrom method of order three for solving fuzzy differential equations. Comput. Methods Appl. Math. **10**, 195–203 (2010). https://doi.org/10.2478/cmam-2010-0011
45. Hussian, E., Suhhiem, M.: Numerical solution of fuzzy partial differential equations by using modified fuzzy neural networks. Br. J. Math. Comput. Sci. **12**, 1–20 (2016). https://doi.org/10.9734/BJMCS/2016/20504
46. Hussian, E., Suhhiem, M.: Numerical solution of fuzzy differential equations based on Taylor series by using fuzzy neural networks. Adv. Math. **11**, 4080–4092 (2015)
47. Streeter, V., Wylie, E., Benjamin, E.: Fluid Mechanics, 4th edn. McGraw-Hill Book Company
48. Hazewinkel, M.: Oscillator harmonic. Springer. ISBN

Fuzzy Feature Representation with Bidirectional Long Short-Term Memory for Human Activity Modelling and Recognition

Gadelhag Mohmed[(✉)], David Ada Adama, and Ahmad Lotfi

School of Science and Technology, Nottingham Trent University,
Clifton Lane, Nottingham NG11 8NS, UK
gadelhag.mohmed2016@my.ntu.ac.uk

Abstract. There has been an increased interest in the development of models to identify and predict human activities. However, the sparsity of the data gathered from the sensory devices in an ambient living environment creates the challenge of representing activities accurately. Also, such data usually comprise arbitrary lengths of dimensions. Recurrent Neural Networks (RNNs) are one of the widely used algorithms in sequential modelling due to their ability to handle the arbitrary lengths of data. In an attempt to address the above challenges, this paper proposes a method of fuzzy feature representation with Bidirectional Long Short-Term Memory (Bi-LSTM) for human activities modelling and recognition. To obtain optimal feature representation, sensor data are fuzzified and the membership degrees represent the selected features which are then applied to the Bi-LSTM model for activity modelling and recognition. The learning capability of the Bi-LSTM allows the model to learn the temporal relationship in sequential data which is used to identify human activities pattern. The learned pattern is then utilised in the prediction of further activities. The proposed method is tested and evaluated using dataset representing Activity of Daily Living (ADL) for a single user in a smart home environment. The obtained results are also compared with existing approaches that are used for modelling and recognising human activities.

Keywords: Human activity recognition · Activity modelling · Fuzzy feature extraction · Bidirectional Long Short-Term Memory

1 Introduction

Human activity modelling and recognition is considered to be a challenging task in Ambient Intelligence (AmI) research due to the fact that human actions are uncertain and unpredictable [4]. The information obtained from unobtrusive sensory devices such as motion sensors [6,12] and camera sensors [2,9], can provide

© Springer Nature Switzerland AG 2020
Z. Ju et al. (Eds.): UKCI 2019, AISC 1043, pp. 15–26, 2020.
https://doi.org/10.1007/978-3-030-29933-0_2

long-term information regarding human activities. Although, a reliable model is required to be able to accurately represent actions and their relationships. Data from various sensors accumulated over a long period of time represent ambiguity as they contain unpredictable and noisy patterns. This will pose a great challenge in the development of data-driven models for extraction and representation of human activity features, for example, time-dependent features.

To represent features of human activities, statistical and dynamic features in the time domain are the most commonly used features [14,19]. They are suited for datasets with little noise. Fuzzy sets and systems have been established to handle the unpredictable uncertainties in using raw data in many practical applications [5]. In this paper, a fuzzy feature representation of human activities is presented. The proposed fuzzy feature representation technique learns the degree of membership functions through the fuzzification of each input variable in the training data to represent the human activities. The membership degrees represent the features used to model and recognise activities. When compared with other feature representation approaches such as deep learning neural networks which are able to learn features from data, the features derived from the fuzzy feature representation can be flexibly constructed to reduce underlying uncertainties in the data [5].

A new approach is proposed in this paper which applies fuzzy feature representation of the sensory data to a Bidirectional Long Short-Term Memory (Bi-LSTM) Neural Networks. The fuzzy feature representation is used to fuzzify the data representing human activities. The membership degrees obtained for each activity in the data are used as features in the proposed model and they are fed into a Bi-LSTM model which learns the sequential relationship over a long-term period of activities data. The use of LSTM networks are suited for modelling temporal relationship in long-term data sequences due to their ability to handle the vanishing gradient problem common in most traditional algorithms [8]. Therefore, a bidirectional approach in which an activity is not only inferred from past knowledge but also from the knowledge of future actions in the given activity context is relevant in modelling and recognition of human activities. This motivates the use of the Bi-LSTM model in this work. The proposed method is suitable for challenging problems involving long-term time-dependent modelling in sequential data-driven applications.

The remainder of this paper is organised as follows: In Sect. 2 a review of the related literature is presented. Section 3 explains the proposed method for fuzzy feature representation and the bidirectional LSTM architecture. Experiments are conducted to obtain data representing different activities. Description of the experiments is presented in Sect. 4 followed by the results presented in Sect. 5. Section 6 concludes the paper and gives a direction of future work.

2   Related Work

The distribution of human activity data influences the process of feature extraction. Many computational intelligence models are developed to recognise activities mainly relying on appropriate features which distinguish activities to classify activities with high accuracy [13]. However, the process of extracting features can be daunting in applications where the distribution of data representing activities is sparse. This can be the case when dealing with binary signals data in which only two groups (or classes) are involved. Examples of such cases are in the use of unobtrusive sensory devices in AmI environments to provide activity information as binary signals [6, 10, 15]. As a result, the classification results will tend to be skewed to some classes due to the data imbalance and this creates challenges in developing reliable models for such activities [14].

Different techniques have been proposed for feature representation of human activities data. A Spectrogram-based approach is proposed in [14] for the representation of features of inertial measurement unit sensor data. The features obtained were tested on a deep LSTM neural network to evaluate the effect of features representation on recognition of activities. Also, among the widely used methods for features representation is the Fuzzy computational approach. In [5], the authors proposed a fuzzy computational approach in a hierarchical method to extract features from one-dimensional input vectors and the features fed to a deep neural network for data classification. A fuzzy temporal windows approach is proposed in [10] to define temporal-sequence representations to aggregate information from binary sensors for real-time recognition of human activities. The methods in [5, 10] have been successful in capturing features which improved the performance of the classification tasks.

In building models for activity recognition, Recurrent Neural Network (RNN) is one of the widely used algorithms due to their ability to learn sequences in data. LSTMs are a type of RNN applied in most cases when the temporal information is considered. In a recent work [11] we proposed a method for modelling human activities with the fusion of LSTM and fuzzy finite state machines. The LSTM algorithm is also used in [1] for learning activity sequences. The algorithm models activities with the advantage of a memory which recalls instances that have occurred previously. Furthermore, an improvement to the LSTM algorithm known as Bi-LSTM [7] was introduced. This learns data sequences by considering both past and future instances. The benefit of the Bi-LSTM algorithm over the LSTM algorithm has resulted in improved performance in activity recognition applications [16–18].

In the framework proposed in this paper, a fuzzy feature representation method is applied for the representation of human activities information obtained from an AmI environment. This approach is used as a means to improve the limited feature set of activities. The features obtained after the fuzzy representation process are used as input to a Bi-LSTM model which is trained for human activity modelling. In the next section, a description of the methodology used in this work is presented.

3   Methodology

The overall framework of the proposed method comprises three stages as illustrated in Fig. 1. Stage one is the data collection process, where the data is

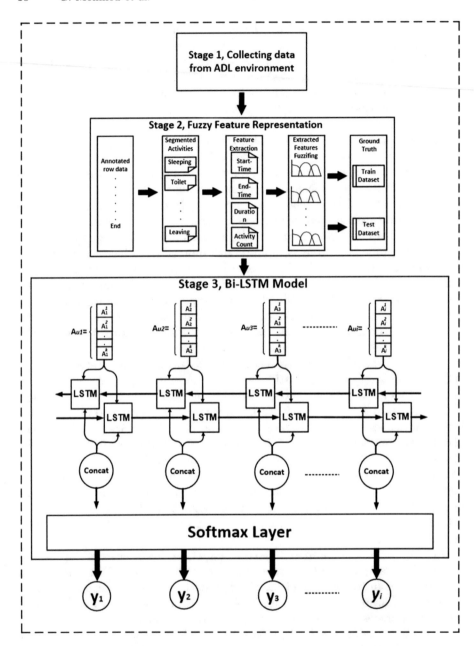

Fig. 1. Proposed framework for fuzzy feature representation and the Bi-LSTM model for human activity modelling and recognition.

gathered using ambient sensory devices such as motion sensors, door entry sensors and pressure sensors. At this stage, the gathered information are time-stamps of binary data representing the ADL. Stage two is the fuzzy feature representation step, which is designed to fuzzify the data and the resulting fuzzy membership degrees are taken as features to be used in modelling and recognition of activities. The final stage in the proposed framework is the Bi-LSTM model for activity modelling and recognition, where the features obtained from stage two are used to train a Bi-LSTM to learn the relationship between the features and their corresponding outputs for modelling and recognising human activities. A detailed description of each stage is given in subsequent sections.

3.1 Data Collection

An Ambient Intelligent environment is often equipped with unobtrusive sensors to gather information representing an occupier's behaviour. Commonly used sensors for data collection are binary sensors including Passive Infrared (PIR) motion sensors, door and window entry sensors, mat pressure sensors to measure bed and sofa occupancy, and electric appliances usage sensor. The recorded information will be transmitted to a central hub to be stored/processed for later use. A representation of sampled data for three days is shown in Fig. 2. More information regarding the used dataset in this research is provided in Sect. 4.

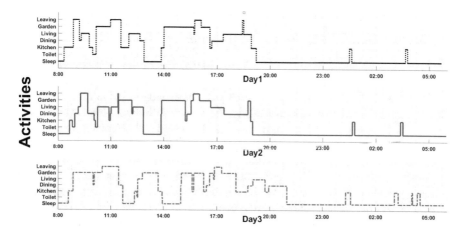

Fig. 2. Multilevel graph showing three days activities obtained from binary sensor signals.

3.2 Fuzzy Feature Representation

Feature extraction is a process applied to the collected activity dataset and it aims to identify unique characteristics in the data [17]. Fuzzy feature representation is used to determine the number of Membership Functions (MFs) to be formed with each input data [16]. By representing each value in the input data with their degree of membership, thus, each value in the input data is represented as the fuzzified values obtained for each MF:

$$u_i(x) = \sum_{x=1}^{j} \{A_{ui}^1(x), A_{ui}^2(x), ..., A_{ui}^k(x)\} \qquad k \neq 0 \qquad (1)$$

where x is the value in the input variable u_i that will be fuzzified, j is the last value in the input value u_i. A and k correspond to the associated labels and the number of associated labels with the MFs respectively, with each value x. The Fuzzy feature presentation process is summarised as follows:

1. Enter the input data to the created fuzzifier algorithm that consists k number of the MFs.
2. Define the degree of fuzziness that corresponds to each associated label A.
3. Determine the maximum degree of fuzziness for the value of x in each iteration.
4. A matrix $f = n \times m$ is created to store the degree of membership for each value x. Where n is the total number of activity instances in the input data u_i, and m is the number of fuzzified values for the input data.
5. Update the matrix f after each iteration.

3.3   Bidirectional Long Short-Term Memory Neural Networks

The Long Short-Term Memory (LSTM) Neural Networks are shown to be a good model to represent sequential forms of data such as the time series data [16]. The LSTM has the ability to store past information by looping inside its architecture, which keeps the information from the previously learned iteration. The LSTM's architecture contains; input gate to handle the input information in the input layer, output gate in the output layer which has a Softmax activation on it, and a forget gate that controls the internal cells to either save or forget the previous information. The main purpose of using the forget gate is to reduce the risk of vanishing gradients. As the LSTM can only get the information from the previous context, a further improvement was made by fusing another LSTM network to give the Bidirectional LSTM (Bi-LSTM) model. The Bi-LSTM model is a combination of two - forward and backwards - LSTMs, which can handle information from both future and past directions.

As the Bi-LSTM is a combination of two directional LSTM's, the input data will be processed firstly in forward direction ($\rightarrow$) and then re-processed again in backward direction ($\leftarrow$). Bi-LSTM's consists of input gate (i_t), forget gate (f_t), output gate (o_t), memory cell (c_t) and the hidden state (h_t) for both forward ($\rightarrow$) and backward ($\leftarrow$) directions at each time step (t). Therefore the hidden state in the Bi-LSTM is calculated based on the combination of the forward hidden state ($h_t^{\rightarrow}$) and the backward hidden state ($h_t^{\leftarrow}$) as formulated in Eq. 2.

$$h_t = h_t^{\rightarrow} \oplus h_t^{\leftarrow} \qquad (2)$$

where the elements of the forward state ($\rightarrow$) are formulated as follows:

$$h_t^{\rightarrow} = o_t^{\rightarrow} \otimes tanh(c_t^{\rightarrow}) \tag{3}$$

$$c_t^{\rightarrow} = f_t^{\rightarrow} \otimes c_{t-1}^{\rightarrow} + i_t^{\rightarrow} \otimes tanh(W_c^{\rightarrow} U_t^{\rightarrow} + V_c^{\rightarrow} h_{t-1}^{\rightarrow} + b_c^{\rightarrow}) \tag{4}$$

$$i_t^{\rightarrow} = \sigma(W_i^{\rightarrow} U_t^{\rightarrow} + V_i^{\rightarrow} h_{t-1}^{\rightarrow} + b_i^{\rightarrow}) \tag{5}$$

$$f_t^{\rightarrow} = \sigma(W_f^{\rightarrow} U_t^{\rightarrow} + V_f^{\rightarrow} h_{t-1}^{\rightarrow} + b_f^{\rightarrow}) \tag{6}$$

$$o_t^{\rightarrow} = \sigma(W_o^{\rightarrow} U_t^{\rightarrow} + V_o^{\rightarrow} h_{t-1}^{\rightarrow} + b_o^{\rightarrow}) \tag{7}$$

on the other hand, the elements of the backward state ($\leftarrow$) are formulated as follows:

$$h_t^{\leftarrow} = o_t^{\leftarrow} \otimes tanh(c_t^{\leftarrow}) \tag{8}$$

$$c_t^{\leftarrow} = f_t^{\leftarrow} \otimes c_{t-1}^{\leftarrow} + i_t^{\leftarrow} \otimes tanh(W_c^{\leftarrow} U_t^{\leftarrow} + V_c^{\leftarrow} h_{t-1}^{\leftarrow} + b_c^{\leftarrow}) \tag{9}$$

$$i_t^{\leftarrow} = \sigma(W_i^{\leftarrow} U_t^{\leftarrow} + V_i^{\leftarrow} h_{t-1}^{\leftarrow} + b_i^{\leftarrow}) \tag{10}$$

$$f_t^{\leftarrow} = \sigma(W_f^{\leftarrow} U_t^{\leftarrow} + V_f^{\leftarrow} h_{t-1}^{\leftarrow} + b_f^{\leftarrow}) \tag{11}$$

$$o_t^{\leftarrow} = \sigma(W_o^{\leftarrow} U_t^{\leftarrow} + V_o^{\leftarrow} h_{t-1}^{\leftarrow} + b_o^{\leftarrow}) \tag{12}$$

where, U_t is the input data at time step t; $U \subset \{u_1, u_2, .., u_k\}$. i, f and o denote the input, forget and output gates respectively. c and y are the cell state and output respectively at time step t. b_j, $j \in \{f, i, o, c\}$ are the bias units for the forget, input and output gates and the memory inputs respectively. W_{ij} is the weight connection between i and j. σ is a logistic sigmoid function and $tanh$ is the tangent hyperbolic function. V is the cell matrix.

Bi-LSTM is used to learn the sequential relations in the given data by storing the information through multiple time steps using the combination of forward and backward states. The next section introduces experiments with the proposed approach, which integrates the learning abilities of the Bi-LSTM model in the temporal data with sequential human activities.

4   Experiments

To evaluate the proposed method, it is applied to a dataset gathered from a real environment representing the ADL of a single user. The dataset used in this experiment represents seven different activities for a single user corresponding to; Sleeping, Toileting, Meal-preparation, Dining, relaxing, Garden and Leaving home. In this work, the sensory devices used in collecting data are listed in Table 1. The fuzzy feature representation stage is applied to clean the collected data by filtering noise and to extract features corresponding to the fuzzy membership degrees of selected variables within the data. In the fuzzy feature representation stage, selected variables from the collected activity data will be used. The selected variables used in this experiment are; activity start time (u_1) - which represents the start time for each activity, activity end time (u_2) - which is the end time for each activity, activity duration (u_3) - which is the time spent for each activity, and the activity count (u_4) - which is the number of activity attempts per hour.

Each value x in the selected variables is fuzzified using Gaussian MFs to transform the data to relevant fuzziness degrees. The membership labels associated with each input variable are represented using the following MFs:

$$U(t) = \begin{cases} u_1(x) \rightarrow \{EM_{u_1}, M_{u_1}, AF_{u_1}, EV_{u_1}, NI_{u_1}\} \\ u_2(x) \rightarrow \{EM_{u_2}, M_{u_2}, AF_{u_2}, EV_{u_2}, NI_{u_2}\} \\ u_3(x) \rightarrow \{VS_{u_3}, SH_{u_3}, ME_{u_3}, LO_{u_3}, VLO_{u_3}\} \\ u_4(x) \rightarrow \{VRU_{u_4}, RU_{u_4}, MU_{u_4}, HU_{u_4}, VHU_{u_4}\} \end{cases} \tag{13}$$

where each value x in the selected input features $\{u_1, u_2, u_3, u_4\}$ is represented with 5 MFs as follows:

- The MFs representing activity start time for the input variable u_1 are represented as $\{EM_{u_1}, M_{u_1}, AF_{u_1}, EV_{u_1}, NI_{u_1}\}$. Where EM, M, AF, EV and NI are MF labels corresponding to Early Morning, Morning, Afternoon, Evening and Night respectively.
- The MFs representing activity end time for the input variable u_2 are represented as $\{EM_{u_2}, M_{u_2}, AF_{u_2}, EV_{u_2}, NI_{u_2}\}$. Where EM, M, AF, EV and NI are the MF labels corresponding to Early Morning, Morning, Afternoon, Evening and Night respectively.

Table 1. List of sensors used for collecting activity data to measure the presence of a single user.

Sensor	No. of sensors	Purpose of use
Passive Infra Red (PIR)	6	Detecting the movement
Door on/off switches	2	Detecting doors' opening and closing
Mat Pressure sensor	1	Measuring bed and sofa occupancy
Electricity Consumption Plugs	2	Measuring electricity consumption

- The MFs representing activity duration for the input variable u_3 are represented as $\{VS_{u_3}, SH_{u_3}, ME_{u_3}, LO_{u_3}, VLO_{u_3}\}$. Where VS, SH, ME, LO and VLO are MF labels corresponding to Very Short, Short, Medium, Long and Very long respectively.
- The MFs representing activity duration for the input variable u_4 are represented as $\{VRU_{u_4}, RU_{u_4}, MU_{u_4}, HU_{u_4}, VHU u_4\}$. Where VRU, RU, MU, HU and VHU are the MF labels corresponding to Very Rare Usage, Rare Usage, Medium Usage, Heavy Usage and Very Heavy Usage respectively.

In the next step, the Bi-LSTM model will be trained with the fuzzy represented features to learn the sequential relationships between the input data (activity data) and their corresponding output (activity labels).

5   Results

This paper focuses on modelling and recognising human activities using fuzzy feature representation with Bi-LSTM model. Based on the explanation provided

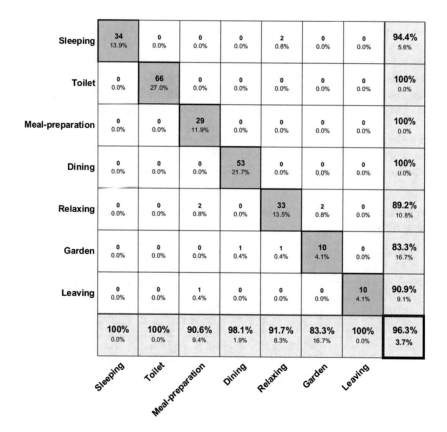

Fig. 3. Confusion matrix plot for the obtained results by using Bi-LSTM.

in the previous sections, the Bi-LSTM model is employed to model and recognise the ADL for a single user within a smart home environment. This section presents the results obtained with the proposed framework using the experimental dataset. As it is reported in some previous researches [3], datasets that represent human activities are usually not balanced because of the uncertainty associated with the way humans behave in their daily activities. In this case, some activities will be more dominant than others. Therefore, if the dominant activities are well modelled and recognised the overall accuracy for the whole system will be high even though the other activities are not well modelled and recognised. Based on that, each activity will be evaluated separately and then the performance over the whole system will be computed. Figure 3 shows the results obtained from the conducted experiment. This shows the confusion matrix illustrating the recall and precision scores for each activity as well as over the whole system. The information given in the confusion matrix is explained in the list as follows:

- The rows represent the output activities and columns represent the target activities. The activities are named as *Sleeping, Toilet, Meal-preparation, Dining, Relaxing, Garden and Leaving home*.
- The diagonal cells from the upper left to the lower right illustrate the activities that are correctly recognised.
- The off-diagonal cells show the incorrectly recognised activities.
- The last column in the right represents the precision for each activity.
- The last row in the bottom shows the recall for each activity.
- The bottom right cell represents the accuracy over the whole model.

In order to evaluate and emphasise the proposed method, the results obtained in the proposed framework employing the Bi-LSTM model are compared with the results obtained using the standard LSTM and the Support Vector Machine

Table 2. Recall and precision for each activity obtained based on Bi-LSTM, standard LSTM and SVM.

Activities	Bi-LSTM		LSTM		SVM	
	Recall	Precision	Recall	Precision	Recall	Precision
Sleeping	100%	94.4%	91.7%	87.50%	91.7%	68.8%
Toilet	100%	100%	100%	90.9%	100%	78.9%
Meal-preparation	90.6%	100%	83.1%	78.3%	58.3%	89.2%
Dining	98.1%	100%	100%	83.3%	77.3%	85.0%
Relaxing	91.7%	89.2%	76.9%	50%	50%	72.2%
Garden	83.3%	83.3%	75%	NaN	NaN	15.2%
Leaving	100%	90.9%	35.9%	42.6%	NaN	NaN
Overall accuracy	**96.3%**		80.8%		74%	

(SVM) with the same dataset. The comparison between the performance of Bi-LSTM and the other two methods has been made for each activity in terms of recall, precision and the overall accuracy of the model as presented in Table 2.

6   Conclusion

In this paper, a model for fuzzy feature representation of ADLs with a focus on modelling and recognising activities is proposed. Fuzzy feature representation method makes it possible for the extraction of unique features from sparse features set of ADL and capture underlying uncertainties. To achieve the aim of this research, a Bi-LSTM model is used. The Bi-LSTM model is used because of its unique capability of modelling sequential data from past and future observations and this is beneficial in human ADL applications such as trend analysis. The results obtained show a high performance of the Bi-LSTM model on recognition of different activities compared with the results that are obtained from a standard LSTM and SVM methods for the same dataset. For future work, more experiments would be conducted with the proposed method in order to evaluate the robustness in modelling ADLs. This will be compared with state-of-the-art methods for feature representation.

References

1. Adama, D.A., Lotfi, A., Langensiepen, C.: Key frame extraction and classification of human activities using motion energy. In: Advances in Computational Intelligence Systems, pp. 303–311. Springer (2019)
2. Adama, D.A., Lotfi, A., Langensiepen, C., Lee, K., Trindade, P.: Human activity learning for assistive robotics using a classifier ensemble. Soft Comput. **22**(21), 7027–7039 (2018)
3. Benmansour, A., Bouchachia, A., Feham, M.: Modeling interaction in multi-resident activities. Neurocomputing **230**, 133–142 (2017)
4. Cook, D.J., Augusto, J.C., Jakkula, V.R.: Ambient intelligence: technologies, applications, and opportunities. Pervasive Mob. Comput. **5**(4), 277–298 (2009)
5. Deng, Y., Ren, Z., Kong, Y., Bao, F., Dai, Q.: A hierarchical fused fuzzy deep neural network for data classification. IEEE Trans. Fuzzy Syst. **25**(4), 1006–1012 (2017)
6. Gochoo, M., Tan, T.-H., Liu, S.-H., Jean, F.-R., Alnajjar, F.S., Huang, S.-C.: Unobtrusive activity recognition of elderly people living alone using anonymous binary sensors and dcnn. IEEE J. Biomed. Health Inform. **23**(2), 693–702 (2019)
7. Graves, A., Schmidhuber, J.: Framewise phoneme classification with bidirectional LSTM and other neural network architectures. Neural Netw. **18**(5), 602–610 (2005)
8. Hochreiter, S., Schmidhuber, J.: Long short-term memory. Neural Comput. **9**(8), 1735–1780 (1997)
9. Ke, S.-R., Thuc, H., Lee, Y.-J., Hwang, J.-N., Yoo, J.-H., Choi, K.-H.: A review on video-based human activity recognition. Computers **2**(2), 88–131 (2013)
10. Medina-Quero, J., Zhang, S., Nugent, C., Espinilla, M.: Ensemble classifier of long short-term memory with fuzzy temporal windows on binary sensors for activity recognition. Expert. Syst. Appl. **114**, 441–453 (2018)

11. Mohmed, G., Lotfi, A., Pourabdollah, A.: Long, short-term memory fuzzy finite state machine for human activity modelling. In: The 12th PErvasive Technologies Related to Assistive Environments Conference (PETRA). ACM, New York (2019)

12. Mohmed, G., Lotfi, A., Pourabdollah, A.: Human activities recognition based on neuro-fuzzy finite state machine. Technologies **6**(4), 110 (2018)

13. Rashidi, P., Mihailidis, A.: A survey on ambient-assisted living tools for older adults. IEEE J. Biomed. Health Inform. **17**(3), 579–590 (2013)

14. Steven Eyobu, O., Han, D.: Feature representation and data augmentation for human activity classification based on wearable imu sensor data using a deep LSTM neural network. Sensors **18**, 9 (2018)

15. Tan, T.-H., Gochoo, M., Jean, F.-R., Huang, S.-C., Kuo, S.-Y.: Front-door event classification algorithm for elderly people living alone in smart house using wireless binary sensors. IEEE Access **5**, 10734–10743 (2017)

16. Yulita, I.N., Fanany, M.I., Arymurthy, A.M.: Fuzzy clustering and bidirectional long short-term memory for sleep stages classification. In: International Conference on Soft Computing, Intelligent System and Information Technology (ICSIIT), pp. 11–16. IEEE (2017)

17. Yulita, I.N., Fanany, M.I., Arymuthy, A.M.: Bi-directional long short-term memory using quantized data of deep belief networks for sleep stage classification. Procedia Comput. Sci. **116**, 530–538 (2017)

18. Zhao, Y., Yang, R., Chevalier, G., Xu, X., Zhang, Z.: Deep residual bidir-LSTM for human activity recognition using wearable sensors. Math. Probl. Eng. **2018**, 7316954, 13 (2018)

19. Zhu, J., San-segundo, R., Pardo, J.M.: Feature extraction for robust physical activity recognition. Hum. Centric Comput. Inf. Sci. **7**(1), 1–16 (2017)

Single Frame Image Super Resolution Using ANFIS Interpolation: An Initial Experiment-Based Approach

Muhammad Ismail[1,2], Jing Yang[1,3], Changjing Shang[1(✉)], and Qiang Shen[1]

[1] Department of Computer Science, Faculty of Business and Physical Sciences,
Aberystwyth University, Aberystwyth, Ceredigion, Wales, UK
{isi,jiy6,cns,qqs}@aber.ac.uk

[2] Department of Computer Science, Sukkur IBA University, Sukkur, Sindh, Pakistan
ismail@iba-suk.edu.pk

[3] School of Computer Science, Nortwestern Polytechnical University,
Xi'an 710072, China

Abstract. Image super resolution is a classical problem in image processing. Different from most of the existing super resolution algorithms that work on sufficient training data, in this work, a new super resolution method is proposed to handle the situation where the training data is insufficient by the use of ANFIS (Adaptive Network-based Fuzzy Inference System) interpolation. ANFIS interpolation aims to interpolate an effective ANFIS given only sparse data in the problem area of interest, with the assistance of two well trained ANFISs in the neighbourhood areas. The interpolated ANFIS constructs mappings from low resolution images to high resolution ones, which provides an effective mechanism for further inference of high resolution images from given low resolution ones. Experimental results indicate that the proposed approach entails improved super resolution performance for situations where there is a shortage of training data.

Keywords: Image super resolution · Insufficient training data · ANFIS interpolation

1 Introduction

Single frame image Super Resolution (SR) is the process of generating a High Resolution (HR) image using one single Low Resolution (LR) image of the same scene as input. Such techniques play an important role in devising high-level image processing tasks (such as classification, recognition and tracking problems) where direct analysis using LR images may be difficult. There are different approaches to single frame image SR. Amongst them learning-based methods which induce the mappings from LR to HR images have gained significant attention over the past decades [1–4].

In an effort to learn the relationship between an LR and an HR image, one approach is to partition the entire training data set available into a cluster of

© Springer Nature Switzerland AG 2020
Z. Ju et al. (Eds.): UKCI 2019, AISC 1043, pp. 27–40, 2020.
https://doi.org/10.1007/978-3-030-29933-0_3

feature spaces, leading to the development of multiple linear mappings [2]. To enable learning of nonlinear mappings, deep convolutional neural network (CNN) can be employed, which represents the state of the art techniques in SR [3]. However, the training of such mappings using CNN is rather time consuming while it is also very difficult to tune the huge number of parameters in the network. Besides, this approach requires substantial amount of training data to work, whilst the resulting mappings are not easy to interpret despite their accuracy. Fuzzy inference systems (FIS) provide an alternative way for learning non-linear mappings using a set of fuzzy rules [4], making both the training process and the learned mappings interpretable.

Fuzzy Inference Systems are popular tools for reasoning with imprecise knowledge. A FIS mainly consists of a knowledge base (aka., rule base) and an inference engine. There are several types of inference engines available to implement FISs, but Mamdani [5] and TSK [6] are the two most commonly used for solving real world problems. In particular, ANFIS (Adaptive Network-based Fuzzy Inference System) [7] has proven to be very powerful for generating non-linear mappings from a given input space to an output space. Based on this observation, the TSK-type ANFIS has been utilised in [8] to learn multiple ANFIS mappings for single frame image SR. Nonetheless, as with most of the existing SR algorithms this method works upon the assumption that there exists sufficient training data. As such, how to perform image SR on insufficient (or sparse) training data remains a challenging problem.

Independent of whether Mamdani or TSK inference method is used, for a problem involving sparse knowledge, no rule would be fired if a given observation does not overlap with any rule antecedent in the rule base. In order to handle such situations where sparse data is present, Fuzzy Rule Interpolation (FRI) techniques can be exploited. FRI was originally proposed in [9] which could achieve appropriate conclusions using a sparse rule base, through manipulating neighbouring rules that do not directly match an observation. Since then, a lot of research has been done on sparse rule bases using Mamdani inference. Particularly, in [10,11], the scale and move based transformation technique was introduced for fuzzy interpolative reasoning. The aforementioned and almost all of the many subsequent developments (e.g., [12–20]) were however, focussed on the use of the Mamdani method until very recently. The latest approach as reported in [21], however, works on TSK models. As such, this seminal algorithm is not pure data-driven.

Inspired by this observation, a novel ANFIS interpolation method has been proposed in [22], with initial evaluations using synthetic data and standard (but small scale) benchmark real-world data. The method constructs an ANFIS without requiring a large number of training data, which is very useful for situations where only sparse training data is available. It works by interpolating a group of fuzzy rules, with the assistance of two ANFISs trained in the neighbourhood of the problem area at hand. This paper further develops such work, through applying it to a much more challenging real problem, that of image SR where the training image pixels are insufficient. The work demonstrates the efficacy of this approach for improving the otherwise poor performance of the original ANFISs that are constructed with limited training data.

The rest of the paper is organised as follows. For academic completeness, Sect. 2 presents an overview of the problem of single frame image super resolution. Section 3 details the proposed ANFIS Interpolation based SR approach, in both descriptive and pictorial form, along with its training and testing phases. Experimental investigation is reported and discussed in Sect. 4, including the experimental setup, performance criteria and experimental results. Section 5 concludes the paper.

2   Single Frame Image Super Resolution

Image resolution reflects how much detail is contained in an image; an image with higher resolution means that more details are captured. High resolution images are desired in various applications, but images may only be obtained with Low Resolution due to poor image sensors, budget limitations or other practical reasons. Image Super Resolution techniques are therefore, developed to improve the resolution of LR images using computational algorithms. Typically, image SR algorithms can be divided into two classes: multi-frame image SR and single-frame image SR. Multi-frame image SR approaches [23,24] require multiple low resolution images of the same scene, which are aligned on the sub-pixel level. Obviously, they are restricted to the situations where multiple LR images of the same scene are available. Besides, their performance greatly relies on the accuracy of the alignment algorithms. Single frame image SR methods [1,25] are proposed to overcome the disadvantages of the multi-frame image SR approaches, generating the HR image using only one given LR image.

In single frame SR problems, the observed LR image $\mathbf{X}$ is assumed to be the down sampled version of the HR image $\mathbf{Z}$:

$$\mathbf{X} = H\mathbf{Z} \tag{1}$$

Here, H represents the down sampling operator. Single frame SR is therefore, the inverse problem of the above image degradation equation. For a given LR input $\mathbf{X}$, infinitely many LR images $\mathbf{Z}$ satisfy Eq. (1), thus super-resolution is an extremely ill-posed inverse problem. In order to handle such a problem, other information is needed. In learning based methods [2–4], for instance, an extra database is used, consisting of a large amount of image pairs with both high and low resolutions. Based on such an additional database the learning algorithms learns the inverse relationships between the LR and HR images. This leads to the state of the art results in addressing single frame SR problems.

As the machine learning techniques are developing, various learning based SR methods have been proposed, including a range of techniques capable of producing learned mappings that simulate the underlying relationship between the LR and HR images. In particular, linear mappings were proposed to describe such mapping relationship. Whilst it is very simple to implement, such a method suffers from inaccurate mapping results. This raises the requirement of developing more accurate, non-linear mappings. In the literature, non-linear mappings learned by fuzzy rules or deep CNN are the most commonly applied. Amongst the existing techniques, the image SR method as reported in [8] is of direct relevance to this work, which learns the non-linear mappings using an ANFIS.

3   Proposed Approach

In this section a novel approach for image SR is proposed using ANFIS interpolation. The description is divided into two parts: training phase and testing phase. In the first phase, although a large number of LR-HR pixel pairs are fed into the learning system in general, for certain mapping properties there is a lack of sufficient training data (and hence, the need for interpolation). In the second part, LR images are fed to a learned ANFIS to compute the corresponding mapped HR image. Thus, it is the first phase that is scientifically more challenging.

3.1   Training Phase Using ANFIS Interpolation

The proposed training process is summarised in Algorithm 3.1, while its flowchart is shown in Fig. 1. Suppose that to train an ANFIS model, 75 Bitmap images are used as the training set. LR-HR patch pairs of a size of 9 by 9 are extracted from the training data set. Then they are converted into LR-HR pixel vector pairs P. Based on their respective pixel values, P is partitioned into

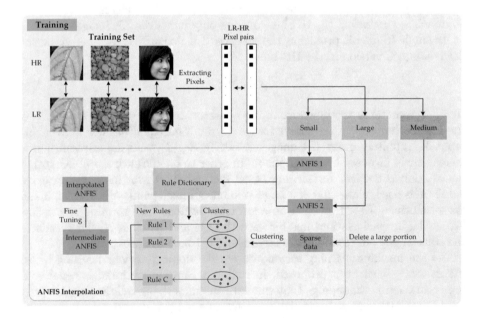

Fig. 1. Flowchart of proposed training phase.

three subsequent categories 'small' pixels, 'medium' pixels and 'large' pixels, as depicted in the upper half of Fig. 1.

Suppose that the 'large' and 'small' pixels in the training set are of sufficient amounts, while the 'medium' pixels are insufficient. From this, for the 'small' and 'large' pixels, two ANFISs (denoted as $\mathcal{A}_s$ and $\mathcal{A}_l$) can be trained using the standard ANFIS training procedure. However, the shortage of 'medium' pixels will lead to very poor performance if the same ANFIS training process is directly applied. In order to improve the accuracy of the learned ANFIS under sparse training data situations, this work proposes an ANFIS interpolation mechanism to construct the required ANFIS $\mathcal{A}_m$ for the sparse 'medium' pixels. That is, the proposed ANFIS interpolation approach utilises the two well trained ANFISs $\mathcal{A}_s$ and $\mathcal{A}_l$ as source ANFISs to support the generation of a target ANFIS $\mathcal{A}_m$. This training process is detailed in the following subsections.

(1) Generating a Rule Dictionary. A rule dictionary is constructed firstly to store the extracted fuzzy rules from the learned source ANFISs $\mathcal{A}_s$ and $\mathcal{A}_l$, which is to be subsequently used for generating an intermediate ANFIS. From a given ANFIS, a set of fuzzy production rules can be extracted. For the present application of image SR, it may be assumed that the ith rule can be expressed in the following format:

$$R_i : if\ x\ is\ A_i,\ then\ z_i = p_i x + r_i \tag{2}$$

where x denotes the input LR pixel, A_i is the corresponding fuzzy set value of the pixel, z_i represents the output of the ith rule (which contributes to the final outcome of the HR image being constructed), and r_i is a constant coefficient within the linear combination of the rule consequent.

The rule dictionary $D = \{D_a, D_c\}$ is generated by reorganising the above extracted fuzzy rules, with the antecedent part D_a and the consequent part D_c each collecting all the antecedents and consequents of those rules. Suppose that two source ANFISs consist of n_1 and n_2 rules, respectively, then there will be totally $N = n_1 + n_2$ rules in the rule dictionary. Thus, D_a consists of the antecedent parts of all the rules:

$$D_a = \{A_1\ A_2\ \cdots\ A_N\} \tag{3}$$

and the consequent part D_c consists of the consequents of the rules:

$$D_c = \begin{bmatrix} p_1\ p_2\ \cdots\ p_N \\ r_1\ r_2\ \cdots\ r_N \end{bmatrix} \tag{4}$$

where each column denotes the linear coefficients in the consequent part of a certain rule.

(2) Interpolating an Intermediate ANFIS. Having obtained the above rule dictionary, the small number of 'medium' pixel pairs $\{(x, z)\}$ are divided into C clusters using the K-means algorithm (although if preferred, any other numeric value-based clustering method may be used as the alternative to perform clustering). For the centre of each cluster, a new fuzzy rule is interpolated. Then, by aggregating all interpolated rules, an intermediate ANFIS results [8].

For each cluster C_k, compute its centre, resulting in $c^{(k)}$. Given the previously obtained antecedent part rule dictionary D_a, the first step then, is to select K closest rule antecedent $A^i \in D_a, i = 1, \ldots, K$ with respect to $c^{(k)}$. This is done on the basis of a distance metric, say for simplicity, $d^i = d(A^i, c^{(k)}) = |Rep(A^i) - c^{(k)}|$, where $Rep(A^i)$ stands for the representative value of the fuzzy set A^i [10]. The K rule antecedents $\{A^i\}$ with the smallest distances d^i are chosen, whose index set is denoted by $\mathcal{K}$.

From this, the next step is set to find the best reconstruction weights for the chosen closest rules. This is achieved by solving the following optimisation problem under the constraint that the sum of all the weights equals to 1:

$$w^{(k)} = \min_{w^{(k)}} ||c^{(k)} - \sum_{i \in \mathcal{K}} Rep(A^i) w_i^{(k)}||^2, \ s.t. \ \sum_{i \in \mathcal{K}} w_i^{(k)} = 1 \tag{5}$$

where $w_i^{(k)}$ denotes the relative weighting of A^i. The solution of this constrained least square problem is as follows:

$$w^{(k)} = \frac{G^{-1} \mathbf{1}}{\mathbf{1}^T G^{-1} \mathbf{1}} \tag{6}$$

where $G = (c^{(k)} \mathbf{1}^T - Y)^T (c^{(k)} \mathbf{1}^T - Y)$ is a defined Gram matrix, $\mathbf{1}$ is a column vector of ones, and the columns of Y are the selected rule antecedents.

Following the standard fuzzy rule interpolation technique as per [10], the weights $w^{(k)}$ are applied onto both the antecedent part and the consequent part in interpolating a new rule to summarise the kth cluster:

$$R_k : if \ x \ is \ A^k, \ then \ z_k = p_k x + r_k \tag{7}$$

where the parameters are generated by:

$$A^k = \sum_{i \in \mathcal{K}} w_i^{(k)} A^i, \ p_k = \sum_{i \in \mathcal{K}} w_i^{(k)} p_i, \ r_k = \sum_{i \in \mathcal{K}} w_i^{(k)} r_i \ k = 1, 2, \cdots, C. \tag{8}$$

3.2   Testing Phase

The phase run to test the implemented approach is illustrated in Fig. 2. From left to right in this figure, bicubic interpolation is applied on the input LR image by the scale factor s to make an image of a larger size.

As the proposed ANFIS model works on raw pixel values LR pixels are extracted from the enlarged image and partitioned into 3 categories based on their pixel values: 'small', 'medium' and 'large'. The testing process runs through the three corresponding trained ANFISs: ANFIS $\mathcal{A}_s$, ANFIS $\mathcal{A}_l$ and the interpolated ANFIS $\mathcal{A}_m$. Those 'small' and 'large' pixel values are fed into $\mathcal{A}_s$ and $\mathcal{A}_l$ respectively while $\mathcal{A}_m$ gets those in the 'medium' range of pixel values. Thus,

---

Algorithm 3.1. Training Phase using ANFIS Interpolation

Input:
   1. HR natural image data set $\{\mathbf{Z}^h\}$
   2. Scale factor s

Step 1: Generating Database
   1. Generate LR images $\{\mathbf{X}^l\}$ from $\{\mathbf{Z}^h\}$ by rate of s;
   2. Extract pixels from LR-HR image pairs $\{(\mathbf{X}^l, \mathbf{Z}^h)\}$ to form
   pixel vector pairs $\mathbf{P} = (\mathbf{x}, \mathbf{z})$;
   3. Divide pixel pairs $\mathbf{P}$ into 'small' pixels, 'medium' pixels and
   'large' pixels using K-Means algorithm;

Step 2: Training ANFIS for 'small' and 'large' pixels
          with standard ANFIS training procedure
   1. Train first ANFIS $\mathcal{A}_\mathbf{s}$ using 'small' pixels;
   2. Train second ANFIS $\mathcal{A}_\mathbf{l}$ using 'large' pixels;

Step 3: Interpolating ANFIS for 'medium' pixels
   1. Construct rule dictionary by extracting rules from $\mathcal{A}_\mathbf{s}$ and $\mathcal{A}_\mathbf{l}$;
   2. Construct intermediate ANFIS $\mathcal{A}_\mathbf{m}$ for 'medium' pixels by
   interpolating rules using $\mathcal{A}_\mathbf{s}$ and $\mathcal{A}_\mathbf{l}$;
   3. Fine tune intermediate ANFIS $\mathcal{A}_\mathbf{m}$

Output:
   Three learned ANFISs $\mathcal{A}_\mathbf{s}$, $\mathcal{A}_\mathbf{m}$ and $\mathcal{A}_\mathbf{l}$

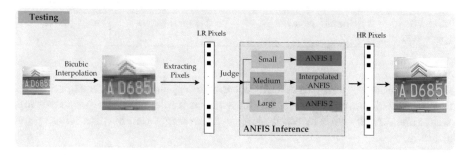

Fig. 2. Flowchart of testing phase.

the trained ANFIS-based Inference takes LR pixel values as input and converts them into HR pixel values, which are subsequently converted into the required HR image. This testing phase is presented in Algorithm 3.2.

With regard to different types of image pixels, three domains are defined: Source Domain 1, Source Domain 2 and Target Domain. As indicated previously, the source domains contain 'small' and 'large' vectors respectively, whilst Target Domain comprises sparse 'medium' pixel vectors.

For comparative analysis purpose, three models are employed, namely: Model 1, Model 2 and Model 3, as illustrated in Table 1. These models run over the same source domain 1 and source domain 2 but a different target domain, which is to be identified. Model 1, which is an ideal (reference) model, performs ANFIS using whole data in the target domain with no need to apply any interpolation

Algorithm 3.2. Testing Phase

Input:
    1. Low resolution image $\mathbf{X}$
    2. Scale factor s

    1. Pre-processing: Upscale LR image $\mathbf{X}$ by scale factor s
    using bicubic interpolation;
    2. **for** each pixel of upscaled LR image:
        1) Determine type of current pixel ('small', 'medium' or 'large');
        2) Inference using corresponding ANFIS ($\mathcal{A}_{\mathbf{s}}$, $\mathcal{A}_{\mathbf{m}}$ and $\mathcal{A}_{\mathbf{l}}$);
        end for
    3. Integrate HR pixels to form HR image $\mathbf{Z}^0$;
    4. Post-processing: Apply iterative back projection:
        $\mathbf{Z}^{t+1} = \mathbf{Z}^t + \lambda * I(\mathbf{X} - D(\mathbf{Z}^t))$

Output:
    Estimated high resolution image $\hat{\mathbf{Z}}$

Table 1. Model specification

Models	Target domain	Interpolation	Source domain 1	Source domain 2
Model 1 (reference model)	Whole data	No	Small pixels	Large pixels
Model 2	Sparse data	No	Small pixels	Large pixels
Model 3 (proposed model)	Sparse data	Yes	Small pixels	Large pixels

technique. Model 2 performs ANFIS without interpolation with the sparse data in the target domain. However, Model 3 which implements the proposed approach, runs ANFIS with interpolation over the target domain involving sparse data only.

4   Experimental Results

This section presents and discusses the results of experimental comparison.

4.1   Experimental Setup

The number of images used for training and testing are 75 and 5 respectively. During the training phase, HR images are down sampled using the scale factor s to generate their respective LR images using the same factor s. Note that this scale factor s is an important input parameter as per Algorithm 3.1. In this initial experimental investigation, its value is empirically set to 2 or 3, dividing the experiments into two cases as shown in Tables 2 and 3 respectively.

Note that the present work is developed for situations with sparse image training data, especially for the cases where in the target domain a large number of training data is missing whilst sufficient training data in neighbourhood source domains are available. Such cases can be common in different data sets. However, in this experimental investigation, in order to have ground truth to compare the performances of different approaches, for the target domain a large portion of 'medium' pixels are randomly removed to artificially create a sparse data.

4.2 Performance Criteria

The PSNR and SSIM indices are employed to quantitatively evaluate the performance of the three models. The PSNR (Peak Signal-to-Noise Ratio) index assesses the model results by computing the difference between the ground truth HR image $\mathbf{Z}$ and the estimated HR image $\hat{\mathbf{Z}}$, which is calculated as follows:

$$MSE = \frac{\| \mathbf{Z} - \hat{\mathbf{Z}} \|_F^2}{MN}, \quad PSNR = 10 \log_{10}(\frac{255^2}{MSE}) \tag{9}$$

where M, N are the image length and image width respectively, and $\| \cdot \|_F$ stands for the Frobenius norm of matrix. The SSIM (Structure SIMilarity) index reflects the degree of structural similarity between the ground truth HR image and the estimated HR image. SSIM is defined by the following:

$$SSIM = \frac{4\mu_{\mathbf{Z}}\mu_{\hat{\mathbf{Z}}}\sigma_{\mathbf{Z},\hat{\mathbf{Z}}}}{(\mu_{\mathbf{Z}}^2 + \mu_{\hat{\mathbf{Z}}}^2)(\sigma_{\mathbf{Z}}^2 + \sigma_{\hat{\mathbf{Z}}}^2)} \tag{10}$$

with $\mu_{\mathbf{Z}}$ and $\mu_{\hat{\mathbf{Z}}}$ denote the mean value of the ground truth and that of the estimated HR image, respectively; $\sigma_{\mathbf{Z}}$ and $\sigma_{\hat{\mathbf{Z}}}$ denote their corresponding standard deviations; and $\sigma_{\mathbf{Z},\hat{\mathbf{Z}}}$ is the covariance of $\mathbf{Z}$ and $\hat{\mathbf{Z}}$. The range of the SSIM index is $[0, 1]$. Generally, a lager PSNR or SSIM means a better result.

4.3 Results and Discussion

Again, experimental results for cases where the scale factor is 2 and 3 are shown in Tables 2 and 3 respectively.

Table 2. Results of three models with scale factor being 2

Image	Index	Models		
		Model 1	Model 2	Model 3
Child	PSNR/SSIM	35.186/0.948	34.549/0.940	34.859/0.945
Butterfly	PSNR/SSIM	28.545/0.927	28.329/0.919	28.417/0.925
Hat	PSNR/SSIM	32.483/0.9103	32.144/0.888	32.289/0.902
House	PSNR/SSIM	32.652/0.914	32.438/0.896	32.599/0.908
Airplane	PSNR/SSIM	30.798/0.940	30.630/0.936	30.707/0.939
Average	PSNR/SSIM	31.933/0.928	31.619/0.916	31.774/0.924

To support analysing and comparing the performances of the three trained ANFIS models, five popular LR images (Child, Butterfly, Hat, House, and Airplane) are used. Generally speaking, the results of using scale factor of 2 are much better than those of 3. In particular, the average values of PSNR with the scale factor being 2 are 3.53 dB up to 4.163 dB higher than those with the scale factor being 3, and the average SSIM values with scale factor 2 are 0.084 up to 0.102 higher than those with scale factor 3.

Importantly, the differences in the average values of PSNR and SSIM between Model 1 (the reference) and Model 3 (the proposed approach) are approximately 0.11 dB and −0.004, respectively. This implies that the values of Model 3 are very similar to those of the reference model. Particularly, the results given in Table 2 (with scale factor being 2) indicate that the difference of average PSNR values between Model 1 and Model 3 is 0.159 dB, and that the difference of average SSIM values between these two models is only 0.001. Both differences are very minor. However, the difference between Model 2 and Model 1 is 0.314 dB which is much higher than 0.155 (of the difference between Models 1 and 3). Examining the results more closely, it can be seen that the SSIM values of Model 2 in relation to Model 1 are particularly lower for the Hat and House images as compared to the respective values of Model 3.

Table 3. Results of three models with scale factor being 3

Image	Index	Models		
		Model 1	Model 2	Model 3
Child	PSNR/SSIM	30.149/0.849	28.215/0.804	30.052/0.846
Butterfly	PSNR/SSIM	24.523/0.820	23.793/0.786	24.449/0.815
Hat	PSNR/SSIM	29.483/0.827	28.934/0.799	29.404/0.818
House	PSNR/SSIM	29.919/0.858	29.246/0.831	29.853/0.845
Airplane	PSNR/SSIM	27.503/0.864	27.094/0.852	27.440/0.861
Average	PSNR/SSIM	28.316/0.844	27.456/0.814	28.240/0.837

The trend of results as shown in Table 3 with the scale factor being 3 is similar. The difference of average PSNR values between Model 1 (the reference model) and Model 3 (the proposed model) is merely 0.076 dB, and the difference of average SSIM values between them is 0.007 which is again quite similar, as with the case when the scale factor is 2. Nevertheless, the difference of average PSNR values between Model 2 and Model 1 is 0.861 dB, which is much higher than 0.076 (between model 1 and model 3). This is also reflected by the comparatively much higher difference of 0.785 dB, regarding the same performance index between

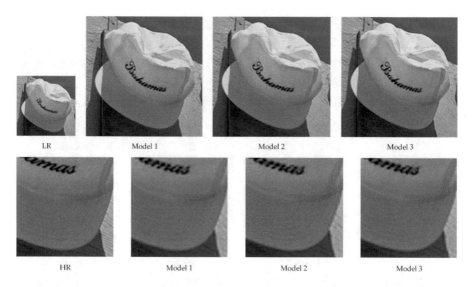

Fig. 3. Visual results of image Hat when scale factor is 2 – First line: LR to HR output images. Second line: Detailed HR images (From left to right: original HR image used as ground truth, estimated HR image using Model 1, Model 2 and Model 3).

Model 3 and Model 2. As a matter of fact, the individual SSIM values of different images such as Child, Butterfly and Hat of Model 2 are quite lower than their respective values between Models 1 and 3.

As an example for qualitative illustration, Figs. 3 and 4 show the LR images of Hat and Child and their respective HR output images through running all three models, when the scare factors are 2 and 3 respectively. For comparison purpose, one patch taken from the Hat images and two different patches extracted from the Child image are also shown in Figs. 3 and 4. It can be observed from these figures that in the estimated HR images using Model 2, there are obvious noise and bad edges, whilst the results of the proposed model (Model 3) are much closer the reference model (Model 1). This demonstrates that the proposed interpolated ANFIS model significantly improves over the poor performance of Model 2 that is directly trained with insufficient data.

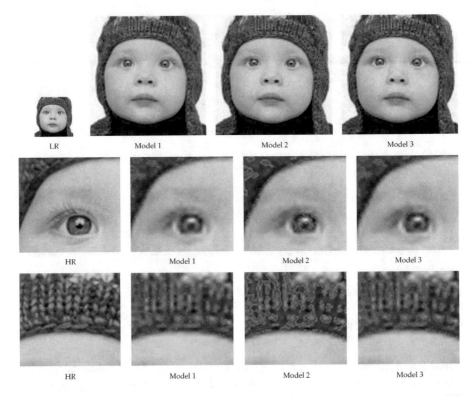

Fig. 4. Visual results of image Child when scale factor is 3 – First line: LR to HR output images. Second and third lines: Detailed HR images.

5   Conclusion

This paper has offered a novel approach for single frame image SR, via generating ANFISs through rule interpolation. It is developed for situations with sparse image training data. The approach works at the pixel level, by dividing given LR image pixels into categories which are referred to as source domains, where sufficient training data are available, or target domains, where a large number of training data is missing. Then, the ANFIS in a target domain is learned through interpolating the ANFISs trained with data from its closest neighbourhood source domains. This study extends the prior work of [22] which considered sparse numerical function approximation only, and that of [8] where multiple ANFISs were mapped with the patches of an LR image to obtain the final HR image. The performance of the proposed approach has been compared to the reference model which is fed with original data, showing that it provides similar results as those attainable by the reference model while significantly improves upon the outcomes achievable by the ANFISs directly trained with sparse data without interpolation.

As an initial attempt on the development of an ANFIS-based image super resolution mechanism, and also for simplicity, the proposed approach directly works on raw image pixels. However, much research has shown that image features are usually more representative than raw image pixels. Therefore, it would be potentially beneficial to replace images' pixels with extracted features instead as input variables of the ANFISs, in an effort to strengthen the effectiveness of this work. In addition, the proposed approach is currently compared to the results of using the standard ANFIS training algorithm only. Apart from ANFIS, there exist other mapping learning techniques (e.g., [2,4]) for image SR. Thus, further experimental evaluation and comparison with such alternative approaches, using sparse training data, forms an interesting piece of future work. Furthermore, as widely recognised in the literature, the use of less but more informative features may do better than utilising all extracted features [26], integrating feature selection techniques (e.g., [27,28]) into such further developments may boost their efficiency.

References

1. Yang, J., Wright, J., Huang, T.S., Ma, Y.: Image super-resolution via sparse representation. IEEE Trans. Image Process. **19**(11), 2861–2873 (2010)
2. Zhang, K., Tao, D., Gao, X., Li, X., Xiong, Z.: Learning multiple linear mappings for efficient single image super-resolution. IEEE Trans. Image Process. **24**(3), 846–861 (2015)
3. Dong, C., Loy, C.C., He, K., Tang, X.: Image super-resolution using deep convolutional networks. IEEE Trans. Pattern Anal. Mach. Intell. **38**(2), 295–307 (2015)
4. Purkait, P., Pal, N.R., Chanda, B.: A fuzzy-rule-based approach for single frame super resolution. IEEE Trans. Image Process. **23**(5), 2277–2290 (2014)
5. Mamdani, E.H.: Application of fuzzy logic to approximate reasoning using linguistic synthesis. In: Proceedings of the Sixth International Symposium on Multiple-Valued Logic, pp. 196–202. IEEE Computer Society Press (1976)
6. Takagi, T., Sugeno, M.: Fuzzy identification of systems and its applications to modeling and control. IEEE Trans. Syst. Man Cybern., 1–132 (1985)
7. Jang, J.-S.: ANFIS: adaptive-network-based fuzzy inference system. IEEE Trans. Syst. Man Cybern. **23**, 665–685 (1993)
8. Yang, J., Shang, C., Li, Y., Shen, Q.: Single frame image super resolution via learning multiple ANFIS mappings. In: 2017 IEEE International Conference on Fuzzy Systems (FUZZ-IEEE), pp. 1–6 (2017)
9. Kóczy, L., Hirota, K.: Approximate reasoning by linear rule interpolation and general approximation. Int. J. Approx. Reason. **9**, 197–225 (1993)
10. Huang, Z., Shen, Q.: Fuzzy interpolative reasoning via scale and move transformations. IEEE Trans. Fuzzy Syst. **14**, 340–359 (2006)
11. Huang, Z., Shen, Q.: Fuzzy interpolation and extrapolation: a practical approach. IEEE Trans. Fuzzy Syst. **16**, 13–28 (2008)
12. Li, F., Shang, C., Li, Y., Yang, J., Shen, Q.: Fuzzy rule based interpolative reasoning supported by attribute ranking. IEEE Trans. Fuzzy Syst. **26**, 2758–2773 (2018)
13. Chen, C., MacParthalain, N., Li, Y., Price, P., Quek, C., Shen, Q.: Rough-fuzzy rule interpolation. Inf. Sci. **351**, 1–17 (2016)

14. Chen, S., Ko, Y.: Fuzzy interpolative reasoning for sparse fuzzy rule-based systems based on α?-cuts and transformations techniques. IEEE Trans. Fuzzy Syst. **16**(6), 1626–1648 (2008)

15. Chen, S., Ko, Y., Chang, Y., Pan, J.: Weighted fuzzy interpolative reasoning based on weighted increment transformation and weighted ratio transformation techniques. IEEE Trans. Fuzzy Syst. **17**(6), 1412–1427 (2009)

16. Yang, L., Shen, Q.: Adaptive fuzzy interpolation. IEEE Trans. Fuzzy Syst. **19**(6), 1107–1126 (2011)

17. Yang, L., Shen, Q.: Closed form fuzzy interpolation. Fuzzy Sets Syst. **225**, 1–22 (2013)

18. Chen, S., Chang, Y., Pan, J.: Fuzzy rules interpolation for sparse fuzzy rule-based systems based on interval type-2 Gaussian fuzzy sets and genetic algorithms. IEEE Trans. Fuzzy Syst. **21**(3), 412–425 (2013)

19. Jin, S., Diao, R., Quek, C., Shen, Q.: Backward fuzzy rule interpolation. IEEE Trans. Fuzzy Syst. **22**(6), 1682–1698 (2014)

20. Naik, N., Diao, R., Shen, Q.: Dynamic fuzzy rule interpolation and its application to intrusion detection. IEEE Trans. Fuzzy Syst. **26**(4), 1878–1892 (2018)

21. Li, J., Qu, Y., Shum, H.P., Yang, L.: TSK inference with sparse rule bases. In: Advances in Computational Intelligence Systems, pp. 107–123. Springer (2017)

22. Yang, J., Shang, C., Li, Y., Li, F., Shen, Q.: Generating ANFISs through rule interpolation: an initial investigation. In: UK Workshop on Computational Intelligence, pp. 150–162. Springer (2018)

23. Li, X., Hu, Y., Gao, X., Tao, D., Ning, B.: A multi-frame image super-resolution method. Signal Process. **90**(2), 405–414 (2010)

24. Irani, M., Peleg, S.: Super resolution from image sequences. In: 10th International Conference on Pattern Recognition, pp. 115–120. IEEE (1990)

25. Glasner, D., Bagon, S., Irani, M.: Super-resolution from a single image. In: Proceedings of the IEEE International Conference on Computer Vision, pp. 349–356. IEEE (2009)

26. Jensen, R., Shen, Q.: Are more features better? EEE Trans. Fuzzy Syst. **17**(6), 1456–1458 (2009)

27. Diao, R., Shen, Q.: Two new approaches to feature selection with harmony search. In: 2010 IEEE International Conference on Fuzzy Systems (FUZZ-IEEE), pp. 1–7 (2010)

28. Jensen, R., Tuson, A., Shen, Q.: Finding rough and fuzzy-rough set reducts with SAT. Inf. Sci. **255**, 100–120 (2014)

L_1-Induced Static Output Feedback Controller Design and Stability Analysis for Positive Polynomial Fuzzy Systems

Aiwen Meng[1(✉)], Hak-Keung Lam[2], Liang Hu[3], and Fucai Liu[4]

[1] Yanshan University, Qinhuangdao 066004, Hebei, China
Aiwen_Meng@126.com
[2] Kings College London, London WC2B 4BG, UK
hak-keung.lam@kcl.ac.uk
[3] De Montfort University, Leicester LE1 9BH, UK
liang.hu@dmu.ac.uk
[4] Yanshan University, Qinhuangdao 066004, Hebei, China
lfc@ysu.edu.cn

Abstract. The aim of this paper is to study the control synthesis and stability and positivity analysis under L_1-induced performance for positive systems based on a polynomial fuzzy model. In this paper, not only the stability and positivity analysis are studied but also the L_1-induced performance is ensured by designing a static output feedback polynomial fuzzy controller for the positive polynomial fuzzy (PPF) system. In order to improve the flexibility of controller implementation, imperfectly matched premise concept under membership-function-dependent analysis technique is introduced. In addition, although the static output feedback control strategy is more popular when the system states are not completely measurable, a tricky problem that non-convex terms exist in stability and positivity conditions will follow. The nonsingular transformation technique which can transform the non-convex terms into convex ones successfully plays an important role to solve this puzzle. Based on Lyapunov stability theory, the convex positivity and stability conditions in terms of sum of squares (SOS) are obtained, which can guarantee the closed-loop systems to be positive and asymptotically stable under the L_1-induced performance. Finally, in order to test the effectiveness of the derived theory, we show an example in the simulation section.

Keywords: Positive polynomial fuzzy-model-based (PPFMB) control systems · Static output feedback control · Stability analysis · Sum of squares (SOS) · L_1-induced performance

1 Introduction

Positive systems attract more and more attention from researchers. Due to the unique characteristics of positive systems, for example, the system states always

© Springer Nature Switzerland AG 2020
Z. Ju et al. (Eds.): UKCI 2019, AISC 1043, pp. 41–52, 2020.
https://doi.org/10.1007/978-3-030-29933-0_4

stay in the positive quadrant with the non-negative initial conditions, some scholars begin to investigate this kind of systems from different points of view. In [1], the controller synthesis for positive linear systems with bounded controls was investigated. Considering that the states of positive systems are not obtained completely in some cases, the authors in [2] proposed two iterative algorithms for solving static output feedback (SOF) stabilization problem for LTI multi-input multi-output systems. The work in [3] devotes to the stability of continuous-time positive switched linear systems. Although these works have provided a theoretical foundation for the study of control synthesis and stability analysis of positive systems, these results only work for positive linear systems. However, in practice, many actual systems are complex positive nonlinear systems. Due to the complexity of positive nonlinear systems, many current results for positive linear systems cannot be directly employed. Therefore, it is worth a try to study the control synthesis for positive nonlinear systems.

Designing a state feedback controller for a positive nonlinear system is easy to achieve, but it is more meaningful to design a SOF controller by thinking about the following two aspects: (1) in many cases, the state feedback controller is not available because it is hard to get all of the state information; (2) the implementation cost of state feedback controller is comparatively high. Hence, we focus on investigating SOF controllers for positive nonlinear systems in this paper. Nevertheless, the non-convex terms which are led by system matrix will generally set up a huge barrier for stability and positivity analysis [4]. As a result, how to realize the transformation of non-convex conditions is a challenging problem to be solved. Moreover, as we all know that in order to study different performance of systems, different performance indexes are given, such as, H_∞ control and H_2 control. H_∞ norm is obtained in L_2 signal space, but it cannot express the features of practical positive systems naturally. Relatively speaking, L_1-norm can describe the features of positive systems more accurately, because L_1-norm provides the sum of the values of the components. For example, when the meaning of values is the number of animals, it is more appropriate to use the L_1-norm than others.

Compared with Takage-Sugeno (T-S) fuzzy model, polynomial fuzzy model has more advantages: Firstly, polynomial fuzzy model allows polynomials in the system matrices and the membership functions (MFs) [5], which means a wider range of nonlinear systems can be expressed by this kind of fuzzy model. Secondly, imperfectly matched premise concept under membership-function-dependent analysis technique is introduced to improve the flexibility of controller implementation. Furthermore, some relaxed methods can be introduced into the stability analysis to reduce the conservativeness [6,7]. Considering the above two points, we choose the polynomial fuzzy model to express complex positive nonlinear systems. Nevertheless, positive polynomial fuzzy systems are different with general polynomial fuzzy systems because this kind of systems need all subsystem matrices to be Metzler matrices and all the elements of input matrix as well as the output matrix are non-negative [8,9]. In addition, the positivity of the closed-loop positive polynomial fuzzy control systems also should be ensured, which leads to most of the previous results based on general polynomial fuzzy systems cannot be employed directly. Thereby, the work in this paper is very interesting but also challenging.

The authors in [10] had studied the positivity and stability analysis for a positive polynomial fuzzy system. After a little trial, more efforts were made to design a polynomial fuzzy controller (PFC) based on state feedback control technique in [11]. Considering that it is hard to obtain the full information of state variables in many cases, the authors had a try to design the SOF controllers for positive polynomial fuzzy systems in [4]. However, the L_1-induced performance of positive polynomial fuzzy systems has not yet been taken into account. Noting the importance of robustness that deals with uncertainty exhibiting in practical systems, it is vital to study the L_1-induced SOF control for positive polynomial fuzzy systems. So far, as is known to the authors, there are no results related to this topic, which also gives us a big incentive to do this work.

Within this paper, the PFC is designed based on the SOF control technique. The L_1-induced performance and the stability and positivity analysis of the positivity polynomial fuzzy control systems are carried out. In order to obtain feasible solutions, on the one hand, the non-convex terms in the stability conditions need to be dealt with so that the SOSTOOL [7] can be employed. On the other hand, the convex positive conditions should be ensured so that the positivity of the closed-loop systems can be achieved. Fortunately, nonsingular transformation technique facilitates the work in this paper. The convex SOS-based stability and positivity conditions are derived in terms of Lyapunov stability theory [12]. A simulation example is employed to verify the effectiveness of this approach.

The remainder of the paper consists of the following sections. In Sect. 2, we mainly introduce some important notations, and show the positive polynomial fuzzy model as well as the PFC based on the SOF control strategy. In Sect. 3, in terms of the Lyapunov stability theory and L_1-induced performance index, the stability and positivity analysis under the L_1-induced performance of the closed-loop positive polynomial fuzzy control system are derived. In Sect. 4, we demonstrate the validity of the L_1-induced SOF control scheme for positive polynomial fuzzy systems. In Sect. 5, a conclusion is given.

2 Preliminaries

2.1 Notation

Throughout this paper, the following notations are employed [13]. The monomial in $\mathbf{x}(t) = [x_1(t), \ldots, x_n(t)]^T$ is defined as $x_1^{d_1}(t), \ldots, x_n^{d_n}(t)$, where d_k, $k \in \{1, \ldots, n\}$, is a non-negative integer. The degree of a monomial is defined as $d = \sum_{k=1}^{n} d_k$. A polynomial $\mathbf{p}(\mathbf{x}(t))$ is shown as finite linear combination of monomials with real coefficients. If a polynomial $\mathbf{p}(\mathbf{x}(t))$ can be expressed as $\mathbf{p}(\mathbf{x}(t)) = \sum_{j=1}^{m} \mathbf{q}_j(\mathbf{x}(t))^2$, where m is a non-zero positive integer and $\mathbf{q}_j(\mathbf{x}(t))$ is a polynomial for all j, it can be concluded that $\mathbf{p}(\mathbf{x}(t)) \geq 0$ is a SOS. For a matrix $\mathbf{N} \in \Re^{m \times n}$ where n_{rs} denotes the element located at the r-th row and s-th column, the expressions $\mathbf{N} \succeq 0$, $\mathbf{N} \succ 0$, $\mathbf{N} \preceq 0$ and $\mathbf{N} \prec 0$ mean that each element n_{rs} is non-negative, positive, non-positive and negative, respectively. $\mathbf{Q}(\mathbf{x}) = \text{diag}(x_1, \ldots, x_n)$ represents $\mathbf{Q}(\mathbf{x})$ is a diagonal matrix with all of the diagonal elements being $x_1, \ldots, x_n$.

2.2   Positive Polynomial Fuzzy Model

A nonlinear positive system is approximated by p polynomial fuzzy rules and the i-th fuzzy rule is shown as follows:

$$\text{Rule } i: \text{IF } f_1(\mathbf{x}(t)) \text{ is } M_1^i \text{ AND} \cdots \text{AND } f_\Psi(\mathbf{x}(t)) \text{ is } M_\Psi^i$$

$$\text{THEN } \begin{cases} \dot{\mathbf{x}}(t) = \mathbf{A}_i(\mathbf{x}(t))\mathbf{x}(t) + \mathbf{B}_i(\mathbf{x}(t))\mathbf{u}(t) + \mathbf{B}_{i\omega}\tilde{\mathbf{w}}(t), \\ \mathbf{z}(t) = \mathbf{D}_i(\mathbf{x}(t))\mathbf{x}(t) + \mathbf{E}_i(\mathbf{x}(t))\mathbf{u}(t) + \mathbf{E}_{i\omega}\tilde{\mathbf{w}}(t), \end{cases} \tag{1}$$

where $f_l(\mathbf{x}(t))$, is the premise variable, Ψ is a positive integer; M_l^i is the fuzzy set of the i-th rule corresponding to the function $f_l(\mathbf{x}(t))$; $\mathbf{x}(t) \in \Re^n$ is the system state vector; $\mathbf{u}(t) \in \Re^m$ is the input vector; $\tilde{\mathbf{w}}(t) \in \Re^p$ is the disturbance signal; $\mathbf{z}(t) \in \Re^q$ is the and controlled output; $\mathbf{A}_i(\mathbf{x}(t))$, $\mathbf{B}_i(\mathbf{x}(t))$, $\mathbf{B}_{i\omega}$, $\mathbf{D}_i(\mathbf{x}(t))$, $\mathbf{E}_i(\mathbf{x}(t))$ and $\mathbf{E}_{i\omega}$ are the system matrices with right dimensions.

The dynamics of the positive polynomial fuzzy system is expressed as follows:

$$\begin{cases} \dot{\mathbf{x}}(t) = \sum_{i=1}^{p} w_i(\mathbf{x}(t))(\mathbf{A}_i(\mathbf{x}(t))\mathbf{x}(t) + \mathbf{B}_i(\mathbf{x}(t))\mathbf{u}(t) + \mathbf{B}_{i\omega}\tilde{\mathbf{w}}(t)), \\ \mathbf{z}(t) = \sum_{i=1}^{p} w_i(\mathbf{x}(t))(\mathbf{D}_i(\mathbf{x}(t))\mathbf{x}(t) + \mathbf{E}_i(\mathbf{x}(t))\mathbf{u}(t) + \mathbf{E}_{i\omega}\tilde{\mathbf{w}}(t)), \\ \mathbf{y}(t) = \mathbf{C}\mathbf{x}(t), \end{cases} \tag{2}$$

where $\sum_{i=1}^{p} w_i(\mathbf{x}(t)) = 1, w_i(\mathbf{x}(t)) \geq 0 \forall i$, $w_i(\mathbf{x}(t))$ is the normalized grade of membership; $\mathbf{y}(t) \in \Re^l$ is the measurement output; $\mathbf{C}$ is the output matrix.

Definition 1 [1]. *A system is deemed to be positive if the initial condition* $\mathbf{x}(0) = \mathbf{x}_0 \succeq 0$ *holds and the corresponding trajectory* $\mathbf{x}(t) \succeq 0$ *for all* $t \geq 0$ *is satisfied.*

Definition 2 [1]. *A matrix* $\mathbf{M}$ *is called a Metzler matrix if its off-diagonal elements are non-negative:* $m_{rs} \succeq 0$, $r \neq s$.

Lemma 1 [14,15]. *System* (2) *is a positive system if* $\mathbf{A}_i(\mathbf{x}(t))$ *is a Metzler matrix,* $\mathbf{B}_i(\mathbf{x}(t)) \succeq 0$, $\mathbf{B}_{i\omega} \succeq 0$, $\mathbf{D}_i(\mathbf{x}(t)) \succeq 0$, $\mathbf{E}_i(\mathbf{x}(t)) \succeq 0$, $\mathbf{E}_{i\omega} \succeq 0$ *and* $\mathbf{C} \succeq 0$.

Under zero initial conditions, the L_1-induced performance of the positive polynomial fuzzy system (2) is defined as follows:

$$||\mathbf{z}(t)||_{L_1} < \gamma ||\tilde{\mathbf{w}}(t)||_{L_1}. \tag{3}$$

2.3   Polynomial Fuzzy Controller Design

In the following, a c-rule PFC in terms of the SOF control scheme is designed:

$$\text{Rule } j : \text{IF } g_1(\mathbf{y}(t)) \text{ is } N_1^j \text{ AND} \ldots \text{AND } g_\Omega(\mathbf{y}(t)) \text{ is } N_\Omega^j$$
$$\text{THEN } \mathbf{u}(t) = \mathbf{v}\mathbf{K}_j(\mathbf{y}(t))\mathbf{y}(t), \tag{4}$$

where $g_\beta(\mathbf{y}(t))$ is the premise variable, Ω is a positive integer; N_β^j is the fuzzy set of j-th rule corresponding to the function $g_\beta(\mathbf{y}(t))$; $\mathbf{K}_j(\mathbf{y}(t)) \in \Re^{1 \times l}$ is the feedback gain to be determined; $\mathbf{v} \in \Re^{m \times 1}$ is a given vector.

Taking $\mathbf{y}(t)$ into the PFC (4), we have:

$$\mathbf{u}(t) = \sum_{j=1}^{c} m_j(\mathbf{y}(t))\mathbf{v}\mathbf{K}_j(\mathbf{y}(t))\mathbf{y}(t) = \sum_{j=1}^{c} m_j(\mathbf{y}(t))\mathbf{v}\mathbf{K}_j(\mathbf{y}(t))\mathbf{C}\mathbf{x}(t), \tag{5}$$

where $\sum_{j=1}^{c} m_j(\mathbf{y}(t)) = 1, m_j(\mathbf{y}(t)) \geq 0, \forall j$, $m_j(\mathbf{y}(t))$ is the normalized grade of membership.

Remark 1. The vector $\mathbf{v}$ is a non-zero constant vector which is given in advance by users. This method can facilitate the introduction of the nonsingular transformation technique which works effectively for solving the non-convex problem.

Remark 2. So far, some authors kept a watchful eye on unstable open-loop general systems [11,16], which means only the positivity of the closed-loop systems require to be ensured. However, the work in [15] ensured both open-loop systems and closed-loop systems to be positive. In this paper, we focus on the latter case that the positivity of both polynomial fuzzy model and closed-loop positive polynomial fuzzy control system should be maintained.

For simplicity, $w_i(\mathbf{x}(t)), m_j(\mathbf{y}(t)), \mathbf{x}(t)$ and $\mathbf{y}(t)$ will be abbreviated as $w_i(\mathbf{x})$, $m_j(\mathbf{y})$, $\mathbf{x}$ and $\mathbf{y}$, respectively.

3   Stability and Positivity Analysis

In this section, we will give the stability and positivity analysis under the L_1-induced performance for the closed-loop positive polynomial fuzzy control system based on the Lyapunov theory and the L_1-induced performance index. Convex SOS-based conditions will be derived by employing some useful techniques to solve non-convex terms.

3.1   Closed-Loop Positive Polynomial Fuzzy Control Systems

Based on (2) and (5), we have the following closed-loop positive polynomial fuzzy control system:

$$\begin{cases} \dot{\mathbf{x}} = \sum_{i=1}^{p} \sum_{j=1}^{c} w_i(\mathbf{x})m_j(\mathbf{y})\Big((\mathbf{A}_i(\mathbf{x}) + \mathbf{B}_i(\mathbf{x})\mathbf{v}\mathbf{K}_j(\mathbf{y})\mathbf{C})\mathbf{x} + \mathbf{B}_{i\omega}\tilde{\mathbf{w}}\Big), \\ \mathbf{z} = \sum_{i=1}^{p} \sum_{j=1}^{c} w_i(\mathbf{x})m_j(\mathbf{y})\Big((\mathbf{D}_i(\mathbf{x}) + \mathbf{E}_i(\mathbf{x})\mathbf{v}\mathbf{K}_j(\mathbf{y})\mathbf{C})\mathbf{x} + \mathbf{E}_{i\omega}\tilde{\mathbf{w}}\Big), \\ \mathbf{y} = \mathbf{C}\mathbf{x}. \end{cases} \tag{6}$$

3.2   Stability Analysis of Closed-Loop Positive Polynomial Fuzzy Control Systems

The following Lyapunov function candidate [15] is chosen to study the stability of the closed-loop positive polynomial fuzzy control system (6):

$$V(t) = \lambda^T \mathbf{x}, \tag{7}$$

where $\lambda = [\lambda_1, \dots, \lambda_n]^T \succ 0$ is a vector to be determined.

Then we have the time derivative of $V(t)$ as follows:

$$
\begin{aligned}
\dot{V}(t) &= \lambda^T \dot{\mathbf{x}} \\
&= \sum_{i=1}^{p} \sum_{j=1}^{c} w_i(\mathbf{x}) m_j(\mathbf{y}) \lambda^T \Big(\mathbf{B}_{i\omega} \tilde{\mathbf{w}} + (\mathbf{A}_i(\mathbf{x}) + \mathbf{B}_i(\mathbf{x}) \mathbf{v} \mathbf{K}_j(\mathbf{y}) \mathbf{C}) \mathbf{x} \Big).
\end{aligned} \tag{8}
$$

In the following, the L_1-induced performance will be taken into consideration, and the L_1-induced performance index is given:

$$J = \int_0^\infty \|\mathbf{z}\|_{L_1} - \gamma \|\tilde{\mathbf{w}}\|_{L_1} dt, \tag{9}$$

Combining stability theory and performance index, the equality (9) can be dealt with as following:

$$
\begin{aligned}
J &= \int_0^\infty \|\mathbf{z}\|_{L_1} - \gamma \|\tilde{\mathbf{w}}\|_{L_1} + \dot{V} - \dot{V} dt = \int_0^\infty \sum_{k=1}^{q} \mathbf{z} - \gamma \sum_{k=1}^{p} \tilde{\mathbf{w}} + \dot{V} dt - V(\infty) \\
&= \int_0^\infty \mathbf{I}_1^T \mathbf{z} - \gamma \mathbf{I}_2^T \tilde{\mathbf{w}} + \dot{V} dt - V(\infty),
\end{aligned} \tag{10}
$$

where $\mathbf{I}_1 \in \Re^q$ and $\mathbf{I}_2 \in \Re^p$ are vectors with all of the elements being 1.

The term $V(\infty)$ is equal to 0 when $t \to \infty$, then by taking $\mathbf{z}$ and (8) into (10), we have

$$
\begin{aligned}
J &= \int_0^\infty \mathbf{I}_1^T \mathbf{z} - \gamma \mathbf{I}_2^T \tilde{\mathbf{w}} + \dot{V} dt \\
&= \int_0^\infty \mathbf{I}_1^T \Big(\sum_{i=1}^{p} \sum_{j=1}^{c} w_i(\mathbf{x}) m_j(\mathbf{y}) (\mathbf{E}_{i\omega} \tilde{\mathbf{w}} + (\mathbf{D}_i(\mathbf{x}) + \mathbf{E}_i(\mathbf{x}) \mathbf{v} \mathbf{K}_j(\mathbf{y}) \mathbf{C}) \mathbf{x}) \Big) \\
&\quad - \gamma \mathbf{I}_2^T \tilde{\mathbf{w}} + \lambda^T \Big(\sum_{i=1}^{p} \sum_{j=1}^{c} w_i(\mathbf{x}) m_j(\mathbf{y}) (\mathbf{B}_{i\omega} \tilde{\mathbf{w}} + (\mathbf{A}_i(\mathbf{x}) + \mathbf{B}_i(\mathbf{x}) \mathbf{v} \mathbf{K}_j(\mathbf{y}) \mathbf{C}) \mathbf{x}) \Big) dt \\
&= \int_0^\infty \sum_{i=1}^{p} \sum_{j=1}^{c} w_i(\mathbf{x}) m_j(\mathbf{y}) \Big((\mathbf{I}_1^T \mathbf{E}_{i\omega} - \gamma \mathbf{I}_2^T + \lambda^T \mathbf{B}_{i\omega}) \tilde{\mathbf{w}} \\
&\quad + (\mathbf{I}_1^T (\mathbf{D}_i(\mathbf{x}) + \mathbf{E}_i(\mathbf{x}) \mathbf{v} \mathbf{K}_j(\mathbf{y}) \mathbf{C}) \\
&\quad + \lambda^T (\mathbf{A}_i(\mathbf{x}) + \mathbf{B}_i(\mathbf{x}) \mathbf{v} \mathbf{K}_j(\mathbf{y}) \mathbf{C})) \mathbf{x} \Big) dt.
\end{aligned} \tag{11}
$$

In order to facilitate the analysis, we define:

$$\mathbf{Q}_{1ij}(\mathbf{x}, \mathbf{y}) = \mathbf{I}_1^T \mathbf{E}_{i\omega} - \gamma \mathbf{I}_2^T + \lambda^T \mathbf{B}_{i\omega}, \tag{12}$$

$$\mathbf{Q}_{2ij}(\mathbf{x}, \mathbf{y}) = \mathbf{I}_1^T (\mathbf{D}_i(\mathbf{x}) + \mathbf{E}_i(\mathbf{x})\mathbf{v}\mathbf{K}_j(\mathbf{y})\mathbf{C}) + \lambda^T (\mathbf{A}_i(\mathbf{x}) + \mathbf{B}_i(\mathbf{x})\mathbf{v}\mathbf{K}_j(\mathbf{y})\mathbf{C}). \tag{13}$$

Based on (11), the inequality $J < 0$ holds if $\mathbf{Q}_{1ij}(\mathbf{x}, \mathbf{y}) \prec 0$ and $\mathbf{Q}_{2ij}(\mathbf{x}, \mathbf{y}) \prec 0$ for all i and j are satisfied. However, we find that the term $\lambda^T \mathbf{B}_i(\mathbf{x})\mathbf{v}\mathbf{K}_j(\mathbf{y})\mathbf{C}$ is non-convex. To solve the non-convex problem, we assume that $\mathbf{B}_i(\mathbf{x}) = \mathbf{B}$ for all i. Inspired by the nonsingular transformation technique in [17], we have:

$$\bar{\mathbf{B}} = \mathbf{B}\mathbf{v}, \mathbf{\Gamma} = \begin{bmatrix} (\bar{\mathbf{B}}^T \bar{\mathbf{B}})\bar{\mathbf{B}}^T \\ \mathrm{ortc}(\bar{\mathbf{B}}) \end{bmatrix}, \tag{14}$$

where $\mathrm{ortc}(\bar{\mathbf{B}}) \in \Re^{(n-1) \times n}$ represents the orthogonal complement of $\bar{\mathbf{B}}$ which satisfies the following equality:

$$\mathbf{\Gamma}\bar{\mathbf{B}} = [1 \quad ; \quad \mathbf{0}_{n-1}] \tag{15}$$

where $\mathbf{0}_{n-1}$ means that the elements in $\mathbf{0}_{n-1}$ are all zero.

Therefore, the non-convex term $\lambda^T \mathbf{B}_i(\mathbf{x})\mathbf{v}\mathbf{K}_j(\mathbf{y})\mathbf{C}$ can be dealt with further:

$$\lambda^T \mathbf{B}_i(\mathbf{x})\mathbf{v}\mathbf{K}_j(\mathbf{y})\mathbf{C} = \lambda^T \mathbf{\Gamma}^{-1} \mathbf{\Gamma}\mathbf{B}\mathbf{v}\mathbf{K}_j(\mathbf{y})\mathbf{C} = p_1 \mathbf{K}_j(\mathbf{y})\mathbf{C} = \mathbf{Z}_j(\mathbf{y})\mathbf{C}, \tag{16}$$

where $\lambda^T \mathbf{\Gamma}^{-1} = \mathbf{p}^T$ and $p_1 \mathbf{K}_j(\mathbf{y}) = \mathbf{Z}_j(\mathbf{y}) \in \Re^{1 \times l}$; as the first element in $\mathbf{p}^T$, p_1 is a positive value and the detailed proof can refer to [4].

Inducing (16) and $\lambda^T = \mathbf{p}^T \mathbf{\Gamma}$ into (13), we can obtain

$$\mathbf{I}_1^T \mathbf{D}_i(\mathbf{x}) + \mathbf{I}_1^T \mathbf{E}_i(\mathbf{x})\mathbf{v}\mathbf{Z}_j(\mathbf{y})/p_1 \mathbf{C} + \mathbf{p}^T \mathbf{\Gamma} \mathbf{A}_i(\mathbf{x}) + \mathbf{Z}_j(\mathbf{y})\mathbf{C} \prec 0. \tag{17}$$

It can be seen that there still is a non-convex term in (17). In this case, we can get around this obstacle by dividing the inequality (17) into two parts. Then the inequality (17) can be ensured if the following two inequalities are satisfied:

$$\mathbf{I}_1^T \mathbf{E}_i(\mathbf{x})\mathbf{v}\mathbf{Z}_j(\mathbf{y})/p_1 \mathbf{C} \prec 0, \tag{18}$$

$$\mathbf{I}_1^T \mathbf{D}_i(\mathbf{x}) + \mathbf{p}^T \mathbf{\Gamma} \mathbf{A}_i(\mathbf{x}) + \mathbf{Z}_j(\mathbf{y})\mathbf{C} \prec 0. \tag{19}$$

Due to $p_1 > 0$, so we can multiply both sides of (18) by p_1, then we have

$$\mathbf{I}_1^T \mathbf{E}_i(\mathbf{x})\mathbf{v}\mathbf{Z}_j(\mathbf{y})\mathbf{C} \prec 0. \tag{20}$$

Based on the above analysis, the convex stability conditions can be derived. However, not only should the stability conditions be guaranteed, but also the positivity conditions need to be satisfied. Hence, in the following, we will analyze the positivity conditions. From the closed-loop positive polynomial fuzzy systems (6), we can get the positivity conditions as follows:

$$\mathbf{A}_i(\mathbf{x}) + \mathbf{B}\mathbf{v}\mathbf{K}_j(\mathbf{y})\mathbf{C} \quad \text{is a Metzler,}$$

$$\mathbf{D}_i(\mathbf{x}) + \mathbf{E}_i(\mathbf{x})\mathbf{v}\mathbf{K}_j(\mathbf{y})\mathbf{C} \succeq 0. \tag{21}$$

Recall the Definition 2 and take $\mathbf{K}_j(\mathbf{y}) = \mathbf{Z}_j(\mathbf{y})/p_1$ into (21), we can obtain

$$p_1 a_{irs}(\mathbf{x}) + \mathbf{b}_r \mathbf{v} \mathbf{Z}_j(\mathbf{y}) \mathbf{c}_s \succeq 0, \forall r \neq s$$
$$p_1 \mathbf{D}_i(\mathbf{x}) + \mathbf{E}_i(\mathbf{x}) \mathbf{v} \mathbf{Z}_j(\mathbf{y}) \mathbf{C} \succeq 0, \tag{22}$$

where $a_{irs}(\mathbf{x})$ is in the r-th row and s-th column of $\mathbf{A}_i(\mathbf{x})$, $\mathbf{b}_r$ is the r-th row of $\mathbf{B}$ and $\mathbf{c}_s$ is the s-th column of $\mathbf{C}$.

Based on the analysis above, we can summarize the results in Theorem 1.

Theorem 1. *The positive polynomial fuzzy model (2) can be controlled to be asymptotically stable and positive by the SOF PFC (6) under L_1-induced performance if there exist vectors $\mathbf{Z}_j(\mathbf{y}) \in \Re^{1 \times l}$, $\mathbf{p} \in \Re^n$ such that the following SOS-based positivity and stability conditions are satisfied:*

$$p_1 a_{irs}(\mathbf{x}) + \mathbf{b}_r \mathbf{v} \mathbf{Z}_j(\mathbf{y}) \mathbf{c}_s \text{ is SOS } \forall \, r \neq s, i, j; \tag{23}$$

$$v^T \Big(diag\big(p_1 \mathbf{D}_i(\mathbf{x}) + \mathbf{E}_i(\mathbf{x}) \mathbf{v} \mathbf{Z}_j(\mathbf{y}) \mathbf{C}\big)\Big) v \text{ is SOS}; \tag{24}$$

$$v^T \Big(diag\big(\mathbf{p}^T \mathbf{\Gamma} - \epsilon_1 \mathbf{I}\big)\Big) v \text{ is SOS}; \tag{25}$$

$$-v^T \Big(diag\big(\mathbf{I}_1^T \mathbf{E}_i(\mathbf{x}) \mathbf{v} \mathbf{Z}_j(\mathbf{y}) \mathbf{C} + \epsilon_2(\mathbf{x}) \mathbf{I}\big)\Big) v \text{ is SOS } \forall \, i, j; \tag{26}$$

$$-v^T \Big(diag\big(\mathbf{I}_1^T \mathbf{D}_i(\mathbf{x}) + \mathbf{p}^T \mathbf{\Gamma} \mathbf{A}_i(\mathbf{x}) + \mathbf{Z}_j(\mathbf{y}) \mathbf{C} + \epsilon_3(\mathbf{x}) \mathbf{I}\big)\Big) v \text{ is SOS } \forall \, i, j; \tag{27}$$

$$-v^T \Big(diag\big(\mathbf{I}_1^T \mathbf{E}_{i\omega} - \gamma \mathbf{I}_2^T + \mathbf{p}^T \mathbf{\Gamma} \mathbf{B}_{i\omega} + \epsilon_4 \mathbf{I}\big)\Big) v \text{ is SOS } \forall \, i, j; \tag{28}$$

where $v \in \Re^n$ is an arbitrary vector independent of $\mathbf{x}$ and $\mathbf{y}$; $\epsilon_1 > 0$ and $\epsilon_4 > 0$ are predefined scalars and $\epsilon_2(\mathbf{x}) > 0$ and $\epsilon_3(\mathbf{x}) > 0$ for $\mathbf{x} \neq 0$ are predefined scalar polynomials. The feedback gain is $\mathbf{K}_j(\mathbf{y}) = \mathbf{Z}_j(\mathbf{y})/p_1$.

Remark 3. The conditions (23) and (24) are the positivity conditions which are used to guarantee the positivity of the closed-loop positive polynomial fuzzy systems. The conditions (25), (26), (27) and (28) are employed to ensure the stability under the L_1-induced performance.

4   Simulation Example

In this section, a simulation example is given to validate the theory in this paper.

4.1   Scenario

The 3-rule positive polynomial fuzzy model is shown as follows:

$$\mathbf{x} = [\, x_1 \; x_2 \,]^T, \mathbf{v} = [\, 1 \,], \mathbf{C} = [\, 1 \; 0 \,], \mathbf{B}_1 = \mathbf{B}_2 = \mathbf{B}_3 = \begin{bmatrix} 1 \\ 1 \end{bmatrix},$$

$$\mathbf{A}_1(x_1) = \begin{bmatrix} 0.11 & 1 \\ 0.55 + 0.5a & -0.85 - 0.12\mathbf{x}_1^2 + 0.11\mathbf{x}_1 \end{bmatrix},$$

$$\mathbf{A}_2(x_1) = \begin{bmatrix} 0.14 & 1.2 \\ 0.72 + 0.5a & -1.37 - 0.24\mathbf{x}_1^2 + 0.25\mathbf{x}_1 \end{bmatrix},$$

$$\mathbf{A}_3(x_1) = \begin{bmatrix} 0.19 & 1.5 \\ 0.85 + 0.5a & -1.9 - 0.42\mathbf{x}_1^2 + 0.31\mathbf{x}_1 \end{bmatrix},$$

$$\mathbf{D}_1(x_1) = \begin{bmatrix} 0.45 + 0.5b & 1.52 + 0.12\mathbf{x}_1^2 + 0.31\mathbf{x}_1 \\ 1.6 & 0.38 \end{bmatrix},$$

$$\mathbf{D}_2(x_1) = \begin{bmatrix} 0.63 + 0.5b & 1.09 + 0.15\mathbf{x}_1^2 + 0.24\mathbf{x}_1 \\ 1.1 & 0.22 \end{bmatrix},$$

$$\mathbf{D}_3(x_1) = \begin{bmatrix} 0.89 + 0.5b & 2.11 + 0.18\mathbf{x}_1^2 + 0.11\mathbf{x}_1 \\ 2.3 & 0.65 \end{bmatrix},$$

$$\mathbf{B}_{1\omega} = \begin{bmatrix} 1.4 \\ 0.16 \end{bmatrix}, \mathbf{B}_{2\omega} = \begin{bmatrix} 1.5 \\ 0.35 \end{bmatrix}, \mathbf{B}_{3\omega}(x_1) = \begin{bmatrix} 1.8 \\ 0.44 \end{bmatrix},$$

$$\mathbf{E}_1 = \begin{bmatrix} 1.41 \\ 0.46 \end{bmatrix}, \mathbf{E}_2 = \begin{bmatrix} 1.65 \\ 0.15 \end{bmatrix}, \mathbf{E}_3 = \begin{bmatrix} 1.98 \\ 0.04 \end{bmatrix},$$

$$\mathbf{E}_{1\omega} = \begin{bmatrix} 1.54 \\ 1.06 \end{bmatrix}, \mathbf{E}_{2\omega} = \begin{bmatrix} 2.25 \\ 0.55 \end{bmatrix}, \mathbf{E}_{3\omega} = \begin{bmatrix} 0.18 \\ 0.84 \end{bmatrix},$$

where a and b are constant scalars.

The open-loop system is a positive system because the off-diagonal elements in $\mathbf{A}_i(x_1), i \in \{1,2,3\}$ are non-negative; the elements of $\mathbf{D}_i(x_1), i \in \{1,2,3\}, \mathbf{B}_{i\omega}, \mathbf{E}_i, \mathbf{E}_{i\omega}, \mathbf{C}$ and $\mathbf{B}$ are non-negative. In addition, $\mathbf{v}$ is a given vector and chosen arbitrarily in the simulation satisfying $\mathbf{Bv} = \bar{\mathbf{B}} = [1; \quad 1] \succeq 0$. The disturbance signal is $\tilde{\mathbf{w}}(t) = 4.5e^{-t}|cos(2t)|$.

A 2-rule PFC is designed. The MFs of the model and the controller are same as the ones in the simulation section in [4], which are illustrated in Fig. 1. Theorem 1 in this paper is verified with $1 \leq a \leq 10$ at the interval of 1 and $2 \leq b \leq 12$ at the interval of 1. And the scalars ϵ_1, $\epsilon_2(\mathbf{x})$, $\epsilon_3(\mathbf{x})$ and ϵ_4 are set as 0.001, the highest degree of $\mathbf{Z}_j(x_1)$ is 2.

Table 1. The feedback gains $K_j(x_1)$, λ and γ obtained by Theorem 1.

(a, b)	$K_j(x_1)$	λ	γ
$a = 1$	$K_1(x_1) = 1.6444e^{-17}x_1^2 - 0.9545$	$p_1 = 42.8049$	44.6147
$b = 2$	$K_2(x_1) = -1.3681e^{-17}x_1^2 - 0.9545$	$p_2 = 4.5212$	
$a = 6$	$K_1(x_1) = -1.19703e^{-15}x_1^2 - 2.2172$	$p_1 = 505.1437$	532.2583
$b = 7$	$K_2(x_1) = 4.7696e^{-15}x_1^2 - 2.2172$	$p_2 = 35.8998$	
$a = 7$	$K_1(x_1) = 2.3482e^{-12}x_1^2 - 2.7222$	$p_1 = 49.2535$	51.4175
$b = 9$	$K_2(x_1) = -3.8775e^{-12}x_1^2 - 2.7222$	$p_2 = 4.9574$	
$a = 10$	$K_1(x_1) = -1.1706e^{-12}x_1^2 - 3.4783$	$p_1 = 83.7958$	87.8538
$b = 12$	$K_2(x_1) = -1.8539e^{-12}x_1^2 - 3.4783$	$p_2 = 7.2983$	

4.2   Feasibility Analysis

From the Fig. 2, it makes clear that the SOF PFC has the ability to achieve the stability and positivity under the L_1-induced performance for an open-loop

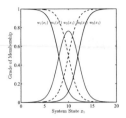

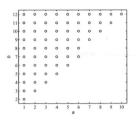

Fig. 1. Membership functions of positive polynomial fuzzy model and SOF PFC.

Fig. 2. Stability region given by Theorem 1

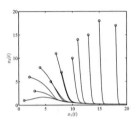

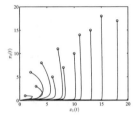

Fig. 3. Phase plot of the states x_1 and x_2 for $a = 1, b = 2$ for the open-loop system.

Fig. 4. Phase plot of the states x_1 and x_2 for $a = 1, b = 2$ for the closed-loop system.

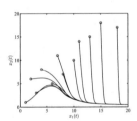

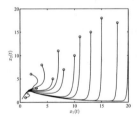

Fig. 5. Phase plot of the states x_1 and x_2 for $a = 6, b = 7$ for the open-loop system.

Fig. 6. Phase plot of the states x_1 and x_2 for $a = 6, b = 7$ for the closed-loop system.

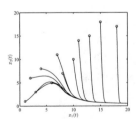

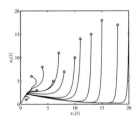

Fig. 7. Phase plot of the states x_1 and x_2 for $a = 7, b = 9$ for the open-loop system.

Fig. 8. Phase plot of the states x_1 and x_2 for $a = 7, b = 9$ for the closed-loop system.

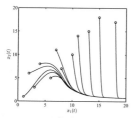

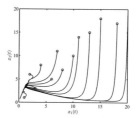

Fig. 9. Phase plot of the states x_1 and x_2 for $a = 10, b = 12$ for the open-loop system.

Fig. 10. Phase plot of the states x_1 and x_2 for $a = 10, b = 12$ for the closed-loop system.

unstable positive polynomial fuzzy system. Meanwhile, to further confirm the correctness of this method, we pick out several feasible points (a, b) to check their phase plots of x_1 and x_2 obeying various initial conditions. In Fig. 2, we choose $(1, 2)$, $(6, 7)$, $(7, 9)$ and $(10, 12)$, respectively. λ, γ and $\mathbf{K}_j(x_1)$ corresponding to these feasible points (a, b) are checked as well and appear in Table 1. The phase plots of these points can refer to Figs. 3, 4, 5, 6, 7, 8, 9 and 10, where Figs. 3, 5, 7 and 9 are the phase plots of these points for open-loop systems, and Figs. 4, 6, 8 and 10 are the phase plots of these points for closed-loop systems. From Figs. 3, 4, 5, 6, 7, 8, 9 and 10, we can see that the open-loop system is positive but unstable, while the closed-loop system is asymptotically stable and positive.

According to the phase plots, the positive polynomial fuzzy system can be driven to the origin. Therefore, it is clear that the unstable open-loop positive polynomial fuzzy systems can be controlled to be stable and positive under L_1-induced performance by the SOF PFC.

5 Conclusion

In this paper, under L_1-induced performance, the SOF control synthesis and stability analysis for closed-loop positive polynomial fuzzy control systems have been studied. The non-convex terms in the stability conditions have been transformed into convex ones so that the MATLAB third-party toolbox SOSTOOL can be used to obtain the feasible solution. The SOF strategy has been introduced in the PFC design so that the unstable open-loop positive systems can be controlled to be asymptotically stable and positive. Meanwhile, taking the L_1-induced performance into consideration, the SOS-based stability and positivity conditions have been obtained in terms of the Lyapunov stability theory. An example has been given to demonstrate the correctness of the theorem, in addition, the stability region and the phase plots of some feasible points have been shown in this paper.

References

1. Rami, M.A., Tadeo, F.: Controller synthesis for positive linear systems with bounded controls. IEEE Trans. Circuits Syst. **54**(2), 151–155 (2007)
2. Bhattacharyya, S., Patra, S.: Static output-feedback stabilization for MIMO LTI positive systems using LMI-based iterative algorithms. IEEE Control Syst. Lett. **2**(2), 242–247 (2018)
3. Ju, Y., Zhu, X., Sun, Y.: Stability analysis of continuous-time positive switched linear systems. In: Proceedings of 18th International Conference on Control, Automation and Systems, pp. 1062–1065 (2018)
4. Meng, A., Lam, H.K., Yu, Y., Li, X., Liu, F.: Static output feedback stabilization of positive polynomial fuzzy systems. IEEE Trans. Fuzzy Syst. **26**(3), 1600–1612 (2018)
5. Lam, H.K.: A review on stability analysis of continuous-time fuzzy-model-based control systems: From membership-function-independent to membership-function-dependent analysis. Eng. Appl. Artif. Intell. **67**, 390–408 (2018)
6. Tsai, S., Jen, C.: H_∞ stabilization for polynomial fuzzy time-delay system: a sum-of-squares approach. IEEE Trans. Fuzzy Syst. **26**(6), 3630–3644 (2018)
7. Lam, H.K.: Polynomial fuzzy model-based control systems: stability analysis and control synthesis using membership function dependent techniques. Springer, Switzerland (2016)
8. Steentjes, T.R.V., Doban, A.I., Lazar, M.: Feedback stabilization of positive nonlinear systems with applications to biological systems. In: Proceedings of 2018 European Control Conference, pp. 1619–1624 (2018)
9. Zhu, S., Pan, F., Feng, J.: Fuzzy filtering design for positive T-S fuzzy systems with Markov jumping parameters. In: Proceedings of 2018 Australian and New Zealand Control Conference, pp. 1–4 (2018)
10. Li, X., Liu, C., Lam, H.K., Liu, F., Zhao, X.: Stability analysis of fuzzy polynomial positive systems with time delay. In: Proceedings of International Conference on Fuzzy Theory and Its Applications, pp. 24–28 (2014)
11. Li, X., Lam, H.K., Liu, F., Zhao, X.: Stability and stabilization analysis of positive polynomial fuzzy systems with time delay considering piecewise membership functions. IEEE Trans. Fuzzy Syst. **25**(4), 958–971 (2017)
12. Prajna, S., Papachristodoulou, A., Parrilo, P.A.: Introducing SOSTOOLS: a general purpose sum of squares programming solver. In: Proceedings of 41st IEEE Conference on Decision and Control, vol. 1, pp. 741–746 (2002)
13. Prajna, S., Papachristodoulou, A., Wu, F.: Nonlinear control synthesis by sum of squares optimization: a Lyapunov-based approach. In: Proceedings of the 5th Asian Control Conference, vol. 1, pp. 157–165 (2004)
14. Farina, L., Rinaldi, S.: Positive Linear Systems: Theory and Applications. Wiley, New York (2000)
15. Zhang, J., Han, Z., Zhu, F., Huang, J.: Brief paper: feedback control for switched positive linear systems. IET Control Theory Appl. **7**(3), 464–469 (2013)
16. Min, M., Shuqian, Z., Chenghui, Z.: Static output feedback control for positive systems via LP approach. In: Proceedings of 31st Chinese Control Conference, pp. 1435–1440 (2012)
17. Lam, H.K.: Stabilization of nonlinear systems using sampled-data output-feedback fuzzy controller based on polynomial-fuzzy-model-based control approach. IEEE Trans. Syst. Man Cybern. B. Cybern. **42**(1), 258–267 (2012)

Curvature-Based Sparse Rule Base Generation for Fuzzy Interpolation Using Menger Curvature

Zheming Zuo[1], Jie Li[2], and Longzhi Yang[1(✉)]

[1] Department of Comptuer and Informaiton Sciences, Northumbria University,
Newcastle upon Tyne NE1 8ST, UK
{zheming.zuo,longzhi.yang}@northumbria.ac.uk
[2] School of Computing & Digital Technologies, Teesside University,
Middlesbrough, UK
jie.li@tees.ac.uk

Abstract. Fuzzy interpolation improves the applicability of fuzzy inference by allowing the utilisation of sparse rule bases. Curvature-based rule base generation approach has been recently proposed to support fuzzy interpolation. Despite the ability to directly generating sparse rule bases from data, the approach often suffers from the high dimensionality of complex inference problems. In this work, a different curvature calculation approach, i.e., the Menger approach, is employed to the curvature-based rule base generation approach in an effort to address the limitation. The experimental results confirm better efficiency and efficacy of the proposed method in generating rule bases on high-dimensional datasets.

Keywords: Fuzzy interpolation · Rule base generation ·
Sparse rule base · Menger curvature · High-dimensional data

1 Introduction

Fuzzy inference systems are built upon the fuzzy sets and fuzzy logic theory to provide a mapping mechanism that maps system input spaces to output spaces. A typical fuzzy inference system consists of two components: a rule base and an inference engine. A number of inference engines have been proposed, with the Mamdani inference and the TSK inference being the most widely applied. The TSK fuzzy inference is able to produce crisp outputs directly, as the polynomials are employed in the rule consequences. In contrast, the Mamdani fuzzy model is more intuitive and suitable for coping with linguistic inputs, thereby, to produce the fuzzy outputs. Common to both fuzzy inference approaches, Mamdani and TSK approach, a dense rule base, in which the entire input domain is fully covered, is required to support the fuzzy inference.

A fuzzy rule base can usually be generated in one of two ways: knowledge-driven, which generates fuzzy rule bases from expert knowledge, or data-driven that extracts rule bases from existing data. The knowledge-driven approaches

Z. Ju et al. (Eds.): UKCI 2019, AISC 1043, pp. 53–65, 2020.
https://doi.org/10.1007/978-3-030-29933-0_5

essentially are a representation of the human expertise in the format of fuzzy rules that require full understanding of the problems by human experts. Recognising that the expert knowledge may not always be available, data-driven approaches were proposed, which extract fuzzy rules from a set of training data. Such data-driven approaches are commonly built upon a large quantity of existing data to target the dense rule bases used by the conventional fuzzy inference engines.

Fuzzy rule interpolation (FRI), which was initially proposed in [1], relaxes the requirement of dense rule bases from conventional fuzzy inference systems [1]. When a given input does not overlap with any rule antecedent, certain conclusions can still be obtained by means of interpolation. A number of fuzzy interpolation approaches have been developed, such as [1–8], which have been successfully applied to deal with real-world problems, such as cybersecurity [9–11], smart home control system [12], network QoS management system [13], personalised exoskeleton control [14], and job planning system [15]. Fuzzy interpolation is not only able to enhance fuzzy inference over a sparse rule base but also able to help in system complexity reduction by removing the rules that can be approximated by others [16]. To reduce the complexity of such rule bases, various rule base reduction approaches have been developed [16–18]. Nevertheless, those approaches usually generate a dense rule base first, which is followed by the removing of redundancy rules based on certain similarity measures. As a consequence, such operations are likely to lead to an extra computational cost.

This paper proposes a different curvature-based data-driven rule base generation approach [19,20] for FRI which not only directly generates sparse rule bases from the given training datasets but also copes well with high-dimensional data. Fundamentally, the majority of existing fuzzy interpolation approaches are fuzzy extensions of crisp linear interpolation. Based on this, the 'flat' or 'straight' regions of a data pattern can be easily approximated by its surroundings. The curvature value of a part of the data pattern in a way indicates the straightness of the given region of the data, where the lower curvature value represents the flatness and vice verse. Based on this observation, the proposed system first determines and removes the 'flat' or 'straight' regions of the given data by calculating the curvature values, thus only extracts rules from the selected regions that have high-curvature values. The proposed approach is evaluated by two experiments with promising results generated.

The rest of the paper is structured as follows: Sect. 2 revisits the related background theories, including the Transformation-based fuzzy interpolation approach and the conventional curvature-based rule base generation approach. Section 3 presents the proposed system. Section 4 demonstrates and evaluates the proposed system. Section 5 concludes the paper and suggests probable future developments.

2   Background

Recent development of rule base generation has been reported, with compact sparse rule bases targeted [20]. This approach is developed based on the concept

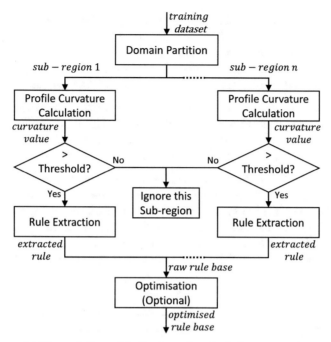

(a) The work flow of the curvature-based rule base generation

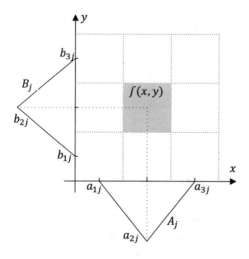

(b) Fuzzy sets extraction

Fig. 1. Curvature-based rule base generation

of profile curvature values of different parts of the data pattern. In particular, the method divides the problem domain into a number of components (sub-regions). As the 'flat' or 'straight' parts of the pattern can be approximated by linear fuzzy interpolation techniques, only the parts with higher curvature values are selected for fuzzy rules extraction. The approach is illustrated in Fig. 1(a).

Domain Partition: Given a training dataset, which is distributed in a 3-dimensional space (2-inputs and signal output), the input domain is equally partitioned into $a \times b$ grid areas, where a and $b \in \mathbb{N}$ indicate the number of partitions on a horizontal axis and a vertical axis, respectively.

Curvature Value Calculation: The profile curvature values are usually used in geospatial analysis, which represents the steepest downward gradient for a given direction [21]. The profile curvature values are used in this approach to help indicate the importance of each sub-region. Given a sub-region $f(x, y)$, the profile curvature, k_p, is the rate at which a surface slope, S, changes whilst moving in the direction of $grad(f)$, which can be calculated by the directional derivative:

$$D_{(\hat{n})}(F) = \triangledown F \cdot \hat{n}. \tag{1}$$

The directional derivative refers to the rate at which any given scalar field, $F(x, y)$, is changing as it moves in the direction of some unit vector, $\hat{n}$, such as $\hat{n} = -(\triangledown f / S)$, where S is the slope defined as the magnitude of the gradient vector and is a scalar field:

$$S(x, y) = |\triangledown f| = \sqrt{f_x^2 + f_x^2}, \tag{2}$$

where $\triangledown f = (f_x, f_y, 0)$ denotes the gradient of this surface, which is a 2D vector that points in the steepest uphill and downhill directions. From here, the profile curvature values, k_p, can be expressed as:

$$k_p = -S^{-1}(\triangledown S \cdot \triangledown f), \tag{3}$$

In order to calculate the overall linearity of a sub-region, eight directions of profile curvature values, which are defined from the centre of the sub-region to the four corners and the central points of the four edges, are defined, such as $k_{p_i}, i = \{1, 2, \cdots, 8\}$. That is, the final profile curvature value takes the maximum value of the eight directional curvature values: $k_p = \max(k_{p_i})$, $i = \{1, 2, \cdots, 8\}$.

Rule Extraction: Given a curvature threshold θ, if the curvature value of a sub-region is greater than θ, the corresponding sub-region will be selected to form a fuzzy rule. In this approach, each selected sub-region is represented by one fuzzy rule. For simplicity, only isosceles triangular fuzzy sets are employed in this approach, each of which can be precisely represented as $A = (a_1, a_2, a_3)$, where a_2 is the core and (a_1, a_3) is the support of the fuzzy set. In this approach, the core of the fuzzy set is set to the centre of the sub-region, and the support of the fuzzy set is equal to twice the span of the corresponding sub-region. Given a selected sub-region $f(x, y)$, the extracted fuzzy sets are illustrated in Fig. 1(b). The raw rule base can then be constructed from all extracted rules.

Optimisation: The generated raw rule base can be employed for results generation. An optional process, the rule base optimisation, can also be applied in the end to fine-tune the parameters by adopting a genetic optimisation algorithm, such as a Genetic Algorithm (GA) [22], in order to increase the performance. The details of the optimisation process are omitted here, as it is not the main focus of this paper.

The profile curvature is based on the directional derivative, which usually used in 3D surface. As a consequence, the profile curvature-based rule base generation approach is limited on three-dimensional problems, which comprise two inputs and signal output.

3 Menger Curvature-Based Rule Base Generation

The original curvature-base rule base generation approach as introduced in Sect. 2 is extended in this section by deploying the concept of the Menger Curvature [23], thereby to relief the limitation of the original approach by allowing to handle the high-dimensional data instances.

3.1 Menger Curvature

The Menger Curvature (MC) measures the curvature of a triple of points in n-dimensional Euclidean space $\mathbb{E}^n$ which is the reciprocal of the radius of the circle that passes through the three points [23]. In this work, only plane curves, that is, only two-dimensional problems, are considered. Assume that $p_1(x_1, y_1), p_2(x_2, y_2), p_3(x_3, y_3)$ are three points in a two-dimensional space $\mathbb{E}^2$ and p_1, p_2, p_3 are not collinear, as depicted in Fig. 2, the MC on p_2 can be defined by:

$$MC(p_1, p_2, p_3) = \frac{1}{R} = \frac{2\sin(\varphi)}{||p_1, p_3||}, \qquad (4)$$

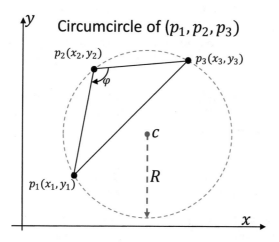

Fig. 2. The Menger Curvature of a triple of points on two-dimensional space

where R represents the $||p_1, p_3||$ denoting the Euclidean distance between p_1 and p_3, and φ is the angle made at the p_2-corner of the triangle spanned by p_1, p_2, p_3, which can be obtained by the Law of Cosines:

$$cos(\varphi) = \frac{||p_1, p_2||^2 + ||p_2, p_3||^2 - ||p_1, p_3||^2}{2 \cdot ||p_1, p_2||^2 \cdot ||p_2, p_3||^2}. \tag{5}$$

Note that the MC on point p_1 and p_3 will not be calculable, due to the boundary points.

3.2 Extended Rule Base Generation

Given a high-dimensional complex inference problem, the same procedure as detailed in Sect. 2 is used for rule base generation, except the step of curvature value calculation. In particular, a high-dimensional data instance is first evenly partitioned to a number of hypercubes. Then, each hypercube is broken down into a set of two-dimensional (2D) problems with one input and a single output. From there, the MC is applied to each data point of the obtained 2D problems, thus to calculate the mean of MC of each 2D problem. Finally, the curvature of each generated hypercube can be determined by the weighted average of the curvature values of its corresponding 2D problems. The hypercubes with higher curvature values are then selected to contribute to the fuzzy rule base generation.

Given a high-dimensional problem denoted as $\mathbb{P}_{n+1}(n > 2)$, assume it contains n inputs features $\mathbb{X} = \{x_1, \dots, x_n\}$, and single output feature y, which have been evenly partitioned into $m_1 \times \cdots \times m_n$ hypercubes, where $m_i(1 \leq i \leq n)$ denotes the number of partitions in the input domain of x_i. Taking a hypercube H_i, which contains h_i data instance, as an example, its curvature value can be computed by the following steps:

Step 1. *Hypercube break down*: Given a hypercube H_i, which contains n input features and single output feature, it is first broken down into n 2D planes, denoted as $P_j^i(1 \leq j \leq n)$, which is implemented by combining each input feature and the single output feature. As a result, the given high-dimensional hypercube H_i can be represented by a set of corresponding decomposed 2D planes.

Step 2. *Mean value determination for the Menger Curvature of each 2D*: Suppose that a decomposed 2D plane P_j^i contains h_i data instances, the mean of MC of P_j^i can be determined by:

$$MC_j^i = \frac{1}{h_i - 2} \sum_{k=1}^{h_i-2} mc_k^i, \tag{6}$$

where mc_k^i represents the Menger Curvature value on the kth data point in the 2D plane P_j^i, which can be calculated by Eq. 4. Note that the MCs on the two boundary points are not calculable.

Step 3. *Curvature value determiunation for each hypercube*: The curvature value C_i of corresponding hypercube H_i can be obtained by averaging all decomposed 2D planes as:

$$C_i = \frac{1}{n} \sum_{j=1}^{n} MC_j^i, \tag{7}$$

where n represents the number of features, MC_j^i is the mean of Menger Curvature of ith decomposed 2D plane, which can be obtained by Eq. 6. Note that the dimensionality reduction techniques, such as Principal Component Analysis (PCA), Fuzzy rough feature selection (FRFS) [24], and Linear Discriminant Analysis (LDA) [25], could be applied in this step in helping identify the most relevant features. In this case, a weighted average method can be employed to calculate the curvature of each hypercube. However, this mechanism will remain as a piece of future work.

Step 4. *Raw Rule Base Generation*: Based on the obtained curvature values of hypercubes, the important rules for FRI can be identified. Similar to the original rule base generation approach as detailed in Sect. 2, given a threshold ϕ, the hypercube with a curvature value higher than the given threshold will be selected to contribute to the generation of a fuzzy rule, which can be expressed as:

$$R_r : \textbf{ IF } x_1 \text{ is } A_1^r, x_2 \text{ is } A_2^r, \ldots, x_n \text{ is } A_n^r \textbf{ THEN } y = B_r, \tag{8}$$

where $A_i, i = \{1, \ldots, n\}$, is a triangular fuzzy sets, which can be precisely represented as $A_i = (a_{1i}^r, a_{2i}^r, a_{3i}^r)$, and its extraction procedure is outlined in Fig. 1 and introduced in Sect. 2. The final fuzzy rule base is constructed by comprising all extracted important rules. The generated rule base could be dense or sparse, depends on the given training dataset. In the case of sparse rule base, the T-FRI will be employed to generate inference results.

4 Experimentation

The proposed system was evaluated in this section. In particular, two experiments have been carried out, a mathematical model and a real-world application.

4.1 Illustrative Example

The problem considered in [19, 20] is re-considered in this work for an illustrative example. This problem is given below:

$$f(x_1, x_2) = \sin\left(\frac{x_1}{\pi}\right) \sin\left(\frac{x_2}{\pi}\right), \tag{9}$$

which takes two inputs x_1 ($x_1 \in [10, 30]$) and x_2 ($x_2 \in [10, 30]$), and produces a single output $y = f(x_1, x_2)$ ($y \in [-1, 1]$), as illustrated in Fig. 3(a).

The problem domain is equally partitioned into 20 grid areas, which consequently results in a total 400 cubes, as shown in Fig. 3(a). Assume that only

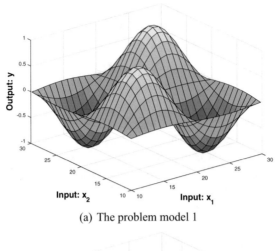

(a) The problem model 1

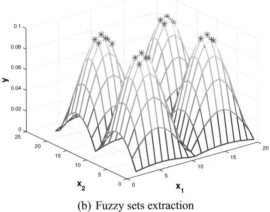

(b) Fuzzy sets extraction

Fig. 3. Curvature-based rule base generation

Table 1. Data instances in H_1 and corresponding menger curvature value

No. (j)	x_1^1	x_2^1	y^1	$MC_j^1(x_1, y)$	$MC_j^1(x_2, y)$
1	10	10	0.0017	N/A	N/A
2	10.1	10.1	0.0054	0.0024	0.0024
3	10.2	10.2	0.0110	0.0034	0.0034
4	10.3	10.3	0.0187	0.0044	0.0044
5	10.4	10.4	0.0282	0.0054	0.0054
6	10.5	10.5	0.0397	0.0064	0.0064
7	10.6	10.6	0.0531	0.0074	0.0074
8	10.7	10.7	0.0683	0.0084	0.0084
9	10.8	10.8	0.0852	0.0094	0.0094
10	10.9	10.9	0.1038	N/A	N/A

Table 2. Curvature values of the cubes

No.	1	2	3	4	5	6	7	8	9	10	11	12	13	14	15	16	17	18	19	20
1	0.0015	0.0026	0.0035	0.0040	0.0042	0.0039	0.0032	0.0022	0.0010	0.0003	0.0016	0.0027	0.0036	0.0041	0.0041	0.0038	0.0031	0.0020	0.0008	0.0009
2	0.0123	0.0219	0.0295	0.0342	0.0353	0.0328	0.0269	0.0184	0.0081	0.0029	0.0136	0.0231	0.0303	0.0345	0.0352	0.0322	0.0259	0.0171	0.0067	0.0071
3	0.0211	0.0382	0.0520	0.0607	0.0628	0.0580	0.0471	0.0318	0.0139	0.0049	0.0235	0.0402	0.0535	0.0614	0.0626	0.0569	0.0453	0.0296	0.0115	0.0124
4	0.0272	0.0497	0.0686	0.0809	0.0840	0.0771	0.0618	0.0412	0.0179	0.0063	0.0302	0.0525	0.0707	0.0819	0.0837	0.0755	0.0594	0.0383	0.0147	0.0162
5	0.0305	0.0562	0.0783	**0.0930**	0.0967	0.0884	0.0703	0.0464	0.0200	0.0071	0.0339	0.0593	0.0807	**0.0942**	**0.0963**	0.0865	0.0674	0.0430	0.0165	0.0186
6	0.0312	0.0577	0.0805	**0.0959**	**0.0998**	**0.0911**	0.0722	0.0476	0.0205	0.0072	0.0348	0.0609	0.0831	**0.0971**	**0.0993**	0.0891	0.0693	0.0441	0.0169	0.0196
7	0.0295	0.0543	0.0754	0.0893	**0.0929**	0.0850	0.0677	0.0448	0.0194	0.0069	0.0328	0.0573	0.0777	**0.0904**	**0.0925**	0.0832	0.0650	0.0416	0.0160	0.0194
8	0.0252	0.0458	0.0629	0.0739	0.0766	0.0705	0.0568	0.0380	0.0166	0.0059	0.0280	0.0483	0.0647	0.0748	0.0763	0.0691	0.0546	0.0353	0.0137	0.0180
9	0.0180	0.0323	0.0438	0.0509	0.0527	0.0487	0.0398	0.0270	0.0119	0.0042	0.0199	0.0340	0.0450	0.0515	0.0525	0.0478	0.0383	0.0251	0.0098	0.0152
10	0.0082	0.0146	0.0196	0.0226	0.0234	0.0217	0.0179	0.0122	0.0054	0.0019	0.0091	0.0154	0.0201	0.0229	0.0233	0.0213	0.0172	0.0114	0.0045	0.0108
11	0.0029	0.0052	0.0070	0.0081	0.0083	0.0077	0.0064	0.0044	0.0019	0.0007	0.0033	0.0055	0.0072	0.0081	0.0083	0.0076	0.0061	0.0041	0.0016	0.0052
12	0.0136	0.0243	0.0327	0.0379	0.0392	0.0363	0.0298	0.0203	0.0090	0.0032	0.0150	0.0255	0.0336	0.0383	0.0390	0.0357	0.0287	0.0189	0.0074	0.0012
13	0.0221	0.0400	0.0545	0.0637	0.0660	0.0609	0.0494	0.0333	0.0146	0.0052	0.0245	0.0421	0.0561	0.0644	0.0657	0.0597	0.0475	0.0309	0.0120	0.0073
14	0.0278	0.0509	0.0703	0.0830	0.0862	0.0790	0.0633	0.0421	0.0183	0.0065	0.0309	0.0537	0.0724	0.0840	0.0858	0.0774	0.0608	0.0391	0.0151	0.0125
15	0.0307	0.0566	0.0790	**0.0939**	**0.0977**	0.0892	0.0709	0.0468	0.0202	0.0071	0.0342	0.0598	0.0814	**0.0951**	**0.0973**	0.0873	0.0680	0.0434	0.0166	0.0163
16	0.0311	0.0575	0.0803	**0.0956**	**0.0994**	**0.0908**	0.0720	0.0474	0.0204	0.0072	0.0347	0.0607	0.0828	**0.0968**	**0.0990**	0.0888	0.0691	0.0440	0.0168	0.0186
17	0.0291	0.0534	0.0741	0.0878	**0.0912**	0.0835	0.0667	0.0442	0.0191	0.0068	0.0324	0.0564	0.0764	0.0889	**0.0909**	0.0818	0.0640	0.0410	0.0158	0.0196
18	0.0244	0.0443	0.0608	0.0713	0.0739	0.0680	0.0549	0.0368	0.0161	0.0057	0.0271	0.0467	0.0625	0.0721	0.0736	0.0667	0.0528	0.0342	0.0132	0.0194
19	0.0168	0.0302	0.0409	0.0475	0.0491	0.0455	0.0371	0.0252	0.0111	0.0039	0.0187	0.0318	0.0420	0.0480	0.0489	0.0446	0.0358	0.0235	0.0092	0.0179
20	0.0068	0.0121	0.0162	0.0187	0.0193	0.0179	0.0148	0.0101	0.0045	0.0016	0.0075	0.0127	0.0166	0.0189	0.0192	0.0176	0.0142	0.0094	0.0037	0.0150

10 data instances are contained in H_1, as listed in the second, third and fourth columns of Table 1, the curvature values of H_1 can be obtained by following steps:

Step 1: To break down H_1 into two 2D plane $P_1(x_1^1, y^1)$ and $P_2(x_2^1, y^1)$.

Step 2: To calculate the menger curvature values for each data point of two decomposed 2D planes by Eq. 4, which are listed in fifth and sixth columns of Table 1, and obtain the mean of menger curvature values of two 2D planes, which are 0.0059 and 0.0059, respectively.

Step 3: To determine the curvature value of cube H_1 by average two mean of menger curvature values of two 2D planes, which is $C_i = 0.0059$. The curvature values of the rest of cubes can be obtained as the same way, which are listed in Table 2, and also visualised in Fig. 3(b).

Given a threshold $\phi = 0.09$, which is identified by human expert knowledge, curvature values of 23 cubes are greater than ϕ, as shown as bold in Table 2, and also are marked as $*$ (i.e. the selected tubes) in Fig. 3(b). Therefore, those 23 cubes are selected to extract to the fuzzy rules. As a result, the generation rule base comprises with total 23 rules, which exactly same as the rule base generated by profile curvature-based approach provided in [19]. The aim of this illustrative example is to demonstrate the working procedure of the proposed rule base generation method, and its competitive ability will be presented in the next experiment by involving a real-world scenario.

4.2   Real-World Scenario

In this experiment, a dataset, which was derived from images in the Mammographic Image Analysis Society (MIAS) database [26] and used for breast cancer risk assessment, is used for evaluation purpose. The dataset includes a set of Medio-Lateral Oblique (MLO) left and right mammogram of 161 women (in total 322 data samples). Each data sample has been pre-processed and represented by 280 features, as well as to be labelled into one of six classes, which are indicated by a proportion of dense breast tissue method in breast cancer risk assessment [27].

In this experiment, all 280 features of the MIAS dataset are selected as the inputs, the six integer numbers $(1-6)$, which indicate the six classes, are used as a single output to construct a high-dimensional dataset. The proposed curvature-based rule base generation approach is used to generate a sparse rule base, and the T-FRI inference approach is applied to produce the inference result. The 10-fold cross-validation strategy was employed for system evaluations. In order to evaluate the performance of the proposed approach, eight commonly used classification approaches, including Gaussian Naive Bayes [9], Naive Bayes [9], k-Nearest Neighbours (k-NN) [28], Multi-functional nearest-neighbour classification (MFNN) [29], Logistic Regression [9], Random Forest [9], Adaptive Boosting (AdaBoost) [9], and Linear Support Vector Machine (Linear SVM) [9], were also implemented with the same experimental setup. Note that the min-max normalisation method was applied for all experimentations for noise reduction. The accuracy of the classification results for each class obtained by different approaches including the proposed one are listed in Fig. 4, which confirmed the competitive ability of the proposed system.

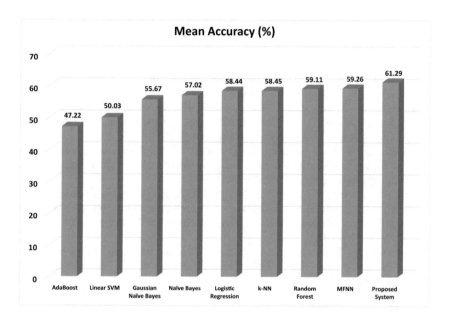

Fig. 4. Experimental results comparison

5 Conclusion

The existing curvature-based rule base generation approach has been modified using a different curvature calculation approach in this work for more efficient fuzzy interpolation in the setting of handling high-dimensional data, in addition to reducing the space and time complexity with a sparse fashion. Though promising experimental results have been obtained, a faster version of this method is expected as a possible piece of future work, which can be achieved, for example, by investigating the computation of the second-order derivatives in a more efficient way. Besides, it is also worthwhile to investigate how the curvature values can be directly used in helping generate TSK style rule bases.

References

1. Kóczy, L.T., Hirota, K.: Approximate reasoning by linear rule interpolation and general approximation. Int. J. Appox. Reason. **9**(3), 197–225 (1993)
2. Huang, Z., Shen, Q.: Fuzzy interpolative reasoning via scale and move transformations. IEEE Trans. Fuzzy Syst. **14**(2), 340–359 (2006)

3. Huang, Z., Shen, Q.: Fuzzy interpolation and extrapolation: a practical approach. IEEE Trans. Fuzzy Syst. **16**(1), 13–28 (2008)
4. Shen, Q., Yang, L.: Generalisation of scale and move transformation-based fuzzy interpolation. J. Adv. Comput. Intell. Intell. Inform. **15**(3), 288–298 (2011)
5. Li, J., Qu, Y., Shum, H.P.H., Yang, L.: TSK inference with sparse rule bases. In: Proceedings of UK Workshop on Computational Intelligence, pp. 107–123 (2016)
6. Yang, L., Shen, Q.: Adaptive fuzzy interpolation. IEEE Trans. Fuzzy Syst. **19**(6), 1107–1126 (2011)
7. Yang, L., Shen, Q.: Closed form fuzzy interpolation. Fuzzy Sets Syst. **225**, 1–22 (2013)
8. Yang, L., Chao, F., Shen, Q.: Generalized adaptive fuzzy rule interpolation. IEEE Trans. Fuzzy Syst. **25**(4), 839–853 (2016)
9. Zuo, Z., Li, J., Anderson, P., Yang, L., Naik, N.: Grooming detection using fuzzy-rough feature selection and text classification. In: Proceedings of IEEE International Conference on Fuzzy Systems, pp. 1–8 (2018)
10. Elisa, N., Li, J., Zuo, Z., Yang, L.: Dendritic cell algorithm with fuzzy inference system for input signal generation. In: Proceedings of UK Workshop on Computational Intelligence, pp. 203–214 (2018)
11. Zuo, Z., Li, J., Wei, B., Yang, L., Chao, F., Naik, N.: Adaptive activation function generation for artificial neural networks through fuzzy inference with application in grooming text categorisation. In: Proceedings of IEEE International Conference on Fuzzy System (2019)
12. Li, J., Yang, L., Shum, H.P.H., Sexton, G., Tan, Y.: Intelligent home heating controller using fuzzy rule interpolation. In: Proceedings of UK Workshop on Computational Intelligence (2015)
13. Li, J., Yang, L., Fu, X., Chao, F., Qu, Y.: Dynamic QoS solution for enterprise networks using TSK fuzzy interpolation. In: Proceedings of IEEE International Conference on Fuzzy Systems, pp. 1–6 (2017)
14. Yin, K., Xiang, K., Pang, M., Chen, J., Anderson, P., Yang, L.: Personalised control of robotic ankle exoskeleton through experience-based adaptive fuzzy inference. IEEE Access **7**, 72221–72233 (2019)
15. Yang, L., Li, J., Chao, F., Hackney, P., Flanagan, M.: Job shop planning and scheduling for manufacturers with manual operations. Expert Syst. e12315 (2018)
16. Koczy, L.T., Hirota, K.: Size reduction by interpolation in fuzzy rule bases. IEEE Trans. Syst., Man, Cybern. **27**(1), 14–25 (1997)
17. Li, J., Shum, H.P.H., Fu, X., Sexton, G., Yang, L.: Experience-based rule base generation and adaptation for fuzzy interpolation. In: Proceedings of IEEE International Conference on Fuzzy Systems, pp, 102–109 (2016)
18. Tao, C.-W.: A reduction approach for fuzzy rule bases of fuzzy controllers. IEEE Trans. Syst., Man, Cybern. B. Cybern. **32**(5), 668–675 (2002)
19. Tan, Y., Shum, H.P.H., Chao, F., Vijayakumar, V., Yang, L.: Curvature-based sparse rule base generation for fuzzy rule interpolation. J. Intell. Fuzzy Syst. **36**(5), 4201–4214 (2019)
20. Tan, Y., Li, J., Wonders, M., Chao, F., Shum, H.P.H., Yang, L.: Towards sparse rule base generation for fuzzy rule interpolation. In: Proceedings of IEEE International Conference on Fuzzy Systems, pp. 110–117 (2016)
21. Peckham, S.D.: Profile, plan and streamline curvature: a simple derivation and applications. In: Proceedings of Geomorphometry, vol. 4, pp. 27–30 (2011)
22. Li, J., Yang, L., Qu, Y., Sexton, G.: An extended Takagi-Sugeno-Kang inference system (TSK+) with fuzzy interpolation and its rule base generation. Soft Comput. **22**(10), 3155–3170 (2018)

23. Léger, J.-C.: Menger curvature and rectifiability. Ann. Math. **149**, 831–869 (1999)
24. Jensen, R., Shen, Q.: New approaches to fuzzy-rough feature selection. IEEE Trans. Fuzzy Syst. **17**(4), 824–838 (2008)
25. Tharwat, A., Gaber, T., Ibrahim, A., Hassanien, A.E.: Linear discriminant analysis: a detailed tutorial. AI Commun. **30**(2), 169–190 (2017)
26. Suckling, J., Parker, J., Dance, D., Astley, S., Hutt, I., Boggis, C., Ricketts, I.: Mammographic Image Analysis Society (MIAS) database v1.21 (2015). https://www.repository.cam.ac.uk/handle/1810/250394/
27. Boyd, N.F., Byng, J.W., Jong, R.A., Fishell, E.K., Little, L.E., Miller, A.B., Lockwood, G.A., Tritchler, D.L., Yaffe, M.J.: Quantitative classification of mammographic densities and breast cancer risk: results from the Canadian National Breast Screening Study. J. Natl. Cancer Inst. **87**(9), 670–675 (1995)
28. Aha, D.W., Kibler, D., Albert, M.K.: Instance-based learning algorithms. Mach. Learn. **6**(1), 37–66 (1991)
29. Qu, Y., Shang, C., Parthaláin, N.M., Wu, W., Shen, Q.: Multi-functional nearest-neighbour classification. Soft Comput. **22**(8), 2717–2730 (2018)

Intelligence in Robotics

A Method of Intention Estimation for Human-Robot Interaction

Jing Luo[1,2], Chao Liu[2], Ning Wang[3], and Chenguang Yang[3(✉)]

[1] South China University of Technology, Guangzhou 510640, China
[2] CNRS-University of Montpellier, UMR 5506, 161 Rue Ada,
34095 Montpellier, France
[3] Bristol Robotics Laboratory, UWE, Bristol BS16 1QY, UK
cyang@ieee.org

Abstract. Dynamics of human wrist play an important role in human-robot interaction. In this paper, we develop a novel method to classify the human wrist's motion and to recognize its stiffness profile. In the proposed method, an integrated framework of linear discriminant analysis and extreme learning machine is developed to evaluate the intention of the wrist. Specifically, linear discriminant analysis is used to classify gestures of the wrist. Based on the result of classification, extreme learning method is use to construct a regression model of the stiffness. The experimental results are demonstrated the effectiveness of the proposed method.

Keywords: Human-robot interaction · Linear discriminant analysis ·
Extreme learning machine · Intention estimation

1 Introduction

Nowadays, the use of robots has enabled to increased working efficiency and reduced economic costs in many areas such as manufacturing, robotic minimally invasive surgery (MIS), telemedicine, etc. [1, 2]. However, the robotic systems have not yet achieve full automation in dangerous or unstructured working environments, the systems still require humans to transmit the motion commands to complete the tasks [3, 4]. Therefore, the study of human factor has been a hot research topic in human robot interaction (HRI). However, the impact of human wrist in the process of HRI has rarely been studied in literature. In many scenarios, the interaction performance is largely affected by the wrist motion and its biomechanical properties. Recently, the electromyography (EMG) signals have been used in HRI to recognize the humans' arm activities. Generally, the EMG signals reflect the comprehensive influence of motor unit action potentials (MAUPs) of the muscle fiber. Surface electromyography (sEMG) signal can be used as the reflection of the muscle activation and motion intention by using machine learning techniques.

In order to extract the effective information from the recorded sEMG signals, many researchers propose a lots of feature extraction algorithms. The purpose of the feature extraction is to obtain the compact and informative features from the raw EMG signals. In [5], the authors utilized time domain (TD) features, frequency domain (FD) features, and time-frequency domain (TFD) features to study the impacts of the various domain feature for the EMG-based prosthetic hand. In [6], Jahromi et al. presented a survey and made a cross comparison for the different feature extraction algorithms. They found that the TD features were most efficient for EMG decomposition. A top and slope (TAS) feature extraction method was proposed for human motion detection by using EMG signals. Compared with the TD feature, the proposed TAS feature provided higher accuracy for locomotion mode detection [7]. Doulah et al., used discrete wavelet transform (DWT) coefficients as the features from the given EMG signals for the purposes of neuromuscular disease classification in [8].

After useful features can be extracted from EMG signals and based on that motion patterns could be classified, a lot of techniques have been used to quantitatively estimate interaction intention information or properties like joint angle, forces, impedance and torques. In [9], Ajoudani et al. presented a novel tele-impedance framework to transfer the human impedance information including postural and stiffness profiles to the robot based on the sEMG signal. Kiguchi et al. proposed a EMG-based impedance controller to recognize human' motion intention and to control the power-assist exoskeleton robot [10]. In [11], the authors proposed a state-space EMG model to estimate the continuous motion of human. A two-step hybrid method based on load and sEMG signals was developed to estimate elbow joint angle in HRI [12]. The above algorithms achieved satisfactory performance to estimate human' intention.

However, the aforementioned existing works treat motion classification and intention estimation/calculation separately. Therefore, no unified framework has been proposed to systematically analyze both at the same time. To remove the above-mentioned constrains, we propose a new unified framework to quantitatively evaluate the intention of the wrist both in motion classification and stiffness estimation in this work. The new framework utilizes LDA algorithm to classify the wrist motion and to obtain output of motion state discriminator for the purpose of intention estimation. Then a regression model (ELM) is employed to estimate wrist intention according to the output of motion state discriminator and acquired muscle stiffness. The novelty of this paper lies in that the wrist motion intention can be quantitatively analyzed both in motion recognition and stiffness estimation simultaneously, and the proposed intention estimation framework provides an effective and user-friendly interface between the humans and the robot for their interaction (HRI). The experimental results demonstrated the effectiveness and flexibility of the developed frameworks.

2 Background Knowledge

In this section, we introduce the background knowledge that will be used in our framework development about muscle activation extraction and stiffness acquisition by using sEMG signals.

2.1 Muscle Activation Extraction

In this paper, the collected sEMG signal u_{sEMG} (unit: mV) can be obtained as

$$u_{sEMG} = \frac{1}{N} \sum_{i=1}^{N} \sqrt{u_{raw}^2(i)} \tag{1}$$

where u_{sEMG} are the raw sEMG signals, $i = 1, 2, \ldots, N$ are the sEMG signals detection channels.

By using a moving average filter, one have

$$u_f(k) = \begin{cases} \frac{1}{k} \sum_{j=0}^{N} u_{sEMG} & k < W_f \\ \frac{1}{W_f} \sum_{j=k-W_f}^{N} u_{sEMG} & k > W_f \end{cases} \tag{2}$$

where W_f is the size of the moving window.

Based on (1) and (2), the relationship between the raw sEMG signals u_{raw} and muscle activation $a(k)$ can be presented as [11].

$$a(k) = \frac{e^{A u_f(k)} - 1}{e^A - 1} \tag{3}$$

where $-3 < A < 0$ is the nonlinear shape factor to define the exponential curvature. Muscle activation $a(k)$ indicates the approximative nonlinear essence of the sEMG-force relationship.

2.2 Stiffness Acquisition

Based on the muscle activation extraction, the wrist stiffness can be defined as a linear function [13]

$$K_i = (K_{max} - K_{min}) \frac{(a_i - a_{min})}{(a_{max} - a_{min})} + K_{min} \tag{4}$$

where K_{max} and K_{min} are the maximum and the minimum stiffness based on the sEMG signals. $a_{min} \leq a_i \leq a_{max}$ is the value of muscle activation. Stiffness indicates the variation tendency of the muscle activation and is proportional to the muscle activation according to (4).

3 Proposed Framework

In this section, the proposed framework is presented in Fig. 1, it contains motion classification module and motion-based stiffness regression model module.

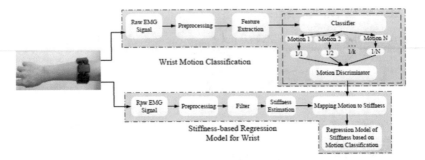

Fig. 1. The proposed framework of wrist motion intention recognition.

Motion Classification Module. The raw sEMG signals are recorded by a MYO armband with 8 channels. We use the classifier to classify the four typical wrist motions after extracting feature of the sEMG signals. The output of this model is the state of motion discriminator. The state of the motion discriminator can be confirmed when there have one and only one motion.

Motion-Based Stiffness Regression Model Module. In this module, we aim to recognize the wrist motion stiffness. According to the state of the motion discriminator in the motion classification module, the corresponding wrist stiffness can be estimated by using the regression model.

3.1   Motion Classification

In this paper, we attempt to classify 4 kinds of wrist motions in the manipulation of the HRI: wrist to the right (WR), wrist to the left (WL), wrist extension (WE), and wrist flexion (WF) as shown in Figs. 2(a)–(d). In this work, we mainly study four wrist rotation motions and for each motion the subject is asked to turn his wrist to the maximum angle possible. The MYO armband are attached to a right forearm as presented in Fig. 2(e). In the motion classification stage (Fig. 3), the input is the recorded sEMG signals. The output is the state of motion discriminator. The motion can be classified via a LDA classifier by using RMS feature.

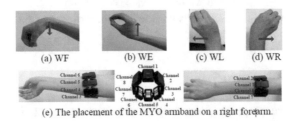

(a) WF   (b) WE   (c) WL   (d) WR

(e) The placement of the MYO armband on a right forearm.

Fig. 2. Four kinds of wrist motions and the placement of the MYO armband on a right forearm.

Data Acquisition and Feature Extraction. The data acquisition device is the MYO armband. The MYO armband can recognize five motions: fist, wave left, wave right, fingers spread, double tap. However, the wrist motions in the experiments are different, thus the gesture detection function provided by the MYO armband is not used in this paper. The sEMG signals are abstracted at a 200 Hz sampling rate with the Bluetooth communication between the MYO armband and a work computer. The MYO armband contains 8 EMG detection electrodes and a 9-axis inertial measurement unit (IMU).

In order to extract the effective features in the motion classification stage, the RMS is used as feature for the recorded sEMG signals.

The RMS feature can be obtained as

$$f_{RMS} = \sqrt{\frac{1}{W}\sum_{i=1}^{W} x_i^2} \tag{5}$$

where W is the length of sampling moving window. x_i is the collected raw sEMG signal for each channel. In the feature extraction session, eight RMS features can be generated in this work.

Training and Classification. After acquisition of the RMS features of the sEMG signal, we utilize a supervised machine learning method (LDA) to train the classifier of wrist motions according to (6)–(10). In this session, each motion trial time is 2 s, and each motion is repeated three times.

In LDA method, we assume that c_k represents the k-th kind of motion type. f_{RMS} indicate the feature parameters of the sEMG signal. The motion type c_k is determined by the feature vector f_{RMS}.

According to the Bayesian formula, the motion type c_k can be expressed as [14]

$$p(c_k|f_{RMS}) = \frac{p(c_k)p(f_{RMS}|c_k)}{p(f_{RMS})} \tag{6}$$

where $p(c_k)$ are the prior probability of the k-th kind of motion type. $p(f_{RMS}|c_k)$ indicate the posterior probability of the k-th kind of motion type with respect to feature vector f_{RMS}. That is to say, $p(c_k|f_{RMS})$ represent the desired classification criterion. When the posterior probability of a certain motion is maximum in comparison with that of other motions, the posterior probability of this certain motion represent the optimal decision of the LDA method. $p(f_{RMS}|c_k)$ are the contingent probability of the k-th kind of motion type with respect to feature vector f_{RMS}, $p(f_{RMS}|c_k)$ can be defined as

$$p(f_{RMS}|c_k) = \frac{1}{\sqrt{(2\pi)^P \det(C)}} \exp\left\{-\frac{1}{2}(f_{RMS} - \mu_k)^T C^{-1}(f_{RMS} - \mu_k)\right\} \tag{7}$$

where P is the number of feature vectors in the training set. μ_k is the mean vector of the motion type c_k. C indicate the covariance matrix of the all motion types. We assume that the contingent probability $p(f_{RMS}|c_k)$ satisfy the multivariate probability distribution.

In order to evaluate the optimal decision of the feature vector f_{RMS} with respect to the corresponding motion (when the $p(c_k|f_{RMS})$ take the maximum value), we take logarithm on both sides of (7), the linear discriminant function can be defined as

$$\delta(k) = f_{RMS}^T \mu_g + c_g \tag{8}$$

With

$$\mu_g = C^{-1}\mu_k \tag{9}$$

$$c_g = -\frac{1}{2}\mu_k^T C^{-1}\mu_k \tag{10}$$

When maximize the linear discriminant function $\delta(k)$, the corresponding motion type k is the optimal decision for the each feature vector f_{RMS}.

3.2   Stiffness-Based Regression

The proposed stiffness regression model is presented in Fig. 3. It can be seen that when the motion is confirmed by the one and only one state of the motion discriminator via the LDA classifier, the ELM method obtains the output state of the state discriminator C^m and the acquired stiffness of the corresponding wrist motion.

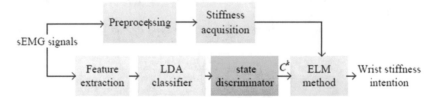

Fig. 3. The motion-based stiffness regression block diagram.

The state discriminator function can be defined as

$$C^m = \left\{ \begin{array}{ll} 1 & \textit{When motion is only confirmed in WR or WL or WE or WF.} \\ 0 & \textit{Others.} \end{array} \right\} \tag{11}$$

Data Processing. In the stiffness regression stage, the data processing procedure is divided into following steps: rectification, squaring, moving average, low pass filter, and envelope.

Regression Model. In Compared with traditional regression algorithms, the ELM method makes the learning algorithm more quickly in the context of ensuring the learning accuracy [15]. In Fig. 4, it can be seen that the ELM structure can be divided into three layer: input layer, hidden layer, and output layer. The input of the regression model is the state of motion discriminator C^m and the calculated stiffness $K_j, j = 1, 2, \ldots, N$ of the corresponding wrist motion according to (4). The output of the regression model is the estimated stiffness S involving the corresponding wrist motion.

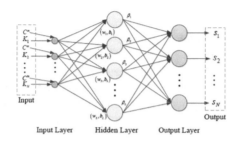

Fig. 4. Architecture of extreme learning machine.

The stiffness regression model S can be expressed by using the ELM algorithm. To this end, a potential mapping for the sample data sets $\left(K_j, C^m\right)_j^N, j = 1, 2, \ldots, N$ based on ELM in the hidden layer between the input $\left(K_j, C^m\right)_j^N$ and output S can be defined as

$$S = \sum_{i=1}^{L} C^m \beta_i f_i(K) = \sum_{i=1}^{L} C^m \beta_i g\left(w_i^T K + b_i\right) \tag{12}$$

where $w_i, i = 1, 2, \ldots, L$ indicate the input weights. $g(\cdot)$ are the activation function. $\beta_i, i = 1, 2, \ldots, L$ represent the output weights. $b_i, i = 1, 2, \ldots, L$ are the biases of the hidden layer. C^m is a parameter of the state discriminator.

In order to acquire the regression model, we should minimize the output error based on a single hidden layer. The minimization problem of output error can be solved as

$$\min_{w,b,\beta} \left\| C^m H \beta - S \right\| \tag{13}$$

With

$$H(w_1, \ldots, w_L, b_1, \ldots, b_L, K_1, \ldots, K_L) = \begin{bmatrix} g\left(w_1^T K_1 + b_1\right) & \cdots & g\left(w_L^T K_1 + b_L\right) \\ \cdots & \cdots & \cdots \\ g\left(w_1^T K_L + b_1\right) & \cdots & g\left(w_L^T K_L + b_L\right) \end{bmatrix} \tag{14}$$

where $H(\cdot)$ is the output of the hidden layer. $S = (S_1, \ldots, S_L)$ indicate the expected output by using ELM algorithm.

When the input weights $W = (w_1, \ldots, w_L)$ and the bias $b = (b_1, \ldots, b_L)$ are random determined, the output of the hidden layer $H(\cdot)$ can be uniquely determined. Then, the problem of training the single hidden layer neural network can be transformed into a problem of solving a linear system. The output weights β can be determined as

$$\beta = C^m H^\dagger S \tag{15}$$

where $H^\dagger$ represent the Moore-Penrose pseudo inverse of the output matrix of the hidden layer H.

Under this framework, starting from the sEMG signal measurement, by using the LDA classifier the state of the motion discriminator can be obtained in motion classification phase. By collecting the motion-based stiffness data, the wrist stiffness with a state discriminator can be estimated via the proposed ELM method. Through combination of the motion classification and the motion-based stiffness regression model, the wrist motion intention can be recognized in a systematic way.

4   Experiments and Results

In the experiments, we aim to recognize the wrist motion and estimate the wrist stiffness intention. In the classification experiment, four types of wrist motions (WR, WL, WE, and WF) are classified. Based on the classification results, the stiffness intention are estimated in the regression experiment.

4.1   Experiment Setup

To evaluate the performance of the proposed framework, the experiment setup is conducted as follows: The MYO armband is mainly used to extract the sEMG signals with 8 channels. MATLAB R2016a and Visual Studio (VS) 2013 are utilized to process the sEMG signals and to train a motion classifier and to recognize the wrist stiffness intention. The software MATLAB R2016a and VS 2013 are running on Windows 7 in a work computer with Intel(R) Core (TM) i7-3770T CPU @ 2.50 GHz and DDR3 1333 MHz 12 GB.

4.2   Performance Metrics

In order to measure experimental performance of the classification and regression, accuracy (ACC), precessing time (PT), root mean square error (RMSE) and coefficient of determination R^2 are used in the experiments.

ACC. This accuracy is used to evaluate the performance of the classifier in the classification experiment.

PT. It is the total time required for each test session to be completed in the classification experiment.

RMSE. RMSE is mainly used to measure the deviation between estimated value and actual value. It is the radication of the mean squared error.

Coefficient of Determination. Coefficient of determination R^2 is used to estimate the error of the regression model. $R^2 \in [0, 1]$ indicates the fitting degree of the regression model. The higher the R^2, the better performance of the regression model will be achieved.

4.3 Wrist Motion Classification

In the classification experiment, two healthy human subjects (age 22–30 years old, 2 males) take part in this experiment. In order to minimize the fatigue of each subject, the subject are asked to performed the each motion repeated twice and the process continued for 2 min and then to rest for half an hour. The sEMG signals are recorded when the targeted motion is reached. We utilize the LDA method to classify the wrist motions. The RMS is used to extract the feature from the recorded sEMG signals for four kinds of motions. The length of the sampling window W for the RMS is set to 400 ms. In our pilot experiment, we found that the classification accuracy with $W = 400$ ms is higher in comparison to others. The sizes of the training set and testing set are 20000*8 and 400*8, respectively. The training set contains four motions, while the testing set contains only one motion.

By using the RMS feature and LDA classifier, we can obtain classification accuracy of the four wrist motions. Table 1 show the classification results by using the RMS feature from the recorded testing sEMG signals to predict the four types of the wrist motions. The averaged accuracy for wrist motions (WR, WL, WE, and WF) classification are 97.50%, 94.75%, 98.75%, and 100%, respectively. From the classification results, they show that the classification output related to the manual annotated label. In the classification results, we notice that the PT for each wrist motion with low cost. The PT for motion WR, WL, WE, and PF are 0.0936 ms, 0.0780 ms, 0.0468 ms and 0.0624 ms, respectively. The LAD method predicts the four kinds of wrist motions in less than 1 ms. The averaged accuracy and averaged procession time for motion classification are $(97.50 + 94.75 + 98.75 + 100)/4$% and $(0.0936 + 0.0780 + 0.0468 + 0.0624)/4$ ms, respectively. Thanks to the inherent predictive nature of EMG signal, with the short computation time, the motion classification can be achieved even before the wrist's physical motion is actually executed and thus enables many other assistive functions in real HRI and simulation training. The classification results indicates that the wrist motions can be successful recognized with high accuracy by using the sEMG signals.

Table 1. The results of wrist gesture classification.

Gesture	WR	WL	WE	WF
WR	**0.9750**	0.025	0	0
WL	0	**0.9475**	0.0525	0
WE	0	0	**0.9875**	0.0125
WF	0	0	0	**1**
PT	0.0936 ms	0.0780 ms	0.0468 ms	0.0624 ms

4.4   Wrist Stiffness Regression Based on Motion Classification Experiment

In the regression experiment, we employ the motion-based ELM method to construct the stiffness regression model for each wrist motion. Wrist stiffness regression experiment is conducted under the circumstance of without weight and with the gesture of holding the joystick. The estimated stiffness is the inherent stiffness of the wrist. The number of the hidden nodes are set as 8000 through a pilot experiment with highest accuracy. The size of the moving window W_f is set as 30. The nonlinear shape factor $A = -0.6891$. In this study, we repeat four times to calculate the wrist stiffness when the corresponding motion is recognized for each motion. The performed time for stiffness estimation is 6–8 s. The sizes of training set and testing set for this study can be founded in Table 2. In this table, the training set and the testing set contain only one kind of data for corresponding wrist motion.

Table 2. The size of training set and testing set for each motion in wrist stiffness experiment.

Data size	WR	WL	WE	WF
Training set	1000*4	1000*4	1000*4	1000*4
Testing set	800*4	600*4	800*4	700*4

We employ a bipolar sigmoid function $g(t)$ as the activation function. We estimate the stiffness for each wrist motion for four times. We can construct the motion-based stiffness regression model by training the four stiffness for each motion. In order to construct the continuous stiffness regression model, we maintain the corresponding wrist motion for a few seconds after recognizing the motion.

According to the (4), the wrist stiffness calculation results are presented in Figs. 5 (a)–(d). In Fig. 5, the red line indicates the estimated wrist stiffness regression intention by using the ELM method, the another four lines are the calculated stiffness for corresponding wrist motion. There are four different regression model for the wrist motion.

It can be seen that the stiffness regression line of motion WR varies between 40 and 60 (in 0–8 s) in Fig. 5(a). In Fig. 5(b), the stiffness regression line varies between 70–90 (in 0–6 s) for motion WL. The stiffness regression lines of motion WE and WF are ranged between 50–80 (in 0–8 s) and 70–90 (in 0–7 s) in Figs. 5(c)–(d), respectively.

In order to evaluate the performance of the stiffness regression model, the RMSE and R^2 are used in the regression experiment. In Table 3, the RMSE of regression model for WR is 0.3630. The RESE of regression model for WL, WE, and WF are 0.7015, 1.0387, and 0.4188, respectively. It can be concluded that the error of regression model for WR between the estimated stiffness and observed stiffness is minimal in comparison with that of others. The R^2 of regression model (ELM) for WR, WL, WE, and WF are 0.9737, 0.8827, 0.8269, 0.9650, respectively. Based on (18), we can conclude that the value of R^2 for WR is minimal. That is to say, the performance of regression model in condition of WR is best. Considering together RMSE and R^2, we can find that a positive correlation relationship among the performance of regression model and the value of RMSE and R^2.

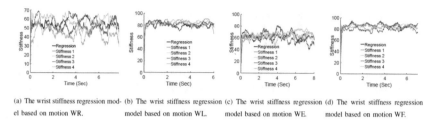

(a) The wrist stiffness regression model based on motion WR.   (b) The wrist stiffness regression model based on motion WL.   (c) The wrist stiffness regression model based on motion WE.   (d) The wrist stiffness regression model based on motion WF.

Fig. 5. The wrist stiffness regression model based on wrist motions.

Table 3. The size of training set and testing set for each motion in wrist stiffness experiment.

Regression error	WR	WL	WE	WF
RMSE	**0.3630**	0.7015	1.0387	0.4188
R^2	0.9737	0.8827	0.8269	0.9650

5   Conclusion

This paper combined the LDA method and motion-based ELM method for the purposes of wrist motion recognition and wrist stiffness estimation by using the MYO armband. The LDA classifier classified four types of wrist motions with RMS feature and the motion-based ELM regression model estimated the wrist stiffness intention for each corresponding motion. An average accuracy of 97.75% was achieved in the wrist classification and an average precessing time of 0.0702 ms in the motion classification stage. In the stiffness regression stage, the proposed regression model effectively estimated the motion stiffness intention. The experimental results verified the performance of the proposed framework. To conclude, the developed novel framework with wrist motion recognition and wrist estimation provides a secure and accurate interface for the human operator in HRI. And this work presents the first effort to provide a unified framework to analyze and quantify human motion based on sEMG measurement.

Funding information. This work was partially supported by National Nature Science Foundation (NSFC) under Grants 61861136009 and 61811530281 Engineering and Physical Sciences Research Council (EPSRC) under Grant EP/S001913.

References

1. Hang, S., et al.: Improved human-robot collaborative control of redundant robot for teleoperated minimally invasive surgery. IEEE Robot. Autom. Lett. **4**(2), 1447–1453 (2019)
2. Hang, S., Sandoval, J., Makhdoomi, M., et al.: Safety-enhanced human-robot interaction control of redundant robot for teleoperated minimally invasive surgery. In: 2018 IEEE International Conference on Robotics and Automation (ICRA), pp. 6611–6616. IEEE (2018)

3. Luo, J., et al.: Enhanced teleoperation performance using hybrid control and virtual fixture. Int. J. Syst. Sci. **50**(3), 451–462 (2019)
4. Luo, J., et al.: A task learning mechanism for the telerobots. Int. J. Humanoid Robot. **16**(2), 1950009 (2019)
5. Geethanjali, P., Ray, K.: A low-cost real-time research platform for EMG pattern recognition-based p°rosthetic hand. IEEE/ASME Trans. Mechatron. **20**(4), 1948–1955 (2015)
6. Jahromi, M.G., Parsaei, H., Zamani, A., Stashuk, D.W.: Cross comparison of motor unit potential features used in EMG signal decomposition. IEEE Trans. Neural Syst. Rehabil. Eng. **26**(5), 1017–1025 (2018)
7. Ryu, J., Lee, B.-H., Kim, D.-H.: sEMG signal-based lower limb human motion detection using a top and slope feature extraction algorithm. IEEE Signal Process. Lett. **24**(7), 929–932 (2017)
8. Doulah, A.S.U., Fattah, S.A., Zhu, W.-P., Ahmad, M.O., et al.: Wavelet domain feature extraction scheme based on dominant motor unit action potential of EMG signal for neuromuscular disease classification. IEEE Trans. Biomed. Circuits Syst. **8**(2), 155–164 (2014)
9. Ajoudani, A., Tsagarakis, N., Bicchi, A.: Tele-impedance: teleoperation with impedance regulation using a body–machine interface. Int. J. Robot. Res. **31**(13), 1642–1656 (2012)
10. Kiguchi, K., Hayashi, Y.: An EMG-based control for an upper-limb power-assist exoskeleton robot. IEEE Trans. Syst., Man, Cybern. Part B (Cybernetics) **42**(4), 1064–1071 (2012)
11. Han, J., Ding, Q., Xiong, A., Zhao, X.: A state-space EMG model for the estimation of continuous joint movements. IEEE Trans. Ind. Electron. **62**(7), 4267–4275 (2015)
12. Tang, Z., Yu, H., Cang, S.: Impact of load variation on joint angle estimation from surface EMG signals. IEEE Trans. Neural Syst. Rehabil. Eng. **24**(12), 1342–1350 (2016)
13. Yang, C., et al.: Haptics electromyography perception and learning enhanced intelligence for teleoperated robot. IEEE Trans. Autom. Sci. Eng. **99**, 1–10 (2018)
14. Jing, P., Heisterkamp, D.R., Dai, H.K.: LDA/SVM driven nearest neighbor classification. IEEE Trans. Neural Netw. **14**(4), 940–942 (2003)
15. Huang, G.B., Zhou, H., Ding, X., Zhang, R.: Extreme learning machine for regression and multiclass classification. IEEE Trans. Syst. Man Cybern. B Cybern. **42**(2), 513–529 (2012)

A Hybrid Human Motion Prediction Approach for Human-Robot Collaboration

Yanan Li[1(✉)] and Chenguang Yang[2]

[1] Department of Engineering and Design, University of Sussex,
Brighton BN1 9RH, UK
yl557@sussex.ac.uk
[2] Bristol Robotics Laboratory, University of the West of England,
Bristol BS16 1QY, UK

Abstract. Prediction of human motion is useful for a robot to collaborate with a human partner. In this paper, we propose a hybrid approach for the robot to predict the human partner's motion by using position and haptic information. First, a computational model is established to describe the change of the human partner's motion, which is fitted by using the historical human motion data. The output of this model is used as the robot's reference position in an impedance control model. Then, this reference position is modified by minimizing the interaction force between the human and robot, which indicates the discrepancy between the predicted motion and the real one. The combination of the prediction using a computational model and modification using the haptic feedback enables the robot to actively collaborate with the human partner. Simulation results show that the proposed hybrid approach outperforms impedance control, model-based prediction only and haptic feedback only.

Keywords: Human-robot collaboration · Robot control ·
Impedance control

1 Introduction

Human-robot collaboration is becoming a popular research topic as it provides new possibilities in various applications [1], including tele-operation [12], heavy load sharing [2], co-assembly [9], etc. To allow human-robot collaboration by physical interaction, impedance control has been successfully implemented and developed in the past decades [3,5]. While early works on impedance control mainly treat a robot as a passive follower by ignoring its autonomy, state-of-the-art research shows significant benefits of taking the robot's autonomy into account and thus suggests a shared control paradigm. Such a paradigm requires the robot to make decisions by using its own sensing and recognition capabilities, which naturally include prediction of human partner's behaviours. Like in

© Springer Nature Switzerland AG 2020
Z. Ju et al. (Eds.): UKCI 2019, AISC 1043, pp. 81–91, 2020.
https://doi.org/10.1007/978-3-030-29933-0_7

human-human collaboration, the robot may sense, recognize and thus predict the human partner's behaviours through various interfaces, such as visual signs [13] and haptic feedback [6]. However, the performance of human-robot collaboration with the current interfaces is still far from expected as in a level comparable to human-human collaboration.

In this paper, we aim to improve human-robot collaboration performance by studying a typical scenario, where the human partner guides the robot through direct physical interaction. This abstract scenario represents applications such as object loading and offloading and the difficulty lies in the unknown human partner's motion intention. Many approaches have been developed to address this problem in the literature, which can be mainly categorized into two groups. The first group is based on an assumption that the human motion follows a certain criterion. In [4, 11], human motion is assumed to follow the minimal jerk criterion. In [13], human's translational motion is assumed to follow a straight line so linear interpolation of human motion is implemented. In [15], the autoregressive integrated moving average (ARIMA) model is used for prediction of human motion. The second group utilizes the information of interaction force, which is deemed to reflect the discrepancy between the actual motion and the desired one. In [6, 7], a human arm dynamics model is considered to estimate the human's desired motion using the interaction force. While the performance of the first group of approaches may be subject to model overfitting, the second group is based on haptic feedback so a response delay is inevitable.

Based on above discussions, it is interesting to combine the two groups of approaches together, resulting in a hybrid approach that predicts the human motion using a feedforward model and corrects the prediction using haptic feedback. In this paper, a computational model is developed as the feedforward model to predict human's future motion. Its parameters are updated by observing the error between the predicted motion and the actual one. The predicted motion is set as the robot's reference position, which is further refined by including the haptic feedback. Theoretical analysis based on the Lyapunov theory shows that the robot's motion perfectly tracks the human's desired motion. Comparative simulations confirm the effectiveness of this approach and demonstrate its performance better than traditional impedance control, model-based prediction only and haptic feedback only.

The rest of this paper is organized as follows. The human-robot collaboration scenario is elaborated in Sect. 2, with introduction of the system dynamics and impedance control. The proposed hybrid human motion prediction approach and robot's reference position design are presented in Sect. 3, with theoretical analysis. Simulation results are given in Sect. 4 and conclusions are summarized in Sect. 5.

2   Human-Robot Collaboration System

The human-robot collaboration scenario includes one robot arm and one human arm, where the human hand holds the end-effector of the robot arm. This simplified scenario can be found in an application where the robot carries an object while the human guides the robot along a desired trajectory.

By treating the force applied by the human as a system input, we write the system dynamics as below

$$M(x)\ddot{x}(t) + C(x,\dot{x})\dot{x}(t) + G(x) = u + f \tag{1}$$

where $x(t)$ is the position at the interaction point, $M(x)$ is the robot's inertia matrix, $C(x,\dot{x})$ is the Coriolis and centrifugal force matrix, $G(x)$ is the gravitational force vector, u is the robot's control input transformed from the motor torques at each joint and f is the force applied by the human. Note that the dynamic model is given in the Cartesian space.

As introduced in the Introduction, impedance control has been widely used for human-robot collaboration. In particular, let us design the robot's control input as

$$u = C(x,\dot{x})\dot{x}(t) + G(x) - f + M^{-1}(x)[\ddot{x}_r - M_d^{-1}(D_d\dot{e} + K_d e - f)] \tag{2}$$

where M_d, D_d and K_d are the desired inertia, damping and stiffness matrices, respectively, x_r is the robot's reference trajectory and $e = x - x_r$ is the robot's trajectory tracking error. By substituting Eq. (2) into Eq. (1), we can obtain the desired impedance model

$$M_d\ddot{e} + D_d\dot{e} + K_d e = f \tag{3}$$

Remark 1. The control input in Eq. (2) requires the knowledge of the matrices $M(x)$, $C(x,\dot{x})$ and $G(x)$. This has been extensively studied in the literature using adaptive control [10], NN control [8], iterative learning control [14] etc, so will not be detailed in this paper.

If we set the robot's reference position $x_r = 0$ and stiffness matrix $K_d = 0$, the impedance model in Eq. (3) becomes $M_d\ddot{x} + D_d\dot{x} = f$ suggesting that the robot will passively follow the human motion. This is an approach that has been used in many works. However, as the robot does not move actively, the human partner has to apply an extra force to compensate for the robot's dynamics, which is undesired especially when the robot has a large M_d. Conversely, when $K_d \neq 0$ and x_r is designed properly according to the human's desired motion, the robot will move to x_r actively so that the human effort can be reduced, which is the main goal of this work.

3   Approach

In this section, we will first present two methods to predict the human partner's motion. Then, the robot's reference position will be designed based on the predicted motion, with the theoretical analysis showing the properties of the proposed approach.

3.1   Prediction of Human Motion

In order to predict the human motion, we employ a simplified human arm model as below

$$f = K_h(x - x_h) \tag{4}$$

where K_h is the human's control gain and x_h is the human's desired position that is unknown to the robot. Suppose that x_h is a continuous trajectory planned with certain acceleration and velocity profiles. Then, without loss of generality, it can be represented as

$$\dot{x}_h(t) = w_0 + w_1 x(t) + w_2 x(t - T) + w_3 x(t - 2T) + \ldots + w_m x(t - (m-1)T)$$

where T is the time step and w_i, $i = 0, 1, \ldots, m$ is the weight determined by the acceleration and velocity profiles of x_h. For example, if x_h has a constant acceleration, we have $m = 2$ and w_i, $i = 0, 1, 2$ can be obtained by using Euler integration. For analysis simplicity, we rewrite x_h in a form of vectors, i.e.

$$\dot{x}_h(t) = W^T X(t) \tag{5}$$

where $W = [w_0, w_1, \ldots, w_m]^T$ and $X(t) = [1, x(t), \ldots, x(t - (m-1)T)]^T$. Note that W is unknown to the robot, indicating unknown human's desired motion. Thus, it is approximated by

$$\dot{\hat{x}}_h(t) = \hat{W}^T X(t) - L \tilde{x}_h(t - T) \tag{6}$$

where $\hat{W}$ is the estimate of W, L is a positive definite matrix and

$$\tilde{x}_h(t - T) = \hat{x}_h(t - T) - x_h(t - T) = \hat{x}_h(t - T) - x(t) \tag{7}$$

The following update law is designed to update $\hat{W}$:

$$\dot{\hat{W}} = -\tilde{x}_h^T(t - T)X(t) - \beta f^T X(t) \tag{8}$$

where $\beta > 0$ and the second term on the right hand side is to compensate for the coupling effect by the haptic feedback that will be elaborated later.

With the predicted x_h, we design the robot's reference trajectory as

$$\dot{x}_r = \dot{\hat{x}}_h - \alpha f + L \tilde{x}_h(t - T) \tag{9}$$

where α is a positive scalar. Note that the above update law of the reference trajectory includes two parts: the first part is $\dot{\hat{x}}_h$ obtained by predicting human motion and the second is $-\alpha f$ which is the haptic feedback to correct the prediction. The last term $L\tilde{x}_h(t - T)$ is used to compensate for the coupling effect as will be shown in the convergence analysis.

3.2   Convergence Analysis

To investigate the convergence of the proposed method, let us consider a Lyapunov function candidate

$$J = J_W + J_h + J_e + J_f \tag{10}$$

where

$$J_W = \frac{1}{2\beta}\text{trace}(\tilde{W}^T \tilde{W}), \ J_h = \frac{1}{2\beta}\tilde{x}_h^T \tilde{x}_h,$$

$$J_e = \frac{1}{2}(\dot{e}^T M_d \dot{e} + e^T K_d e), \ J_f = \frac{1}{2}(x - x_h)^T K_h^T (x - x_h) \tag{11}$$

By taking the time derivative of J_W, we have

$$\dot{J}_W = \frac{1}{\beta}\text{trace}(\tilde{W}^T \dot{\tilde{W}}) = \frac{1}{\beta}\text{trace}(\tilde{W}^T \dot{\hat{W}}) \tag{12}$$

where we have used the fact that $\dot{W} = 0$. By substituting the update law in Eq. (8) into the above equation, we have

$$\dot{J}_W = -\text{trace}[\tilde{W}^T(\frac{1}{\beta}\tilde{x}_h^T(t - T)X(t) + f^T X(t))] \tag{13}$$

By taking the time derivative of J_h, we have

$$\dot{J}_h = \frac{1}{\beta}\tilde{x}_h^T \dot{\tilde{x}}_h \tag{14}$$

By deducting Eq. (6) by Eq. (5), we have

$$\dot{\tilde{x}}_h(t) = \tilde{W}^T X(t) - L\tilde{x}_h(t - T) \tag{15}$$

where $\tilde{W} = \hat{W} - W$. Substituting the above equation into Eq. (14), we have

$$\dot{J}_h = \frac{1}{\beta}\tilde{x}_h^T[\tilde{W}^T X(t) - L\tilde{x}_h(t - T)] \tag{16}$$

By considering Eqs. (13) and (16) and assuming that T is small, we have

$$\dot{J}_W + \dot{J}_h = -f^T \tilde{W}^T X - \frac{1}{\beta}\tilde{x}_h^T L\tilde{x}_h \tag{17}$$

By taking the time derivative of J_e, we have

$$\dot{J}_e = \dot{e}^T(M_d \ddot{e} + K_d e) \tag{18}$$

By considering the impedance model (3), we have

$$\dot{J}_e = \dot{e}^T(-D_d \dot{e} + f) = -\dot{e}^T D_d \dot{e} + \dot{e}^T f \tag{19}$$

By taking the time derivative of J_f, we have

$$\dot{J}_f = (x - x_h)^T K_h^T (\dot{x} - \dot{x}_h) \tag{20}$$

By considering Eqs. (4) and (9), we have

$$\dot{J}_f = f^T (\dot{x} - \dot{\hat{x}}_h + \dot{\tilde{x}}_h) = f^T (\dot{x} - \dot{x}_r - \alpha f + L\tilde{x}_h + \dot{\tilde{x}}_h)$$
$$= -\alpha f^T f + f^T \dot{e} + f^T (L\tilde{x}_h + \dot{\tilde{x}}_h) \tag{21}$$

By substituting Eq. (15) into the above equation, we have

$$\dot{J}_f = -\alpha f^T f + f^T \dot{e} + f^T \tilde{W}^T X \tag{22}$$

By considering Eqs. (17), (19) and (22), we have

$$\dot{J} = \dot{J}_W + \dot{J}_h + \dot{J}_e + \dot{J}_f = -\frac{1}{\beta}\tilde{x}_h^T L\tilde{x}_h - \dot{e}^T D_d \dot{e} - \alpha f^T f \leq 0 \tag{23}$$

which indicates that $\tilde{x}_h$, $\dot{e}$ and f converge to 0 when $t \to \infty$. In other words, the desired human motion x_h is obtained as $\hat{x}_h$, the actual trajectory of the robot x tracks the reference x_r and the human does not need to apply a force to move the robot, i.e. $f = 0$.

4   Simulations

In this section, we conduct comparative simulations to illustrate the effectiveness of the hybrid approach. In particular, we consider that the robot's dynamics are governed by the impedance model in Eq. (3) with the parameters: $M_d = 5\,\text{kg}$, $D_d = 100\,\text{Ns/m}$ and $K_d = 100\,\text{N/m}$. The human model is given in Eq. (4) with the parameters: $K_h = 100\,\text{N/m}$ and two types of x_h are considered in two cases. In the first case, $\dot{x}_h = 0.1\,\text{m/s}$ indicating the human's desired motion is a straight line with a constant speed. In the second case, $\ddot{x}_h = 0.1\sin(t)\,\text{m}^2/\text{s}^2$ indicating a motion with time-varying acceleration.

The number of the historical data in Eq. (5) is set as $m = 5$, the gain matrix in Eq. (6) is $L = 5$, the updating rate in Eq. (8) is $\beta = 0.0001$ and $\alpha = 0.1$ in Eq. (9). For comparison, the following four methods are considered: (i) impedance control: the stiffness K_d and the robot's reference position x_r are set as zero; (ii) prediction only: α in Eq. (9) is set as zero; (iii) haptic feedback only, Eq. (9) is modified to $\dot{x}_r = \alpha f + L\tilde{x}_h(t - T)$; and (iv) hybrid approach.

The robot's trajectories and the force applied by the human with four methods are shown in Figs. 1, 2, 3 and 4, respectively. From Fig. 1, it is clearly found that with impedance control the robot's trajectory lags behind the human's desired trajectory and thus requires the human to apply a relatively large force to move the robot. In comparison, in Fig. 2 we find that the robot is set a reference trajectory x_r with the prediction of the human's desired trajectory, so that it can actively move to x_r. However, as x_r goes beyond x_h, a negative force is generated indicating the human having to slow down the robot's motion, although the magnitude of this force is smaller than that in Fig. 1. By considering the haptic feedback only, the robot is able to better track x_h and thus requires less force from the human, as shown in Fig. 3. In comparison, by combining both the prediction and haptic feedback, the tracking force is further improved and the human force drops to around zero, as illustrated in Fig. 4.

With the same parameters except changing the human's desired motion to be with time-vary acceleration, the simulation results with four methods are shown in Figs. 5, 6, 7 and 8, respectively. These results further confirm the moderate performance of impedance control, effects of prediction and haptic feedback and finally the effectiveness of the proposed hybrid approach.

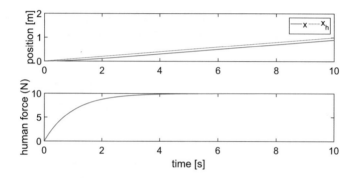

Fig. 1. Case 1 with impedance control: human's desired trajectory and robot's actual trajectory (top); human's force (bottom)

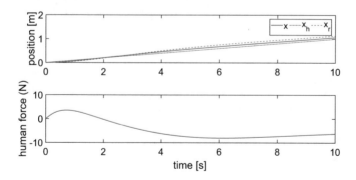

Fig. 2. Case 1 with prediction only: human's desired trajectory, robot's reference trajectory and robot's actual trajectory (top); human's force (bottom)

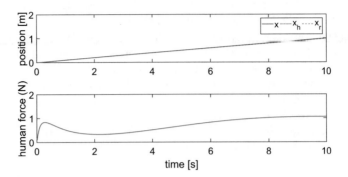

Fig. 3. Case 1 with haptic feedback only: human's desired trajectory, robot's reference trajectory and robot's actual trajectory (top); human's force (bottom)

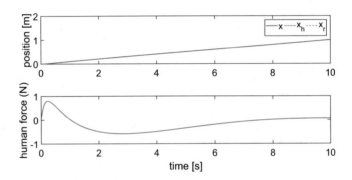

Fig. 4. Case 1 with hybrid approach: human's desired trajectory, robot's reference trajectory and robot's actual trajectory (top); human's force (bottom)

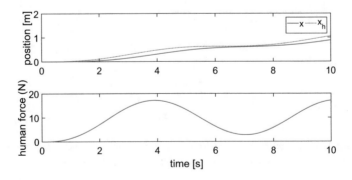

Fig. 5. Case 2 with impedance control: human's desired trajectory and robot's actual trajectory (top); human's force (bottom)

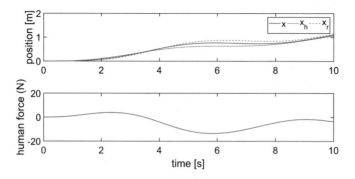

Fig. 6. Case 2 with prediction only: human's desired trajectory, robot's reference trajectory and robot's actual trajectory (top); human's force (bottom)

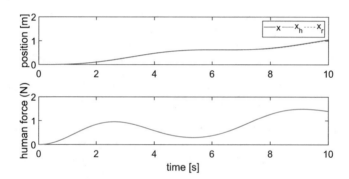

Fig. 7. Case 2 with haptic feedback only: human's desired trajectory, robot's reference trajectory and robot's actual trajectory (top); human's force (bottom)

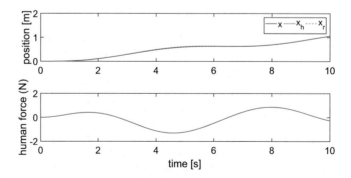

Fig. 8. Case 2 with hybrid approach: human's desired trajectory, robot's reference trajectory and robot's actual trajectory (top); human's force (bottom)

5   Conclusions

Prediction of the human's desired motion and using haptic feedback both can improve human-robot collaboration in the sense of actively following the human and reducing human effort. While they have been separately studied in the literature, it is verified in the paper that their combination yields a better performance. The derivations of the hybrid approach are presented and it is tested and compared to other approaches using simulations. This approach shows potential advantages in applications where the robot is expected to move according to the human's intention, such as human-robot co-manipulation and robotic exoskeleton for assisting human movement. Our future works will focus on implementation of this approach in these applications.

References

1. Ajoudani, A., Zanchettin, A.M., Ivaldi, S., Albu-Schäffer, A., Kosuge, K., Khatib, O.: Progress and prospects of the human-robot collaboration. Auton. Robot. **42**(5), 957–975 (2018). https://doi.org/10.1007/s10514-017-9677-2
2. Al-Jarrah, O.M., Zheng, Y.F.: Arm-manipulator coordination for load sharing using compliant control. In: Proceedings of the 1996 IEEE International Conference on Robotics and Automation, pp. 1000–1005, April 1996
3. Albu-Schäffer, A., Ott, C., Hirzinger, G.: A unified passivity based control framework for position, torque and impedance control of flexible joint robots. Int. J. Robot. Res. **26**(1), 23–29 (2007)
4. Corteville, B., Aertbelien, E., Bruyninckx, H., Schutter, J.D., Brussel, H.V.: Human-inspired robot assistant for fast point-to-point movements. In: IEEE International Conference on Robotics and Automation, pp. 3639–3644 (2007)
5. Hogan, N.: Impedance control: an approach to manipulation. J. Dyn. Syst. Meas. Control Trans. ASME **107**(1), 1–24 (1985)
6. Li, Y., Ge, S.S.: Human-robot collaboration based on motion intention estimation. IEEE/ASME Trans. Mechatron. **19**(3), 1007–1014 (2014). https://doi.org/10.1109/TMECH.2013.2264533. http://ieeexplore.ieee.org/document/6545352/
7. Li, Y., Ge, S.S.: Force tracking control for motion synchronization in human-robot collaboration. Robotica **34**(6), 1260–1281 (2016). https://doi.org/10.1017/S0263574714002240
8. Li, Y., Ge, S.S., Zhang, Q., Lee, T.H.: Neural networks impedance control of robots interacting with environments. IET Control Theory Appl. **7**(11), 1509–1519 (2013). https://doi.org/10.1049/iet-cta.2012.1032
9. Li, Y., Tee, K.P., Yan, R., Chan, W.L., Wu, Y.: A framework of human-robot coordination based on game theory and policy iteration. IEEE Trans. Robot. **32**(6), 1408–1418 (2016). https://doi.org/10.1109/TRO.2016.2597322. http://ieeexplore.ieee.org/document/7548305/
10. Lu, W.S., Meng, Q.H.: Impedance control with adaptation for robotic manipulations. IEEE Trans. Robot. Autom. **7**(3), 408–415 (1991)
11. Maeda, Y., Hara, T., Arai, T.: Human-robot cooperative manipulation with motion estimation. In: Proceedings 2001 IEEE/RSJ International Conference on Intelligent Robots and Systems. Expanding the Societal Role of Robotics in the the Next Millennium (Cat. No.01CH37180), vol. 4, pp. 2240–2245 (2001). https://doi.org/10.1109/IROS.2001.976403

12. Passenberg, C., Buss, M.: A survey of environment-, operator-, and task-adapted controllers for teleoperation systems. Mechatronics **20**(7), 787–801 (2010). https://doi.org/10.1016/J.MECHATRONICS.2010.04.005. https://www.sciencedirect.com/science/article/abs/pii/S0957415810000735

13. Rubagotti, M., Taunyazov, T., Omarali, B., Shintemirov, A.: Semi-autonomous robot teleoperation with obstacle avoidance via model predictive control. IEEE Robot. Autom. Lett. **4**(3), 2746–2753 (2019). https://doi.org/10.1109/LRA.2019.2917707

14. Wang, D., Cheah, C.C.: An iterative learning-control scheme for impedance control of robotic manipulators. Int. J. Robot. Res. **17**(10), 1091–1099 (1998)

15. Wang, Y., Sheng, Y., Wang, J., Zhang, W.: Optimal collision-free robot trajectory generation based on time series prediction of human motion. IEEE Robot. Autom. Lett. **3**(1), 226–233 (2018). https://doi.org/10.1109/LRA.2017.2737486

A Robotic Chinese Stroke Generation Model Based on Competitive Swarm Optimizer

Quanfeng Li[1], Chao Fei[1,2]([✉]), Xingen Gao[1], Longzhi Yang[3], Chih-Min Lin[4], Changjing Shang[2], and Changle Zhou[1]

[1] Department of Artificial Intelligence, School of Informatics, Xiamen University, Xiamen 361005, Fujian, People's Republic of China
fchao@xmu.edu.cn
[2] Aberystwyth University, Aberystwyth SY23 3DB, Wales, UK
[3] Northumbria University, Newcastle NE1 8QH, UK
[4] Yuan Ze University, Taoyuan, Taiwan

Abstract. The process of neural network based robotic calligraphy involves a trajectory generation process and a robotic manipulator writing process. The writing process of robotic writing cannot be expressed by mathematical expression; therefore, the conventional gradient back-propagation method cannot be directly used to optimize trajectory generation system. This paper alternatively explores the possibility of using competitive swarm optimizer (CSO) algorithm to optimize the neural network used in the robotic calligraphy system. In this paper, a variational auto-encoder network (VAE) including an encoder and a decoder is used to establish the trajectory generation model. The training of the VAE is divided into two steps. In Step 1, the decoder part of VAE network is trained by using the gradient descent method to extract the features of the input strokes. In the second step, the first encoder is used to obtain the image features directly as the input of the decoder, and the writing sequence of stroke trajectory points is obtained directly by the decoder. CSO is applied to train the decoder of VAE. Then the writing sequence is sent to the robot manipulator for writing. Experiments show that the strokes generated by this method can achieve similar but slightly different strokes from the training samples, so that the stroke writing diversity can be retained by VAE. The results also indicate the potential in autonomous action-state space exploration for other real-world applications.

Keywords: Robotic calligraphy · Competitive swarm optimizer · Variational auto-encoder network

1 Introduction

The research on robotic Chinese calligraphy is a bridge building connections between robotics and arts [1,11]; however, it is a challenging task to design

© Springer Nature Switzerland AG 2020
Z. Ju et al. (Eds.): UKCI 2019, AISC 1043, pp. 92–103, 2020.
https://doi.org/10.1007/978-3-030-29933-0_8

Chinese calligraphy robotic systems. This special challenge represents a new development direction to further advance the robotic field by better replicating human intelligence in machines. The challenge can be expressed as two technical difficulties. First, a robotic calligraphy system must convert an input Chinese stroke or letter image into robot motion sequences. In conventional approaches, such conversions were usually established by using pre-defined stroke database or manually designed by human engineers. For instance, [9,10] matches writing trajectories with pre-defined stroke database. Excessive arrangement of writing results and reference strokes were considered only in [6]. Other projects [8,12] designed the evaluation method of robot Chinese character calligraphy by manually analyzing the geometric characteristics of calligraphy. But such methods require a lot of manpower. The above works seriously constrains writing diversity. A promising solution to increasing writing diversity is to use probability distributions to represent stroke trajectories. Therefore, several work suggested using deep neural networks to implement the probability distributions [2].

Second, the current neural network based robotic calligraphy systems adopted various gradient-based methods for system training. However, these methods contain a low-efficiency limitation. In particular, if the network size becomes very large, the computational cost will increase tremendously. Other researchers suggested to applied optimization search methods for the training of large-size networks [3]; however, current optimization search algorithms also poorly performed on high-dimensional optimization problems. Therefore, in order to apply deep neural networks for robotic calligraphy, an optimization algorithm with fast convergence speed must be adopted in this work.

This paper proposes a new robotic Chinese calligraphy system, which applies a variational auto-encoder (VAE) [7] is applied as the data-driven writing result evaluation subsystem. The VAE firstly extracts stroke features which encode stroke images as low dimensional codes. Then, each written image is compared with the whole dataset, and the well-written strokes are used as candidates for reconstruction with no or very marginal loss of information. The comparison between the input image and the reconstruction provides an effective means for writing result evaluation, which is used to support the automatic development of optimal writing trajectory models. Competitive swarm optimizer, (CSO) [3] is adapted in this work to obtain optimal writing trajectory models with the support of the CAE-based evaluation subsystem.

The remainder of this paper is organized as follows: Section 2 serves as a brief introduction to the autoencoder network and competitive swarm optimizer. Section 3 specifies the proposed system, which allows a calligraphy robot to automatically learn to write strokes with high quality. Section 4 presents the experimental set up and discusses the experimental results. Section 5 concludes the paper and points out important future work.

2   Background

The main work of the method proposed in this paper is to use VAE (Variational Auto-Encoder) to directly obtain the stroke trajectory generation model.

Then the competitive swarm optimizer (CSO) algorithm is used to optimize the decoder parameters of VAE. This section introduces these basic technologies.

2.1   Variational Auto-Encoder Network

A traditional auto-encoder is a fully connected neural network, which consists of two parts: encoder and decoder. The encoder maps high-dimensional data to latent codes (low-dimensional), while decoders reconstruct initial data from potential code. Auto-encoder is usually used for dimension reduction or feature learning. The variational auto-Encoder network (VAE) is an upgraded version of auto-encoder. The structure of VAE is similar to that of auto-encoder, which also consists of encoder and decoder [4]. The VAE encoder and decoder are briefly described as follows:

Encoder: The encoder consists of convolution and pooling layers. A convolution layer convolutes the input image with a parameter filter (convolution core) to generate features mapping from the input images. The output X^k of the k-th feature map of a single channel is expressed as:

$$X^k = \sigma(X^{k-1} \otimes W^k + b^k) \tag{1}$$

where $\sigma(\cdot)$ is an activation function (usually a non-linear function) representing a 2D convolution operation; W^k and b^k represent the weight and bias of the k-th convolution kernel, respectively.

The pooling layer divides the input feature map into several rectangular regions and outputs the aggregation of each rectangular region. In this method, the network consists of two output values of the encoder, namely, the mean and covariance of the image.

Decoder: The decoder network used in this method is composed of full connection layers. The output Y^k of the k-th layer is expressed as:

$$Y^k = \sigma(V^k Y^{k-1} + c^k) \tag{2}$$

where $\sigma(\cdot)$ denotes an activation function; V^k and c^k are the weights and biases of the k-the full connection layer. In this method, the output value of the decoder is a series of stroke writing trajectory points.

Loss Function: The most commonly used objective function is the sum of the mean square error (MSE) between the input data x and the reconstructed data y, the KL divergence of the encoder output and the standard Gaussian distribution:

$$Loss = \frac{1}{2N} \sum_{i=1}^{N} (x_i - y_i)^2 + KL[\mathbb{N}(\mu(x), \sum(x) \| \mathbb{N}(0, I))] \tag{3}$$

where $\mu(x)$ represents the average value of the output of the encoder; (x) represents the covariance of the output of the encoder; $N(0, I)$ denotes the standard Gauss distribution. All parameters of VAE are conveniently optimized by back-propagation algorithm.

2.2 Competitive Swarm Optimizer

Competitive swarm Optimizer (CSO), inspired by particle swarm optimization (PSO), is a variant of PSO. In CSO, particle updating does not involve either the best individual location of each particle or the best global or neighborhood location. Instead, a pairwise competition mechanism is introduced. In the mechanism, the lost particles will update their position by learning from the more competitive particles (also known as winners).

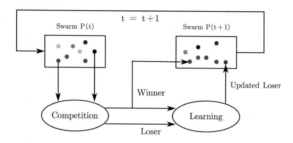

Fig. 1. Basic procedure of CSO

The procedure of CSO algorithm is shown in Fig. 1. First, two or more particles are randomly selected in $SwarmP(t)$ to form a competition pair, and the winner and loser are determined according to their adaptability in the environment. The winner keeps the original position unchanged, and the loser updates its position by learning from the winner. Then, the winners and losers form the $(T+1)$-th generation population ($SwarmP(t+1)$), and then iterate the process of competition and learning.

3 Proposed Method

3.1 Overview

The proposed system allows calligraphy robots to learn to write high-quality Chinese characters and strokes. The technical process of the system is shown in Fig. 2. It includes a stroke path generation system based on variational self-encoder and a writing system of a calligraphy robot.

The general process of the proposed method is as follows:

(1) Pre-training of the VAE encoder. Stroke training images are input into VAE, and the output is the image generated by VAE. The parameters of the network are optimized by minimizing the mean square error of the input and output images, and the KL divergence of the output distribution and the multivariate standard Gaussian distribution. After training, only the weight of the encoder is used to generate stroke trajectories.

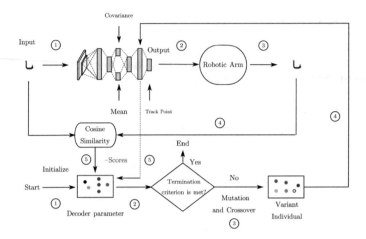

Fig. 2. An overview of the proposed writing framework for robots. CSO algorithm is used to optimize VAE to generate stroke trajectory sequence and the stroke is written by robot system.

(2) Using CSO algorithm to optimize the parameters of VAE decoder. As shown in Fig. 2, the whole process can be divided into two synchronous subprocesses. Step 1, the training samples are input into the pre-trained encoder; and outputs are the corresponding mean and covariance of the sample. Step 2: the mean and covariance are used as the input of the decoder, then the decoder outputs the sequence of strokes trajectory points. Step 3: the manipulator receives the sequence of strokes trajectory points then begins to write. Step 4: the robot vision system captures each written result as an image. Step 5: the cosine similarity between the sample image and the image written by the robot arm is calculated, and the negative cosine similarity score is used as the fitness value of the CSO algorithm.

For CSO procedure, Step 1: the population of VAE decoder parameters is initialized as the dimension of particles in the population. Step 2: According to the fitness value, CSO must determine whether the termination condition of the algorithm is reached. Step 3: if the termination condition is not reached, according to the fitness value of each particle, the evolution and selection of particles can be used to obtain the next generation population (i.e. new parameters of the decoder). Step 4: the latest parameters of the decoder are updated. Step 5: the new generation population with the previous generation population is generated for a new round of evolution and selection.

Figure 3 shows the final generation model after training. It contains only the decoder part of the variational autoencoder and the writing system of the calligraphy robot. By inputting the Gaussian noise, $z \backsim N(0|I)$, and stroke type, c, the calligraphy robot can write the corresponding Chinese character strokes.

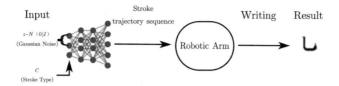

Fig. 3. Decoder of VAE

3.2   Implementation of Variational Auto-Encoder

The first task of the variational auto-encoder module is to extract the stroke features of each stroke in the encoder, and then reconstruct the image by the decoder. The variational auto-encoder used in this method has six hidden layers. Table 1 shows the important parameters of the variational auto-encoder network. The encoder of VAE consists of two convolution layers, two max-pooling layers and two full connection layers. The decoder consists of two full connection layers. Because the dimensionality of the training data set is relatively low, no pooling layer is used in this VAE architecture. The first convolution layer of the encoder contains 10 convolution cores, each of which is 5×5 in size. Each convolution core generates a feature map with a resolution of 24×24 pixels. The first max-pooling layer of the encoder contains 10 convolution cores, each of which is 2×2 in size. Each convolution core generates a feature map with a resolution of 12×12 pixels. The second convolution layer contains 20 convolution cores, each of which is 5×5 in size. Each convolution core generates a feature map with a resolution of 8×8 pixels. The second max-pooling layer of the encoder contains 20 convolution cores, each of which is 2×2 in size. Each convolution core generates a feature map with a resolution of 4×4 pixels. The three full-connection layers contain two full-connection layers for potential variables of output samples. Potential variables here refer to the mean and covariance of the sample. The decoder consists of only two full connection layers.

3.3   Cosine Similarity Evaluation of Strokes

The quality of the generated strokes is measured by calculating the similarity between the input strokes and strokes written by the robot. Since all the images used in this work are grayscale, cosine similarity can be used to score strokes:

$$Score(x) = \frac{x \cdot \hat{x}}{\|x\| \cdot \|\hat{x}\| + \alpha} \tag{4}$$

where x denotes the strokes image vector; $\hat{x}$ denotes the strokes image vector of VAE output; α is a positive number close to 0.

Table 1. Table type styles

Layer	Parameter	Dimension
Input	N/A	$1 \times 28 \times 28$
Conv. 1	10 5×5 kernels*	$10 \times 24 \times 24$
Max-pooling 1	10 2×2 kernels	$10 \times 12 \times 12$
Conv. 2	20 5×5 kernels*	$20 \times 8 \times 8$
Max-pooling 2	10 2×2 kernels	$20 \times 4 \times 4$
Fully-connected 1	100 neurons	100
Fully-connected-μ	30 neurons	30
Fully-connected-σ	30 neurons	30
DeFully-connected 1	100 neurons	100
DeFully-connected 2	28×28 neurons	28×28
Output	N/A	$1 \times 28 \times 28$

*$stride = 1$ and $pad = 0$

3.4 Stroke Writing Trajectory Generation Model Based on CSO Algorithms

In this work, the Gaussian distribution is used to represent the writing trajectory model of strokes. The model here is actually the decoder part of VAE. We use CSO algorithm to optimize the parameters of VAE decoder.

The population structure and main algorithm flow of CSO algorithm used in this method are summarised as follows:

(1) Population structure. The population is composed of N_t particles. In this experiment, the number of particles is set to 250. t denotes the number of iterations. Each particle contains its position information and velocity information. Vectors $x(t) \in R^d$ and $v(t) \in R^d$ are used to represent the spatial position and velocity of particles, respectively. The vector dimension, D, is used to represent the number of parameters of VAE decoder. The values of each dimension of each dimensional position vector are bounded, and the upper and lower limitations must be specified simultaneously:

$$x_i^s < x_{j,i}(t) < x_i^m; \quad j = 1, 2, \cdots, d; i = 1, 2, \cdots, N_t \tag{5}$$

where x_i^s and x_i^m represent the minimum and maximum values of the first dimension of each position vector in the population, i.e. the minimum and maximum values of the first parameter of VAE decoder.

(2) Initialization. Once the location initialization boundary is established, a positive generator can be used to assign a value to each parameter of each vector within a specified range. The specific operation is defined as:

$$x_{j,i}(0) = x_i^s + rand(0, 1)(x_i^m - x_i^s) \tag{6}$$
$$v_{j,i}(0) = 0 \tag{7}$$

where Rand (0,1) is a uniformly distributed sampling on [0,1]. Usually every dimension of individual velocity vector is initialized to 0.

(3) Evolution. Among the N_t particles, two bodies are randomly selected for matching to form a $N_t/2$ pair of competitors. By comparing their fitness values, which particle can be determined to be the winner and which one to be the loser. This method makes the winner with low adaptability. Then the loser updates his position by learning from the winner. Assuming the individuals $x_1(t)$ and $x_2(t)$ are a pair of competitors in the t-th iteration, then the evolutionary operation can be described as:

$$x_w(t) = \begin{cases} x_1(t), & \text{if } f(x_1(t)) \leq f(x_2(t)) \\ x_2(t), & \text{otherwise} \end{cases} \tag{8}$$

$$v_l(t+1) = R_1(t)v_l(t) + R_2(t)(x_w(t) - x_l(t)) + \phi R_3(t)(\bar{x}(t) - x_l(t)) \tag{9}$$

$$x_l(t+1) = x_l(t) + v_l(t+1) \tag{10}$$

where $x_w(t)$, $x_l(t)$, $v_l(t)$ denote the winner position vector, the loser position vector and velocity vector in $x_1(t)$ and $x_2(t)$ respectively. $f(\cdot)$ is the objective function to minimize, called fitness function. The utility function used in this experiment is the negative cosine similarity. $R_1(t)$, $R_2(t)$, and $R_3(t)$ denote the three randomly generated vectors in the competition and learning process of the first iteration. $\bar{x}(t)$ is the average position of all particles in the t-th iteration. ϕ is the parameter of the influence of control.

After all the losers in a group learn to update their positions from the winners, we will enter the next generation of evolution. Evolution is repeated until termination conditions are met. For example, the best value is found or the number of iterations reaches a predefined maximum.

3.5 Robotic System

The calligraphy robot system used in this work is shown in Fig. 4a. The calligraphy robot system consists of a four-axis robot arm and a camera mounted on a bracket. A brush is installed at the end effector of the robot arm. Writing works within the range of the arm. The whiteboard placed in front of the robot is the writing area of the robot. After the robot arm has finished writing the brush, the brush will return to the position of the initialization definition. Then the camera with the manipulator system captures the written strokes as images.

4 Experimentation

The experiment mainly involves two training processes: (1) Train the encoder part of VAE by using the traditional VAE training method; and (2) Train the process of the overall framework. Dividing the experiment into two training processes can greatly improve the efficiency of training. In this experiment, a Chinese character stroke training set shown in Fig. 4c was used to train.

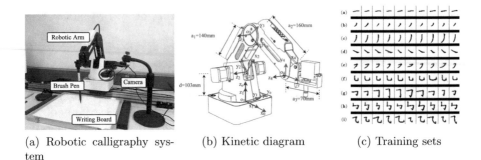

(a) Robotic calligraphy sys-   (b) Kinetic diagram   (c) Training sets
tem

Fig. 4. Robotic system and training sets

4.1   Training Sets

The training data set used in this experiment contains nine different Chinese character strokes. The data set contains 4,500 gray-scale images with a resolution of 28 × 28 and a total of 500 samples for each image. Some training data were shown in Fig. 4c. From top to bottom, these strokes are (a) horizontal; (b) short left-falling; (c) long left-falling; (d) right-falling; (e) horizontal and left-falling; (f) vertical and turn-right hook; (g) horizontal,fold and hook; (h) vertical,fold and curved-hook; and (i) horizontal and curved-hook.

4.2   Evaluation of Stroke Generation Model

When the pre-training of the encoder is completed, the average and covariance of the output of the encoder were used as the input of the second step training. Then, the decoder outputs a sequence of stroke trajectory points. The stroke trajectory sequence was inputted into the calligraphy robot system, and the corresponding stroke image was obtained through the writing process described in Sect. 3.4. The negative cosine similarity between the input sample image and the image written by the manipulator is used as the fitness function of the CSO algorithm to optimize the decoder parameters.

Nine models were trained in this step. For each model, trajectory points (T), population size (N_t), maximum algebra (G), and CSO algorithm control average position parameters (ϕ) were set to 10, 250, 2,400, and 0.25, respectively.

Figure 5 shows the training process of nine stroke models. In this figure, the fitness, i.e., the negative cosine similarity between the sample and the writing results, was steadily declining during the 2,400 generations. Solid points in the graph represent the average fitness of all individuals in each generation. With the algebraic growth of training, the fitness range becomes smaller and smaller. Note that all the training errors can drop to a very low level. This situation proves CSO algorithm can successfully optimize the decoder of VAE.

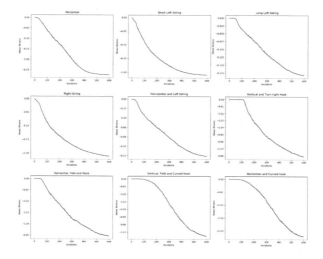

Fig. 5. Error curves during training

4.3   Writing Results

The final writing results of the nine strokes are shown in Fig. 6. Each stroke contains 16 images. These strokes were written by the calligraphy robot using the optimal stroke trajectory distribution. In this figure, all kinds of written strokes own the similar shapes of the strokes used in the training set. These similar shapes indicated the effectiveness of the proposed system. Moreover, the results of the same stroke writing are slightly different to each other; such situation illustrates that the calligraphy robot can learn to write the stroke flexibly, rather than simply copy the stroke trajectory.

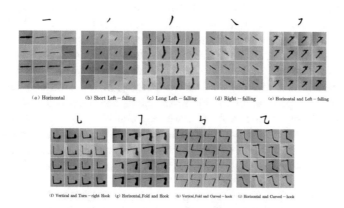

Fig. 6. Writing results of the nine strokes

Table 2. Frechet inception distance measurement

Stroke writing results	FID values
Stroke 1	49.04
Stroke 2	24.89
Stroke 3	63.58
Stroke 4	34.30
Stroke 5	45.24
Stroke 6	73.34
Stroke 7	97.95
Stroke 8	98.28
Stroke 9	105.89

We also use the Frechet Inception Distance (FID) method [5] to measure whether the written results of the proposed method are close to the pre-training data set. The FID value of a smaller data set indicates that the written result is closer to the data set. The measurement results are shown in Table 2. Each score was averaged by 10 experiments. Strokes 2 and 4 can obtain small FID values; since the structures of these two strokes are relatively simpler than that of the rest strokes. The scores of rest stroke are not very low; this situation shows that the proposed method does not simply retain the consistency of the writing results with the training set, but it can make the writing results more diverse.

5   Conclusion

This paper presents a new learning method for robotic calligraphy. The proposed method used VAE to generate stroke trajectory sequence; the training of the VAE involved the traditional backpropagation algorithm to optimized the encoder of VAE, and also adopted the CSO algorithm to optimize VAE decoder. Nine Chinese stroke datasets are used to evaluate the proposed system. Writing results showed that the proposed system successfully learns high-quality and diverse strokes. Therefore, the proposed method effectively solved the problem that gradient descent cannot be used for training in robot-related tasks. There is still room for improvement in this work. First, the proposed system can only learn Chinese strokes. In the future, we can consider to learn stroke collocation, and then develop a system for write complete Chinese characters. Second, the writing sequence of writing strokes has been not taken into account in the sampling of stroke trajectory points, this will be further studied in the future.

Acknowledgment. This work was supported by the National Natural Science Foundation of China (No. 61673322, 61673326, and 91746103), the Fundamental Research Funds for the Central Universities (No. 20720190142), Natural Science Foundation of Fujian Province of China (No. 2017J01128 and 2017J01129), and the European Union's Horizon 2020 research and innovation programme under the Marie Sklodowska-Curie grant agreement (No. 663830).

References

1. Chao, F., Huang, Y., Zhang, X., Shang, C., Yang, L., Zhou, C., Hu, H., Lin, C.: A robot calligraphy system: from simple to complex writing by human gestures. Eng. Appl. Artif. Intell. **59**, 1–14 (2017). https://doi.org/10.1016/j.engappai.2016.12.006

2. Chao, F., Lv, J., Zhou, D., Yang, L., Lin, C.M., Shang, C., Zhou, C.: Generative adversarial nets in robotic Chinese calligraphy. In: 2018 IEEE International Conference on Robotics and Automation, pp. 1104–1110 (2018)

3. Cheng, R., Jin, Y.: A competitive swarm optimizer for large scale optimization. IEEE Trans. Cybern. **45**(2), 191–204 (2015)

4. Gao, X., Zhou, C., Chao, F., Yang, L., Lin, C.M., Xu, T., Shang, C., Shen, Q.: A data-driven robotic Chinese calligraphy system using convolutional auto-encoder and differential evolution. Knowl.-Based Syst. (2019)

5. Heusel, M., Ramsauer, H., Unterthiner, T., Nessler, B., Hochreiter, S.: GANs trained by a two time-scale update rule converge to a local nash equilibrium, pp. 6626–6637 (2017). http://papers.nips.cc/paper/7240-gans-trained-by-a-two-time-scale-update-rule-converge-to-a-local-nash-equilibrium.pdf

6. Mueller, S., Huebel, N., Waibel, M., Dandrea, R.: Robotic calligraphy - learning how to write single strokes of Chinese and Japanese characters. In: IEEE/RSJ International Conference on Intelligent Robots and Systems, pp. 1734–1739 (2013). https://doi.org/10.1109/IROS.2013.6696583

7. Pu, Y., Zhe, G., Henao, R., Xin, Y., Li, C., Stevens, A., Carin, L.: Variational autoencoder for deep learning of images, labels and captions (2016)

8. Wei, L., Song, Y., Zhou, C.: Computationally evaluating and synthesizing Chinese calligraphy. Neurocomputing **135**(C), 299–305 (2014)

9. Yao, F., Shao, G., Yi, J.: Extracting the trajectory of writing brush in Chinese character calligraphy. Eng. Appl. Artif. Intell. **17**(6), 631–644 (2004). https://doi.org/10.1016/j.engappai.2004.08.008

10. Yao, F., Shao, G., Yi, J.: Trajectory generation of the writing brush for a robot arm to inherit block style Chinese character calligraphy techniques. Adv. Robot. **18**(3), 331–356 (2004). https://doi.org/10.1163/156855304322972477

11. Zeng, H., Huang, Y., Chao, F., Zhou, C.: Survey of robotic calligraphy research. CAAI Trans. Intell. Syst. **11**(1), 15–26 (2016)

12. Zhe, M., Su, J.: Aesthetics evaluation for robotic Chinese calligraphy. IEEE Trans. Cogn. Dev. Syst. **9**(1), 80–90 (2017)

Deep Learning Approaches

A Two-Stream CNN Framework for American Sign Language Recognition Based on Multimodal Data Fusion

Qing Gao[1,2,3], Uchenna Emeoha Ogenyi[4], Jinguo Liu[1,2(✉)], Zhaojie Ju[1,2,4], and Honghai Liu[4]

[1] State Key Laboratory of Robotics, Shenyang Institute of Automation, Chinese Academy of Sciences, Shenyang 110016, China
liujinguo@sia.cn
[2] Institutes for Robotics and Intelligent Manufacturing, Chinese Academy of Sciences, Shenyang 110169, China
[3] University of Chinese Academy of Sciences, Beijing 100049, China
[4] School of Computing, University of Portsmouth, Portsmouth PO1 3HE, UK

Abstract. At present, vision-based hand gesture recognition is very important in human-robot interaction (HRI). This non-contact method enables natural and friendly interaction between people and robots. Aiming at this technology, a two-stream CNN framework (2S-CNN) is proposed to recognize the American sign language (ASL) hand gestures based on multimodal (RGB and depth) data fusion. Firstly, the hand gesture data is enhanced to remove the influence of background and noise. Secondly, hand gesture RGB and depth features are extracted for hand gesture recognition using CNNs on two streams, respectively. Finally, a fusion layer is designed for fusing the recognition results of the two streams. This method utilizes multimodal data to increase the recognition accuracy of the ASL hand gestures. The experiments prove that the recognition accuracy of 2S-CNN can reach 92.08% on ASL fingerspelling database and is higher than that of baseline methods.

Keywords: Hand gesture recognition · CNN · Multimodal data fusion

1 Introduction

At present, with the development of computer technology and artificial intelligence (AI), the HRI has evolved from robot-centered methods to human-centered methods. The purpose is to achieve more convenient, natural and coordinated interaction between humans and robots, and give full play to the advantages of people and robots to achieve higher work efficiency. The new methods of HRI mainly include hand gestures, voice and EEG [1]. Among them, the vision-based hand gesture interaction is very suitable for HRI because of its intuitive, natural, and non-contact characteristics. So, it has been paid much attention and in-depth research by many scholars.

© Springer Nature Switzerland AG 2020
Z. Ju et al. (Eds.): UKCI 2019, AISC 1043, pp. 107–118, 2020.
https://doi.org/10.1007/978-3-030-29933-0_9

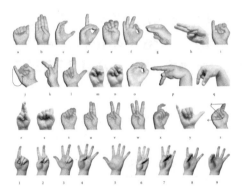

Fig. 1. ASL hand gesture dataset

But the vision-based hand gesture interaction is not yet mature now. It has mainly three difficulties: (1) better and more natural interactive hand gestures; (2) data processing to remove interference from noise and background; (3) the designs of hand gesture feature extractor and classifier.

The design of hand gesture sets in HRI is crucial. Simple and natural interactive hand gestures make it easier for operators to interact with robots. We chose the ASL hand gesture set [2], which contains 26 different hand gestures, including 24 static hand gestures and 2 dynamic hand gestures. These hand gestures are shown in Fig. 1. We can choose several hand gestures from them for HRI and use different alphabetic hand gestures to represent the corresponding interactive hand gestures. It is very convenient for the operator to learn and use.

Illumination changes and background can seriously affect the recognition accuracy of vision-based hand gesture recognition. The hand segmentation can effectively remove the background. For RGB images, skin color-based hand segmentation is often used, but the objects in the background with similar skin colors can seriously affect the effect of the method. The depth image can avoid the influence of illumination changes, and it is more convenient to remove the background. So, the depth images are utilized in this paper.

Traditional hand gesture recognition methods mainly extract the shallow layer features of hand gestures. These features are handcrafted, such as SIFT and HOG [3]. But they cannot get good performance for complex hand gesture recognition. With the development of deep learning, more and more works have focused on applying deep neural networks to hand gesture recognition. Among the deep learning methods, CNN is mainly used for image recognition [4]. Typical CNN are mainly VGG, GoogLeNet, Inception, ResNet and MobileNet. Each of these methods has its own characteristics and combining these methods can improve the efficiency of image recognition.

In this paper, a two-stream CNN framework is designed for the recognition of ASL hand gestures. The contributions are mainly as follows:

- A two-stream CNN framework is proposed. It uses two CNN feature extractors to extract gesture RGB and depth features and a fusion layer to fuse the recognition results. Multimodal data features are utilized to increase the hand gesture recognition accuracy.

- In the data processing part, a simple and convenient data enhancement method is proposed by combining the advantages of RGB and depth images.
- In the result fusion part, a fusion layer is designed and connected to the network. The weights of the fusion can be obtained through training.

The rest of this paper is organized as follows. Related work of vision-based hand gesture recognition is introduced in Sect. 2. The 2S-CNN proposed in this paper is introduced in Sect. 3, which includes network framework, data preparation, data enhancement, CNN feature extractor, fusion method. The experiments and analyzes are shown in Sect. 4. Section 5 is conclusion remark and future work.

2   Related Work

Traditional vision-based hand gesture recognition algorithms include dynamic time warping (DTW), artificial neural network (ANN) and hidden Markov model (HMM) [5]. But each of them has some shortcomings. With the development of deep learning in the field of image recognition, more and more scholars have applied deep learning methods to the research of vision-based hand gesture recognition and have achieved certain research results. For example, Oyebade K used the CNN to identify 24 ASL hand gestures and achieved a high recognition accuracy. But the gesture database used is too small and the image background is single [6]. Jawad Nagi used a CNN with Max-Pooling layers to classify six hand gestures and used them to control a robot. He wore gloves to segment and classify hand gestures simply. But this method is not universal [7]. Youngwook Kim used DCNN to identify 10 gestures of radar-acquired Micro-Doppler Signatures and achieved a high recognition accuracy. But the disadvantage is that the image background is single [8]. Takayoshi Yamashita designed a deep convolutional neural network with bottom-up structure for hand gesture recognition. However, this method used two-dimensional grayscale images, which may not be suitable for some complicated three-dimensional hand gestures [9]. There are a lot of interference information for hand gesture recognition in images, such as complex background, illumination changes and noise. In-depth researches are still needed to achieve robust hand gesture recognition.

Depth images can effectively reduce the effects of background and illumination changes. At present, with the development of depth sensors like Kinect, many scholars introduce depth images into hand gesture recognition, or fuse depth images and RGB images to achieve a higher recognition accuracy. For example, Qing Gao used a parallel CNNs to fuse depth information with RGB information, which improved the recognition accuracy of the ASL hand gesture [10]. C. Jose L. Flores used two CNN architectures with different amounts of layers and parameters per layer to classify 24 alphabet hand gestures of sign language of Peru (LSP) [11]. Zhenyuan Zhang proposed a HandSense network, which extracts features of RGB and depth hand gesture images through two

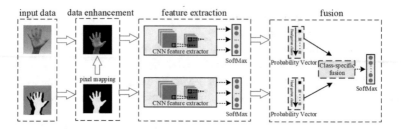

Fig. 2. 2S-CNN framework

parallel 3D-CNNs, and then fused the features and classified hand gestures by a SVM classifier [12]. Although there are many work to study multi-modal information fusion for hand gesture recognition, some work still need further research, such as how to make full use of RGB and depth information, which CNN network is more effective, and how to efficiently fuse information.

The recognition of ASL hand gestures belongs to a fine-grained image recognition task. This paper makes full use of the advantages of hand gesture RGB and depth information by fusing the two kinds of information. Efficient hand gesture segmentation, feature extraction and recognition methods are designed. The proposed 2S-CNN can effectively improve the recognition accuracy of ASL hand gestures.

3   Two-Stream CNN Framework (2S-CNN)

3.1   Network Framework

The RGB images of hand gestures can express various performance information such as the colors and shapes of the hands. While the depth images of hand gestures can express the spatial information of the hands. So, combining RGB with depth images can take advantage of more hand gesture information to increase hand gesture recognition accuracy. The proposed 2S-CNN adopts a two-stream CNN framework [13]. One channel is used to process the RGB images, and a CNN is used to extract the performance features of the hand gestures. The other channel is used to process the depth image, and another CNN is used to extract the 3D space features of the hand gestures. Finally, the outputs of the two channels are fused using class-specific fusion method to achieve the final prediction of hand gestures. It can get more feature information of hand gestures to achieve a robust prediction and better performance. Its network framework is mainly divided into four parts, which are input data, data enhancement, feature extraction, and fusion. Its framework is shown in Fig. 2.

The main ideas of the 2S-CNN are shown as follows:

- Augment the amount of input data to prevent over-fitting when train the network and improve the generalization ability of the network.

- In the data enhancement part, a simple and efficient hand segmentation method is designed by using the characteristics of the depth images to remove background and noise from the RGB and depth images.
- In the feature extraction part, adopt a more efficient CNN feature extractor.
- In the fusion part, a fusion layer is designed and connected into the network. Then, the fusion weight can be obtained through training.

The details of the method are described below.

3.2 Data Preparation

In this paper, the ASL fingerspelling database is used as training and testing database. It contains both ASL hand gesture RGB and depth images. There are 24 static hand gestures, representing 24 English letters (except J and Z because they are represented by dynamic hand gestures). Each hand gesture includes 5000 hand gesture images which are 2500 RGB and 2500 depth images collected by 5 objects in different backgrounds. In order to more efficiently segment hand and train the data, the RGB and depth images should be aligned firstly, and then, make the data augmentation.

Image Alignment. The depth and the RGB images in the ASL database are not aligned, which affects the subsequent hand segmentation operation. So, we need to align the RGB and depth images first. The alignment equations are shown as follows [14]:

$$Z_{rgb} * p^{rgb} = R * Z_d * p_d + T \tag{1}$$

$$R - K_{ryb} * R_{d2ryb} * K_d^{-1} \tag{?}$$

$$T = K_{rgb} * T_{d2rgb} \tag{3}$$

where rgb means RGB image, d means depth image. $Z * p$ represents the mapping relationship of the homogeneous 3D points ($P = [XYZ1]^T$) in the respective camera coordinate systems to the pixel coordinates ($p = [uv1]^T$) on the respective images. K_{rgb} and K_d are internal parameters for color camera and depth camera of Kinect. Set R_{rgb} and T_{rgb} as the external parameters of the color camera, and R_d and T_d as the external parameters of the depth camera. So, the equations of rigid body transformation matrices (R_{d2rgb} and T_{d2rgb}) of the two cameras can be get as follows:

$$R_{d2rgb} = R_{rgb} * R_d^{-1} \tag{4}$$

$$T_{d2rgb} = T_{rgb} - R_{d2rgb} * T_d \tag{5}$$

Data Augmentation. The increase of training data can help to avoid overfitting and improve the generalization of the network. For data enhancement, we use the following enhancement methods: (a) flip horizontal; (b) translation transformation: pan the image by ±5 pixels; (c) rotation transformation: rotate the image by ±10°.

3.3   Data Enhancement

There is some interference information in the data of ASL fingerspelling database, such as background in RGB images and noise in depth images, which will affect the accuracy of hand gesture recognition. Because the depth images can effectively distinguish hands and background, depth images are utilized to segment hands first, and then segment RGB images by mapping the segmentation pixels. Suppose the pixel with the closest distance value in the depth image appears in the hand area (it's true in ASL fingerspelling database). Therefore, the following equation can be used to segment the depth images.

$$d_{w,h} = \begin{cases} d_{w,h} & if d_{w,h} \geq d_{min} - \triangle d \\ 0 & if d_{w,h} < d_{min} - \triangle d \end{cases} \tag{6}$$

where w and h represent the width and height of the image, respectively. $w, h \in [0, 200]$. $d_{w,h}$ represents the depth value of the image at the coordinate point (w, h). d_{min} is the minimum value in the depth image. $\triangle d$ represents the difference between the depth value and the minimum depth value d_{min}, set $\triangle d = 2$. So, the pixel whose depth value is in the range $[d_{min}, d_{min} - \triangle d]$ defaults to the hand gesture. The remaining pixels are regarded as background and their values are set to 0. Then, the pixel coordinates of the hand gesture in the depth map are mapped onto the color image, and the segmented hand gesture on color image is obtained.

Since the sizes of the ASL fingerspelling database images are different, all image sizes are converted into 299 × 299 pixels, which can be performed at the input layer of the network.

3.4   CNN Feature Extractor

Feature extraction of hand gesture images is the key to hand gesture recognition. A more efficient CNN feature extractor can better extract hand gesture features and help to improve the recognition accuracy of hand gestures. In this part, the Inception-ResNet v2 is chosen as the CNN feature extractor by analyzing a variety of typical CNN structures [15]. Inception-ResNet v2 combines the advantages of Inception and ResNet. It can train deeper networks with better feature extraction and avoid overfitting. Its network structure is shown in Fig. 3. Where the network structure of the three Inception-resnets in Fig. 3 are shown in Fig. 4.

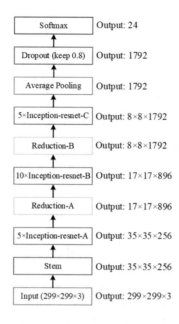

Fig. 3. Inception-ResNet v2 structure

3.5 Fusion Method

It can be seen from Fig. 2 that in the proposed 2S-CNN, the hand gesture predicted probability scores can be obtained from the input RGB and depth images through CNN feature extractors and softmax classifiers. After that, the results from the two streams need to be fused. A fusion layer is designed and connected after the network. So, the weight of the fusion can be obtained through training. Firstly, the form of outputs is converted into probability vectors. Set P_1 and P_2 denote the probability vectors obtained from the RGB and depth streams, respectively. Where $P_1, P_2 \in R^{1 \times N}$, and N denotes the number of classifications of the ASL hand gestures. Then, set $\omega_i = \omega_1^i, \omega_2^i, \cdots, \omega_N^i (i = 1, 2)$ indicates the corresponding fusion weight. Where ω_j^i represents the $j-$th class fusion weight in the $i-$th stream. Finally, set P_f represents the vector through the fusion of P_1 and P_2, then P_f is expressed as

$$P_f \cdot \omega = P_1 \odot \omega_1 + P_2 \odot \omega_2 = [P_1^1, \cdots, P_1^N, P_2^1, \cdots, P_2^N] \begin{bmatrix} \omega_1^1 & 0 & \cdots & 0 \\ 0 & \omega_1^2 & \cdots & 0 \\ \vdots & \vdots & \ddots & \vdots \\ 0 & 0 & \cdots & \omega_1^N \\ \omega_2^1 & 0 & \cdots & 0 \\ 0 & \omega_2^2 & \cdots & 0 \\ \vdots & \vdots & \ddots & \vdots \\ 0 & 0 & \cdots & \omega_2^N \end{bmatrix} \quad (7)$$

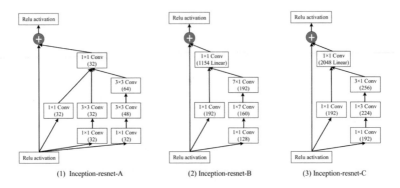

Fig. 4. Inception-resnet structure

where $\omega \in R^{2N \times N}$ is a matrix composed by the hand gesture class fusion weights. $\odot$ represents the element-wise product.

The purpose of Eq. (7) is to learn the optimal hand gesture class fusion weights to achieve the best fusion effect. The network is finally classified by a softmax layer. To avoid over-fitting of the training, a rectified linear unit (ReLU) is integrated into the loss function of the softmax layer. The loss function is given as follows:

$$\omega = \min_{\omega} loss(p, y; \omega) + \max(0, \omega) \tag{8}$$

where $loss(\cdot)$ is the original loss function of the softmax layer, y represents the ground-truth labels, and $\max(0, \omega)$ is the ReLU result of.

Finally, the recognized hand gesture c is:

$$c = \arg \max P_f \tag{9}$$

4   Experiment Results and Discussion

4.1   Train

The training process uses Transfer Learning, which can effectively reduce training time and steps. The Inception_resnet_v2 model trained under ImageNet database is used as the pre-training model, freeze all network layer parameters except softmax layer. Then, change the neuron number in the softmax layer to 24. After that, retrain the model on the ASL fingerspelling database.

In the process of training, the data is divided into batches, which can improve the efficiency of training. Set the batch size N to 24. The gradient solution method in this experiment uses Stochastic gradient descent (SGD). In the SGD, ω is updated by the linear combination of the negative gradient and the last weight update value V_t. The iteration equation is shown as follows:

$$V_{t+1} = \mu V_t - \alpha \bigtriangledown L(\omega_t) \tag{10}$$

$$\omega_{t+1} = \omega_t + V_{t+1} \tag{11}$$

where α is the learning rate of the negative gradient. μ is the weight of the last gradient value and it is used to weight the effect of the previous gradient direction on the current gradient direction. These two parameters can get the best results through tuning. t represents the current number of iterations. The adjustment method of the learning rate selects the step uniform distribution strategy, which can make the network quickly converge in the early stage and reduce the oscillation in the later stage. Its calculation equation is shown as follows:

$$\alpha = \alpha_0 \times \gamma^{(t/s)} \tag{12}$$

where α_0 is the initial learning rate and is set to 0.001. γ is the adjustment parameter and is set to 0.1. s represents the iteration length of the adjustment learning rate and is set to 10000. That is, when the current iteration number t reaches an integral multiple of 10000, the learning rate is adjusted. The total number of training steps is 30000.

4.2   Validation

The verification process of the experiment adopts the "half-half" method. That is, half of the ASL fingerspelling database is used as training data and the other half is used as testing data. Data augmentation is performed only on the training data, and data alignment and data enhancement are performed on the test data.

To verify the superiority of the Inception-ResNet v2, the experimental results are compared with that of other typical networks. Four networks, VGG-19, ResNet152, Inception v4 and Inception-Resnet v2, are used as CNN feature extractors for 2S-CNN. After training, 4 ASL gesture recognition models are obtained. These four models are used to verify their accuracies and speeds on the testing data. The results are shown in Table 1. The experiments are carried out under the Ubuntu16.04 system. The deep learning framework uses Tensorflow and the GPU chose GTX1060.

Table 1. The accuracies and speeds of four models

CNN feature extractor	Speed(ms)	Accuracy(%)	GPU
VGG-19 [16]	36	80.86	GTX1060
ResNet152 [17]	38	83.80	GTX1060
Inception v4 [15]	**30**	85.32	GTX1060
Inception-ResNet v2 [15]	33	**92.08**	GTX1060

As can be seen from Table 1, the speeds of the four ASL gesture recognition models are all within 40ms, and the recognition accuracies are all above 80%.

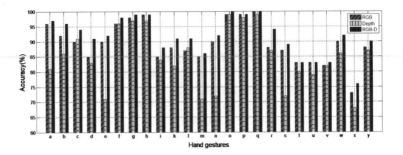

Fig. 5. ASL hand gesture recognition accuracy comparison chart

Where the model that uses ResNet152 as the CNN feature extractor gets the slowest speed (38 ms). The model that uses Inception v4 as the CNN feature extractor gets the fastest speed (30 ms). The model that uses Inception-ResNet v2 as the CNN feature extractor gets the highest accuracy (92.08%). And the model that uses VGG-19 as the CNN feature extractor gets the lowest accuracy (80.86%). Therefore, it can prove that the Inception-ResNet v2 can extract hand gesture features efficiently and help improve the recognition accuracy of the model. And the recognition speed of the model can be real-time. So, it can be applied as a gesture recognition model to the real-time HRI system.

In addition, in order to verify the effectiveness of the proposed 2S-CNN network for fusing RGB and depth features, we compared the hand gesture recognition results of the 2S-CNN with single-stream frameworks using RGB and depth, respectively. These three models all use Inception-ResNet v2 as the CNN feature extractors. The comparison of the hand recognition accuracies of the three models for each ASL gesture is shown in Fig. 5.

As can be seen from Fig. 5, the average recognition accuracies of the ASL hand gestures by the RGB-stream framework and depth-stream framework are 89.7% and 84.8%, respectively. The ASL hand gesture recognition accuracy of the 2S-CNN is 92.1%. Its average recognition accuracy increases by 2.4% and 7.3% compared to the RGB-stream and depth-stream, respectively. And its recognition accuracy of each ASL hand gesture exceeds 75%. Therefore, it can be proved that the proposed 2S-CNN framework can effectively improve the hand gesture recognition accuracy.

To further verify the superiority of our method, the experimental result is compared with some baseline methods. The comparison results are shown in Table 2.

It can be seen from Table 2 that the proposed 2S-CNN has a higher recognition accuracy for ASL hand gestures than the main baseline methods. Therefore, the validity and superiority of the method are proved.

Table 2. Comparison of the recognition accuracy

Method	Accuracy (%)
GF-RF [18]	75
ESF-MLRF [19]	87
RF-JP [20]	59
RF-JA+C [21]	90
2S-CNN	**92**

5   Conclusion Remark and Future Work

In this paper, a two-stream CNN framework combining RGB and depth information is proposed for increase the recognition of ASL hand gestures. The framework extracts the RGB and depth features of the hand gesture through two
CNN streams, respectively, and fuses the recognition results. The validity and
superiority of the proposed method are verified by comparison of experimental
results. The contributions of this paper mainly include as follows:

– A two-stream CNN framework is designed to fuse multiple hand gesture information to improve the recognition accuracy of hand gestures.
– Effective processing and enhancement of ASL data facilitates subsequent hand
  gesture feature extraction and recognition.
– The fusion layer is designed and connected to the network. By doing so, the
  optimal fusion weight can be obtained through network training.

Currently, this method is only applied to the recognition of static hand gestures. We know that the application of dynamic hand gesture recognition is more
important in HRI, so the application of the idea in this paper can be transferred
to dynamic hand gesture recognition in the future work.

Acknowledgment. Research supported in part by the Key Research Program of
the Chinese Academy of Sciences under Grant Y4A3210301, in part by the Research
Fund of China Manned Space Engineering under Grant 050102, in part by the Natural
Science Foundation of China under Grant 51775541, 51575412, 51575338 and 51575407,
in part by the EU Seventh Framework Programme (FP7)-ICT under Grant 611391, in
part by the Research Project of State Key Lab of Digital Manufacturing Equipment &
Technology of China under Grant DMETKF2017003, in part by National Key R&D
Program Projects 2018YFB1304600.

References

1. Goodrich, M.A., Schultz, A.C.: Human crobot interaction: a survey. Found. Trends
   in Hum. Comput. Interact. **1**(3), 203–275 (2008)
2. Liu, J., Luo, Y., Ju, Z.: An interactive astronaut-robot system with gesture control.
   Comput. Intell. Neurosci. **2016**, 7845102 (2016)

3. Rautaray, S.S., Agrawal, A.: Vision based hand gesture recognition for human computer interaction: a survey. Artif. Intell. Rev. **43**(1), 1–54 (2015)
4. LeCun, Y., Bengio, Y., Hinton, G.: Deep learning. Nature **521**(7553), 436 (2015)
5. Wang, T., Li, Y., Hu, J., Khan, A., Liu, L., Li, C., Ran, M.: A survey on vision-based hand gesture recognition. In: International Conference on Smart Multimedia, pp. 219-231. August 2018
6. Oyedotun, O.K., Khashman, A.: Deep learning in vision-based static hand gesture recognition. Neural Comput. Appl. **28**(12), 3941–3951 (2017)
7. Nagi, J., Ducatelle, F., Di Caro, A.G., Cirean, D., Meier, U., Giusti, A., Gambardella, L.M.: Max-pooling convolutional neural networks for vision-based hand gesture recognition. In: Conference on 2011 IEEE International In Signal and Image Processing Applications (ICSIPA), pp. 342-347 (2011)
8. Kimm, Y., Toomajian, B.: Hand gesture recognition using micro-Doppler signatures with convolutional neural network. IEEE Access **4**, 7125–7130 (2016)
9. Yamashita, T., Watasue, T.: Hand posture recognition based on bottom-up structured deep convolutional neural network with curriculum learning. In: Image Processing (ICIP), 2014 IEEE International Conference on, pp. 853-857 (2014)
10. Gao, Q., Liu, J., Ju, Z., Li, Y., Zhang, T., Zhang, L.: Static hand gesture recognition with parallel CNNs for space human-robot interaction. In: International Conference on Intelligent Robotics and Applications, pp. 462-473 (2017)
11. Flores, C.J.L., Cutipa, A.G., Enciso, R.L.: Application of convolutional neural networks for static hand gestures recognition under different invariant features. In: 2017 IEEE XXIV International Conference on Electronics, Electrical Engineering and Computing (INTERCON), pp. 1-4 (2017)
12. Zhang, Z., Tian, Z., Zhou, M.: HandSense: smart multimodal hand gesture recognition based on deep neural networks. Journal of Ambient Intelligence and Humanized Computing, 1-16 (2018)
13. Hao, S., Wang, W., Ye, Y., Nie, T., Bruzzone, L.: Two-stream deep architecture for hyperspectral image classification. IEEE Trans. Geosci. Remote Sens. **56**(4), 2349–2361 (2018)
14. Zhang, Z.: Microsoft kinect sensor and its effect. IEEE Multimedia **19**(2), 4–10 (2012)
15. Szegedy, C., Ioffe, S., Vanhoucke, V., Alemi, A.A.: Inception-v4, inception-resnet and the impact of residual connections on learning. AAAI **4**, 12 (2017)
16. Simonyan, K., Zisserman, A. : Very deep convolutional networks for large-scale image recognition. arXiv preprint arXiv:1409.1556 (2014)
17. He, K., Zhang, X., Ren, S., Sun, J.: Identity mappings in deep residual networks. In: European Conference on Computer Vision, pp. 630-645 (2016)
18. Pugeault, N., Bowden, R.: Spelling it out: real-time ASL fingerspelling recognition. In: 2011 IEEE International Conference on, Computer Vision Workshops (ICCV Workshops), pp. 1114-1119 (2011)
19. Kuznetsova, A., Leal-Taix, L. , Rosenhahn, B.: Real-time sign language recognition using a consumer depth camera. In: Proceedings of the IEEE International Conference on Computer Vision Workshops, pp. 83-90 (2013)
20. Keskin, C., Kraç, F., Kara, Y.E., Akarun, L.: Real time hand pose estimation using depth sensors. In Consumer depth cameras for computer vision, 119-137 (2013)
21. Dong, C., Leu, M.C., Yin, Z.: American sign language alphabet recognition using microsoft Kinect. In: Proceedings of the IEEE Conference on Computer Vision and Pattern Recognition Workshops, pp. 44-52 (2015)

DeepSwarm: Optimising Convolutional Neural Networks Using Swarm Intelligence

Edvinas Byla and Wei Pang[(✉)]

Department of Computing Science, University of Aberdeen, Aberdeen AB24 3UE, UK
e.byla.15@aberdeen.ac.uk, pang.wei@abdn.ac.uk

Abstract. In this paper we propose DeepSwarm, a novel neural architecture search (NAS) method based on Swarm Intelligence principles. At its core DeepSwarm uses Ant Colony Optimization (ACO) to generate ant population which uses the pheromone information to collectively search for the best neural architecture. Furthermore, by using local and global pheromone update rules our method ensures the balance between exploitation and exploration. On top of this, to make our method more efficient we combine progressive neural architecture search with weight reusability. Furthermore, due to the nature of ACO our method can incorporate heuristic information which can further speed up the search process. After systematic and extensive evaluation, we discover that on three different datasets (MNIST, Fashion-MNIST, and CIFAR-10) when compared to existing systems our proposed method demonstrates competitive performance. Finally, we open source DeepSwarm (https://github.com/Pattio/DeepSwarm) as a NAS library and hope it can be used by more deep learning researchers and practitioners.

Keywords: Ant Colony Optimization · Neural Architecture Search

1 Introduction

In recent years it has become increasingly challenging for human engineers to manually design deep neural architectures for specific tasks. This is mainly due to the following two facts: (1) modern deep neural architectures tend to be very complex with a lot of layers and hyperparameters; (2) one architecture might perform well on one dataset or on one type of problems but poorly on others. These two factors have resulted in a boom of research that tries to develop methods that can automate the design of neural architectures, the so-called neural architecture search [22].

In this paper we propose a novel neural architecture search method based on Swarm Intelligence (SI). To start with, we focus on Convolutional Neural Networks (CNN) [13], one of the most commonly used deep neural architectures. To discover new CNN architectures our method uses Ant Colony Optimization

© Springer Nature Switzerland AG 2020
Z. Ju et al. (Eds.): UKCI 2019, AISC 1043, pp. 119–130, 2020.
https://doi.org/10.1007/978-3-030-29933-0_10

(ACO) [5]. The motivation for using SI for NAS is due to the fact that SI possesses many appealing properties that could be helpful when dealing with NAS problems. This includes fault tolerance, decentralisation, scalability and ability to share and combine the knowledge, just to name a few. In particular, ACO has few distinct characteristics that make it naturally fit into the NAS domain: ACO is good at solving discrete problems which can be represented as graphs and it can easily adapt to dynamic environment (changing graph). Another significant motivating factor to use SI is the fact that the majority of its methods have not been explored in the context of NAS.

The novel contributions of this research are summarised as follows:

- We show that ACO can be used to effectively optimise CNNs.
- We use heuristic information when performing NAS based on ACO.
- We dynamically change the graph size and progressively search for the architectures when performing NAS based on ACO.

The rest of the paper is organised as follows: Sect. 2 presents related work; Sect. 3 introduces our proposed method; Sect. 4 presents the evaluation of our method; and Sect. 5 concludes the paper and explores possible future directions.

2   Related Work

Neural Architecture Search (NAS) is an automated process that aims to discover the best performing neural network architectures for a specific problem. Even though NAS research goes back as far as three decades [16], it has attracted new attention in recent years with the rapid development of deep learning, significant improvements in hardware, and growing interest of the machine learning community. Furthermore, even with this renewed interest from many deep learning researchers and practitioners it still seems that most of the existing NAS research predominantly focuses on using Evolutionary Algorithms [15,19,21], Bayesian Optimisation [4,10], and Reinforcement Learning [1,24,25]. However, considering most of these approaches require huge amounts of computational resources, some new work which tries to reduce the computational costs have emerged [8,17,18]. For example, in [18] the authors proposed to use large computational graph which stores all the weights, and they reported that sharing these weights among child models could be 1000 times less computationally expensive than standard NAS approaches.

To the best of our knowledge, ACO was first applied to NAS problem in 2014 [20], and in their work ACO was used to optimise feed-forward neural networks. Furthermore, in their work the authors discovered that reusing the weights of the best solution can further improve the performance of their method. In 2015 ACO was used to optimise the structure of deep recurrent neural networks [3], where the authors try to address the problem of predicting general aviation flight data. The authors reported that using ACO they could achieve better prediction performance for airspeed, altitude, and pitch compared with the previous best published results. Finally, in more recent work [7], ACO was used to optimise long short-term memory recurrent neural networks, and they achieved an increase in prediction accuracy, while also reducing the number of trainable weights by 55%.

It is noted that another relevant work to our research is the Progressive Neural Architecture Search (PNAS) approach [14]: similar to PNAS, the system proposed in this paper explores enormous CNN search space by using small incremental steps. In [14] the authors concluded that PNAS can achieve the same level of performance as the previous NAS approach [25] while being 8 times faster in terms of the required total computational time.

3   DeepSwarm

In this section we first present the details of the proposed DeepSwarm, and then we give the overall workflow.

As mentioned before, DeepSwarm search for new architectures in the order of increasing complexity similar to PNAS. At the beginning of a NAS task, DeepSwarm creates an internal graph which contains only the input node. Then a specified number of ants are generated. Next, one by one each ant is placed on the input node. After being placed on the input node each ant uses the Ant Colony System (ACS) [6] selection rule to select one of the available nodes in the next layer of CNN, and the ACS selection rule is as follows:

$$
s = \begin{cases} \arg\max_{u \in J_k(r)}\{[\tau(r,u)] \cdot [\eta(r,u)]^\beta\}, & \text{if } q \leq q_0 \quad \text{(exploitation)}. \\ S, & \text{otherwise} \quad \text{(biased exploration)}, \end{cases} \tag{1}
$$

In the above $\tau(r,u)$ denotes the pheromone amount on the edge that goes from node r to node u and $\eta(r,u)$ denotes the heuristic value associated with the edge going from node r to node u. Furthermore, $J_k(r)$ denotes a set of nodes that are available to visit from node r. The value of q is a random number uniformly distributed over $[0 \ldots 1]$. Parameters $q_0 \in (0,1]$ and $\beta \in (0, \inf)$ control the algorithm's greediness and the relative importance of heuristic information. Finally, S is a random variable selected according to the probabilistic distribution defined by Eq. (2):

$$
p_k(r,s) = \begin{cases} \dfrac{[\tau(r,s)] \cdot [\eta(r,s)]^\beta}{\sum_{u \in J_k(r)} [\tau(r,u)] \cdot [\eta(r,u)]^\beta}, & \text{if } s \in J_k(r). \\ 0, & \text{otherwise}. \end{cases} \tag{2}
$$

Once a node is selected the system checks if this node already exists in the graph at the depth of the selection. If this node is a new one which does not exist in the graph, it is added to the graph as a neighbour node to the previous node (i.e., the node where the ant was before the selection) so the subsequent ants can exploit the pheromone information. After an ant selects a particular node it also performs the same selection rule as defined by Eqs. (1) and (2) to select the attributes of that node (i.e. filter size, kernel size). When the selection is completed the node is added to the ant's path. Once an ant reaches the current maximum allowed depth, its path is transformed into a neural network architecture which then gets evaluated. Furthermore, after an ant finishes a walk it performs ACS local pheromone update as defined by Eq. (3) for each edge it has used:

$$\tau(r, s) \longleftarrow (1 - \rho) \cdot \tau(r, s) + \rho \cdot \tau_0 \tag{3}$$

In the above, parameter ρ denotes the pheromone decay factor and parameter τ_0 is the initial pheromone value. This local update rule decays pheromone values so the other ants can be encouraged to explore other paths. After all ants are evaluated the best ant is found (the ant which found the architecture with the highest accuracy). This best ant then performs the ACS global pheromone update as defined by Eq. (4), which increases the pheromone values for the edges found in the best path.

$$\tau(r, s) \longleftarrow (1 - \alpha) \cdot \tau(r, s) + \alpha \cdot \Delta\tau(r, s), \tag{4}$$

where

$$\Delta\tau(r, s) = \begin{cases} C_{gb}, & \text{if } (r, s) \in \text{ global-best-tour.} \\ 0, & \text{otherwise.} \end{cases} \tag{5}$$

Here parameter α controls pheromone evaporation and its range is $(0, 1)$. C_{gb} is the cost of the global best tour (the best model accuracy). After the graph's current maximum allowed depth is increased, a new population of ants is generated. This cycle is repeated until the maximum depth (specified by the user) is reached. An illustrative example of NAS performed by DeepSwarm can be seen in Fig. 1, and the pseudocode is given in Algorithm 1.

We list several interesting outcomes of using ACO as a search strategy as follows: (1) weight reusability is straightforward to implement: we find the longest common sub-path in the graph and reuse the best weights from that sub-path, (2) the search space can be explored progressively as ants can adapt to the dynamic environment (when we expand the graph from depth n to $n + 1$ we do not lose the information which was gathered up to depth $n + 1$), and (3) because ACO uses domain-specific heuristics (Eqs. (1) and (2)) domain experts can easily provide their own knowledge to speed up the search further.

We finally point out the differences of DeepSwarm compared with previous work on ACO for optimising neural networks as well as PNAS [14] as mentioned in Sect. 2. First, DeepSwarm uses dynamic graphs and performs the search layer by layer progressively, and this is similar to PNAS [14]; while previous work [3,7,20] used static graphs and tried to search complete paths, for instance, in [20], each ant constructs a complete neural network at each iteration. Second, DeepSwarm differs from PNAS [14] as it uses heuristic information and individual searchers (ants) share information with each other; while PNAS employed a deterministic search strategy. Third, DeepSwarm reused the weights of the partial neural networks trained previously in order to improve the training speed, while most of the previous work did not reuse the previously trained weights except for [20].

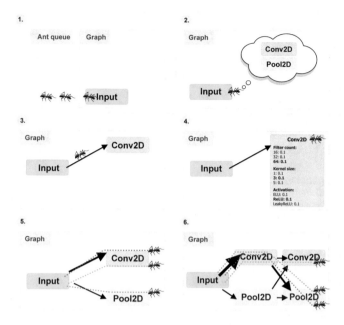

Fig. 1. An overview of the NAS process of DeepSwarm. (1) The ant is placed on the input node. (2) The ant checks what transitions are available. (3) The ant uses the ACS selection rule to choose the next node. (4) After choosing the next node the ant selects the node's attributes. (5) After all ants finished their tour the pheromone is updated. (6) The maximum allowed depth is increased and the new ant population is generated. **Note**: Arrow thickness indicates the pheromone amount, meaning that thicker arrows have more pheromone.

4   Experiments

For the experimental design, three different datasets were chosen: (1) MNIST [12], (2) Fashion-MNIST [23], and (3) CIFAR-10 [11]. Each of these three datasets is quite different from the others and requires different CNN architectures to achieve the best results. As a result the combination of them is a good way to test the algorithm's robustness and performance. In order to evaluate our proposed method the baselines taken from [10] were used. All of our tests were carried out in the Google Colab environment (1x Tesla K80 GPU) [9] using a MacBook Pro (Early 2015 model) to interact with this environment. Note that even though in [10] they ran each method only for 12 hours, they used NVIDIA GeForce GTX 1080 Ti GPU, which according to a few benchmarks is approximately 2–3 times faster than our selected Tesla K80 GPU. This is the reason why we are not going to constrain our runs to 12 h.

Algorithm 1. DeepSwarm

Function search():
  graph = Graph() // build graph containing only the input node
  while *graph.current_depth < max_depth* **do**
    ants = generate_ants()
    best_ant = find_best(ants)
    graph.global_pheromone_update(best_ant)
    graph.increase_depth()
  return best_ant

Function generate_ants():
  ants = []
  for *i = 0* **to** *ant_count* **do**
    ant = Ant()
    ant.path = generate_path()
    ant.evaluate()
    ants.append(ant)
    graph.local_pheromone_update(ant)
  return ants

Function generate_path():
  current_node = graph.input_node
  path = [current_node]
  for *i = 0* **to** *current_max_depth* **do**
    if *current_node.neighbours* ⟵ ∅ **then**
      break
    current_node = aco_select_rule(current_node.neighbours)
    path.append(current_node)
  completed_path = complete_path(path) // completes the path if needed
  return path

Function aco_select(*neighbours*):
  foreach *neighbour ∈ neighbours* **do**
    probability = neighbour.pheromone × neighbour.heuristic
    probabilities ← probability
    denominator += probability
  if *random.uniform(0, 1) ≤ greediness* **then**
    max_index = probabilities.index(max(probabilities))
    return neighbours[max_index]
  probabilities = probabilities / denominator
  neighbour_index = wheel_selection(probabilities)
  return neighbours[neighbour_index]

4.1   Evaluation Procedure

When evaluating the system the following procedure was followed: (1) create a new Google Colab instance, (2) import the source code of the library, (3) split the training set 90-10 to training and validation sets, (4) run the algorithm until the max depth is reached, (5) take the best found network, (6) for CIFAR-10 dataset apply standard data augmentation (random horizontal flips, rotation and scaling) to the training data, (7) train the best found network for additional 50 epochs on the augmented data, (8) load the weights which showed the best performance on the validation set during those 50 epochs, and (9) evaluate the network with these best weights on the testing data.

4.2   Ant Count

The ant count (the number of ants used during search) is one of the most important hyperparameters in DeepSwarm. This is because it is a trade-off between the performance of the final model and the run-time of the algorithm. In order to find a good trade-off, we ran multiple tests by exponentially increasing the ant count. Furthermore, we split the results into two parts: before and after the final training. Before the final training is a part where DeepSwarm finds potentially the best model and after the final training is the part where the best found model is trained for an additional 50 epochs on augmented data. The reason for this choice is that the results before the final training can reflect the real implications that the ant count has on the error rate, whereas the results after the final training can show how the ant count can affect the generalisation. This follows from the fact that before the final training the models are trained on the same data, whereas during the final training the models are trained on the augmented data which can show how well they can learn. The results before the final training are presented in Fig. 2, the results after the final training can be seen in Fig. 3, and the run time is shown in Fig. 4.

Looking at the results one can see that changing the ant count from 1 to 2 had a significant impact on the error rate. This finding was to be expected because when only one ant exists both exploration and exploitation must suffer. The exploration suffering is associated with the fact that the ant can only explore one architecture per depth, meaning that only a small subset of available architectures will be explored. The exploitation degradation occurs because at each depth acquired knowledge scales only linearly, for example, at depth 3 the ant will only know about 2 other architectures. Furthermore, having only one ant will result in rather greedy behaviour where the same ant will explore the same sub-tree in the graph and will only rarely explore the parallel sub-trees. We further noticed that even though doubling the ant count almost doubles the run time, it will not always result in drastically improved performance. For example, when we increased the ant count from 4 to 8 ants the run time increased from 7 hours to 18 hours, while the average error rate decreased only by 0.13%. The most drastic changes in the error rate happened when the ant count was changed from 1 to 2 (3.11% decrease) and from 8 to 16 (2.1% decrease). However, due

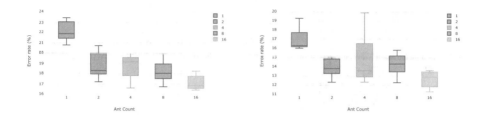

Fig. 2. The error rate on the CIFAR-10 dataset before the final training across five separate trials.

Fig. 3. The error rate on the CIFAR-10 dataset after the final training across five separate trials.

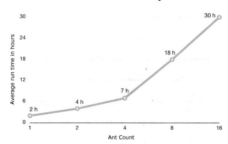

Fig. 4. The average run time (across five different trials) in hours for different ant counts on the CIFAR-10 dataset.

to the computational restrictions we did not test ant counts beyond 16 which means that there might be even bigger performance improvements when going beyond 16 ants.

4.3   Greediness

Another important hyperparameter of DeepSwarm is greediness. As mentioned in Sect. 3, the greediness is used in Eq. (1) to decide how greedy each ant should be. As greediness can be defined in the range from 0.0 to 1.0, we test the greediness with its value increases from 0 to 1 at a step size of 0.25. Furthermore, similarly to the ant count, the results were divided into before and after the final training. The results before the final training are shown in Fig. 5 and the results after the final training are shown in Fig. 6.

Looking at the results it seems that when selecting the greediness for the algorithm one should never go to extremes as this will most likely result in poor performance. The more general insight we gathered from the results was that selecting the greediness values which were close towards the middle (0.5) resulted in the best performance. The reason why the extremely greedy ants perform poorly is as follows: at the beginning of the search they base their search purely on the heuristic information and then, once the pheromone is laid on the graph, all of them will reuse the same path, therefore generating the same architecture. Furthermore, the local pheromone update rule will not help

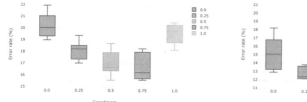 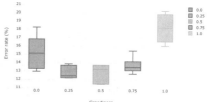

Fig. 5. The error rate on the CIFAR-10 dataset before the final training across five separate trials.

Fig. 6. The error rate on the CIFAR-10 dataset after the final training across five separate trials.

here because once the pheromone evaporates these greedy ants will use the same heuristics which will result in the same paths being chosen again. In contrast, the ants with no greediness will always base each of their decisions only on the wheel selection without exploiting the gathered information (as the first part in Eq. 1 is always skipped) and because during the path generation an ant needs to make a lot of these decisions (choosing the next node and each attribute), a substantial part of them will be random, which will result in a poor performance. Another interesting observation was that the greedy models tend to generalise worse than the less greedy ones. For example, the average error rate difference before the final training between 0.25 greediness and 0.75 greediness was 1.32% (18.12% and 16.80% respectively), but after the final training, the difference was -0.83% (12.89% and 13.72% respectively). Furthermore, we noticed that the greediness had some impact on the average network depth, for example, the best architectures which were found using no greediness, were on average five layers deeper than the ones which were found using 1.0 greediness. As a result of that, these less greedy architectures had more regularisation and feature extraction. We believe that this could be the reason why these less greedy architectures were generalising better during the final training.

4.4 Accuracy

In order to compare the performance of DeepSwarm with that of other methods, we report the average and best performance achieved during the five separate runs on three different datasets. These results are shown in Table 1. From these results we can see that on the MNIST dataset from all of the methods Deep-Swarm showed the best performance. When compared with the straightforward methods (random and grid search [2]) DeepSwarm showed a significantly lower error rate (1.79%, 1.68% versus 0.46%). On the Fashion-MNIST dataset, Deep-Swarm achieved the lowest error rate and once again proved to be superior to the straightforward methods which had almost a two times bigger error rate (11.36%, 10.28% versus 6.75%). Finally, on the CIFAR-10 dataset, even though DeepSwarm managed to find the architecture with the lowest error rate (11.31%), on average its performance was not as good as some other methods. Overall on all of the three datasets, DeepSwarm still produced very competitive and promising results. To see the best architectures discovered by DeepSwarm please visit following external resource[1].

Table 1. The error rates on the CIFAR-10 dataset.

Method	MNIST	Fashion-MNIST	CIFAR-10
RANDOM	1.79%	11.36%	16.86%
GRID	1.68%	10.28%	17.17%
SPMT	1.36%	9.62%	14.68%
SMAC	1.43%	10.87%	15.04%
SEAS	1.07%	8.05%	12.43%
NASBOT	N/A	N/A	12.30%
AutoKeras BFS	1.56%	9.13%	13.84%
AutoKeras BO	1.83%	7.99%	12.90%
AutoKeras BFS	0.55%	7.42%	**11.44%**
DeepSwarm Average	**0.46%**	**6.75%**	12.70%
DeepSwarm Best	0.39%	6.44%	11.31%

4.5   Discussion

Even though there exists a NAS approach developed by Google Brain [25] which can achieve better results than DeepSwarm on the CIFAR-10 dataset, we think that it would be not fair to compare our work with theirs for the following reasons: (1) they used 400 GPUs (also their GPUs were much more powerful than the one used in our experiments) for 4 days, (2) they used skip and add connections which are not implemented into DeepSwarm yet. We also point out that as they did not open source their code, it is not easy for us to test their approach in our environment to compare the performance difference. Nevertheless, based on the results seen in Sect. 4.4 DeepSwarm proved to be a competitive approach against already existing NAS methods. However, there is still some work that needs to be done in order to further improve DeepSwarm. We think that the two main components that can be added in the future are skip and add nodes. Adding these two components would allow DeepSwarm to search for more complex architectures which in turn could substantially improve the overall learning performance. Finally, we list the main advantages of DeepSwarm compared with other existing NAS systems as follows:

- DeepSwarm offers competitive performance. As shown in Sect. 4.4, on all 3 datasets DeepSwarm can achieve comparable or better results than the other NAS systems.
- DeepSwarm can look for diverse structures. DeepSwarm does not enforce a specific structure, which allows it to find novel and interesting architectures.

---

[1]   https://edvinasbyla.com/assets/images/best-architectures.pdf.

- DeepSwarm can offer fast search. As mentioned earlier, DeepSwarm is built to search for architectures progressively and has a mechanism to reuse the old weights which boosts its performance.
- DeepSwarm allows the users to provide heuristic information which can further speed up the search process.
- DeepSwarm is easy to use. To start the neural architecture search a user just needs to write a few lines of code (see detailed instructions on DeepSwarm's GitHub page).
- DeepSwarm is easy to be further developed and extended. As we open source DeepSwarm and share it with the wider machine learning community, other researchers can further develop and extend DeepSwarm.

5   Conclusion and Future Work

In this paper we presented DeepSwarm and demonstrated that Swarm Intelligence can be used to effectively tackle NAS problems. After evaluating Deep-Swarm we discovered that when compared to other similar methods it can show competitive performance. Furthermore, we open source DeepSwarm[2] and share it with the community, and we hope more people will benefit from it and further develop it.

The main contribution of this work is to show that ACO can be used to effectively search for optimal CNN architectures. Our second contribution is to demonstrate that domain expert knowledge can be successfully incorporated into ACO based NAS. The final contribution of this work is to show that progressive architecture search approach can be applied to ACO based NAS methods.

For future work we propose to explore the following directions: (1) implement skip and add connections which would allow ants to look for more complex architectures, (2) try to use ACO to perform cell based search (similar to [25]) rather than the full architecture search, (3) compare conventional search method with the progressive search when ACO is applied to NAS problem, and (4) explore ACO in other deep learning contexts i.e. find which neurons to drop in the dropout layer.

References

1. Baker, B., Gupta, O., Naik, N., Raskar, R.: Designing neural network architectures using reinforcement learning. arXiv preprint arXiv:1611.02167 (2016)
2. Bergstra, J., Bengio, Y.: Random search for hyper-parameter optimization. J. Mach. Learn. Res. **13**, 281–305 (2012)
3. Desell, T., Clachar, S., Higgins, J., Wild, B.: Evolving deep recurrent neural networks using ant colony optimization. In: European Conference on Evolutionary Computation in Combinatorial Optimization, pp. 86–98. Springer (2015)
4. Domhan, T., Springenberg, J.T., Hutter, F.: Speeding up automatic hyperparameter optimization of deep neural networks by extrapolation of learning curves. IJCAI **15**, 3460–8 (2015)

---

[2] https://github.com/Pattio/DeepSwarm.

5. Dorigo, M.: Optimization, learning and natural algorithms. PhD Thesis, Politecnico di Milano (1992). https://ci.nii.ac.jp/naid/10000136323/en/
6. Dorigo, M., Gambardella, L.M.: Ant colony system: a cooperative learning approach to the traveling salesman problem. IEEE Trans. Evol. Comput. **1**(1), 53–66 (1997)
7. ElSaid, A., Jamiy, F.E., Higgins, J., Wild, B., Desell, T.: Using ant colony optimization to optimize long short-term memory recurrent neural networks. In: Proceedings of the Genetic and Evolutionary Computation Conference, pp. 13–20. ACM (2018)
8. Elsken, T., Metzen, J.H., Hutter, F.: Simple and efficient architecture search for convolutional neural networks. arXiv preprint arXiv:1711.04528 (2017)
9. Google: https://colab.research.google.com
10. Jin, H., Song, Q., Hu, X.: Efficient neural architecture search with network morphism. arXiv preprint arXiv:1806.10282 (2018)
11. Krizhevsky, A., Nair, V., Hinton, G.: The cifar-10 dataset. http://www.cs.toronto.edu/kriz/cifar.htmlp. 4 (2014)
12. LeCun, Y.: The mnist database of handwritten digits. http://yann.lecun.com/exdb/mnist/ (1998)
13. LeCun, Y., Bengio, Y., Hinton, G.: Deep learning. Nature **521**(7553), 436 (2015)
14. Liu, C., Zoph, B., Neumann, M., Shlens, J., Hua, W., Li, L.J., Fei-Fei, L., Yuille, A., Huang, J., Murphy, K.: Progressive neural architecture search. In: Proceedings of the European Conference on Computer Vision (ECCV), pp. 19–34 (2018)
15. Miikkulainen, R., Liang, J., Meyerson, E., Rawal, A., Fink, D., Francon, O., Raju, B., Shahrzad, H., Navruzyan, A., Duffy, N., et al.: Evolving deep neural networks. In: Artificial Intelligence in the Age of Neural Networks and Brain Computing, pp. 293–312. Elsevier (2019)
16. Miller, G.F., Todd, P.M., Hegde, S.U.: Designing neural networks using genetic algorithms. ICGA **89**, 379–384 (1989)
17. Negrinho, R., Gordon, G.: Deeparchitect: Automatically designing and training deep architectures. arXiv preprint arXiv:1704.08792 (2017)
18. Pham, H., Guan, M.Y., Zoph, B., Le, Q.V., Dean, J.: Efficient neural architecture search via parameter sharing. arXiv preprint arXiv:1802.03268 (2018)
19. Real, E., Moore, S., Selle, A., Saxena, S., Suematsu, Y.L., Tan, J., Le, Q., Kurakin, A.: Large-scale evolution of image classifiers. arXiv preprint arXiv:1703.01041 (2017)
20. Salama, K., Abdelbar, A.M.: A novel ant colony algorithm for building neural network topologies. In: International Conference on Swarm Intelligence, pp. 1–12. Springer (2014)
21. Suganuma, M., Shirakawa, S., Nagao, T.: A genetic programming approach to designing convolutional neural network architectures. In: Proceedings of the Genetic and Evolutionary Computation Conference, pp. 497–504. ACM (2017)
22. Wistuba, M., Rawat, A., Pedapati, T.: A survey on neural architecture search (2019)
23. Xiao, H., Rasul, K., Vollgraf, R.: Fashion-MNIST: a novel image dataset for benchmarking machine learning algorithms. arXiv preprint arXiv:1708.07747 (2017)
24. Zoph, B., Le, Q.V.: Neural architecture search with reinforcement learning. arXiv preprint arXiv:1611.01578 (2016)
25. Zoph, B., Vasudevan, V., Shlens, J., Le, Q.V.: Learning transferable architectures for scalable image recognition. In: Proceedings of the IEEE Conference on Computer Vision and Pattern Recognition, pp. 8697–8710 (2018)

Single-Grasp Detection Based on Rotational Region CNN

Shan Jiang[1(✉)], Xi Zhao[1], Zhenhua Cai[1], Kui Xiang[1], and Zhaojie Ju[2]

[1] Wuhan University of Technology, Wuhan, Hubei, China
jswhut@163.com
[2] University of Portsmouth, Portsmouth, UK

Abstract. Object grasp detection is foundational to intelligent robotic manipulation. Different from typical object detection tasks, grasp detection tasks need to tackle the orientation of the graspable region in addition to localizing the region since the ground truth box of the grasp detection is arbitrary-oriented in the grasp datasets. This paper presents a novel method for single-grasp detection based on rotational region CNN (R^2CNN). This method applies a common Region Proposal Network (RPN) to predict inclined graspable region, including location, scale, orientation, and grasp/non-grasp score. The idea is to deal with the grasp detection as a multi-task problem that involves multiple predictions, including predict grasp/non-grasp score, the inclined box and its corresponding axis-align bounding box. The inclined non-maximum suppression (NMS) method is used to compute the final predicted grasp rectangle. Experimental results indicate that the presented method can achieve accuracies of 94.6% (image-wise splitting) and 95.6% (object-wise splitting) on the Cornel Grasp Dataset, respectively. This method outperforms state-of-the-art grasp detection models that only use color images.

Keywords: Robotic Grasp · Convolutional Neural Network · Region Proposal Network · Faster-RCNN · Rotational Region CNN

1 Introduction

With the advance of robotics and relevant applications in industry and daily life, researchers pay more attention to robotic grasp which is foundational to robotic manipulation. When humans try to grasp an object for the first time, they can perceive, think, and probably figure out how to grasp it effectively. However, grasping a novel object is relatively challenging for robots as this task involves many subjects, including computer vision, robot kinematics, control science and path planning. Nowadays, most of robotic grasp schemas highly rely on the predefined programs, which simply depend on repeating a series of predetermined basic motions. Obviously, such schemas subject to a lack of generalization and robustness if objects or environments are varying. Therefore, practical robotic grasp manipulation requires more intelligent and robust strategies.

For the task of grasp detection, many studies [1–7, 17–19] focused on predicting grasp rectangles. Compared with single grasp point, a grasp rectangle contains more information including the graspable region as well as the orientation information. This

© Springer Nature Switzerland AG 2020
Z. Ju et al. (Eds.): UKCI 2019, AISC 1043, pp. 131–141, 2020.
https://doi.org/10.1007/978-3-030-29933-0_11

orientation indicates a proper opening direction for parallel robotic gripper. Consequently, the existence of the orientation makes the problem of grasp detection distinguishable from other general object detection algorithms.

This paper aims to propose a new method to improve grasp rectangle detection using images for practical real-time robotic grasp. To achieve this goal, the authors employ convolutional neural network (CNN) and regional proposal network (RPN) to explore feasible models that can effectively determine inclined grasp region. By combining inspirations from some state-of-the-art methods [8–10], a deep learning method is presented to detect the grasp region of objects for accurate robotic manipulation. Experimental results based on the Cornel Grasp Dataset are discussed as well as future work to improve the presented method.

The major contributions of this paper are summarized as below:

(1) A new grasp detection method, which is specifically designed for inclined grasp region, is presented. This method considers the grasp detection problem as a triple-task problem by adding the axis-align box as well. It outperforms existing grasp detection methods that only use image information.
(2) The smaller anchor scales is added to cover the tiny objects and the inclined non-maximum suppression (inclined NMS) is adopted to select the optimal grasp rectangle.

2   Related Work

As an elemental manipulation, the task of robotic grasp has been extensively studied [11]. Most research can be divided into two categories: heuristic method and machine learning methods.

Jiang et al. [1] adopted manually designed two-step color feature to achieve the detection result with 85.5% accuracy. Dogar and Goldfeder [12, 13] used full physical simulation given 3D models to predict correct grasps. The results are not very satisfying whereas the process was very time-consuming. Traditional heuristic methods [1, 12–15] for feature extraction are not suitable enough for robotic grasp detection.

With the advance of computer vision and deep learning, some researchers studied the robotic grasp problem using those new technologies, which significantly improved both accuracy and computational speed. Lenz et al. [2] proposed a two-stage cascaded system for detecting robot grasps, wherein a small deep network was used to generate some potential rectangles and then a larger deep network was used to select the top-ranked rectangle from these candidates. Redmon et al. [3] conducted the detection based on RGB-D images with a neural network inspired by the AlexNet to address the same problem as the former researchers. Kumra et al. [4] used two 50-layer deep convolutional residual neural networks in parallel for RGB-D feature extraction, one for RGB feature and the other for depth information. Then these features were fused and fed into the detection network. Their research showed that the use of deep residual layers can extract better features from the input images than the ordinary convolutional layers. Guo et al. [5] associated each grid cell with several reference rectangles in different scales and ratios and then these reference rectangles were refined to their corresponding predicted rectangles. In follow-up study [6], they proposed a hybrid system that combined the vision and tactile information for robotic grasping. Furthermore, a new THU grasp dataset which contained the visual, tactile and grasp configuration information was collected, and the results showed that the tactile data can help improve the accuracy for grasp detection. Although the result was relatively outstanding, it was hard and too complex to conduct the grasp experiment as it required plenty of experiment instruments. Chu et al. [7] presented a system that can be applied to both single-grasp and multi-grasp detection situation. In their research, they converted the problem of regression for grasp orientation into a problem of classification. The system quantitated the orientation into 19 categories and assigned the predicted rectangles to the corresponding classes. The prediction accuracy on RGB model was 94.4% and 95.5% on image-wise and object-wise splitting, respectively.

3   Rotated-RCNN for Single-Grasp Detection

3.1   Grasp Configuration

Similar to the previous work [1–7, 17–19], the five-dimensional grasp configuration is applied to this paper. The configuration consists of both position information and orientation information. The ground truth rectangle is defined as:

$$G = \{x, y, w, h, \theta\} \tag{1}$$

wherein, the (x, y) represents the coordinate of the center point, w and h mean the width and height of the rectangle respectively. The angle θ symbolizes the angle between the rectangle and the horizontal x-axis and the gripper can get close the object in this direction. Figure 1 shows an example of the grasp configuration.

Different from normal object detection which the bounding box is axis-align, the ground-truth box of grasp detection is normally inclined. So how to tackle the orientation problem is the key for grasp detection.

Fig. 1. Five-dimensional grasp configuration

3.2   Network Architecture

Different from the horizontal ground truth box, grasp rectangles are normally arbitrary-orientated since the object is placed on the platform randomly. Under the circumstances, the Rotated-RCNN is applied to tackle the orientation problem in the grasp detection problem. Figure 2 shows the whole network architecture of the grasp detection system. The input images pass through multiple convolutional layers to produce the feature maps. The RPN is used to generate axis-align bounding boxes that encircle the graspable region. Considering about the diversity of the sizes and aspect ratios of the grasp rectangles, a smaller anchor scale is applied to the model. Experimental result shows the smaller anchor is effective in the grasp detection.

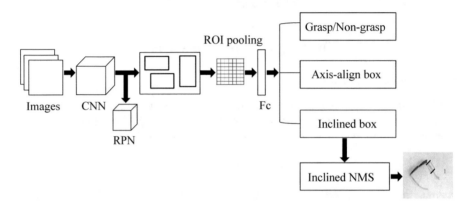

Fig. 2. The whole architecture of the grasp detection network

Two fully-connected layers are used for classification and regression in parallel. The region proposal produced by RPN is classified as grasp or non-grasp. The inclined bounding box and its associated axis-aligned bounding box get refined as well. The regression loss of the axis-align bounding box is considered in the whole loss function, the evaluation of [8] confirmed the effectiveness of this idea.

As the subject of the grasp detection is oriented rectangle, the inclined non-maximum suppression is utilized to post-process detection candidates as to obtain the output grasp rectangle.

3.3   Loss Function

The loss function in this paper contains two parts, the loss of the region proposal network L_{RPN} and the loss of grasp configuration detection L_{GCD}. The RPN generates axis-align region proposals that encircle the inclined graspable region. The loss function of the RPN consists of the classification loss and regression loss, defines as:

$$L_{RPN}(p_i, t_i) = \sum_i L_{RPN\_cls}(p_i, p_i^*) + \lambda_1 \sum_i p_i^* L_{RPN\_reg}(t_i, t_i^*) \tag{2}$$

wherein, L_{RPN_cls} defines the log loss of the proposal classification and L_{RPN_reg} is the smooth L1 loss of the proposal regression. p_i is the probability of the anchor belonging to the foreground. The ground truth label p_i^* is 1 if the anchor is positive and is 0 if the anchor is negative [9]. The regression loss is calculated only when the anchor is assigned to the foreground. t_i represents the four-dimensional coordinate vector of the predicted axis-align bounding box and t_i^* means that of the horizontal box rotated from the inclined ground truth box.

The loss function of the grasp configuration detection consists of three parts: the grasp/non-grasp classification loss, the loss of the axis-align box that encircles the graspable region and the loss of the inclined box. The loss function defines as:

$$L_{GCD}(\rho_i, \beta_i, \delta_i) = \sum_i L_{GCD\_cls}(\rho_i, \rho_i^*) + \lambda_2 \sum_i \rho_i^* L_{GCD\_regh}(\beta_i, \beta_i^*)$$
$$+ \lambda_3 \sum_i \rho_i^* L_{GCD\_regi}(\delta_i, \delta_i^*) \tag{3}$$

wherein, L_{GCD_cls} means the log loss of the grasp classification. Grasps are labeled as 1 and others are assigned background. The parameter $\beta = (\beta_x, \beta_y, \beta_w, \beta_h)$ means the predicted regression for axis-align bounding box for the graspable class and β^* means the true regression target. The parameter $\delta = (\delta_x, \delta_y, \delta_w, \delta_h, \delta_\theta)$ means the predicted regression vector for the inclined bounding box and δ^* is the corresponding ground truth grasp bounding box vector. Similarly, the regression loss is considered only when the proposal is assigned to the graspable class. The parameter λ_2 and λ_3 are used to balance these three kinds of loss. It should be noted that in all the experiments of this paper λ_2 and λ_3 are set 1.

The total loss for the end-to-end training of the grasp detection defines as:

$$L_{Total} = L_{RPN} + L_{GCD} \tag{4}$$

4 Experiment

4.1 Dataset

In order to evaluate the performance of the network compared with existing studies, the network is trained and tested on the standard Cornel Grasp Dataset. The Cornel Grasp Dataset applied to this research contains 885 images of 244 graspable objects and each object in these images is associated with several positive grasp rectangles and negative rectangles. In this research, the positive rectangles are defined as ground-truth box.

The five-fold cross-validation is conducted in the experiments and the dataset is divided in two different ways:

(1) Image-wise splitting divides the images into training set and validation set at random. This aims to test the generalization ability of the network to the new position and orientation of an object.

(2) Object-wise splitting divides the dataset at object instance level. The training and test dataset does not share the images of the same instance. This method aims to test the generalization ability of the network to the novel object.

In practice, both splitting methods give comparable performance. This may due to the similarity between different objects in the dataset [3].

4.2   Data Preprocessing

The training process for the deep neural network needs a large amount number of labeled data. Since the amount of the Cornel Grasp Dataset is insufficient, a process of data augmentation is required before the dataset is fed into the network. Several methods of data augmentation such as rotation, translation and crop have been adopted before the experiments. Firstly, a region of 320 * 320 pixels is center cropped from the original image. Secondly, the cropped region is padded with 50 pixels in both x and y directions. Then the padded image is translated with random pixel between −50 and 50 pixels in both x-axis and y-axis. Then the rotated image is rotated with a random angle between 0° to 360°. Lastly, the image is resized to 512 * 512 pixels. Each original image extends to 25 images after the augmentation and these processed images will be sent to the input of the network. The label of the ground truth box is transformed as well.

4.3   Training

To improve the efficiency of the training process and avoid overfitting, the transfer learning is applied to this research. The network is initialized by the ResNet-50 pre-trained on the ImageNet.

The model is based on the GPU version of TensorFlow framework with cuda-8.0 and cudnn-5.1.0 package. The whole network is trained end-to-end for 200 epochs and the whole training and test process runs on a single NVIDIA GTX1080Ti. The initial learning rate is 0.001 with a weight decay of 0.0005 and the momentum of 0.9.

4.4   Evaluation Metric for Detection

The point metric and rectangle metric [3, 4] are the most popular evaluation metrics in grasp detection on the Cornel Grasp Dataset. For the point metric, the distances between the center point of ground-truth grasps and center point of predicted grasp are considered. If any of these distances is less than the predefined threshold, the predicted grasp is regarded as a correct prediction.

Obviously, the point metric cannot comprehensively evaluate predicted grasp. This kind of metric does not evaluate the size and orientation of the predicted grasp and thus may overestimate the performance of the algorithm for grasp detection.

In this paper, the rectangle metric is chosen as the evaluation metric. In this metric, the predicted grasp is regarded to be correct if it satisfies both conditions:

(1) The angle difference between the predicted grasp and the ground-truth grasp is within 30°.
(2) The Jaccard index of the ground-truth grasp and the predicted grasp is larger than 0.25.

The Jaccard index is defined as:

$$J(G, G^*) = \frac{Area(G \cap G^*)}{Area(G \cup G^*)} \tag{5}$$

The Jaccard index is similar to the Intersection over Union (IoU) threshold [7] for object detection. The G means the area of the top-ranked predicted grasp rectangle in this algorithm and G^* denotes the area of the ground-truth rectangle. $G \cap G^*$ is the intersection of these two rectangles and $G \cup G^*$ denotes the union of these two rectangles. Note that as the ground-truth grasp rectangles cannot be labeled exhaustively, the Jaccard index is 25% rather than 50% used in the normal object detection. A rectangle with the right orientation that only overlaps by 25% with one of ground truth boxes can still be considered as a reliable prediction. All the experiments are performed using this kind of rectangle metric.

5 Results and Discussion

Different from many methods adopted in the image augmentation for training dataset, the test dataset only uses the center crop. The model is evaluated by the metric mentioned above.

The result of self-comparison of the proposed algorithm with different parameters shows in Tables 1 and 2 shows the comparison of this model and other previous works on the Cornel Grasp Dataset with the same evaluation metric. It should be noted that all of the results only consider about the single grasp of the object. In other words, the inclined bounding box with the highest confidence is set as the output grasp rectangle. It is clear that smaller anchor scale and inclined NMS can improve detection accuracy. In Table 2, the result shows the proposed model outperforms previous works with RGB images. On image-wise splitting, the accuracy is up to 94.8%, which is 0.4% higher than the up-to-date 94.4% accuracy [7] in grasp research. While on the object-wise splitting, the detection accuracy is 95.6%, which is 0.1% higher than the 95.5% accuracy of Chu's [7] work. Both Chu's work [7] and this approach are generated on the basis of the Faster-RCNN, wherein Chu's work converts the problem of the regression over the orientation to the problem of discretization orientation classification and this paper deals with this problem in a continuous manner. Moreover, this paper applies smaller anchor to improve the accuracy at the cost of little additional runtime.

Table 1. Result of network with different parameters

Anchor scale	Inclined NMS	Prediction accuracy (%)	
		Image-wise	Object-wise
(8, 16, 32)	No	92.5	94.2
	Yes	93.4	95.1
(4, 8, 16, 32)	No	93.2	94.8
	Yes	94.8	95.6

Table 2. Single grasp evaluation

Approach	Prediction accuracy (%)		Speed fps
	Image-wise	Object-wise	
Jiang et al. [1]	60.5	58.3	0.02
Lenz et al. [2]	73.9	75.6	0.07
Redmon et al. [3]	88.0	87.1	3.31
Wang et al. [18]	81.8	N/A	7.10
Asif et al. [19]	88.2	87.5	-
Kumra et al. [4]	89.2	88.9	16.03
Mahler et al. [20]	93.0	N/A	~1.25
Guo et al. [5]	93.2	89.1	-
Chu et al. (Res50 RGB) [7]	**94.4**	**95.5**	8.33
Chu et al. (Res50 RGB-D) [7]	96.0	96.1	8.33
Ours (Res50 RGB)	*94.8*	*95.6*	7.25

As shown in Fig. 3, some positive rectangles were generated from the grasp detection system. The top row shows the ground truth grasp rectangles which are obtained from the Cornel Grasp Dataset. The red line in these pictures indicates the gripper's orientation. As shown in the picture, the number and the size of the grasp rectangles are varying and some of the rectangles are even small. Therefore, it is necessary to add smaller size to the anchor scale. The second row displays the top-ranked inclined grasp rectangle predicted by the detection system. The last row reveals all the inclined rectangles output from the detection system. The black line in the

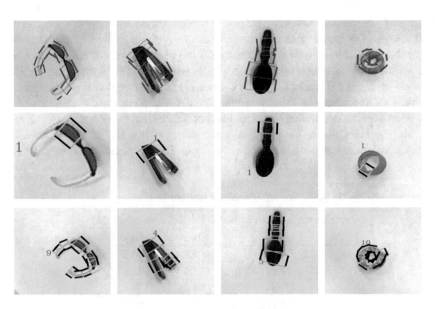

Fig. 3. Some positive examples from the network

picture of the second and last rows means the gripper's orientation and the number in these pictures represents the number of the output rectangles. The results indicate that the detection network can accurately predict position and orientation of the grasp rectangles.

Besides, some incorrect grasp rectangles are shows in Fig. 4. The first row is the ground truth box and the other row shows the wrong results. It should be noted that the wrong results here means that the inclined rectangle does not meet the rectangle metric mentioned above. Although the left two pictures in the second row are assigned to be incorrect, as the grasp rectangle cannot be labeled completely, these outputs can be thought as proper prediction as well.

Fig. 4. Some incorrect prediction from the network

6 Conclusion

In this paper, a robust and accurate robotic grasp detection method based on CNN and RPN is presented. The architecture of the network is adapted from the R^2CNN which was originally designed to detect inclined scene text, which redefines the meaning of the network and shows the generalization of the network when it is comes to the grasp problem. Many modifications have been made to solve the grasp detection tasks, and the presented method is verified to effectively improve the accuracy of the grasp detection. The experimental results show that this novel network achieves an accuracy of 94.6% (image-wise splitting) and 95.6% (object-wise splitting), respectively. The network outperformed previous work with the same evaluation metric. Granted, the computational speed of the algorithm is not satisfactory enough and further work need to be conducted on the problem of shortening time as well as the practical grasp manipulation.

References

1. Jiang, Y., Moseson, S., Saxena, A.: Efficient Grasping from RGB-D images: Learning using a new rectangle representation. In: 2011 IEEE International Conference on Robotics and Automation, pp. 3304–3311. IEEE (2011)
2. Lenz, I., Lee, H., Saxena, A.: Deep learning for detecting robotic grasps. Int. J. Robot. Res. 34(4–5), 705–724 (2015)
3. Redmon, J., Angelova, A.: Real-time grasp detection using convolutional neural networks. In: 2015 IEEE International Conference on Robotics and Automation (ICRA), pp. 1316–1322. IEEE (2015)
4. Kumra, S., Kanan, C.: Robotic grasp detection using deep convolutional neural networks. In: 2017 IEEE/RSJ International Conference on Intelligent Robots and Systems (IROS), pp. 769–776. IEEE (2017)
5. Guo, D., Sun, F., Kong, T., Liu, H.: Deep vision networks for real-time robotic grasp detection. Int. J. Adv. Robot. Syst. 14(1) (2016)
6. Guo, D., Sun, F., Liu, H., Kong, T., Fang, B., Xi, N.: A hybrid deep architecture for robotic grasp detection. In: 2017 IEEE International Conference on Robotics and Automation (ICRA), pp. 1609–1614. IEEE (2017)
7. Chu, F.J., Xu, R., Vela, P.A.: Real-world multi-object, multi-grasp detection. IEEE Robot. Autom. Lett. 3(4), 3355–3362 (2018)
8. Jiang, Y., Zhu, X., Wang, X., et al.: R2CNN: rotational region CNN for orientation robust scene text detection. arXiv preprint arXiv:1706.09579 (2017)
9. Ren, S., He, K., Girshick, R., Sun, J.: Faster R-CNN: towards real-time object detection with region proposal networks. In: Advances in Neural Information Processing Systems, pp. 91–99 (2015)
10. He, K., Zhang, X., Ren, S., Sun, J.: Deep residual learning for image recognition. In: Proceedings of the IEEE Conference on Computer Vision and Pattern Recognition, pp. 770–778 (2016)
11. Bohg, J., Morales, A., Asfour, T., Kragic, D.: Data-driven grasp synthesis—a survey. IEEE Trans. Robot. 30(2), 289–309 (2013)
12. Dogar, M., Hsiao, K., Ciocarlie, M., Srinivasa, S.: Physics-Based Grasp Planning Through Clutter, pp. 78–85 (2012)
13. Goldfeder, C., Ciocarlie, M., Dang, H., Allen, P.K.: The Columbia Grasp Database (2008)
14. Miller, A.T., Knoop, S., Christensen, H.I., Allen, P.K.: Automatic grasp planning using shape primitives, pp. 1824–1829 (2003)
15. Piater, J.H.: Learning visual features to predict hand orientations (2002)
16. Girshick, R.: Fast R-CNN. In: Proceedings of the IEEE International Conference on Computer Vision, pp. 1440–1448 (2015)
17. Zhang, H., Zhou, X., Lan, X., Li, J., Tian, Z., Zheng, N.: A real-time robotic grasp approach with oriented anchor Box. arXiv preprint arXiv:1809.03873 (2018)
18. Pinto, L., Gupta, A.: Supersizing self-supervision: learning to grasp from 50k tries and 700 robot hours. In: 2016 IEEE International Conference on Robotics and Automation (ICRA), pp. 3406–3413. IEEE (2016)
19. Watson, J., Hughes, J., Iida, F.: Real-world, real-time robotic grasping with convolutional neural networks. In: Annual Conference Towards Autonomous Robotic Systems, pp. 617–626. Springer, Cham (2017)

20. Wang, Z., Li, Z., Wang, B., Liu, H.: Robot grasp detection using multimodal deep convolutional neural networks. Adv. Mech. Eng. **8**(9) (2016)
21. Asif, U., Bennamoun, M., Sohel, F.A.: RGB-D object recognition and grasp detection using hierarchical cascaded forests. IEEE Trans. Robot. **33**(3), 547–564 (2017)
22. Mahler, J., Liang, J., Niyaz, S., et al.: Dex-Net 2.0: deep learning to plan robust grasps with synthetic point clouds and analytic grasp metrics. arXiv preprint arXiv:1703.09312 (2017)

Interpreting the Filters in the First Layer of a Convolutional Neural Network for Sleep Stage Classification

Gulrukh Turabee$^{(\boxtimes)}$, Yuan Shen$^{(\boxtimes)}$, and Georgina Cosma$^{(\boxtimes)}$

Nottingham Trent University, Nottingham, UK
gulrukh.turabee2018@my.ntu.ac.uk, {yuan.shen,georgina.cosma}@ntu.ac.uk

Abstract. Sleep stage classification is the categorisation of Electroencephalogram (EEG) epoch into different sleep stages. Various supervised and unsupervised models have been developed for sleep stage classification. Emphasis of those models has been on classifying sleep stages using deep learning models such as the Convolutional Neural Network (CNN), however, very limited work exists on interpreting those CNN filters learned from EEG data in a supervised manner. This paper focuses on investigating and interpreting the output filters of the first CNN layer of the DeepSleepNet model, which is a model developed for automatic sleep stage scoring based on raw Single-Channel EEG. Experiments were carried out using a public benchmark dataset, namely the Sleep EDF Database. Spectral properties of both EEG epoch (input) and the learned filters obtained from the first CNN layer were compared. Results showed similar spectral properties between sleep EEG patterns and the learned filters which were obtained from the first CNN layer, and these findings suggest that 'sleep stage'-defining EEG patterns are associated with certain learned CNN filters.

Keywords: Sleep stage classification · Deep learning ·
Electroencephalogram data

1 Introduction

An Electroencephalogram (EEG) is a tool used to investigate electrical activity of the brain. EEG is one of the strongest brain imaging modalities with high practical value in clinical neurology. It tracks brain wave patterns in the cerebral cortex and aids in detecting the abnormal or pathological brain states such as epilepsy, seizures, dizziness, and head injuries. This test is conducted by attaching a metal disc to the scalp with wires. The metal discs are called Electrodes, and evaluate the electrical impulses of the brain and send them to computer in order to record the results [1].

EEG recordings have been widely used in sleep research. The normal sleep process begins with a sleep stage called Non-Rapid Eye Movement (NREM) during which the eyes do not move back and forth and brain wave activity slows

© Springer Nature Switzerland AG 2020
Z. Ju et al. (Eds.): UKCI 2019, AISC 1043, pp. 142–154, 2020.
https://doi.org/10.1007/978-3-030-29933-0_12

down considerably, followed by the Rapid Eye Movement (REM) sleep stage. During the REM stage, brain waves are most active and eye movements are rapid from side to side. A normal person during sleep goes through multiple sleep cycles at night in approximately 90 min between REM and NREM [1]. Different sleep manuals, such as the American Academy of Sleep Medicine (AASM) are used by sleep experts for the categorization of sleep stages. This process of classifying consecutive epochs of EEG data (Electrical Activity of Brain) into wake, REM and NREM is called Sleep Stage Scoring [4]. NREM sleep stage is further divided into 4 sleep stages. Starting from sleep stage 1 and 2, also known as 'Light Sleep', during which muscle movement and brain waves begin to slow down is followed by sleep stage 3, well known as 'Deep Sleep'. Transition between REM and NREM occurs during sleep stage 3 and brain wave activity becomes the slowest. Subsequently, the NREM sleep stage 4 begins during which the deep sleep continues [6]. Visual scoring of sleep stages is time consuming and requires considerable work by sleep experts. Hence, there is a need to automate the sleep staging classification process [4].

Many hand engineering techniques and well-known machine learning algorithms have been used for automatic sleep stage scoring. For instance, the Convolutional Neural Network (CNN) and Recurrent Neural Network (RNN) particularly, Long Short-Time Memory (LSTM) have recently been explored by researchers for the task of sleep stage classification and achieved high accuracy of 80% and above [2,7]. Section 2 provides more details about the kind of models mentioned above. The literature review revealed that the models proposed for sleep stage classification perform well, however, these models provide limited explanation regarding the ways in which sleep stage scoring is associated with the learned filters of first layer among these models. It may be related to the feature learning capability of CNN which can help the algorithm learn 'sleep stage'-defining EEG patterns from EEG data in the form of CNN filters. To validate this hypothesis, we investigated the *DeepSleepNet* model - one of such sleep stage classification models which utilises multiple CNN layers.

This paper describes an in depth study of the DeepSleepNet model proposed by Supratak et al. [7] for automatic Sleep Stage Scoring based on Raw Single-Channel EEG. The study provides results when working on the first CNN Layer of this model for sleep stage classification. To this end, the learned CNN filters of the first layer are analysed in order to find a connection between the sleep EEG pattern and learned filters of the first CNN layer [7]. The paper is structured as follows. Section 2 provides a literature review of related work; Sect. 3 discusses the experimental methodology which includes details about the data, the architecture of the first part of the DeepSleepNet model, and the methodology which was adopted for investigating the filters in the first layer of the model. Subsequently, Sect. 4 presents the acquired results; and Sect. 5 provides a conclusion and future work. The source-code of the methods described in this paper is available online[1].

---

[1] https://github.com/gcosma/SleepStageCNN.git.

2   Related Work

Biswal et al. [2] proposed an automated sleep staging System, referred to as 'SleepNet', which applies deep neural network techniques on extracted features of multi-channel EEG data obtained from clinical routine Polysomnography (PSG). The architecture of SleepNet comprises of two modules, namely the training and deployment module. The main purpose of the training module is to find the best classification algorithm for EEG data and to classify that data into sleep stages. This is done by taking the multi-channel EEG data as input, extracting features from the data and then finding the classification algorithm to identify the combination of best features (model input) and algorithm (model class) among all the given algorithms. Following the training module is the deployment module, which deploys the best performing algorithm. The deployment module inputs new PSG data and classifies the data into sleep stages automatically using the best performing algorithm [2].

Similarly, Tsinalis et al. [9] proposed a machine learning method to automate the process of sleep stage scoring using stacked sparse autoencoders and time-frequency analysis for hand-crafted feature extraction. According to Tsinalis et al. [9], detection of sleep disruption aids in diagnosing the primary phase of neurodegenerative diseases, such as Alzheimer's disease. This is possible by monitoring the difference of disruption in the sleep pattern of healthy human brains and brains of humans with neurodegenerative disease. An openly available dataset containing EEG of 20 young healthy patients was used. At first, time frequency analysis was performed by extracting features from EEG data using Morlet Wavelets. The reason for preferring time frequency analysis instead of Fast Fourier transform (FFT) was to capture the spectral properties and their interrelations at different time intervals. Then, Stacked Sparse Auto Encoders (SSAEs), a special type of neural network model comprising of multiple layers of autoencoders was used for sleep stage classification. It was observed that the method used either outperformed with overall accuracy of 78% for all subjects or had nearly equivalent performance as of those other traditional methods in all dimensions [9].

Moreover, Tsinalis et al. [8] proposed another sleep stage scoring method using CNN with single channel EEG. The rationale behind this was to learn and analyse task-specific filters of CNN layer and compare them with different fold of subjects. After some analysis, it was concluded that sleep stage scoring rules as mentioned in AASM manual were consistent with rules learnt by CNN filters. The Sleep EDF database was used an input. The filters learned were analysed by computing the power spectral density using Fast Fourier Transform (FFT). Then, the mean activation was computed per sleep stage for all epochs. All filters analyzed were visualize and a similarity was observed between learned filter and EEG patterns of different folds. However, it was noted from the visualization results that association was not much accurate using FFT [8].

The model analysed in this paper is the DeepSleepNet [7] model which utilizes CNN and LSTM to automate the process of sleep stage scoring. This model is different from all other current models as it relies on the feature extraction capabilities of deep learning instead of manually extracting features from EEG. The inspiration behind the DeepSleepNet model was to utilize the feature extraction capabilities of deep learning instead of relying on old methods such as spectral density estimate to extract features from EEG for sleep stage classification and prediction. Overall, the DeepSleepNet [7] model has two parts: the first part utilizes eight CNN layers and second part contains two Bidirectional LSTM network. In the first part of the model, the first two layers are those of a CNN with two different filter sizes in order to extract temporal features at two different time scales. The first CNN layer comprises of a small filter size to find temporary information (i.e. the occurrence of certain EEG at different time intervals) and a second CNN layer contains a large filter size for frequency information from EEG (repeating sleep stages per unit time). Temporal and frequency components are essential to acquire in order to extract complementary features of each sleep stage. After two CNNs are combined together, a Bidirectional LSTM encodes temporal information, for instance some rules defining transition from one sleep stage into another sleep stage. For other sleep EEG datasets, this model could be used to predict sleep stage without modifying its model architecture or re-training its parameters. This architecture follows a two-step training algorithm which includes supervised pre-training followed by a supervised fine-tuning step using two different learning rates. The results revealed that the DeepSleepNet model can be used to train different EEG channels without changing the training algorithm and architecture of the model. Undoubtedly, it can be said that its architecture has a good generalisation capacity. However, the DeepSleepNet model, and in particular the first part of the model, has not been explored and explained in much depth. This paper investigates and interprets the filters in the first layer of the Convolutional Neural Network of the DeepSleepNet for sleep stage classification model, to provide an understanding of the hidden aspects of the model.

3   Experimental Methodology

This section describes the data and the first part of the DeepSleepNet architecture that was utilised for the experiments.

3.1   Data: Sleep EDF Database

The EEG data used in this work is extracted from a sleep EEG database (expanded) which is available on *Physionet* Website [7]. This data set comprises of a collection of 197 Polysomnographic Sleep Recordings in EDF format (European Data format) containing EEG (from Fpz-Cz and Pz-Oz Electrode Locations), EMG (Electromyography), EOG (Electrooculography) and event markers along with the Hyponogram.edf files which contain their corresponding sleep patterns. These recordings are classified manually by sleep experts using the 1968 Rechtschaffen and Kales manual into 5 sleep stages as: wake, sleep stage 1, sleep stage 2, sleep stage 3 and REM sleep stage. In addition, information about artifacts such as those caused by head movement is also provided.

A total of 39 subjects have been used as an input. The EDF files for input have been converted into .mat files using MATLAB software. The leave-one-subject-out procedure was employed to compute test accuracy, and evaluations were performed 39 times, where each time the data of 38 subjects was used as training data and the remaining subject's data was used as testing data. Long periods of EEG epochs corresponding to sleep stage 0 (awake) were excluded from the data because the study focused on classification all-night EEG recordings rather than whole day EEG one. Moreover, movement time and unknown stages were also excluded as they did not belong to any of the five sleep stages and sleep stage 3 has been merged with sleep stage 4 as it is currently recommended by the American Academy of Sleep Medicine (AASM) [7,9].

3.2   Architecture of First Part of the DeepSleepNet Model

The architecture of *DeepSleepNet* has two main parts: the 'representation learning' part for classification of EEG epochs regarding their corresponding sleep stages, followed by 'sequence residual learning' for predicting sleep stage of upcoming EEG epochs. This section of the paper introduces the architecture of the first part of the model in depth.

Model Specification of Representation Learning: This part of the model is trained to extract temporal features from 30 second epochs. It has two parallel compartments consisting of CNN and pooling layers for extracting features at two different time scales. To achieve this, the filter used in the first CNN layer of these two compartments is of different size. The small filter with filter size of half sampling rate matches the time scale at which EEG pattern occurs. The large filter with the filter size of four times the sampling rate is used to capture information about frequency components. The sampling rate is taken as 100 Hz. Both sub parts consist of six layers with four CNN layers and two max pooling layers. The specifications of stride sizes, filter sizes, number of filters and pooling sizes of each layer are given in Table 1, where Fs is the Sampling rate and Fig. 1 represent the block of the architecture of first part.

Assume that the EEG recording is divided into N consecutive epochs of a fixed length, say $\{\mathbf{x}_1, \ldots, \mathbf{x}_j, \ldots, \mathbf{x}_N\}$ where $\mathbf{x}_j$ is a vector representing the EEG time series in the j-th epoch. Let $\mathbf{g}$ be the feature vector resulted from applying the function of CNN with small filter size on the j-th EEG epoch then the equation for this can be written as follows:

$$\mathbf{g}_j^s = \mathbf{CNN}_{\theta_s}(\mathbf{x}_j) \tag{1}$$

Where the superscript indicates the size of CNN filters and s stands for small filter size and θ_s are the parameters of that CNN layer. Similarly, the CNN function with large filter size operating on the j-th EEG epoch is given by

$$\mathbf{g}_j^l = \mathbf{CNN}_{\theta_l}(\mathbf{x}_j) \tag{2}$$

Table 1. Specification of the first part of the model

Layer	Specifications	Small filter CNN	Large filter CNN
1st layer	Filter size	Fs/2	Fs *4
	Stride size	Fs/16	Fs/2
	Number of filters	64	64
2nd layer	Pooling size	8	4
	Stride size	Fs/8	Fs/4
3rd, 4th and 5th layer	Filter size	8	6
	Stride size	Fs/16	Fs/2
	Number of filters	128	128
6th layer	Pooling size	4	2
	Stride size	Fs/4	Fs/2

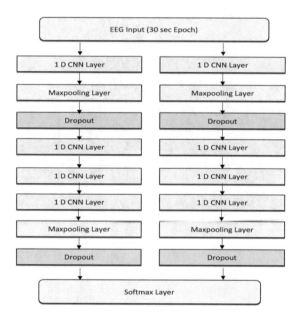

Fig. 1. Block diagram of the architecture of the first part of the DeepSleepNet model [7].

where the subscript l stands for large filter and θ_l are the parameters of CNN with large filter sizes. The output of the last layer of both CNNs (small and large filter size) are combined together as:

$$\mathbf{z}_j = \mathbf{g}_j^s \| \mathbf{g}_j^l \tag{3}$$

Where || denotes the concatenated function of CNN two vectors (that is $\mathbf{g}_j^s$ and $\mathbf{g}_j^l$). The representation of N EEG epochs, that is, $\mathbf{z}_1, \ldots, \mathbf{z}_N$, are then passed to second part of the model containing LSTM for the prediction of sleep stages.

Training and Classification: An overall accuracy of 63% was achieved during the training of first part of the DeepSleepNetmodel model [7]. Figure 2 illustrates the confusion matrix obtained after training the first part of the model for the classification of sleep stages from 1 to 4. The diagonal line represents the correctly classified sleep stages from 1 to 4 accordingly.

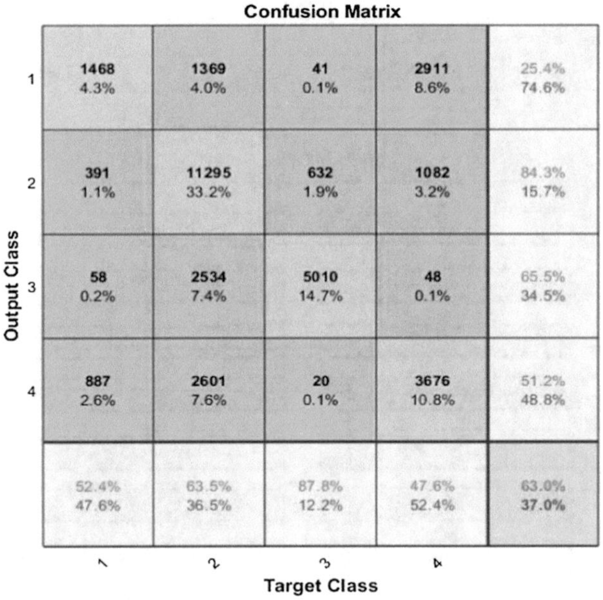

Fig. 2. Confusion matrix

3.3 Methodology for Investigating and Interpreting the Filters in the First Layer of DeepSleepNet

The main objective of this paper is to identify associations between sleep EEG Patterns and learned filters. This objective will be achieved by estimating the spectral power of all 64 CNN filters of small filter size as well as that of selected EEG epochs of sleep stage 2 and 4. This is because EEG time series from these two sleep stages exhibited for distinct EEG patterns. As EEG patterns manifest themselves by oscillation of particular frequency, the spectral density of an EEG epoch with such activities should have a significant peak at the corresponding frequency [5]. The spectral power qualities of both sleep stages and learned filters can help identify the association among them. The step by step experimental methodology carried out is as follows:

1. First, we select a number of sleep stage 2 and sleep stage 4 epochs with profoundly manifested EEG patterns. Only these epochs are used for further analysis. To select these epochs automatically as well as objectively, we used the spectral power in the frequency range of 13 and 15 Hz for selecting sleep stage 2 epochs, and spectral power in the frequency range of 0.5 and 4 Hz for selecting sleep stage 4 epochs.
2. Next, we compute the average output of all 64 CNN filters from the trained first CNN layer, averaged over all selected EEG epochs of sleep stage 2 and those of sleep stage 4 separately. This statistic measures association between a CNN filter and a particular sleep stage. We rank all 64 filters based on this association measure for sleep stage 2 and for sleep stage 4;
3. Following the identification of those top ranked CNN filters, we now compute the spectral density of these filters. Note that these filters are actually short time series. For short time series data, AR model based Yule-Walker estimator outperforms FFT based Welch estimator (see Sect. 4.5);
4. Compared both the results obtained from steps 1 and 3, in order to find the consistency between the spectral density of EEG epochs and that of the top ranked CNN filters.

4   Results

4.1   Prototypical EEG Patterns in Sleep Stage 2

Sleep stage 2, also known as first NREM stage, is a precedence of deeper sleep. During this stage, eye movements slow down along with brain wave activity and certain kind of significant patterns tends to appear during this sleep stage known as 'Sleep Spindles' and 'K Complexes' [10]. Figure 3 illustrates these patterns.

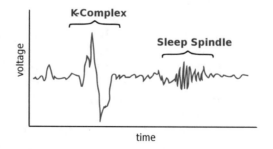

Fig. 3. Sleep spindle and K complex pattern

In order to identify the association between sleep stages and the learned CNN filters, the EEG epochs selected from sleep stage 2 were used because two unique EEG patterns, that is, 'spindles' and 'K complexes' occur in these epochs. Instead of averaging all the epochs from sleep stage 2 over all subjects, as done in [7], some of the best epochs such as those having more prominent K complexes and sleep spindle shapes were averaged. The main motivation to do this was to include only those epochs which have the strongest signal of the underlying brain activities during sleep. Because of the large number of sleep stage 2 epochs over all the subjects, visual identification is not a reasonable option [4]. Therefore, the best approach is to compute the power spectral density (PSD) of sleep stage 2 over all subjects and then set the peak frequency within 13–15 Hz range as the spindle usually peaks at approximately 14 Hz. The threshold value was kept as $PSD^{thres} = 100$, and all the peaks occurring above threshold value were selected Before computing the PSD, the EEG time series was normalized so that it has unit variance. As a result, the area under the spectral density curve is fixed to 1 for all EEG epochs. This can help avoid any false detection which simply resulted from changes in EEG's amplitudes rather than its spectral properties.

4.2   Selection of EEG Epochs with Prominent EEG Patterns via Spectral Analysis

Welch's Power Spectral Density (PSD) estimate was used to select EEG epochs with highest spectral content over the frequency range between 13 Hz and 15 Hz. Let $\{f_1, \ldots, f_n\}$ denote a set of n frequencies. We compute the power spectrum density (PSD) for each of these frequencies and denote the PSD of frequency f_j by P_j, $j = 1, \ldots, n$. Once the PSD is computed, the spectral power of a given frequency band can be derived by

$$P_{f_{low}, f_{upp}} = \sum_{f'_{low} \leq f_j \leq f_{upp}} P_j \tag{4}$$

where f_{low} and f_{upp} stand for the lower and upper bound of the given frequency band. For example, for checking the spindle and/or K complex activities, we compute $P_{13\,Hz, 15\,Hz}$ for all EEG epochs.

It was evident from the results that some of the epochs had higher peak as compared to the other epochs. This is mainly because of more prominent pattern of sleep spindle and K complexes in that particular epoch. A profound peak around 14 Hz was observed in the 128-th epoch derived from the 2nd subject's EEG recording.

Finally, note that out of 39 subjects, only 34 subjects were used as remaining subjects did not had sleep stage 2.

In order to illustrate the results of epoch selection, the spectral density of 6 EEG epochs selected from a single subject (subject 2) are plotted in Fig. 4. Within this subject, these 6 epochs are the one with the highest spectral content in the frequency range between 13 Hz and 15 Hz. Figure 4 showed a clear peak around 14 Hz. One of the epochs (epoch 15) clearly showed a significant K complex pattern (for less than one second interval) after 25th second followed by a spindle between the 25-th and 30-th second. Similarly, for another epoch

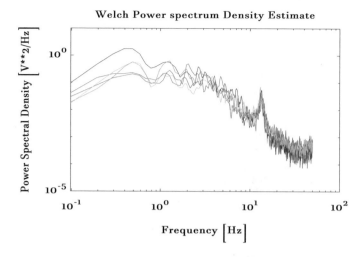

Fig. 4. Welch power spectrum density estimate (subject no. 2-sleep stage 2)

(epoch 282), multiple complexes were found between 5-th and 25-th second of the epoch followed by a sleep spindle.

4.3   Ranking CNN Filters

After having selected EEG epochs in sleep stage 2, the next step is to identify those CNN filters that are associated with sleep stage 2. This was achieved by estimating the activation of individual CNN filters from the first CNN layer. For each of the selected sleep stage 2 epochs, the estimate of its activation level was computed across all 64 filters. For all of those epochs whose spectral density was plotted in Fig. 4, their top ranked filters were found to be 5 and 47. Both filter shapes showed a visible sleep spindle pattern. Among the 64 filters, based on how often they occurred in 170 stage 2 epochs selected from 34 subjects, the top 9 ranked filters were found to be those indexed by 5, 6, 11, 23, 47, 51, 52, 57, 60 and 63 which in total are 9 filters out of 64 filters. The bar chart in Fig. 5 shows the number of times a certain learned CNN filter appeared in the descending order.

4.4   Interpretation of the Top-Ranked CNN Filters

As only 9 learned filters (out of 64) occurred at least once in those 170 epochs of Sleep Stage 2, it is reasonable to hypothesize that the shape of these filters should be similar to EEG patterns such as K complex and spindles. The best way to test this hypothesis, is to estimate the power spectrum of the filters and check whether it possesses peaks around the 14 Hz. Therefore, the power spectral estimate of 9 top ranked filters were estimated by using the power spectral density method named as 'Yule Walker Method', as described in Sect. 4.5.

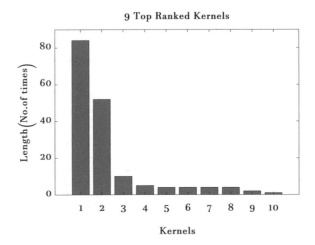

Fig. 5. The occurrence frequencies of the 9 top ranked filters

4.5 Yule-Walker Method

The Yule-Walker method was used to compute the spectral power density of a short time series. Here, a CNN filter is a time series of 50 data points. This time series is too short to apply Welch or FFT methods. This method computes the spectral density by fitting an auto-regressive (AR) model of lower-order to the time series. It does so by minimizing the discrepancy between the theoretical auto-correlation function of the fitted AR model and the empirical auto-correlation function of that time. This formulation leads to the Yule-Walker equations [3]. The order of AR model is set to 6 so as to account for the oscillations of varying periods that could occur in the short time series. As the number of data points is limited to 50, the frequency was specified between 1 to 25 cycles per unit time. The data points were normalized before calculating the PSD using the standard deviation method. Figure 6 shows the results where the high peak at 14 Hz clearly shows a similarity between the learned filters and the epochs.

Figure 4 further illustrates the log representation of Welch PSD of selected top 5 epochs with highest peaks (15, 282, 353, 63 and 9 among 362 epochs) of the 2nd Subject. A clear pattern of peak of these epochs can be seen after 10 Hz. The similarity of learned filters and selected EEG epoch is quite evident by comparing the results obtained from the Welch PSD and Yule-Walker PSD as illustrated in Figs. 4 and 6.

4.6 Spectral Analysis of CNN Filters: Yule-Walker Versus FFT

In order to evaluate the obtained results using Yule-Walker Method, they were compared with results obtained for CNN learned filters using Fast Fourier Transform (FFT). It has been previously experimented by Tsinalis et al. [8] with first

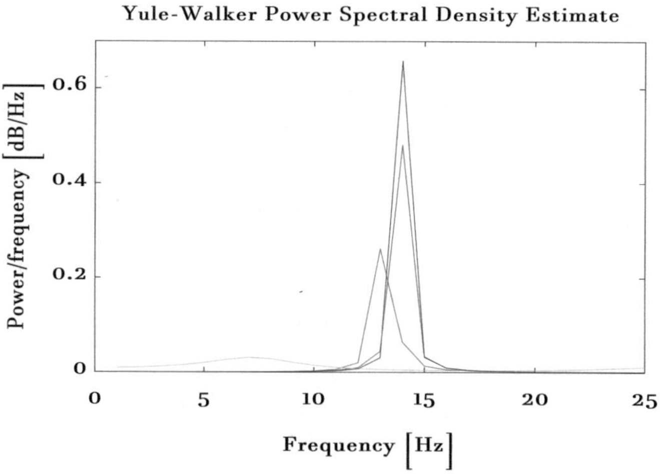

Yule-Walker Power Spectral Density Estimate

Fig. 6. Yule-Walker power spectral density Estimate for top 9 filters, where each line is a filter.

CNN layer learned filters using the same Sleep-EDF database with 100 Hz frequency. However, it was observed that using Yule-Walker method provided more clear structure of filters with accurate peak between 13 and 15 Hz.

5 Conclusion and Future Work

The present study focuses on the interpretation and exploration of sleep stage classification using the DeepSleepNet prediction model for automatic sleep stage scoring based on Raw Single-Channel EEG data. DeepSleepNet utilizes a Convolutional Neural Network (CNN) and Bidirectional LSTM for classification and prediction of sleep stages respectively. The Sleep EDF Database was used in the experiments described in this paper. The main objective of the study was to find the association between sleep EEG patterns and learned filters of the first CNN layer of the DeepSleepNet model. For this, epochs belonging to sleep stage 2 and 4 were used due to occurrence of significant patterns like 'Spindles' and 'Slow waves' among them. Spectral power qualities were estimated and compared for EEG sleep patterns and learned filters for both sleep stages. The Welch theorem and Yule-Walker method were used for obtaining the spectral PSD of EEG sleep patterns and the learned filters respectively. Results demonstrated a similarity between them by identifying a peak around 14 Hz and 2 Hz for sleep stages 2 and 4 respectively. Future work includes interpretation of other CNN layers used in the first part of the model. Moreover, the second part of the DeepSleepNet model utilizing Bidirectional LSTM will also be analyzed.

References

1. Aboalayon, K.A.I., Faezipour, M., Almuhammadi, W.S., Moslehpour, S.: Sleep stage classification using EEG signal analysis: a comprehensive survey and new investigation. Entropy **18**(9), 272 (2016)
2. Biswal, S., Kulas, J., Sun, H., Goparaju, B., Westover, M.B., Bianchi, M.T., Sun, J.: SLEEPNET: automated sleep staging system via deep learning. CoRR abs/1707.08262 (2017)
3. Howarth, R.J.: Dictionary of Mathematical Geosciences: With Historical Notes, pp. 669–671. Springer, Cham (2017)
4. Malafeev, A., Laptev, D., Bauer, S., Omlin, X., Wierzbicka, A., Wichniak, A., Jernajczyk, W., Riener, R., Buhmann, J., Achermann, P.: Automatic human sleep stage scoring using deep neural networks. Front. Neurosci. **12**, 781 (2018)
5. Shen, Y., Olbrich, E., Achermann, P., Meier, P.: Dimensional complexity and spectral properties of the human sleep EEG. Clin. Neurophysiol. **114**(2), 199–209 (2003)
6. Sors, A., Bonnet, S., Mirek, S., Vercueil, L., Payen, J.F.: A convolutional neural network for sleep stage scoring from raw single-channel EEG. Biomed. Signal Process. Control **42**, 107–114 (2018)
7. Supratak, A., Dong, H., Wu, C., Guo, Y.: DeepSleepNet: a model for automatic sleep stage scoring based on raw single-channel EEG. IEEE Trans. Neural Syst. Rehabil. Eng. **25**(11), 1998–2008 (2017)
8. Tsinalis, O., Matthews, P.M., Guo, Y., Zafeiriou, S.: Automatic sleep stage scoring with single-channel EEG using convolutional neural networks, October 2016
9. Tsinalis, O., Matthews, P.M., Guo, Y.: Automatic sleep stage scoring using time-frequency analysis and stacked sparse autoencoders. Ann. Biomed. Eng. **44**(5), 1587–1597 (2016)
10. Vilamala, A., Madsen, K.H., Hansen, L.K.: Deep convolutional neural networks for interpretable analysis of EEG sleep stage scoring. In: 2017 IEEE 27th International Workshop on Machine Learning for Signal Processing (MLSP), pp. 1–6, September 2017

Gradient Boost with Convolution Neural Network for Stock Forecast

Jialin Liu[1], Chih-Min Min Lin[2], and Fei Chao[1(✉)]

[1] Department of Artificial Intelligence, School of Informatics, Xiamen University, Xiamen 361005, Fujian, People's Republic of China
31520171153232@stu.xmu.edu.cn, fchao@xmu.edu.cn
[2] Department of Electrical Engineering, Yuan Ze University, Chung-Li, Tao-Yuan 320, Taiwan
cml@saturn.yzu.edu.tw

Abstract. Market economy closely connects aspects to all walks of life. The stock forecast is one of task among studies on the market economy. However, information on markets economy contains a lot of noise and uncertainties, which lead economy forecasting to become a challenging task. Ensemble learning and deep learning are the most methods to solve the stock forecast task. In this paper, we present a model combining the advantages of two methods to forecast the change of stock price. The proposed method combines CNN and GBoost. The experimental results on six market indexes show that the proposed method has better performance against current popular methods.

Keywords: Ensemble learning · Deep learning · Stock forecast

1 Introduction

The stock price of a company is an important criterion for measuring the actual value of the company. In the stock market, well decision depends on well forecast. Due to the development of computational technology and intelligence technologies. New tools are recently developed to process information on stock forecast. The analysis of financial market movements has been widely studied in the fields of finance, engineering and mathematics in the last decades [15]. Using intelligent technologies on stock prediction has widely spread in recent years.

In the past, the most commonly algorithms for stock forecast in the past are artificial neural network (ANN) and support vector machine (SVM or SVR). In contract to ANN, SVM is a statistical learning method that is widely used in pattern recognition tasks. In 2003, Kim et al. predicted stock price by using a SVR model and proved that the prediction precision of support vector regression model was better than the back propagation (BP) neural network prediction model and case-based reasoning (CBR) [11]. In 2006, Xu et al. came up with a revised Least squares (LS)-SVM model and forecasted Nasdaq Index movement, and model brought satisfactory results [16].

© Springer Nature Switzerland AG 2020
Z. Ju et al. (Eds.): UKCI 2019, AISC 1043, pp. 155–165, 2020.
https://doi.org/10.1007/978-3-030-29933-0_13

In the past several years, there are many business applications, in which the technology of ANN was used. It may be due to the non-linear approximation ability of ANN, and it is frequently used in combination with other methods. Martinex et al. proposed that the neural network to solve the financial forecasting problem use mostly a back propagation algorithm to optimize a multi-layer forward neural network (MLP) with high performance [13].

In recent years, ensemble learning and deep learning have developed quickly in a lot of fields [4,5,8,12,14]. These two methods have their own advantages and disadvantages in solving with stock data. In general, there are two kinds of different views in stock prediction. (i) To obtain the enough information, financial analysis methods must be used to obtain high quality information. In this case, market economy data involves the indices with different characters, which are suitable to be handed by ensemble learning. However, simplifies computation of analysis methods would causes loss of information. (ii) Only use the stock price history to get information. It is possible that all the information is available from the historical behavior of a financial asset as a time series. The time series of stock prices involves enough information and is suitable to be handed by deep learning. However, the time series of stock prices also involves a lot of noise and uncertainties. In order to improve forecast accuracy we desire to use technique indices and retain the sequence structure of stock data to make them complementary to each other. Thus, a model combines the advantages of both ensemble learning and deep learning is the objective of this paper. This paper proposes a model combining the ensemble techniques of extreme gradient boost (XGBoost) [4] and 1D convolution neural network (CNN), called as CGBoost, to obtain a better performance.

In addition, we use a sparse autoencodes (SAEs) [3,9] to process data to reduce noise in the stock price time series, the training implemented by encoding and decoding data and reducing the loss in each iteration. If we only and use the original data to train CGBoost without this process, we will get only a highly overfitting result. Then we also try to training one model on several different market indexes, so as to test whether the propose model can unify data from different market indexes to improve overall performance.

The remainder of this paper contains three sections. Section 2 draws the details of each technique used in this work and how to combine all those techniques as a complete system. Section 3 describes data resources, evaluation and other details about the experiment. The results and analysis of the experiment is also in this section. Finally, Sect. 4 draws conclusion and future work.

2   Methology

In order to generate the deep and invariant features for one-step-ahead stock price forecast, this paper presents a gradient boosting framework with a deep learning for financial time series. The framework uses a deep (CNN) and width (GBoost) learning-based predicting scheme that integrates the architecture of CNN and GBoost. The flow chart of this framework shown in Fig. 1, involves

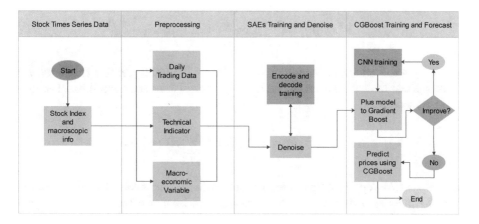

Fig. 1. Stock prediction system contains sparse autoencodes and convolution gradient boost.

three stages: (1) data preprocessing, the clipping and normalizing transform, which are applied to rescale the stock price time series to some scale; (2) adopting of the SAEs, which has a deep architecture trained in an unsupervised manner, combined with 1D CNN; and (3) GBoost, it has 1D CNN to generate the one-step-ahead prediction. Since the first step is related to data descriptions, details the first step are introduced in the Sect. 3. The rest steps are detailed as follows.

2.1   SAEs Training and Denoise

SAEs is a type of deep learning model to reduce dimension and noise of data [3,9]. Since manually adding category tags to data is a very cumbersome process, the machine must learn part of the important features in the sample. By imposing some restrictions on the hidden layer, SAEs can better express the characteristics of the sample in a harsh environment. SAEs has this limitation on the sparseness of the hidden layer.

The sparsity is represented as the activated states of neuron. If the sigmoid function is used as the activation function, and the neuron output value is 0, this situation is regarded as a suppression. The sparsity limit ensures that most of the neuron output is 0 and the state is suppressed. Then, the functions can approximate,

$$\hat{\rho} = \rho, \; \hat{\rho} = \frac{1}{m} \sum_{i=1}^{m} \left[a_j(x^{(i)}) \right] \tag{1}$$

where a_j denotes the activation of the hidden neuron j; $\hat{\rho}$ is the average of the activation; and ρ is a sparsity parameter, usually it is a small value close to 0 (such as $\rho = 0.05$).

In order to achieve this limitation, an additional penalty factor is added to our optimization objective function, which can punish those $\hat{\rho}_j$ has significantly different conditions with ρ in hidden layers, it is given by:

$$\sum_{j=1}^{s_2} \mathrm{KL}(\rho||\hat{\rho}_j) = \sum_{j=1}^{s_2} \rho \log \frac{\rho}{\hat{\rho}_j} + (1-\rho) \log \frac{1-\rho}{1-\hat{\rho}_j}, \tag{2}$$

where s_2 is the number of hidden neurons in the hidden layer; and the index j in turn represents each neuron in the hidden layer. Then, the overall loss function is expressed as:

$$J_{\mathrm{sparse}}(W, b) = J(W, b) + \beta \sum_{j=1}^{s_2} \mathrm{KL}(\rho||\hat{\rho}_j), \tag{3}$$

where $J(W, b)$ represents the reconstruction loss; β controls the weight of the sparsity penalty factor, W and b are weight and bias of neural network, respectively.

Finally, we apply stochastic gradient descend to optimize W and b. In order to minimize $J_{\mathrm{sparse}}(W, b)$. After training SAEs, $\mathrm{a}(x^{(i)}) = a_j(x^{(i)})_j$ is used as the feature of sample $\{x^{(i)}, y^{(i)}\}$.

2.2   CGBoost Training and Forecast

The gradient boost algorithm is an ensemble learning technology. The algorithm generates a prediction model by integrating weak prediction models, such as decision trees. It builds the model in a step-by-step manner like other gradient methods and promotes them by allowing the use of any differentiable loss function.

In the experiments, rather than using original GBoost, we apply the training way in XGBoost [4]. Different from the traditional GBoost method, only the first derivative information is used. XGBoost performs the second-order Taylor expansion on the loss function, and adds the regular term to the objective function to balance the decline of the objective function, so as to avoid overfitting. The objective function of the based learner is given by:

$$Obj^{(t)} \approx \sum_{i=1}^{n} \left[g_i f_t(x_i) + \frac{1}{2} h_i f_t^2(x_i) \right] + \Omega(f_t), \tag{4}$$

where $\Omega(f_t)$ is the L2 regularization $\sum_l \|W_l\|^2$, due to all based estimators are CNNs, W_l denotes the weights of l layer, $g_i = \partial l(y_i, y^{(t-1)_i})/\partial y_i^{(t-1)}$ and $h_i = \partial^2 l(y_i, y^{(t-1)_i})/\partial y_i^{(t-1)^2}$. Because the goal is to predict the real price of stock the loss function can be square loss function. Then, the form is given by:

$$Obj^{(t)} \approx \sum_{i=1}^{n} \left[2(y_i^{(t-1)} - y_i) f_t(x_i) + f_t^2(x_i) \right] + \Omega(f_t), \ t \neq 1 \tag{5}$$

when all based estimators are obtained, the forecast result of x_i is calculated by $F(x_i) = \sum_{t=1}^{T} f_t(x_i)$, where T denotes the number of based estimators.

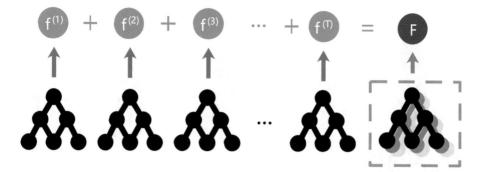

Fig. 2. Ensemble model with neural network based learners.

2.3 1D Residual Network

In this paper, we use 1D residual neural network (resnet) [8], a kind of CNN, within both SAEs and GBoost. Due to the common size of our model we use resnet to train well. However, using resnet still can accelerate training significantly [8].

A structural diagram of a standard 1D CNN is shown in Fig. 3. Each layer receives the output from the previous layer and outputs abstract features. In the training, gradient is back propagated from the output of last layer. The level number of network is larger than a certain number, the gradient vanishing will occur, so as to make deep network to be trained difficultly.

The resnet applies the idea of the "shortcut connections", the idea of cross-layer linking to improve it, in order to prevent the gradient vanishing. The input x is directly passed to the output as the initial result, and the output result is $H(x) = F(x) + x$. If $F(x) = 0$, $H(x)$ will become the identity map, $H(x) = x$. As the network deepening, it still retain the much shallower tunnel. Therefore, the gradient does not decrease as the network deepening.

3 Experiment

The experiments are designed to answer two questions: (1) Can the model combine ensemble learning and deep learning produce more accurate predictions than single deep learning model? (2) Is proposed model able to fit the data from different indices and still improve the performance?

The proposed model compares with the accuracy of WSAEs-LSTM [2], which applied the deep learning model to forecast the stock price series, so as to answer the first question. Following [2], we chose "CSI 300", "DJIA", "Hang Seng", "Nifty 50", "Nikkei 225" and "S&P500" indices as the predict goals. We conducted experiments training one model for each index and training one model for all indexes, and CGBoost and CGBoost6 were applied to denote them respectively. Their results can answer the second question.

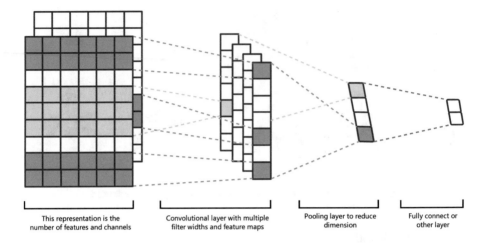

This representation is the number of features and channels

Convolutional layer with multiple filter widths and feature maps

Pooling layer to reduce dimension

Fully connect or other layer

Fig. 3. A standard 1D convolutional neural network diagram.

It is different from Fig. 1 we use a fixed number of base models in CGBoost. The reason is that we train the model on training data and validation data after adjusting hyper-parameter. Thus not validation data can be used to test whether the model is improved or not. Besides, the model is used to predict stock price indirectly by predicting the change rate of price. Based on our experience, this way can get better results.

3.1   Data Descriptions

The data used in this experiment is detailed as follows.

Data Resource. We use the data provided by [2], which was sample from CSMAR[1] and WIND[2]. The sample is from 1st Jul. 2008 to 30th Sep.2016.

Data Features. Three types of feature are chose in our experiment. Following the previous literature the first type of feature includes OHLC variables, which is the price variables (Open, High, Low, and Close price). The second kind of feature is the technical indicators of each index. Each of them is described in Table 1. The final part of inputs is the macroeconomic variable. It is related to stock price. We chose the Interbank Offered Rate and US dollar Index to our system.

Data Divide. Refer to the rule of stock, we cannot use the data from future. Thus we use the first two year as the training set, next three months as the validating data and last three months as the test set. It is divided into four steps to obtain a one-year prediction result for testing. We divide prediction into six years to evaluate accuracy.

---

[1] http://www.gtarsc.com.

[2] http://www.wind.com.cn.

Table 1. The techniques indices and their definition is described in this table.

Name	Definition
MACD	Moving Average Convergence
WVAD	Williams's Variable Accumulation/Distribution
ATR	Average true range
EMA20	20 day Exponential Moving Average
BOLL	Bollinger Band MID
MA5/MA10	5/10 day Moving Average
MTM6/MTM12	6/12 month Momentum
SMI	Stochastic Momentum Index
ROC	Price rate of change
CCI	Commodity channel index

3.2   Evaluate

Following [1,6,7,10], the results were evaluated by "MAPE", "Theil U" and "linear correlation between prediction and real prices" (use R to denote). These indicators are denoted as follows:

$$\text{MAPE} = \frac{1}{N} \sum_{t=1}^{N} \left| \frac{y_t - y_t^*}{y_t} \right| \tag{6}$$

$$\text{R} = \frac{\sum_{t=1}^{N}(y_t - \overline{y_t})(y_t^* - \overline{y_t^*})}{\sqrt{\sum_{t=1}^{N}(y_t - \overline{y_t})^2 \sum_{t=1}^{N}(y_t^* - \overline{y_t^*})^2}} \tag{7}$$

$$\text{Theil U} = \frac{\sqrt{\frac{1}{N} \sum_{t=1}^{N}(y_t - y_t^*)^2}}{\sqrt{\frac{1}{N} \sum_{t=1}^{N}(y_t)^2} + \sqrt{\frac{1}{N} \sum_{t=1}^{N}(y_t^*)^2}} \tag{8}$$

where y_t^* is the forecast of model and y_t is the actual price on time t. N is the number of prediction, in our experiment it is the number of days open in a year. R is different from MAPE and Theil U, if R is larger, the predicting price is similar to the actual value.

3.3   Results

The proposed method has improved results significantly. As show in Tables 2, 3 and 4, both cgboost and cgboost6 have low average predicting error in each year and each index, both MAPE and Theil U, and predicting result has higher linear correlation with actual price than based experiment. This result proved that proposed model can introduce a more accurate prediction than deep learning. Besides, the result of CGBoost6 is much better than that CGBoost, it is also answer the second question.

Table 2. The prediction accuracy in CSI 300 and DJIA indices.

Year	CSI 300 index							DJIA index						
	Year 1	Year 2	Year 3	Year 4	Year 5	Year 6	Average	Year 1	Year 2	Year 3	Year 4	Year 5	Year 6	Average
Panel A. MAPE														
CGBoost6	**0.011**	**0.011**	**0.010**	**0.008**	**0.018**	**0.011**	**0.011**	**0.008**	**0.008**	**0.005**	**0.005**	**0.007**	**0.007**	**0.007**
CGBoost	0.015	0.012	0.014	0.010	0.024	0.014	0.015	0.011	0.010	0.008	0.008	0.011	0.011	0.010
WSAEs-LSTM	0.025	0.014	0.016	0.011	0.033	0.016	0.019	0.016	0.013	0.009	0.008	0.008	0.010	0.011
Panel B. Correlation coefficient														
CGBoost6	**0.974**	**0.971**	**0.978**	**0.978**	**0.992**	**0.975**	**0.978**	**0.972**	**0.977**	**0.994**	**0.983**	**0.962**	**0.976**	**0.977**
CGBoost	0.953	0.963	0.965	0.965	0.988	0.967	0.967	0.949	0.964	0.987	0.956	0.919	0.939	0.952
WSAEs-LSTM	0.861	0.959	0.955	0.957	0.975	0.957	0.944	0.922	0.928	0.984	0.952	0.953	0.952	0.949
Panel C. Theil U														
CGBoost6	**0.007**	**0.007**	**0.007**	**0.005**	**0.013**	**0.008**	**0.008**	**0.006**	**0.005**	**0.003**	**0.003**	**0.005**	**0.004**	**0.004**
CGBoost	0.010	0.008	0.009	0.006	0.016	0.010	0.010	0.008	0.007	0.005	0.005	0.007	0.007	0.007
WSAEs-LSTM	0.017	0.009	0.011	0.007	0.023	0.011	0.013	0.010	0.009	0.006	0.005	0.005	0.006	0.007

Table 3. The prediction accuracy in HangSeng and Nifty 50 indices.

Year	HangSeng index							Nifty 50 index						
	Year 1	Year 2	Year 3	Year 4	Year 5	Year 6	Average	Year 1	Year 2	Year 3	Year 4	Year 5	Year 6	Average
Panel A. MAPE														
CGBoost6	**0.010**	**0.011**	**0.007**	**0.007**	**0.010**	**0.010**	**0.009**	**0.010**	**0.009**	**0.008**	**0.006**	**0.008**	**0.007**	**0.008**
CGBoost	0.014	0.014	0.011	0.009	0.016	0.013	0.013	0.015	0.013	0.013	0.016	0.013	0.011	0.013
WSAEs-LSTM	0.016	0.017	0.012	0.011	0.021	0.013	0.015	0.020	0.016	0.017	0.014	0.016	0.018	0.017
Panel B. Correlation coefficient														
CGBoost6	**0.982**	**0.961**	**0.967**	**0.976**	**0.985**	**0.980**	**0.975**	**0.981**	**0.969**	0.935	**0.997**	0.958	**0.988**	**0.971**
CGBoost	0.963	0.938	0.943	0.954	0.954	0.966	0.953	0.957	0.935	0.873	0.978	0.910	0.972	0.937
WSAEs-LSTM	0.944	0.924	0.920	0.927	0.904	0.968	0.931	0.895	0.927	**0.992**	0.885	**0.974**	0.951	0.937
Panel C. Theil U														
CGBoost6	**0.006**	**0.007**	**0.005**	**0.004**	**0.007**	**0.006**	**0.006**	**0.006**	**0.006**	**0.006**	**0.004**	**0.005**	**0.004**	**0.005**
CGBoost	0.009	0.009	0.007	0.006	0.012	0.008	0.009	0.010	0.008	0.008	0.011	0.008	0.007	0.009
WSAEs-LSTM	0.011	0.010	0.008	0.007	0.018	0.008	0.011	0.013	0.010	0.010	0.009	0.010	0.011	0.011

Table 4. The prediction accuracy in Nikkei 225 and S&P500 indices.

Year	Nikkei 225 index							S&P500 index						
	Year 1	Year 2	Year 3	Year 4	Year 5	Year 6	Average	Year 1	Year 2	Year 3	Year 4	Year 5	Year 6	Average
Panel A. MAPE														
CGBoost6	**0.011**	**0.009**	**0.013**	**0.009**	**0.010**	**0.013**	**0.011**	**0.009**	**0.008**	**0.006**	**0.005**	**0.007**	**0.007**	**0.007**
CGBoost	0.015	0.013	0.019	0.012	0.017	0.017	0.015	0.014	0.012	0.008	0.008	0.011	0.010	0.011
WSAEs-LSTM	0.024	0.019	0.019	0.019	0.018	0.017	0.019	0.012	0.014	0.010	0.008	0.011	0.010	0.011
Panel B. Correlation coefficient														
CGBoost6	**0.966**	**0.977**	**0.994**	0.957	**0.987**	**0.970**	**0.975**	**0.972**	**0.982**	**0.994**	**0.990**	0.953	**0.976**	**0.976**
CGBoost	0.943	0.958	0.990	0.932	0.971	0.957	0.958	0.938	0.966	0.988	0.976	0.892	0.954	0.952
WSAEs-LSTM	0.878	0.834	0.665	**0.972**	0.774	0.924	0.841	0.944	0.944	0.984	0.973	0.880	0.953	0.946
Panel C. Theil U														
CGBoost6	**0.008**	**0.006**	**0.009**	**0.006**	**0.007**	**0.009**	**0.007**	**0.006**	**0.005**	**0.004**	**0.003**	**0.005**	**0.005**	**0.005**
CGBoost	0.010	0.008	0.013	0.007	0.011	0.011	0.010	0.009	0.008	0.005	0.005	0.008	0.006	0.007
WSAEs-LSTM	0.016	0.013	0.013	0.013	0.012	0.012	0.013	0.009	0.010	0.006	0.005	0.008	0.006	0.007

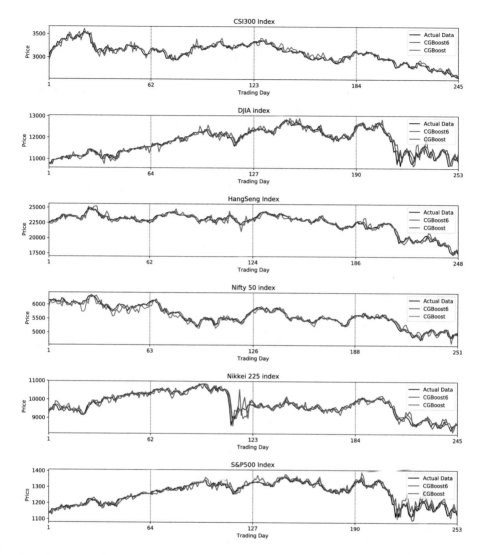

Fig. 4. Shows the actual curves and predicted curves from the our methods for six stock index from 2010.10.01 to 2011.09.30.

Figure 4 shows an example of the Year 1 predicted price from proposed model and the corresponding actual price. CGBoost6 is closer to the actual stock price time series than CGBoost and has lower volatility.

4   Conclusion and Future Work

In this paper we built a new predicting framework to forecast the next day stock price of six stock indices from the financial markets from different country. The process for building this predicting framework is: First, clipping the high value and normalizing the technical index and other feature. Second, using 1D resnet SAEs to denoise and reduce the dimension of features. Last, CGBoost was used to predict the next day price, this is a supervised manner. Our input features include the daily technical indicators, OHLC variables and macroeconomic variables. The main contribution of this paper is attempting to combine 1D resnet with GBoost, a kind of ensemble learning method in stock predict, and prove its performance. Besides, we successfully trains one model on different market and obtain a better prediction on the overall test set.

Future work could focus on increasing the diversity of based estimators. We may try to replace same construction of basic 1D CNNs with several different constructions, in order to improve the performance of CGBoost. Another interesting direction is to use CGBoost in other fields. CGBoost may be applicable to the sequence data including several time series of different features, such as weather forecast, traffic forecast and etc. CGBoost may be able to get better performance in these field.

Acknowledgment. This work was supported by the National Natural Science Foundation of China (No. 61673322, 61673326, and 91746103), the Fundamental Research Funds for the Central Universities (No. 20720190142), Natural Science Foundation of Fujian Province of China (No. 2017J01128 and 2017J01129), and the European Union's Horizon 2020 research and innovation programme under the Marie Sklodowska-Curie grant agreement (No. 663830).

References

1. Altay, E., Satman, M.H.: Stock market forecasting: artificial neural network and linear regression comparison in an emerging market. J. Financ. Manag. Anal. **18**(2), 18 (2005)
2. Bao, W., Yue, J., Rao, Y.: A deep learning framework for financial time series using stacked autoencoders and long-short term memory. PloS one **12**(7), e0180,944 (2017)
3. Bengio, Y., Lamblin, P., Popovici, D., Larochelle, H.: Greedy layer-wise training of deep networks. In: Advances in Neural Information Processing Systems, pp. 153–160 (2007)
4. Chen, T., Guestrin, C.: XGBoost: a scalable tree boosting system. In: Proceedings of the 22nd ACM SIGKDD International Conference on Knowledge Discovery and Data Mining, pp. 785–794. ACM (2016)
5. Diao, R., Chao, F., Peng, T., Snooke, N., Shen, Q.: Feature selection inspired classifier ensemble reduction. IEEE Trans. Cybern. **44**(8), 1259–1268 (2013)
6. Emenike, K.O.: Forecasting Nigerian stock exchange returns: evidence from autoregressive integrated moving average (ARIMA) model. SSRN Electron. J. **2010**, 1–19 (2010)
7. Guo, Z., Wang, H., Liu, Q., Yang, J.: A feature fusion based forecasting model for financial time series. PloS One **9**(6), e101,113 (2014)
8. He, K., Zhang, X., Ren, S., Sun, J.: Deep residual learning for image recognition. In: Proceedings of the IEEE Conference on Computer Vision and Pattern Recognition, pp. 770–778 (2016)

9. Hinton, G.E., Salakhutdinov, R.R.: Reducing the dimensionality of data with neural networks. Science **313**(5786), 504–507 (2006)
10. Hsieh, T.J., Hsiao, H.F., Yeh, W.C.: Forecasting stock markets using wavelet transforms and recurrent neural networks: an integrated system based on artificial bee colony algorithm. Appl. Soft Comput. **11**(2), 2510–2525 (2011)
11. Kim, K.J.: Toward global optimization of case-based reasoning systems for financial forecasting. Appl. Intell. **21**(3), 239–249 (2004)
12. Lawrence, R.: Using neural networks to forecast stock market prices. University of Manitoba 333 (1997)
13. Martinez, L.C., da Hora, D.N., Palotti, J.R.d.M., Meira, W., Pappa, G.L.: From an artificial neural network to a stock market day-trading system: a case study on the BM&F BOVESPA. In: 2009 International Joint Conference on Neural Networks, pp. 2006–2013. IEEE (2009)
14. Nassirtoussi, A.K., Aghabozorgi, S., Wah, T.Y., Ngo, D.C.L.: Text mining of news-headlines for forex market prediction: a multi-layer dimension reduction algorithm with semantics and sentiment. Expert Syst. Appl. **42**(1), 306–324 (2015)
15. Porshnev, A., Redkin, I., Shevchenko, A.: Machine learning in prediction of stock market indicators based on historical data and data from Twitter sentiment analysis. In: 2013 IEEE 13th International Conference on Data Mining Workshops, pp. 440–444. IEEE (2013)
16. Rui-Rui, X., Tian-Lun, C., Cheng-Feng, G.: Nonlinear time series prediction using ls-svm with chaotic mutation evolutionary programming for parameter optimization. Commun. Theor. Phys. **45**(4), 641 (2006)

Urban Village Identification from City-Wide Satellite Images Leveraging Mask R-CNN

Xueyi Wang[2], Tianqi Xie[1], and Longbiao Chen[1(✉)]

[1] Fujian Key Laboratory of Sensing and Computing for Smart City,
School of Informatics, Xiamen University, Xiamen, China
longbiaochen@xmu.edu.cn
[2] University of Hertfordshire, Hertfordshire, UK

Abstract. Urban villages emerge with the rapid urbanization process in many developing countries, and bring serious social and economic challenges to urban authorities, such as overcrowding and low living standards. A comprehensive understanding of the locations and regional boundaries of urban villages in a city is crucial for urban planning and management, especially when urban authorities need to renovate these regions. Traditional methods greatly rely on surveys and investigations of city planners, which consumes substantial time and human labor. In this work, we propose a low-cost and automatic framework to accurately identify urban villages from high-resolution remote sensing satellite imagery. Specifically, we leverage the Mask Regional Convolutional Neural Network (Mask-RCNN) model for end-to-end urban village detection and segmentation. We evaluate our framework on the city-wide satellite imagery of Xiamen, China. Results show that our framework successfully detects 87.18% of the urban villages in the city, and accurately segments their regional boundaries with an IoU of 74.48%.

Keywords: Urban village · Image segmentation · Mask-RCNN ·
Deep learning · Urban computing · Ubiquitous computing

1 Introduction

Urban village refers to the residential area that is lagging behind the pace of development of urbanization, free from the management of modern cities, and with low living standards in the process of urban development [1]. In China, These villages are used to be gathered by the old cottages remained from years ago, and they are commonly inhabited by low income and transient communities. As a results, urban villages continuously suffer from overcrowding and social problems [2]. Accurately identifying the regional boundaries of urban villages is important to the management and planning of the development of urban villages, including real estate reform, road renovation, and sanitation management.

© Springer Nature Switzerland AG 2020
Z. Ju et al. (Eds.): UKCI 2019, AISC 1043, pp. 166–172, 2020.
https://doi.org/10.1007/978-3-030-29933-0_14

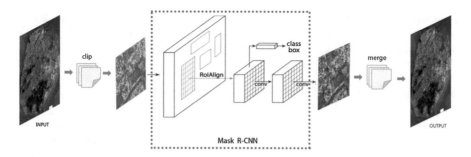

Fig. 1. Framework overview.

In the past, detecting the location and finding the regional boundary of an urban village mainly rely on field surveys of city planners and their local knowledges, which is usually time-consuming and inaccurate for a comprehensive understanding of city-wide urban villages [1]. Recently, the ubiquitousness of high resolution satellite images and the rapid development of deep learning techniques provide us with new opportunities to identifying urban village regions in a *low-cost and automatic* manner. Specifically, Mask Regional Convolutional Neural Networks (Mask-RCNN) has a very good performance in target detection and instance segmentation from images [3]. In this work, we propose an end-to-end framework to detect urban villages and segment their boundaries from city-wide satellite images using the Mask-RCNN architecture. The **main contributions** of this work include:

- To the best of our knowledge, this is the first work on urban village detection and segmentation from satellite imagery, which provides a low-cost alternative for urban planning and management.
- We propose an end-to-end framework to detect and segment urban villages from city-wide satellite images. We first clip a large city-wide satellite image into small patches, and then collect the urban village mask labels using a *crowdsourcing* platform. We train a Mask-RCNN model on a randomly-selected patch set and segment the urban village masks on the left patches. Finally, we merge all the patches to obtain the regional boundaries of all the urban village in the city-wide satellite image.
- We conduct real-world evaluation in Xiamen, China. Results show that our framework successfully detects the urban villages in the city with a precision of 90.67% and a recall of 87.18%, and accurately segments their regional boundaries with an IoU of 74.48%.

2   Framework Overview

The overview of the proposed framework is shown in Fig. 1. First, we clip a large city-wide satellite image into small patches, and mask the urban villages in the patches using a crowdsourcing platform. Then, we train a Mask-RCNN model on a set of randomly-selected patches, and predict the urban village masks for the other patches. Finally, we merge all the patches to obtain the regional boundaries of all the urban villages in the city-wide satellite image.

3   City-Wide Satellite Image Clipping

High-resolution satellite images can be obtained from various geographic information services, such as Google Earth[1]. However, city-wide satellite imageries are usually very large. For example, the satellite image of Xiamen island with a resolution of 0.5 m can be as large as 1.80 GB. Directly processing such a large image for urban village identification is computationally intractable. Therefore, we first clip a city-wide satellite image into small patches for training.

Specifically, we employ the Python Imaging Library (PIL)[2] for satellite imagery clipping, which has proven to be efficient and can preserve the geographical coordinates embedded in the satellite imagery. We determine the size of each patch to be $500 \times 500\,m^2$ squares based on previous studies on the geographic spans of typical urban village [2].

4   Urban Village Mask Labeling

Providing an sufficient number of samples for machine learning tasks is the basis of ensuring model performance [4]. However, labeling urban village masks in each patch is time consuming and requires domain knowledge. Therefore, we exploit the crowdsourcing mechanism to outsource the label masking tasks to the massive qualified crowd workers.

First, we recruit a group of participants with incentives from Xiamen University and train them with background knowledge about urban villages. We then develop a web-based crowdsourcing platform to randomly assign patches to participants. Specifically, each patch is assigned to three participants for cross validation. We integrate an open-source image mask labeling tool *labelme*[3] into the platform to facilitate the masking process. Finally, we obtain the urban village masks for all the patches via the crowdsourcing platform. Figure 2 shows an example of the collected urban village mask labels near a train station.

5   Urban Village Detection and Segmentation

In this step, we train a Mask-RCNN model to detect and segment urban villages from each image patch. Mask-RCNN provides an end-to-end solution to efficiently detects objects in an image while simultaneously generating a high-quality segmentation mask for each instance [3].

---

[1] https://www.google.com/earth.
[2] https://www.pythonware.com/products/pil/.
[3] https://github.com/wkentaro/labelme.

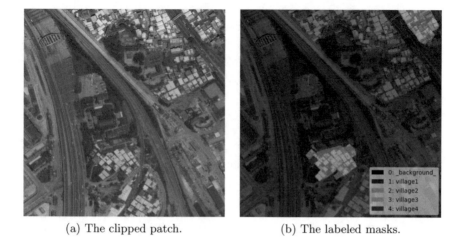

(a) The clipped patch.   (b) The labeled masks.

Fig. 2. An illustrative example of the collected urban village masks labeled by the crowdsourcing participants.

Specifically, we randomly select a small set of patches with urban village masks as the training set, and use them to train a Mask-RCNN model. The selection of the training set size is based on repeated experiments. We then exploit the trained model to detect urban villages and predict their corresponding masks simultaneously. Afterwards, we merge all the patches and masks into one large image to obtain a city-wide view of urban village distribution. We also conduct several auxiliary image processing steps to eliminate unnecessary boundaries between adjacent patches.

6   Evaluations

6.1   Experiment Settings

We evaluate our framework using high-resolution satellite imagery of Xiamen Island from Google Earth. Table 1 shows the details of the collected imagery. We clip the whole imagery into $500 \times 500\,\mathrm{m}^2$ patches and obtain 650 patches. We randomly select 32 patches (5% of all patches) with urban villages to train

Table 1. Imagery description

Items	Specification
Northwest coordinates	[24.561492, 118.064736]
Southeast coordinates	[24.423240, 118.198513]
Geographic span	$13.54\,\mathrm{km} \times 15.39\,\mathrm{km}$
Satellite image resolution	$0.54\,\mathrm{m}$

(a) Baijiacun Village   (b) Zengcuoan Village

Fig. 3. Two examples of identified urban villages.

the Mask-RCNN model, and predict the masks on the other patches. We deploy our framework on a server with an nVIDIA GeForce GTX 1080Ti graphic card and 16 GB RAM.

6.2   Evaluation Metrics

Detection Accuracy: if an urban village in the ground truth has a spatial overlapping with the detected instance, we mark the detection as a hit, and otherwise a miss. Based upon this, the precision and recall are calculated as:

$$precision = \frac{|\{\text{truth instance}\} \cap \{\text{detected instance}\}|}{|\{\text{detected instance}\}|} \tag{1}$$

$$recall = \frac{|\{\text{ground-truth instance}\} \cap \{\text{detected instance}\}|}{|\{\text{ground-truth instance}\}|} \tag{2}$$

Segmentation Accuracy: we adopt the popular Intersection over Union (IoU) metric to evaluate the segmentation accuracy over the city-wide imagery,

$$IoU = \frac{|\{\text{ground-truth pixel}\} \cap \{\text{detected pixel}\}|}{|\{\text{ground-truth pixel}\}|} \tag{3}$$

6.3   Evaluation Results

Figure 4 shows the result of city-wide urban village detection and segmentation in Xiamen Island. Our framework successfully identifies the urban villages with various sizes and locations. For example, Fig. 3 demonstrates two examples of

Fig. 4. Result of city-wide urban village detection and segmentation in Xiamen.

identified urban villages. Specifically, our framework detects 75 urban villages, among which 69 of them are found in the 78 ground truth instances, achieving a detection precision of 90.67% and a recall of 87.18%, respectively. For segmentation accuracy, our framework achieves an IoU of 74.48%, which is quite good for city-wide image segmentation.

7   Conclusion

In this work, we detect the urban villages and segment their geographic boundary from city-wide high-resolution satellite imagery. We propose a framework to exploit the state-of-the-art Mask-RCNN model for instance detection and segmentation in an end-to-end manner. The proposed framework is evaluated in Xiamen Island and achieves accurate detection and segmentation results. In the future, we plan to explore how the villages boundaries change with time.

References

1. Magnaghi, A.: The Urban Village: A Charter for Democracy and Sustainable Development in the City. Zed Books, London (2005)
2. Brindley, T.: The social dimension of the urban village: a comparison of models for sustainable urban development. Urban Des. **8**(1), 53–65 (2003)
3. He, K., Gkioxari, G., Dollár, P., Girshick, R.: Mask R-CNN. arXiv:1703.06870 [cs], March 2017
4. Bishop, C.M., et al.: Pattern Recognition and Machine Learning, vol. 4. Springer, New York (2006)

Surface Crack Detection Using Hierarchal Convolutional Neural Network

Davis Bonsu Agyemang[(⊠)] and Mohamed Bader[(⊠)]

University of Portsmouth, Portsmouth, Hampshire, UK
up785062@myport.ac.uk, mohamed.bader@port.ac.uk

Abstract. Cracks on surface walls may imply that a building possesses problems with its structural integrity. Evaluating these types of defects needs to be accurate to determine the condition of the building. Currently, the evaluation of surface cracks is conducted through visual inspection, resulting in occasions of subjective judgements being made on the classification and severity of the surface crack which poses danger for customers and the environment as it not being analysed objectively. Previous researchers have applied numerous classification methods, but they always stop their research at just being able to classify cracks which would not be fully useful for professionals such as surveyors. We propose building a hybrid web application that can classify the condition of a surface from images using a trained Hierarchal-Convolutional Neural Network (H-CNN) which can also decipher if the image that is being looked is a surface or not. For continuous improvement of the H-CNN's accuracy, the application will have a feedback mechanism for users to send an email query on incorrectly classified images which will be used to retrain the H-CNN.

1 Introduction

A variety of methods have been used to detect surface cracks in walls. One method is known as "Edge Detection" which detects lines or edges that are present in the image as stated by Amer and Absushaala (2015, p. 1). One of the common edge detection techniques is called the "Sobel Operator" which detects surface cracks by applying a filter (matrix of values) on the image that contains the crack as stated by Amer and Absushaala (2015, p. 1). Machine and deep learning methods have also been applied in this area due to its capabilities to learn the images' pixels containing the defect to classify new images according to da Silva and de Lucena (2018, p. 1).

Classifying a building based on the walls as safe or dangerous is extremely vital for surveyors and their customers. This task requires intensive analysis of each surface wall found within the building as the thickness/thinness and shape of the crack can determine its structural condition which is severe if misinterpreted or incorrectly recorded, leading to possible persecution of the surveyor according to Danso (2018) and Neale (2018). Surveyors currently inspect surfaces manually through visual examination as there is no current tool that could aid them in recording and classifying surface conditions from our knowledge. Due to this implication, a tool must be implemented to mitigate these weaknesses according to Kunal and Killemsetty (2014, p. 64).

© Springer Nature Switzerland AG 2020
Z. Ju et al. (Eds.): UKCI 2019, AISC 1043, pp. 173–186, 2020.
https://doi.org/10.1007/978-3-030-29933-0_15

This paper proposes a solution of creating an hybrid web application using Hierarchical Convolution Neural Networks (H-CNN) consisting of 2 CNNs (Convolutional Neural Networks) in which the first CNN classifies: random objects (not a surface), blank (blank surface), thick cracks, thin cracks the second CNN classifies: horizontal, vertical and diagonal cracks. This was achieved by using the combination of surface crack datasets collected by Özgenel (2018) and SDNET2018 image dataset in Dorafshan et al. (2018) in which the image dataset for home objects from Caltech (2006) was used for the random object's class. The reason why 2 surface crack datasets were used was to train the H-CNN on a variety of surface cracks which the 2 image datasets possessed, this would hopefully help it classify better. We decided to use the Caltech image dataset in Caltech (2006) as it represents the objects that would be present in a building, so when the surveyor accidentally takes a picture of a bottle, for instance, the H-CNN should be able to know that the image is not a surface. From our knowledge, no previous researchers have tried to classify the exact classifies specified for this paper, making our chosen classes novel.

The application will also have the capability to allow surveyors to give a query via the "feedback mechanism" if the application incorrectly classified an image to promote continuous improvements for the accuracy of the application. The user's query will be sent to us via email which will have the attachment of the misclassified image with the information on what classification should have been according to the user. That image will be used to retrain the appropriate H-CNN level to improve the model's accuracy. When building the H-CNN each CNN will be measured against the test accuracies. The whole application will be build using Python, Flask Keras TensorFlow backend for the back-end and the creation for the H-CNN, and HTML (Hypertext Mark-up Language), Bootstrap CSS (Cascading Style Sheet), and JavaScript will be used for the front-end (User interface).

2   Literature Review

2.1   H-CNN

One of the most challenging problems in CNN is when 2 or more classes share visual similarities according to Seo and Shin (2019, p. 331). This is difficult due to CNN being a discriminative neural network, meaning that it distinguishes the correct classification amongst the other classes according to the works of Dai and Wu (2015). So, classifying between an apple and an orange is much challenging than an apple and a car, due to apples and oranges looking similar. An H-CNN (also known as HD-CNN) which solves this problem by first separating the classes that are easier to differentiate from one another for instances an apple and a bus. This process is known as the initial coarse classifier according to Seo and Shin (2019, p. 331). Once the image has passed the coarse classifier is then passed to the fine classifier to generate the final classification.

However, from our research, many researchers who dealt with surface crack detection used 1 deep CNN model which is in the next sub-sections of this chapter. This could be due to the main disadvantages of the H-CNN having to train each CNN within the hierarchy while also having to fine-tune the hyperparameters for each CNN architecture such as deciding the number of convolution filters. This could be time-consuming due to the empirical nature of configuring hyperparameters. H-CNN can only have a maximum of 2 CNNs in its hierarchy as it only uses a coarse and fine classifier category to perform classifications. Nevertheless, this leads to Branch Convolution Neural Network (B-CNN) or Multi-branch Convolutional Neural Network (MB-CNN) which is multi-layered H-CNN that produces multiple output layers probability predictions from the coarse classifier levels to the fine classifier levels and based on the total predictions a final classification can be made with more granular information according to Zhu and Bain (2017, p. 2) and Aslani et al. (2018, pp. 1–2).

When traditional CNNs are only trained on 1 topic for instance fruits dataset (banana, oranges and apple classes), it only knows that these classes exist so when a user inputs an image of a bus it can only classify it based on available classes it was trained upon, meaning that it will classify the bus image based on the class that resembles it the closest. This is problematic when it comes to defect detections in building as in future if CNNs are going to be attached to robots to capture and classify images from places that are too dangerous places for humans to enter such as severely damaged or polluted buildings, it is important that can recognise if an image does not belong to any of its classes otherwise it will give false information to the user. Having an H-CNN, one could add another class that contain random images that do not relate to the topic of the application, so the H-CNN's first layer (coarse classifier) can check if the image is related to the topic for instance fruit if it is not classified it as the random image class. The only issue with this approach is that the developer would need to ensure that none of the images in the random image class dataset resembles the topic the H-CNN wishes to classify. This would require one to manually pre-process the image dataset. Still, one of the future use-cases for H-CNN could be for anomaly detection.

2.2 Related Work

As mentioned before in this paper, neglecting a building's wall condition can pose problems for the surveyor, the customer and the environment. The problems with current literature regarding this area relates to the usability for surveyors and the lack of in-depth of information derived for the classifications made from their proposed methods.

Ellenberg et al. (2014, p. 1788) used a drone to capture images from masonry building walls to detect surface cracks by using MATLAB to apply the edge detection technique: "Prewitt edge detection" which are vertical and horizontal filters containing values that removes the background of the image to only preserve the edges (surface crack) according to Adlakha et al. (2016, pp. 1483–1484). "Percolation" was used to reduce the noise in the image for instance extremely bright images according to Ellenberg et al. (2014, p. 1793). Previously without using drones to capture images, Hu et al. (2012, pp. 597–598) also used "Prewitt edge detection" but combined it with "adaptive threshold" which dynamically makes some pixels values of an image more prominent if these values are above the threshold value, and less prominent if they are below the threshold. This preserves the surface cracks in the image.

One problem Ellenberg et al. (2014, pp. 1793–1794) faced was the quality of their images once the percolation algorithm was applied since it made the surface cracks appear thinner which affected their methods ability to classify accurately.

Both Ellenberg et al. (2014, pp. 1793–1794) and Hu et al. (2012, p. 599) edge detection methods struggled to classify cracks when foreign objects were presents in the image, for example, a surface crack that was near the "edge of a window". As mentioned earlier in this paper, this problem could have been mitigated by using a pooling technique if CNNs were used instead of edge detection as it provides spatial invariance, meaning it is not affected by the position of the surface crack it just needs to be present. Failing to capture a defect due to spatial invariance will lead to unreliable building inspection as Ellenberg et al. (2014, pp. 1793–1794) and Hu et al. (2012, p. 599) would limit how surveyors can take images of surface cracks. Using drones to capture images could also pose issues like bad weather conditions can damage the drone and limit users' control of the drone according to Ellenberg et al. (2014, p. 1794) negatively influencing the quality of the images. This leads to bad quality training dataset and the cost of maintenance or being forced to purchase a new one when the drone is damaged, making this method not cost-effective.

Recently, Hoang (2018a, b, p. 1) combined the edge detection "Otsu method "and "Min-Max Gray Level Discrimination (M2GLD)" to detect surface cracks. The Otsu method automatically finds the best threshold value for an image according to Otsu (1979, p. 66) and M2GLD can reduce the grayscale intensity of the image, making the surface cracks appear darker in the image, and increase the grayscale intensity to make non-surface cracks to appear lighter, enabling the image to be distinguishable according to Hoang (2018a, b, p. 5). Hoang (2018a, b, p. 9) method produced good results as it was able to detect the test images accurately. However, for a user to use their method they would need to fine-tune 2 parameters which are the ratio and the margin parameter. Hoang (2018a, b, p. 9) seemed to assume that users would have background knowledge regarding these 2 parameters which is not the case, this tool may be complex for some users which will result to them not using it.

Kim and Cho (2018, p. 1) use CNN for surface crack detection on walls using the AlexNet architecture consisting of 5 "convolutional layers followed by max-pooling layers, and three fully-connected layers". The CNN was trained on a multi-class dataset containing classes such as "cracked"," joint/edge multiple lines (ML)", "joint/edge single line (SL)", "intact surface" and "plant". The results of this method were out-standing, achieving a test accuracy of 96.64% (3.36% error rate). However, the problem with the works of Kim and Cho (2018, p. 1) and even the researchers men-tioned earlier in this chapter, is that they do not realise that user such as surveyors also need to report to their customer about the severity based on the aesthetics of the surface crack such as shape and thickness/thinness. Hoang's (2018a, b, pp. 1–4) other work in Hoang (2018a, b, pp. 1–4) almost achieves this by treating this as multi-class problem using SVM (Support Vector Machine) algorithm to detect "longitudinal crack", "transverse crack", "diagonal crack", "spall damage", "intact wall" (Hoang 2018a, b, p. 1). SVM creates a hyperplane that is a line which separates each class based on their characteristics ensuring that separation between classes is at the maximal distance possible. Hoang (2018a, b, p. 1) attained a test accuracy of 85.33% (14.67% error rate) using 100 image samples per class. Even though Hoang's work in Hoang (2018a, b, pp. 1–4) method treated the problem as a multi-class task, the researcher did not think about training their model to detect if the image is not a surface at all as the CNN cannot recognise if an image does not belong to its classes. For this, an H-CNN would be required to achieve that functionality.

3 Methodology

3.1 H-CNN Architecture Used to Detect Surface Crack

This section illustrates the H-CNN architecture that will be used for this paper as well as how it will be integrated with the front-end using Python, Flask, HTML, Bootstrap CSS and JavaScript. This architecture uses tradition CNN components such as the convolutional layer, max-pooling and fully connected layer for each H-CNN level. The CNN at each level transforms any uploaded image input to the size of 90×90. The rationale behind this size was to decrease the latency of the computations performed by the trained H-CNN when users upload a new image, as mobile systems have less computational power in comparison to the desktop. Using large image sizes such as 224×224 can increase the accuracy of the CNN according to Wu et al. (2015, p. 2). However, this requires many computational resources which will increase the individual CNNs training time due to having to handle multiple images at a large size. This will also slow down the application's ability to classify when used in real-time as performing the CNN operations takes longer on larger images.

Rectifier Linear Unit (RELU) activation function was used in each convolution layers and hidden layers of the fully-connected layers for the purpose of non-linearity to allow the 2 CNNs to learn complex features within an image.

Since the application handles multiple classes, a SoftMax layer is used within the final fully-connected output layers. SoftMax distributes the probabilities throughout each class in which the class with the highest probability will be CNN's classification result according to Lucke and Sahani (2007, pp. 657–659). Below in Fig. 1 is a depiction of the H-CNN architecture and Tables 1 and 2 shows hyperparameter structure and values used in the CNN levels:

The image classes for CNN1 were:

- Random image: this is any image that is not related to surfaces. This will be used to help the application to detect images that are not related to surfaces for instance a spoon or a shoe.
- Blank: this is a surface which possess no cracks.
- Thin cracks: these are surfaces that contain thin hairline cracks.
- Thick cracks: these are surfaces that contain large thick cracks.

If the CNN1 detects a thin or thick crack in the image, CNN2 will then perform classification to determine the shape of the crack. The image classes of CNN2 are:

- Vertical crack
- Horizontal crack
- Diagonal crack

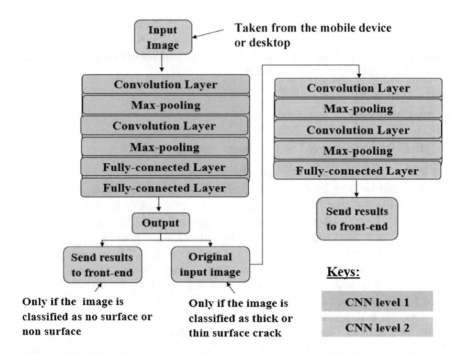

Fig. 1. H-CNN architecture interaction with the surveyor's mobile device or desktop.

Table 1. CNN1 (coarse classifier) parameters are orders as the architecture in Fig. 1.

CIMN1 parameters	Description
Convolution layer	65 convolution filters, size: 5×5, activation function: RELU
Max-pooling	Size: 5×5
Convolution layer	65 convolution filters, size: 5×5, activation function: RELU
Max-pooling	Size: 5×5
Fully-connected layer	Hidden nodes (units): 100, activation: RELU
Fully-connected layer	Hidden nodes (unit): 20, activation: RELU
Output	Activation: softmax

Table 2. CNN2 (fine classifier) parameters are orders as the architecture in Fig. 1.

CNN2 parameters	Description
Convolution layer	32 convolution filters, size: 3×3, activation function: RELU
Max-pooling	Size: 5×5
convolution layer	64 convolution filters, size: 3×3, activation function: RELU
Max-pooling	Size: 5×5
Fully-connected layer	Hidden nodes (units): 50, activation: RELU
Output	Activation: softmax

Below in Fig. 2 shows pseudo code of how the H-CNN will classify images from users:

Classification1 = CNN1 (image)

If Classification1 = = thin or thick crack:

    Classification2== CNN2 (image)

      Print(Classification1+Classification2 + Severity rating)

Else:

    Print(Classification1+ Severity rating)

Fig. 2. Pythonic pseudo code for the H-CNN.

3.2 Training and Testing Process for the CNNs for the H-CNN

The H-CNN was trained on 3 images datasets, containing a total of 920 (230 per class) images for CNN1 and a total of 390 (130 per class) images for CNN2. The batch size of CNN1 was 20 which meant that CNN1 would learn the patterns of 20 images at a time during training (Radiuk 2017, p. 20). CNN1's number of epochs (cycles) was 15 in which in 1 cycle it would learn 20 images from the training set. CNN2 used a batch size of 10 for 10 epochs during the training process. The batch sizes for the test set was the same size for CNN1 and CNN2 as Keras TensorFlow backend simultaneously classifies the images in the test set during training which meant that the epochs were also the same. We chose these batch sizes and epochs to rapidly train the CNNs as in future it needs to be able to retrain on the incorrectly classified images from the "feedback mechanism" at a fast rate so that users do not have to wait for long to use the updated H-CNN. Data augmentation was used on the training set images, which transformed the images to help the CNNs learn the patterns of the images from a different position which is important as the user may take a photo from different angles. The optimizer used for the 2 CNNs was Adam at learning rate 0.001 to speed up the training process, this is important for retraining the H-CNN in the future at a fast rate from the images gathered from the feedback mechanism. The best fully connected layer weights and convolution filter values for each CNN will be saved using keras' "ModelCheckPoint" function. This ensures that the CNN with the highest test accuracy is saved for the H-CNN.

4 H-CNN Test Accuracy Results

Testing the architectures through the performance metrics specified in the introduction was beneficial as it enabled us to see if the CNNS would classify correctly most of the time. Table 3 and Figs. 3 and 4 shows the training and test accuracy and loss for CNN 1 and Table 4 and Figs. 5 and 6 shows the training and test accuracy and loss for CNN 2:

Table 3. CNN1 training and test values.

CNN level 1 training and test values per epoch				
Epochs	Training accuracy	Test accu- racy	Trainning loss	Test loss
1	0.7963	0.7862	0.5787	0.5903
2	0.8822	0.9122	0.3327	0.2673
3	0.9141	0.8844	0.2426	0.3226
4	0.9272	0.9240	0.2017	0.2511
5	0.9349	0.9491	0.1735	0.1617
6	0.9407	0.9088	0.1584	0.2571
7	0.9381	0.9406	0.1658	0.1852
8	0.9502	0.9510	0.1340	0.1437
9	0.9532	0.9558	0.1277	0.1453
10	0.9560	0.9541	0.1196	0.1496
11	0.9524	0.9131	0.1325	0.2667
12	0.9553	0.9606	0.1210	0.1718
13	0.9610	0.9519	0.1040	0.1497
14	0.9646	0.9727	0.0978	0.1084
15	0.9622	0.9713	0.1011	0.1084

Saved with ModelCheckPoint function

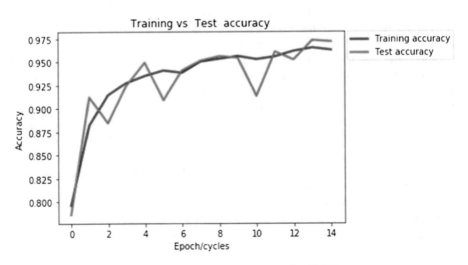

Fig. 3. Training vs test accuracy for CNN1.

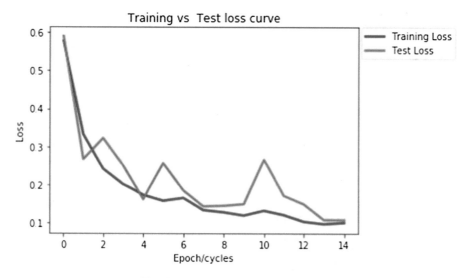

Fig. 4. Training vs test loss for CNN1.

Table 4. CNN2 training and test values.

CNN level 2 training and test values per epoch				
Epochs	Training accuracy	Test accuracy	Trainning loss	Test loss
1	0.5783	0.5972	0.9177	0.8017
2	0.7876	0.7361	0.5494	0.7457
3	0.8423	0.8750	0.4184	0.5386
4	0.8829	0.8889	0.3245	0.3805
5	0.9157	0.9167	0.2511	0.3068
6	0.9328	0.9167	0.2081	0.2634
7	0.9486	0.9444	0.1642	0.2556
8	0.9602	0.9444	0.1335	0.2075
9	0.9684	0.9444	0.1060	0.2357
10	0.9737	0.9444	0.0943	0.2033

Saved with
ModelCheckPoint
function

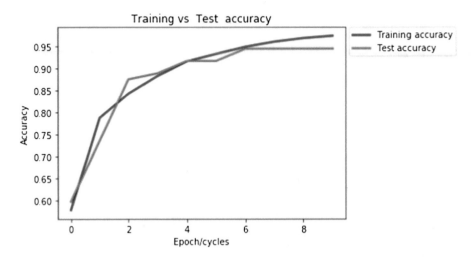

Fig. 5. Training vs test accuracy for CNN2.

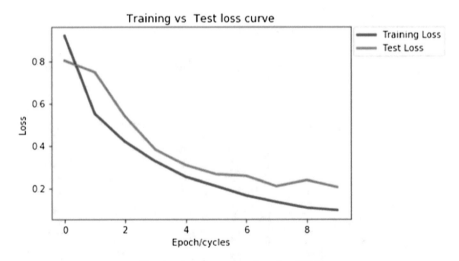

Fig. 6. Training vs test loss for CNN2.

4.1   Discussion

The best test accuracy obtained using the "ModelCheckPoint" function for CNN1 and CNN2 were: 0.9727 (97.3%) in epoch 14 and 0.9444 (94.4%) in epoch 7. As it can be seen in Table 4 from epoch 7 to 10 the test accuracy stops improving but the training accuracy continuously increased for the same epoch range having a standard deviation of 0.0109 (1.09%). The possible causes of the lack of increase for the test accuracy from epoch 7 to 10 in Table 4 could be due to slight overfitting. The test loss at epoch 14 for CNN1 was exceptional as it signified how close the CNN1's classification were from the actual labels of the images in the test set which meant that at epoch 14 CNN1 classified majority of test images correctly with minimal mistakes.

Using the ModelCheckPoint function was advantageous particularly for CNN1 as after the 14th epoch the test accuracy decreased by 0.14% and test loss was also increased by 0.33%, this meant if the ModelCheckPoint function was not used, the final epoch would have been saved instead, losing the opportunity of having a better performing model. The average test accuracy of the H-CNN (combination of CNN1 and CNN2) was 95.85% (4.15% error rate), considering the lack of image samples for CNN at level 2.

5   Surface Crack Detector H-CNN Capabilities

The integration process of the H-CNN to the front end was achieved through the Flask web framework. Flask enabled output of the H-CNN to be made visible in the HTML, CSS and JavaScript front-end. In Fig. 7 shows the classification functionality and feedback mechanism:

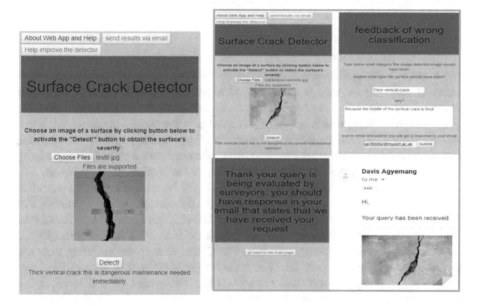

Fig. 7. Demonstration of the Surface Crack Detector classification functionality (orange highlighted screen is the classification functionality and the green highlighted image is the feedback mechanism).

The classification functionality directly accesses the user's mobile phone camera to take the picture but if the user uses the desktop version it will access the file explorer. The feedback mechanism could have been automated in terms of allowing users to update the model without the author and the co-author checking if the H-CNN needed to be retrained based on the user's query. However, the dangerous of this approach is the users can potentially decrease the accuracy of the H-CNN if they decided to place the wrongly classified image in the wrong class for retraining, hence why it needs to be checked before to mitigate that risk. Using the feedback mechanism will enable the surface crack detector to continuously improve its accuracy by learning from its mistakes which is an area that many researchers such as the ones mentioned earlier in this paper have not ventured.

Overall, the application performed exceptionally well in terms of accuracy, but occasionally the application classified an input incorrectly especially if the uploaded image was not focused and the detecting thick or thin cracks was challenging as the way a user took the picture influenced that, hopefully in the future that will be fixed by using an algorithm that can determine depth of the image to determine the width of the crack.

This application has the potential to be beneficial for users such as surveyors, as it does not only provide increased efficiency in evaluating surface conditions, but its use cases can also be to help users who suffer from partial vision problems to still conduct inspections. This is the link for the of the application if one wishes to use the application: https://surfacecrackdetector.herokuapp.com.

6   Conclusion

The benefit of the application is that users such as surveyors will be able to record cracks at rapid rate as they would not need to manually record the conditions. The reason why that benefit is important is that users such as surveyors are required to write a report on a building which is a long process, but if the surface crack detection was used they would be able to quickly take pictures of the building, send the classification results to themselves and go back to the office to elaborate on the classification made by the application for their report. This would improve the work flow of surveyors.

This paper presents the creation of an H-CNN based surface crack detection application. The H-CNN was trained on 3 images datasets, containing a total of 920 (230 per class) images for CNN1 and a total of 390 (130 per class) images for CNN2. Evaluation and analysis were achieved by testing the individual CNNs against its test set to monitor the test accuracy and loss. The CNNs were evaluated using their test accuracies to see if they would be appropriate for the H-CNN. The front-end end was built for the usability for users using Flask, HTML, Bootstrap CSS and JavaScript. It was concluded that the application has many use-cases, and with the "feedback mechanism," this application has the potential to become more accurate over time for continuous improvement.

7 Future Works

The authors hope to implement features such as to classify other defects such as damps and mould in buildings and to have the ability to classify a batch of images at once for users who wish to classify more images. A surveying Master lecturer asked the authors if in future when the application becomes more established that it can be merged with his augmented reality tool which can look at parts of a building such as a wall to inform the user on the last time it was maintained. Combining the surface crack detection application will help his augmented reality tool to see if a surface within a building needs maintenance.

References

Adlakha, D., Adlakha, D., Tanwar, R.: Analytical comparison between sobel and prewitt edge detection techniques. Int. J. Sci. Eng. Res. **7**(1), 1482 (2016)

Amer, H.M., Abushaala, M.A.: Edge detection methods. In: 2015 2nd World Symposium on Web Applications and Networking (WSWAN), p. 1. IEEE, Sousse (2015)

Aslani, S., Dayan, M., Storelli, L., Filippi, M., Murino, V., Rocca, M., Sonaa, D.: Multi-branch convolutional neural network for multiple sclerosis lesion segmentation. Neuroimage 1–2 (2018)

Caltech: Home Objects dataset, 12 December 2006. Caltech: http://www.vision.caltech.edu/pmoreels/Datasets/Home_Objects_06/

Dai, J., Wu, Y.N.: Generative modeling of convolutional neural networks. In: The International Conference on Learning Representations (ICLR 2015), p. 1. The International Conference on Learning Representations (ICLR), San Diego (2015)

Danso, M.: Interview Validate Customer Requirements and Gain Advice. (D. Agyemang, Interviewer), 18 October 2018

Dorafshan, S., Thomas, R.J., Maguire, M.: SDNET2018: an annotated image dataset for non-contact concrete crack detection using deep convolutional neural networks. Data Brief **21**, 1664–1668 (2018)

Ellenberg, A., Kontsos, A., Bartoli, I., Pradhan, A.: Masonry crack detection application of an unmanned aerial vehicle. In: International Conference on Computing in Civil and Building Engineering, Florida, p. 1788 (2014)

Hoang, D.N.: Detection of surface crack in building structures using image processing technique with an improved Otsu method for image thresholding. Adv. Civ. Eng. 1 (2018a)

Hoang, D.N.: Image processing-based recognition of wall defects using machine learning approaches and steerable filters. Comput. Intell. Neurosci. 1 (2018b)

Hu, D., Tian, T., Yang, H., Xu, S., Wang, X.: Wall crack detection based on image processing. In: Third International Conference on Intelligent Control and Information Processing, p. 597. IEEE, Dalian (2012)

Kim, B., Cho, S.: Automated vision-based detection of cracks on concrete surfaces using a deep learning technique. Sensors **18**, 3452 (2018)

Kunal, K., Killemsetty, N.: Study on control of cracks in a structure through visual identification & inspection. IOSR J. Mech. Civil En. **11**, 64 (2014)

Lucke, J., Sahani, M.: Generalized Softmax networks for non-linear component extraction. In: 17th International Conference, pp. 657–659. International Conference on Artificial Neural Networks, Porto (2007

Maggiori, E., Tarabalka, Y., Charpiat, G., Alliez, P.: High-resolution image classification with convolutional. In: IEEE International Geoscience and Remote Sensing Symposium, p. 2. IEEE, Fort Worth (2017)

Neale, S: Capturing Requirements. (D. Agyemang, Interviewer), 12 October 2018

O'Shea, K., Nash, R.: An introduction to convolutional neural networks. arXiv:1511.08458, 9 (2015)

Otsu, N.: A threshold selection method from gray-level histograms. IEEE Trans. Syst. Man Cybern. **9**(1), 66 (1979)

Özgenel, F.Ç.: Concrete Crack Images for Classification, 15 January 2018. mendeley: https://data.mendeley.com/datasets/5y9wdsg2zt/1

Radiuk, M.P.: Impact of training set batch size on the performance of convolutional neural networks for diverse datasets. Inf. Technol. Manag. Sci. **20**, 20–24 (2017)

Seo, Y., Shin, K.: Hierarchical convolutional neural networks for fashion image classification. Expert Syst. Appl. **116**, 328–339 (2019)

Sharma, N., Vibhor, J., Mishra, A.: An analysis of convolutional neural networks for image classification. Procedia Comput. Sci. **132**, 379 (2018)

da Silva, L., de Lucena, S.: Concrete cracks detection based on deep learning image classification. In: Proceedings, p. 1. Molecular Diversity Preservation International (MDPI), Brussels (2018)

Wu, R., Yan, S., Shan, Y., Dang, Q., Sun, G.: Deep image: scaling up image recognition. arXiv, 2 (2015)

Zhu, X., Bain, M.: B-CNN: branch convolutional neural network for hierarchical classification. arXiv:1709.09890 (Preprint), 2 (2017)

Optimisation and Classification

Hyper-parameter Optimisation
by Restrained Stochastic Hill Climbing

Rhys Stubbs$^{(\boxtimes)}$, Kevin Wilson, and Shahin Rostami

Faculty of Science & Technology, Bournemouth University,
Bournemouth BH12 5BB, UK
{i7433085,kwilson,srostami}@bournemouth.ac.uk
https://research.bournemouth.ac.uk/project/ciri/

Abstract. Machine learning practitioners often refer to hyper-parameter optimisation (HPO) as an art form and a skill that requires intuition and experience; Neuroevolution (NE) typically employs a combination of manual and evolutionary approaches for HPO. This paper explores the integration of a stochastic hill climbing approach for HPO within a NE algorithm. We empirically show that HPO by restrained stochastic hill climbing (HORSHC) is more effective than manual and pure evolutionary HPO. Empirical evidence is derived from a comparison of: (1) a NE algorithm that solely optimises hyper-parameters through evolution and (2) a number of derived algorithms with random search optimisation integration for optimising the hyper-parameters of a Neural Network. Through statistical analysis of the experimental results it has been revealed that random initialisation of hyper-parameters does not significantly affect the final performance of the Neural Networks evolved. However, HORSHC, a novel optimisation approach proposed in this paper has been proven to significantly out-perform the NE control algorithm. HORSHC presents itself as a solution that is computationally comparable in terms of both time and complexity as well as outperforming the control algorithm.

Keywords: Hyper-parameter optimisation · Global optimisation ·
Neuroevolution · Artificial neural networks · Random search ·
Stochastic hill climbing

1 Introduction

Neuroevolution (NE), a sub-field of Artificial Intelligence (AI), originated in the 1980s as shown by the work of Montana & Davis [21]; NE involves evolving and adapting Artificial Neural Networks (ANN) by employing Evolutionary Algorithms (EA) such as Genetic Algorithms (GA), a sub-field of Evolutionary Computation (EC), to optimise ANNs as well as solve complex Reinforcement Learning (RL) tasks [33]. As proposed by Yao [42], the Neural Network(s) produced by NE can be described as an Evolutionary Artificial Neural Network (EANN).

© Springer Nature Switzerland AG 2020
Z. Ju et al. (Eds.): UKCI 2019, AISC 1043, pp. 189–200, 2020.
https://doi.org/10.1007/978-3-030-29933-0_16

Traditionally, non-evolving ANN's often employ Gradient-based, Back Propagation algorithms [22,35,43] to learn and optimise performance by adjusting the ANN's hyper-parameters and weights. However, in traditional NE approaches the topology is chosen prior to the employment of the learning algorithm [33] and will not evolve during training. The ANN topology is an important factor to consider when planning to employ an ANN; this is due to the fact that the topology chosen plays a fundamental role in its functionality and performance [11], and therefore it's ability to produce a meaningful output.

Prominent empirical studies [26,33] have shown that EAs can be employed to dynamically evolve the weights as well as the overall topology of ANNs. The GA's used are often categorised depending on the type of evolution achieved. Algorithms that solely evolve the weights are known as conventional Neuroevolution algorithms [13,14,24,33]; whereas, algorithms that evolve the weights and topology are known as Topology and Weight Evolving Artificial Neural Network algorithms (TWEAAN) [33]. Many popular and modern NE algorithms have adopted the TWEANN approach. However, the topics of simplification and complexification are crucial factors that algorithms should consider. The goal of a NE algorithm is to find the best performing solution for a given problem while reducing unnecessary complexity of the topology.

It should be noted that NE is not the only approach used for hyper-parameter optimisation (HPO). It is in fact one of five popular HPO approaches that can be applied to ANN evolution and learning; moreover, NE can be used for optimisation of model parameters and hyper-parameters. Alternative approaches include grid search, random search, Bayesian optimisation and Gradient-based optimisation; the latter of these is most commonly employed and has become the primary approach for Neural Network (NN) parameter optimisation by applying a Gradient Descent and Back-propagation algorithm. The research will focus on the most popular, prominent and currently available NE algorithm implementations in order to evaluate and compare them against alternative hyper-parameter optimisation approaches.

2   Background Study

2.1   Neuroevolution

Neuroevolution is the process of evolving ANNs through evolutionary algorithms [17] and is inspired by the evolution of biological brains [32,33]. Various research papers have proven that a GA can be applied and used to find ANNs that consistently display improved learning speeds in comparison to a typical Feed Forward ANN [40]. NE establishes a fundamentally different approach to learning tasks in contrast to alternative techniques such as Back Propagation with Gradient Descent. NE is a phylogenetic learning approach that focuses on evolving a whole population of solutions, whereas conventional ontogenetic approaches focus on training a single solution [15,36]. Moreover, GAs have been proven to successfully perform tasks such as: connection weight training, architecture design, learning rule adaptation, input feature selection, connection weight initialisation and rule extraction [42].

Neuroevolution can have a number of evolutionary taxonomies, including weights, topology, and learning rules [42]. Moreover, there have been numerous discussions on the methods that evolve topologies; specifically, whether the algorithm should complexify or simplify a NN topology, also known as constructive and destructive algorithms [1,30,42,44]. This would entail adding or removing connections and/or neurons from the NN. In recent years, there has been an increasing amount of literature on NE, specifically, on how to improve the performance of the NNs evolved against increasingly difficult multi-objective problems such as Atari game playing, and complex real-world control/automation problems. Differentiable plasticity, a concept proposed by [20] seeks to address the problem of "learning to learn". It aims to decouple training and learning, as currently agents must be retrained for a different task than the one initially learnt. Despite the initial usage of GAs for NE, recent research has moved towards using Evolution Strategies (ES) for NN training [23], a class of black box optimisation algorithms whose performance, when applied to complex RL problems, rivals that of algorithms such as Q-Learning (DQN) and Policy Gradients (A3C), and which are also highly parallelisable [5,29]. A primary differentiating factor of ES is that solutions are encoded using real numbers [25]. Also, ES uses mutation in a fundamentally different way to GAs and it is achieved through self-adaptation or Covariance Matrix Adaptation (CMA) [38]. However, despite these alternative approaches, researchers at Uber AI Labs [35] have developed Deep GA, a GA for *Deep Neuroevolution*; moreover, their findings show that a simple GA is competitive at completing modern problems that were originally thought of as extremely challenging [38].

2.2 Hyper-parameter Optimisation

Hyper-parameter optimisation is a crucial step of applying a ML algorithm and finding optimal hyper-parameter values manually is often a time consuming and tedious task [10] and it is often referred to as an art form [12]. ML algorithms are rarely hyper-parameter free, and indeed, they are often considered nuisances, however, the process of automatically determining optimal values can be seen as a process of optimisation [31]. HPO is the process of optimising a loss function over a graph-structured configuration space [4]; the aim is to maximise or minimise a given function [39]. Determining appropriate values for the hyper-parameters is fundamental in finding an optimal solution, however, it is a frustratingly difficult task [8–10,18] and the performance of NNs crucially depend on the hyper-parameters used [6,9]. For example, the internal structure is a key factor in determining the efficiency of the NN [27]. Moreover, a major challenge when designing and building a NN is determining the optimal hyper-parameters for the network given the data for the problem at hand [7]. There are a number of popular and widely employed HPO approaches; each has been developed with a different aim and generally improves upon the previously developed approach. However, grid search and manual search are presently the most widely used strategies for HPO [3].

Grid and Random Search. Grid search, also known as a parameter sweep, is the traditional approach used for finding the optimal set of hyper-parameters for a given function; the approach is an exhaustive search that tries all possible combinations to find the optimal value(s) [3]. Due to the fact that all possible values are explored, the approach can guarantee reliable optimisation in low dimensional spaces [3]. However, not necessarily efficiently because it exhaustively tries all possible combinations and suffers from the curse of dimensionality as the number of values grows exponentially with the number of hyper-parameters [2,3].

The random search approach was developed and aimed to reduce the cost of computation and find an alternative to grid search. In principle, the random search approach is very similar to that of grid search, however, instead of all possible value combinations tested, the algorithm stochastically tries values within the search space. A number of empirical studies have shown random search to outperform and be computationally more efficient than grid search at finding an optimal combination of hyper-parameters [3].

3   Experimental Design

The design of this experiment draws on existing research, most notably: [3,4,19, 45]. A considerable amount of literature has been published on grid search, it is widely accepted as a computationally expensive approach. However, in problems where the intrinsic dimensionality is low, it may be an appropriate approach. NE and EAs are stochastic algorithms that often employ manual and evolutionary HPO, however, there are some unanswered questions about the validity of this approach and its ability find optimal hyper-parameters. Therefore, this research aims to integrate and utilise a random search approach for HPO. The question that then naturally arises is whether this approach is reliable, as randomising the optimisation process may arbitrarily produce optimal solutions, which may represent a long-winded process as the computation time required to find an optimal solution increases at each step an optimal solution is not found.

This experiment involves taking a standard NE algorithm and producing five distinct modified versions, as listed in Table 1, and applying them to an Unsupervised Learning problem that involves the NNs learning to target seek. The modifications included in the experiment can be categorised as either an initialisation modification (alias prefix IN) or a run-time optimisation modification (alias prefix RT). The NE algorithm used is *Neataptic* [37], a JavaScript implementation based on NEAT [33]. The random search optimisation modifications will be integrated into multiple distinct copies of the original Neataptic algorithm.

In order to carry out the experiment, each algorithm must be configured as shown in Table 2. Each algorithm will undergo a total of 10000 function evaluations with a sample size of 30 executions for each algorithm. The population size has been determined using a method developed in [34], specifically, the population size $P = N \times 10$ where N is the number of objectives (i.e. 1 in our experiment).

Algorithm 1 is the pseudo-code for the experiment and outlines the fundamental flow of execution for the experiment and the algorithms.

Table 1. Included algorithm descriptions and alias definitions

Algorithm	Alias
Neataptic (no modifications)	Vanilla
Neataptic with initial neuron activation function modification	INAFM
Neataptic with initial neuron bias modification	INBM
Neataptic with initial network topology modification	INTM
Neataptic with run-time activation function optimisation	RTAFO
Neataptic with run-time network topology optimisation	RTNTO

Table 2. Parameter configurations for the GA

Parameter	Value		
Mutation rate	0.3		
Elitism	0.1		
Selection method	Tournament		
Crossover method	Uniform		
Mutation methods	Add neuron	Remove neuron	Add self-connection
	Add connection	Remove connection	Remove self-connection
	Modify weight	Modify bias	Remove recurrent connection
	Add gate	Remove gate	Add recurrent connection
Population size	100		
Generation count	100		

4   Hyper-parameter Optimisation by Restrained Stochastic Hill Climbing

Stochastic hill climbing chooses it's next value at random from the available search-space [28]. The approach introduced in this section, named hyper-parameter optimisation by restrained stochastic hill climbing (HORSHC), proposes an approach to HPO that rivals the manual and evolutionary approach found in the NE algorithm used in our experiment.

The HORSHC process outlined in Algorithm 2 begins by defining a limit, this is the *restrainment* applied to the algorithm. This can be considered another hyper-parameter for optimisation. The algorithm then goes on to stochastically increase or decrease a network's size, doing so until either the limit is reached or performance no longer improves.

Algorithm 1. Hyper-parameter optimisation experiment execution cycle.

```
1:  g ← 0
2:  p ← [size]                                          ▷ in our experiment size = 100
3:  p = INITIALISEPOPULATION(size)
4:  while g ≤ max do                                    ▷ in our experiment max = 100
5:      for network ← 0 to size do
6:          APPLYTOLOSSFUNCTION(network)
7:          f = EVALUATEFITNESS(network)
8:          start = NOW
9:          OPTIMISE(network)
10:         finish = NOW
11:         STORETIMES(finish - start)
12:         STOREFITNESS(f)
13:     end for
14:     g ← g + 1
15: end while
```

Algorithm 2. HORSHC execution cycle

```
1:  limit ← 3
2:  h ←network.score
3:  i ← 0
4:  do
5:      h ←network.score
6:      if p >= 0.5 then                                ▷ p is assigned a random number 0-1
7:          network = SIMPIFLYNETWORK
8:      else
9:          network = COMPLEXIFYNETWORK
10:     end if
11:     APPLYTOLOSSFUNCTION(network)
12:     network.score ← FITNESSFUNCTION
13:     i ← i + 1
14: while network.score > h and i != limit
```

---

5   Numerical Results

Figure 1 depicts the overall performance of each of the initialisation modifications (INAFM, INBM, INTM), compared with the unmodified vanilla version; specifically, the average fitness scores as well as the average of the worst and best performing solutions throughout the 100 generations. Figure 2 does the same for the run-time activation function modification (RTAFO), and Fig. 3 shows the results for the run-time network topology modification (RTNTO).

In order to complement the other findings, the Wilcoxon signed-rank test [41] has been performed using a significant value of 0.05. The test will determine if

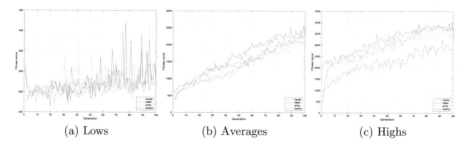

Fig. 1. Experimental results for the initialisation modifications

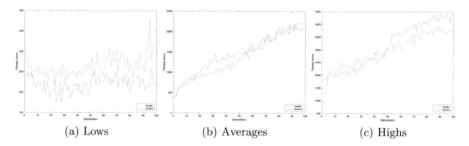

Fig. 2. Experimental results for the activation function modifications

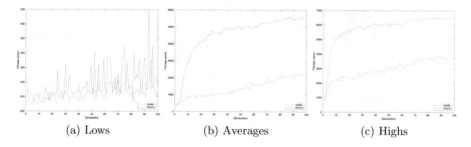

Fig. 3. Experimental results for the topology optimisation modifications

there is a significant statistical difference between the results obtained from the vanilla algorithm and the other algorithms, hypothesis values with a '=' indicates equal performance, '−' indicates inferior performance and a '+' indicates superior performance.

Table 3 shows the average of the worst, mean, and best performing solutions for all algorithm variations across 30 independent samples of 10000 function evaluations in comparison to the vanilla. Despite the results illustrated in Fig. 1 that depicted potential performance increases, none of the results for the initialisation modifications were significantly different to the results produced by the vanilla algorithm. However, results for the run-time modifications show that both RTAFO and RTNTO significantly outperform the vanilla algorithm.

Table 4 shows the Wilcoxon test results for the total times of the full experiment execution, RTNTO represents a significantly inferior algorithm compared to the vanilla algorithm in terms of overall time taken to execute. Whereas, RTAFO had equal performance of that shown by the vanilla algorithm.

Table 3. Wilcoxon test results for the modified algorithms.

	Worst		Mean		Best	
Algorithm	p-value	Hypothesis	p-values	Hypothesis	p-value	Hypothesis
INAFM	0.65176	0 (=)	0.65544	0 (=)	0.51585	0 (=)
INBM	0.26876	0 (=)	0.77657	0 (=)	0.19093	0 (=)
INTM	0.85942	0 (=)	0.95153	0 (=)	0.37074	0 (=)
RTAFO	0.00078147	1 (−)	0.33874	0 (=)	0.00078147	1 (+)
RTNTO	0.56638	0 (=)	1.8626e−08	1 (+)	9.3132e−09	1 (+)

Table 4. Wilcoxon test results for the total time of RTAFO and RTNTO in comparison to the vanilla algorithm.

	Total time		
Algorithm	Mean time (mins)	p-value	Hypothesis
RTAFO	9.1542	0.55611	0 (=)
RTNTO	10.9112	5.7183e−07	1 (−)

6   Conclusion

The experiment carried out during this research examined 3 hyper-parameter initialisation approaches and 2 run-time HPO approaches. The 3 initialisation approaches aimed to better improve the initialisation performance of solutions; whereas, the run-time optimisation approaches aimed to optimise existing solutions in the population. The novel run-time optimisation approaches employed have performed remarkably well in comparison to the vanilla algorithm. It should be remarked that RTNTO has been shown to significantly outperform all other algorithms and was able to do so within a third of the total function evaluations allocated.

With that said, the initialisation algorithms do present some interesting results. Despite neither INAFM, INBM, or INTM displaying superior performance to that of the vanilla algorithm, all 3 were able to produce results that were of equal performance. However, the results show that INAFM has consistently improved the initial performance of the best performing solutions; therefore, it suggests that the selection and crossover mechanisms used by the algorithm are not subsequently producing better performing solutions. Instead,

INAFM is accelerating the initial optimisation process that the other approaches were unable to achieve.

An important question to answer is whether performance can be improved by different permutations of the initialisation approaches; however, this could also result in a further acceleration of the optimisation and not result in higher performing solutions at the end of the process. The method of complexification employed by many modern algorithms ensures that the initial size of the NNs are small, however, INTM disregards this and allows solutions to have potentially larger initial topologies. Depending on the requirements, it may not be advantageous to produce high performing solutions whose topologies are potentially over-complex in comparison to their counterparts.

Moreover, the results of RTNTO have revealed an unanticipated superior performance in comparison to the vanilla algorithm and all other algorithms used in this research. As previously mentioned, RTNTO was able to outperform the other algorithms within a third of the allocated function evaluations. As with INAFM, the RTNTO algorithm was able to not only accelerate the optimisation of solutions, but also increase overall solution performance by 56%. However, despite the significant performance improvement, as with all of the algorithms, RTNTO was unable to increase performance of the whole population as average low scores were significantly lower and did not show signs of improvement during execution.

Similarly, due to the increase in computation time required, RTNTO takes significantly longer to complete than both the vanilla and RTAFO algorithms. Generally speaking, a single, optimal solution is what a researcher/practitioner requires and this inability to remove weak performing solutions and/or execute in the fastest time may not pose a problem. Interestingly, RTNTO was a by-product of another algorithm and was similar in its approach but was unrestricted in terms of how may hidden layers/neurons could be added during a single function evaluation. Despite a lack of statistical evidence, this former algorithm displayed similar performance to that of RTNTO.

Contrasting RTAFO and RTNTO, despite the initial promising performance of INAFM during the initialisation experiments, RTAFO was unable to achieve similar results to that of RTNTO. Comparing RTAFO to the vanilla algorithm, it was able to significantly outperform in terms of average highest fitness for candidate solutions. However, it has become apparent throughout the research that all of the algorithms are unable to improve the population as a whole and RTAFO is no exception to this; in fact, it is the only algorithm that was significantly outperformed by the vanilla algorithm on this basis. Due to this consistency, it opens a question to whether the surrounding components are to blame or whether additional logic is required to eliminate the issue and allow the whole population to improve.

7   Future Work

The results uncovered during this research on the integration of alternative HPO approaches within a Neuroevolution algorithm have revealed 2 probable hypotheses: (1) random initialisation of hyper-parameters has little significance on the final performance of solutions; (2) HORSHC performed significantly better than pure evolutionary and/or manual search strategies for finding high performing solutions. HORSHC is a competitive HPO algorithm and it is proposed that it should be employed for optimising NNs solving single objective problems such as the target seeking problem used in our experiment. However, several questions remain unanswered: (1) how well do the approaches demonstrated in this research perform against a problem with multi/many objectives?; (2) how well do the approaches demonstrated in this research perform on a different set of problems?; (3) Are there permutations of initialisation methods that provide better optimisation results? A natural progression of this work would be to explore the application of the approaches proposed in this research according to questions 1 and 2 as well as the combination of approaches to see if they reveal further performance increases. Furthermore, a further study could assess the effect of activation function optimisation and its significance; [16] performed a similar experiment and introduced HA-NEAT that is analogous to RNAFO. Furthermore, as RTNTO was restricted to 3 modifications, it could be argued that this is another hyper-parameter to tune and increases/decreases may yield better results. Finally, as ES lead the latest research and have fewer hyper-parameters [29], a further study of traditional HPO with ES would be a complementary contribution.

References

1. Angeline, P.J., Saunders, G.M., Pollack, J.B.: An evolutionary algorithm that constructs recurrent neural networks. IEEE Trans. Neural Netw. $5(1)$, 54–65 (1994)
2. Bellman, R.E.: Adaptive Control Processes: A Guided Tour, vol. 2045. Princeton University Press, Princeton (2015)
3. Bergstra, J., Bengio, Y.: Random search for hyper-parameter optimization. J. Mach. Learn. Res. 13, 281–305 (2012)
4. Bergstra, J.S., Bardenet, R., Bengio, Y., Kégl, B.: Algorithms for hyper-parameter optimization. In: Advances in Neural Information Processing Systems, pp. 2546–2554 (2011)
5. Conti, E., Madhavan, V., Such, F.P., Lehman, J., Stanley, K., Clune, J.: Improving exploration in evolution strategies for deep reinforcement learning via a population of novelty-seeking agents. In: Advances in Neural Information Processing Systems, pp. 5027–5038 (2018)
6. Dernoncourt, F., Lee, J.Y.: Optimizing neural network hyperparameters with Gaussian processes for dialog act classification. In: 2016 IEEE Spoken Language Technology Workshop (SLT), pp. 406–413. IEEE (2016)
7. Diaz, G.I., Fokoue-Nkoutche, A., Nannicini, G., Samulowitz, H.: An effective algorithm for hyperparameter optimization of neural networks. IBM J. Res. Dev. $61(4/5)$, 1–9 (2017)
8. Eggensperger, K., Feurer, M., Hutter, F., Bergstra, J., Snoek, J., Hoos, H., Leyton-Brown, K.: Towards an empirical foundation for assessing Bayesian optimization of hyperparameters. In: NIPS Workshop on Bayesian Optimization in Theory and Practice, vol. 10, p. 3 (2013)

9. Eggensperger, K., Hutter, F., Hoos, H., Leyton-Brown, K.: Efficient benchmarking of hyperparameter optimizers via surrogates. In: Twenty-Ninth AAAI Conference on Artificial Intelligence (2015)
10. Feurer, M., Springenberg, J.T., Hutter, F.: Initializing Bayesian hyperparameter optimization via meta-learning. In: Twenty-Ninth AAAI Conference on Artificial Intelligence (2015)
11. Fiesler, E.: Neural Network Topologies. Springer, Boston (1996)
12. Yufeng, G.: The 7 steps of machine learning (2017). https://towardsdatascience. com/the-7-steps-of-machine-learning-2877d7e5548e
13. Gomez, F., Schmidhuber, J., Miikkulainen, R.: Efficient non-linear control through neuroevolution. In: European Conference on Machine Learning, pp. 654–662. Springer (2006)
14. Gomez, F., Schmidhuber, J., Miikkulainen, R.: Accelerated neural evolution through cooperatively coevolved synapses. J. Mach. Learn. Res. 9(May), 937–965 (2008)
15. Gomez, F.J.: Robust non-linear control through neuroevolution. Ph.D. thesis (2003)
16. Hagg, A., Mensing, M., Asteroth, A.: Evolving parsimonious networks by mixing activation functions. In: Proceedings of the Genetic and Evolutionary Computation Conference, pp. 425–432. ACM (2017)
17. Heidrich-Meisner, V., Igel, C.: Neuroevolution strategies for episodic reinforcement learning. J. Algorithms 64(4), 152–168 (2009)
18. Ilievski, I., Akhtar, T., Feng, J., Shoemaker, C.A.: Efficient hyperparameter optimization for deep learning algorithms using deterministic RBF surrogates. In: Thirty-First AAAI Conference on Artificial Intelligence (2017)
19. Larochelle, H., Erhan, D., Courville, A., Bergstra, J., Bengio, Y.: An empirical evaluation of deep architectures on problems with many factors of variation. In: Proceedings of the 24th International Conference on Machine Learning, pp. 473–480. ACM (2007)
20. Miconi, T., Clune, J., Stanley, K.O.: Differentiable plasticity: training plastic neural networks with backpropagation. arXiv preprint arXiv:1804.02464 (2018)
21. Montana, D.J., Davis, L.: Training feedforward neural networks using genetic algorithms. Int. Jt. Conf. Artif. Intell. 89, 762–767 (1989)
22. Morse, G., Stanley, K.O.: Simple evolutionary optimization can rival stochastic gradient descent in neural networks. In: Proceedings of the Genetic and Evolutionary Computation Conference 2016, pp. 477–484. ACM (2016)
23. Rechenber, I.: Optimierung technischer systeme nach prinzipien der biologischen evolution. Ph.D. thesis, Verlag nicht ermittelbar (1970)
24. Risi, S., Togelius, J.: Neuroevolution in games: state of the art and open challenges. CoRR abs/1410.7326 (2014). http://arxiv.org/abs/1410.7326
25. Rostami, S.: Preference focussed many-objective evolutionary computation. Ph.D. thesis, Manchester Metropolitan University (2014)
26. Rostami, S., Neri, F.: Covariance matrix adaptation pareto archived evolution strategy with hypervolume-sorted adaptive grid algorithm. Integr. Comput.-Aided Eng. 23(4), 313–329 (2016)
27. Rostami, S., O'Reilly, D., Shenfield, A., Bowring, N.: A novel preference articulation operator for the evolutionary multi-objective optimisation of classifiers in concealed weapons detection. Inf. Sci. 295, 494–520 (2015)
28. Russell, S.J., Norvig, P.: Artificial intelligence: a modern approach. Pearson Education Limited, Malaysia (2016)

29. Salimans, T., Ho, J., Chen, X., Sidor, S., Sutskever, I.: Evolution strategies as a scalable alternative to reinforcement learning. preprint arXiv:1703.03864 (2017)
30. Siebel, N.T., Botel, J., Sommer, G.: Efficient neural network pruning during neuro-evolution. In: 2009 International Joint Conference on Neural Networks, pp. 2920–2927. IEEE (2009)
31. Snoek, J., Larochelle, H., Adams, R.P.: Practical Bayesian optimization of machine learning algorithms. In: Advances in neural information processing systems, pp. 2951–2959 (2012)
32. Stanley, K.O., D'Ambrosio, D.B., Gauci, J.: A hypercube-based encoding for evolving large-scale neural networks. Artif. life **15**(2), 185–212 (2009)
33. Stanley, K.O., Miikkulainen, R.: Evolving neural networks through augmenting topologies. Evol. Comput. **10**(2), 99–127 (2002)
34. Storn, R.: On the usage of differential evolution for function optimization. In: Proceedings of North American Fuzzy Information Processing, pp. 519–523. IEEE (1996)
35. Such, F.P., Madhavan, V., Conti, E., Lehman, J., Stanley, K.O., Clune, J.: Deep neuroevolution: genetic algorithms are a competitive alternative for training deep neural networks for reinforcement learning. arXiv preprint arXiv:1712.06567 (2017)
36. Togelius, J., Schaul, T., Wierstra, D., Igel, C., Gomez, F., Schmidhuber, J.: Onto-genetic and phylogenetic reinforcement learning. Künstliche Intell. **23**(3), 30–33 (2009)
37. Wagenaartje, T.: wagenaartje/neataptic (2018). https://github.com/wagenaartje/neataptic
38. Wang, L., Feng, M., Zhou, B., Xiang, B., Mahadevan, S.: Efficient hyper-parameter optimization for NLP applications. In: Proceedings of the 2015 Conference on Empirical Methods in Natural Language Processing, pp. 2112–2117 (2015)
39. White Jr., R.: A survey of random methods for parameter optimization. Simulation **17**(5), 197–205 (1971)
40. Whitley, D., Starkweather, T., Bogart, C.: Genetic algorithms and neural networks: optimizing connections and connectivity. Parallel Comput. **14**(3), 347–361 (1990)
41. Wilcoxon, F.: Individual comparisons by ranking methods. Biom. Bull. **1**(6), 80–83 (1945)
42. Yao, X.: Evolving artificial neural networks. Proc. IEEE **87**(9), 1423–1447 (1999)
43. Yao, X., Liu, Y.: A new evolutionary system for evolving artificial neural networks. Trans. Neur. Netw. **8**(3), 694–713 (1997)
44. Yao, X., Liu, Y.: A new evolutionary system for evolving artificial neural networks. IEEE Trans. Neural Networks **8**(3), 694–713 (1997)
45. Young, S.R., Rose, D.C., Karnowski, T.P., Lim, S.H., Patton, R.M.: Optimizing deep learning hyper-parameters through an evolutionary algorithm. In: Proceedings of the Workshop on Machine Learning in High-Performance Computing Environments, p. 4. ACM (2015)

Swarm Inspired Approaches for K-prototypes Clustering

Hadeel Albalawi$^{(\boxtimes)}$, Wei Pang$^{(\boxtimes)}$, and George M. Coghill$^{(\boxtimes)}$

Department of Computing Science, University of Aberdeen, Aberdeen, UK
{r01haal7, pang.wei, g.coghill}@abdn.ac.uk

Abstract. Data clustering is a well-researched area in data mining and machine learning. The clustering algorithms that can handle both numeric and categorical variables have been extensively researched in the recent years. However, the clustering algorithms have a major limitation that converge to a local optima. Therefore, to address this problem this paper has proposed a novel algorithm ABC k-prototypes (Artificial Bee Colony clustering based on k-prototypes) for clustering mixed data. In our proposed approach we use the combination between the distribution centroid and the mean to calculate the dissimilarity between data objects and prototypes. The proposed algorithm is tested on five different datasets taken from the UCI machine learning data repository. The comparative results in the performance measures of the clustering showed that the proposed algorithm outperformed the traditional k-prototypes.

Keywords: K-prototypes · Mixed data · Artificial bee colony

1 Introduction

Data clustering is the basic concept in data mining and machine learning. The problems containing a large number of variables along with a large number of samples make it a useful technique in diverse fields of application. Clustering algorithms can be split into two main categories: hierarchical clustering and partitional clustering. Hierarchal clustering starts with each sample being a separate cluster and later combines them gradually based on a specific similarity measure. For large datasets, hierarchical clustering is not recommended, being computationally expensive. On the other hand, partitional clustering starts from the position that the whole dataset is assumed to be one cluster. It then iteratively splits the data into small sub-sets based on within cluster and between clusters variability. These models include the well-established k-means algorithm (Jain et al. 1999; Aggarwal and Reddy 2013; Shirkhorshidi et al. 2014). The original version of the k-means algorithm is limited to numeric data only, as usually the dissimilarity is calculated using Euclidean distance, which is not suitable for categorical datasets. In the case of mixed data containing both numeric and categorical variables, k-means fails to provide a suitable solution (Cheung and Jia 2013). Data transformations, such as a pre-processing step before k-means, usually result in a loss of information rather than improvement in clustering performance (García-Escudero et al. 2010; Aggarwal and Reddy 2013). Therefore, Huang et al. (1998) proposed the

© Springer Nature Switzerland AG 2020
Z. Ju et al. (Eds.): UKCI 2019, AISC 1043, pp. 201–209, 2020.
https://doi.org/10.1007/978-3-030-29933-0_17

k-prototypes algorithm which extends to the k-means and it is suitable in the situations where the set of variables include both qualitative and quantitative attributes.

An increased interest in swarm intelligence has emerged in recent years. This involved studying and understanding the decentralized behaviors of individual insects and the collectively centralized behaviors of the swarm (Bonabeau et al. 1999). Therefore, the development of swarm intelligence models and techniques is inspired from principles of natural swarm intelligence. Over the years, an enormous amount of research has been studied the different behaviors of social insects in an attempt to utilize the swarm intelligence models and build various artificial swarm systems (Bonabeau et al. 1999). Karaboga (2005) proposed the Artificial Bee Colony (ABC) algorithm to solve numerical optimization problems. This algorithm works on the swarm intelligence principles of foraging behavior of honey bees. Like other swarm based techniques, the ABC has flexibility, robustness and self-organization. Bee colony algorithms represent a clustering problem into an optimization problem that localizes the optimal centroids rather than to find an optimal partition. Clustering techniques in swarm intelligence translate a clustering problem into finding n points in the n-dimensional space that can maximize between clusters variation and minimize variation inside each cluster. Although there exist so many different algorithms on bee swarm for clustering problems, a few studies that address the issue of clustering with both qualitative and quantitative variables. Therefore, in this research we integrate the artificial bee colony with k-prototypes to cluster mixed data.

1.1 Motivation and Aim

The essential task in data mining is to produce a clustering by finding the similarities between data objects. In many applications, the datasets contain both numerical and categorical attributes, so to cluster this type of data we use the k-prototypes algorithm (Ji et al. 2013). One major issue of the k-prototypes algorithm is that it may easily reach local optimal solutions. So in this research we aim to combine the advantages of swarm intelligence algorithm with the k-prototypes algorithm to address this limitation and produce an efficient clustering.

As far as we are aware, little research has been done on swarm intelligence for clustering mixed datasets. In addition, the existing literature shows that the ABC algorithm was previously applied to solve clustering problems consisting of only numeric or only categorical data (Karaboga et al. 2014). Therefore, we decide to explore the potential Artificial Bee Colony algorithm to solve clustering problems, and we believe there is much room to develop new techniques using the existing swarm intelligence approaches. We expect our research will make novel contributions to clustering mixed data considering the mixed datasets are in abundance in various fields. So, this research will not only open new directions for the swarm intelligence but also aim to achieve better clustering performance.

2   Related Work

This section reviews the existing literature on swarm intelligence based clustering algorithms. Data consisting of both numeric and categorical features are abundant in the real-life applications, but the clustering algorithms usually result in local minima (Anderlucci and Hennig 2014). To resolve this problem, a number of researchers proposed different algorithms. Ji et al. (2015) proposed a novel clustering algorithm termed Artificial Bee Colony based on k-Modes (ABC k-Modes) using the ABC approach integrated with k-modes clustering algorithm to cluster categorical data objects. The proposed approach performs one iteration with k-modes algorithm and combine this solution as the initial population solution for ABC clustering algorithm. The proposed algorithm was tested on six real-world categorical datasets and the comparative analysis illustrated that ABC K-modes achieves higher values on the evaluation measures AC, PR, RE, and RI than three other clustering algorithms (k-modes, fuzzy k-modes and genetic k-modes).

Pham et al. (2011) proposed a new algorithm based on bees algorithm and k-prototypes clustering for mixed numeric and categorical dataset named random search with k-prototypes algorithm (RANKPRO). Their algorithm efficiently explores the search space to randomly select new solutions and improves only promising solutions to implement k-prototypes algorithm. The proposed algorithm was applied on different datasets and proved that the proposed algorithm is more efficient than the traditional k-prototypes algorithm. However, it is noted from the results that in connect-4 dataset the efficiency of the k-prototypes algorithm is better than the RANKPRO algorithm. Prabha and Visalakshi (2015), proposed a new algorithm to achieve global optimal clustering solution using the k-prototypes and binary Particle Swarm Optimization algorithms. The efficiency of the proposed algorithm is assessed by Rand Index, Jaccard Index, Entropy, and F-measure on different datasets taken from the UCI benchmark datasets repository. It is observed that the new algorithm performed better than traditional K-Prototype and K-modes algorithms.

3   The Proposed Algorithm (Artificial Bee Colony Based K-Prototypes)

A novel clustering algorithm is presented using k-prototypes and Artificial Bee Colony optimization by further extending the existing ABC k- modes algorithm (Ji et al. 2015). There are three types of artificial honeybees: employed bees, onlookers, and scouts. The clustering solution in ABC optimization stands for the food sources that are the best choices based on stochastic search through a large number of solutions. A good solution's characteristics are the quality, quantity of nectar, and risk in the way to the food source.

Ji et al. (2013) proposed the concept of distribution centroid to represent the center in a cluster for qualitative features. It should be noted that in a cluster the frequency for each value of the categorical attributes is recorded by the distribution centroid. Then, it is combined the mean with distribution centroid to cluster the mixed data.

The following formula is the objective function and it was proposed by Ji et al. (2013), and we implemented it to calculate the dissimilarity between data objects and prototypes of clusters:

$$E(U, Q, S) = \sum_{l=1}^{k} \sum_{i=1}^{n} \sum_{j=1}^{m} u_{il} s_j^\lambda d(x_{ij}, q_{lj}), \qquad (1)$$

where,

- U is the partition matrix $U_{n \times k}$, u_{il} is one element of U, $u_{il} \in \{0, 1\}$ and $\sum_{l=1}^{k} u_{il} = 1$.
- S is the significance values, $0 \leq s_j \leq 1$ and $\sum_{j=1}^{m} s_j = 1$.
- $\lambda > 1$ is an exponent for the significance of s_j.
- q_{lj} is the center of the jth attributes in the cluster l.

According to the ABC k-modes algorithm, to derive an improved candidate solution from the present food source, a fuzzy one-step k-prototypes procedure (OKP) is used (Ji et al. 2015). A fuzzy OKP is an exploitation process used by employed bees and onlooker bees. It is used to search for a neighbor food source based on the current food source and is essentially one iteration step in the search process of the k-prototypes algorithm.

Let f_i be the current food source, then the fuzzy OKP consists of the following two steps:

1. Each data point is assigned to one cluster based on the distance formula proposed on Eq. (1).
2. To update the cluster centroids, both the numeric and categorical features are considered. For categorical part of the centroid, the distribution centroid is used and for numeric data, the mean of the data points in the cluster is used.

3.1 The Pseudo code of ABC k-prototypes Algorithm

Input: The bee colony size N, the maximum cycle number MCN, the number of clusters k, and the parameter that determines the number of trial L.
Output: The best food source.

1. Initialize the population of food sources $P_{fs} = \{f_1, f_2, \ldots, f_H\}$ randomly; for each food source, select k data objects randomly from the dataset X as the prototypes of clusters; set the exploitation numbers of food sources $En_1 = 0$, $En_2 = 0, \ldots, En_H = 0$.
2. Calculate the nectar amounts of the food sources $NA(f_1)$, $NA(f_2)$, $\ldots$ $NA(f_H)$; $E(f_i) = E(U, Q, S)$:

$$NA(f_i) = \frac{1}{E(f_i) + 1} \qquad (2)$$

3. Set $CN = 1$ which is the cycle number.
4. For each employed bee:

    a.   Produce a new food source f_i' from the current food source f_i by using the fuzzy one-step k- prototypes procedure OKP, and set $En_i = En_i + 1$;

    b.   Calculate the nectar amount $NA(f_i')$ for the food source f_i' based on Eq (2);

    c.   If $NA(f_i') > NA(f_i)$, the current food source f_i is replaced by the new food source; otherwise the current food source f_i is retained.

5. Calculate the probability pro_i for each food source f_i:

$$pro_i = \frac{NA(f_i)}{\sum_{j=1}^{H} NA(f_i)} \qquad (3)$$

6. For each onlooker bee

    a.   Pick up one food source f_i as the current food source according to the calculated probabilities;

    b.   Produce a new food source f_i' from the current food source f_i by using OKP, set $En_i = En_i + 1$;

    c.   Calculate the nectar amount of f_i', $NA(f_i')$;

    d.   If $NA(f_i') > NA(f_i)$, the current food source f_i is replaced by the new food source f_i'; otherwise the current food source f_i is retained;

    e.   Update the probability pro_i for each food source f_i based on Eq (3).

7. For each food source f_i, if the exploitation number En_i is no less than L, this food source is abandoned, and the corresponding employed bee becomes a scout.

8. If there exists an abandoned food source f_i:

    a.   Send the scout in the search space to find T candidate food sources $\{f_i^1, f_i^2, \ldots\ldots\ldots f_i^T\}$ by $f_i' = Rand(Dom(x))$ which is the function for selecting k

data points from the dataset randomly.

b.   Calculate the nectar amounts $\{NA\,(f_i^1),\,NA(f_i^2),\ldots\ldots\ldots NA(f_i^T)\}$ of the food sources $\{f_i^1\,f_i^2,\ldots\ldots\ldots f_i^T\}$.

c.   Choose the food source with the highest nectar amount as the new food source f_i', and set $En_i = 0$;

d.   If $NA(f_i') > NA(f_i)$ the current food source f_i is replaced by the new food source f_i';

   otherwise the current food source f_i is retained.

9. $CN = CN+1$;

10. If $CN = MCN$, terminate the algorithm and output the best food source; otherwise go to step 4).

4   Experimental Results

The proposed ABC k-prototypes algorithm was implemented and tested on 5 datasets taken from the UCI Machine Learning Repository (http://archive.ics.uci.edu/ml/datasets.html). The datasets are dermatology, zoo, sponge, adult, and covertype. The covertype is the large dataset, therefore the sample of size 10,000 was randomly chosen for clustering to keep processing time limited. In the performance analysis, we ran the proposed algorithm ABC based k-prototypes 20 trials. Then, we compared the results with the simple k-prototypes. Experimental results proved that the proposed algorithm was better in performance measures compared to the standard k-prototypes algorithm, except for the Sponge dataset where the accuracy measure in the standard k-prototype algorithm was better than the proposed one. Further investigations need to be performed in the future work to reach the root cause. Also, when we ran the proposed algorithm on the large dataset such as covertype, the algorithm takes more time so we will improve our algorithm to solve these problems and to achieve the better results. The following tables shows the details of datasets and the comparison between the proposed algorithm and the traditional k-prototypes. According to the performance measures, the better clustering results determines by the high values of Rand Index, Accuracy Measure, and Precision (Tables 1, 2, 3, 4, 5 and 6).

Table 1. Details of datasets

Dataset	No. of instances	No. of attributes
Dermatology	366	33
Zoo	101	17
Sponge	76	45
Covertype	581012	54
Adult	48842	14

Table 2. The dermatology dataset

The performance measure	ABC k-prototypes	K-prototypes
The average of accuracy	0.7288	0.8143
The average of precision	0.8483	0.1463
The average of rand Index	0.8169	0.6512
Algorithm running time (milliseconds)	20271.0	2.617004

Table 3. The zoo dataset

The performance measure	ABC k-prototypes	k-prototypes
The average of accuracy	0.8311	0.6979
The average of precision	0.8231	0.0970
The average of rand index	0.8629	0.5940
Algorithm running time (milliseconds)	27715.0	2.617004

Table 4. The sponge dataset

The performance measure	ABC k-prototypes	k-prototypes
The average of accuracy	0.6289	0.7383
The average of precision	0.7052	0.1171
The average of rand index	0.8450	0.3997
Algorithm running time (milliseconds)	83082.0	5.046717

Table 5. The covertype dataset

The performance measure	ABC k-prototypes	k-prototypes
The average of accuracy	0.4953	0.03
The average of precision	0.4947	0.08
The average of rand index	0.5889	0.01
Algorithm running time (milliseconds)	5373207.0	30.63

Table 6. The adult dataset

The performance measure	ABC k-prototypes	k-prototypes
The average of accuracy	0.7591	0.0003
The average of precision	0.7601	0.1328
The average of rand index	0.5008	−0.0074
Algorithm running time (milliseconds)	3162864.0	904.1047

5 Conclusion

Data clustering is the area in machine learning research that splits a dataset into an optimal number of segments. Bee colony algorithms pose a clustering problem as an optimization problem that localizes the optimal centroids rather than finding an optimal partition. Artificial bee colony (ABC) optimization is one of the swarm techniques which aims to solve clustering problems. The existing literature shows that the ABC algorithm has previously been applied to solve clustering problems consisting of only numeric or only categorical data. Therefore, we integrate ABC with k-prototypes to cluster mixed data. It is shown that the comparison between the proposed algorithm and standard k-prototypes proves that the ABC k-prototypes algorithm is more efficient than traditional k-prototypes. However, the proposed algorithm takes more time when tested on large datasets so we will improve our algorithm to solve these problems to achieve better results. In addition, to overcome the computational cost, the model can be run in parallel on different machines or different cores to ensure that processing is completed more quickly.

References

Anderlucci, L., Hennig, C.: The clustering of categorical data: a comparison of a model-based and a distance-based approach. Commun. Stat.-Theory Methods **43**(4), 704–721 (2014)

Aggarwal, C.C., Reddy, C.K. (eds.): Data Clustering: Algorithms and Applications. Chapman and Hall/CRC (2013)

Bonabeau, E., Marco, D.D.R.D.F., Dorigo, M., Théraulaz, G., Theraulaz, G.: Swarm Intelligence: from Natural to Artificial Systems, No. 1. Oxford University Press, New York (1999)

Cheung, Y.M., Jia, H.: Categorical-and-numerical-attribute data clustering based on a unified similarity metric without knowing cluster number. Pattern Recognit. **46**(8), 2228–2238 (2013)

García-Escudero, L.A., Gordaliza, A., Matrán, C., Mayo-Iscar, A.: A review of robust clustering methods. Advances Data Anal. Classif. **4**(2–3), 89–109 (2010)

Huang, Z.: Extensions to the k-means algorithm for clustering large data sets with categorical values. Data Min. Knowl. Dis. **2**, 283–304 (1998)

Jain, A.K., Murty, M.N., Flynn, P.J.: Data clustering: a review. ACM Comput. Surv. (CSUR) **31**(3), 264–323 (1999)

Ji, J., Bai, T., Zhou, C., Ma, C., Wang, Z.: An improved k-prototypes clustering algorithm for mixed numeric and categorical data. Neurocomputing **120**, 590–596 (2013)

Ji, J., Pang, W., Zheng, Y., Wang, Z., Ma, Z.: A novel artificial bee colony based clustering algorithm for categorical data. PLoS ONE **10**(5), e0127125 (2015)

Karaboga, D.: An idea based on honey bee swarm for numerical optimization, vol. 200. Technical report-TR06, Erciyes University, Engineering Faculty, Computer Engineering Department (2005)

Karaboga, D., Gorkemli, B., Ozturk, C., Karaboga, N.: A comprehensive survey: artificial bee colony (ABC) algorithm and applications. Artif. Intell. Rev. **42**(1), 21–57 (2014)

Pham, D.T., Suarez-Alvarez, M.M.. Prostov, Y.I.: August. Random search with k-prototypes algorithm for clustering mixed datasets. In: Proceedings of the Royal Society of London A: mathematical, Physical and Engineering Sciences, vol. 467, no. 2132, pp. 2387–2403. The Royal Society (2011)

Prabha, K.A., Visalakshi, N.K.K.: Particle swarm optimization based k-prototype clustering algorithm. IOSR J. Comput. Eng. **17**, 56–62 (2015)

Shirkhorshidi, A.S., Aghabozorgi, S., Wah, T.Y., Herawan, T.: Big data clustering: a review. In: International Conference on Computational Science and Its Applications, pp. 707–720. Springer, Cham, June 2014

A Study of the Necessity of Signal Categorisation in Dendritic Cell Algorithm

Noe Elisa[1], Fei Chao[2], and Longzhi Yang[1($\boxtimes$)]

[1] Department of Computer and Information Sciences, Northumbria University,
Newcastle upon Tyne NE1 8ST, UK
{noe.nnko,longzhi.yang}@northumbria.ac.uk
[2] Cognitive Science Department, School of Information Science and Engineering,
Xiamen University, Xiamen, China
fchao@xmu.edu.cn

Abstract. Dendritic Cell Algorithm (DCA) is a binary classifier in the category of artificial immune systems. During its pre-processing phase, DCA requires features to be mapped into three signal categories including safe signal, pathogenic associated molecular pattern, and danger signal, which is usually referred to as signal categorisation. Conventionally, feature-to-signal mapping is performed either manually or automatically by using dimension reduction or feature selection techniques such as principal component analysis and fuzzy rough set theory. The former has been criticised for its potential over-fitting, whilst the latter may suffer from either the loss of underlying feature meaning or impractical for large and complex datasets. This work therefore investigate the necessity of the signal categorisation process by proposing a DCA without the use of signal categorisation but with generalised context detection functions, where the more complex parameters of these functions are learned using the genetic algorithm. This is followed by a comparative study on twelve well-known datasets; the experimental results show overall better performances in terms of accuracy, sensitivity and specificity compared to the conventional DCAs. This confirms that the signal categorisation phase is not necessary, if the weights of the generalised context detection functions can be optimised.

Keywords: Dendritic cell algorithm · Signal categorisation · Genetic algorithm · Feature-to-signal mapping · Classification

1 Introduction

Dendritic Cell Algorithm (DCA) is an artificial immune system aiming to detect anomalies, which is abstracted from the behavior and functioning of natural dendritic cells (DCs) [1]. The DCA has caught the attention of different researchers

This work has been supported by the Commonwealth Scholarship Commission (CSC-TZCS-2017-717) and Northumbria University, UK.

© Springer Nature Switzerland AG 2020
Z. Ju et al. (Eds.): UKCI 2019, AISC 1043, pp. 210–222, 2020.
https://doi.org/10.1007/978-3-030-29933-0_18

due to its interesting classification performance and several beneficial features for binary classification problems. Usually, the DCA goes through four phases namely, signal categorisation, context detection, context assignment and data labeling, after pre-processing, to process and classify data instances. During pre-processing phase, feature selection is usually employed to select the most significant features from the dataset. Each selected feature is then mapped into one of the three signal categories, either safe signal (SS), pathogenic associated molecular pattern (PAMP), or danger signal (DS). These categories were abstracted from the natural biological immune system. The resulting three signal values are used as the system input to the DCA to perform binary classification.

A manual signal categorisation method was proposed in the seminal work of DCA [1], which uses expert knowledge of the problem domain to map each selected feature to its appropriate signal category. It is simply a brute force permutation process of trying different combination of features for categorisation. This method has been criticised to be undesirable as it may manually over-fit the data to the algorithm [2,3]. Nevertheless, the manual method is application dependent and requires a deep understanding of the problem domain, which is not feasible or practical for general users. Also, if cross-validation is applied with the manual method, the process is required to be performed for every single set, which is time consuming and sometimes difficult or even impossible to achieve.

An automated feature categorisation techniques have been introduced based on principal component analysis (PCA) in the work of [2], to overcome the limitation of manual method. In particular, PCA creates a new set of features with attributes ranked in terms of their variances. Then, the PCA ranking of attributes are mapped into the three signals of the DCA in the order of SS, PAMP and DS, entailing the importance of each signal category to the feature-to-signal mapping [2]. The PCA destroys the meaning of initial features presented in the dataset by generating a new set of features and often produces poor classification performance [3,4].

Fuzzy Rough Set Theory based DCA (FRST-DCA) was proposed for automatic feature selection and signal categorisation based on QuickReduct algorithm [5], to solve the limitations of PCA, in the work of [4]. The FRST-DCA uses the reduct and the core rough set theory concepts to extract the most important features and categorises them to their specific signal types. Firstly, the FRST-DCA selects the attribute having the highest dependency degree in the reduct to form the SS signal. Then, the second attribute to form the reduct is assigned to the PAMP signal while the rest of attributes in the reduct are combined to form the DS signal [4]. The FRST-DCA produces better classification results than PCA, however, it works only on simple datasets and presents an information loss as data should be discretised in advance [3].

Note that the aggregation of the assigned features for each signal is implemented by the simple arithmetic average function; and the resultant signals are used as the inputs for other weighted functions. Therefore, the signal categorisation step may not be necessary if these two functions can be combined into one with a proper set of weights. From this observation, this work proposes a new

DCA without the use of the signal categorisation step, but with more generalised context detection functions. The weights of the generalised functions are determined using the genetic algorithm (GA). Briefly, the proposed approach firstly performs feature selection, then, without signal categorisation, it employs GA to generate the optimal weights associated with the selected features. Secondly, it detects the context of artificial DCs by using the proposed generalised weighted function. After this, the same with the original DCA, it assigns the context to DCs either as normal termed as semi-mature (smDC) or anomalous termed as mature (mDC) based on the concentration of the contexts. This is followed by the labelling of the data using exactly the same majority votes approach. The proposed approach was validated using twelve well known two-class datasets from UCI machine learning repository [6] with overall better performances demonstrated, compared to the conventional DCAs in terms of accuracy, sensitivity and specificity.

The organisation of the rest of this paper is as follows: Sect. 2 briefly provides the background of the danger theory and the dendritic cell algorithm. Section 3 details the proposed approach without the employment of the signal categorisation phase. Section 4 demonstrates the experimentation process, validation and analysis of the results. Finally, Sect. 5 concludes this study and points out the possible future direction.

2   Background

The underpinning biological danger theory model, and the abstracted original DCA are reviewed in this section.

2.1   Biological Danger Theory

Danger theory articulates that immune system rely not only on discrimination between self and foreign molecules but rather reacts to what might cause damage to the body. In biology, the DCs coordinate antigens (e.g. virus) presentation from the external tissues (e.g. skin and lung) and immune system [7]. They produce co-stimulatory molecules (csm) on their cell surface which limit the time they spend sampling the antigens in the tissue. DCs are sensitive to the concentrations of the three signals, including PAMP, DS and SS. In particular, PAMP are abnormal proteins produced by viruses and bacteria which can easily activate immune response; DS are released from the disrupted or stressed cells in the tissue which indicates an anomalous situation but with lower confidence than PAMP; and SS are produced by normal cell death process in the tissue, which is an indicator of normal cell behavior.

Usually, DCs exist in one of the three states depending on the concentrations of SS, PAMP or DS signals in the tissue: (1) Immature DCs (iDCs), which are found in tissues in their pure state where they still collect antigens (i.e.; normal proteins or something foreign). The concentration of the signals of the collected antigens causes iDC to move to a full-mature or semi-mature state; (2) Mature

DCs (mDCs), which are transformed from iDCs when they are exposed to a greater quantity of either PAMP or DS than SS which causes immune reaction; (3) Semi-mature DCs (smDCs), which are transformed to smDCs when they expose to more SS than PAMP and DS which causes immune tolerance.

2.2   The Original Dendritic Cell Algorithm

The development of the original DCA was inspired by the danger theory and the functioning of natural DCs [1]. Firstly, feature selection process is applied with the DCA to select the most informative features, which is followed by four phases, including signal categorisation, context detection, context assignment and labeling as discussed below.

Signal Categorisation. The selected features are categorised into either SS, PAMP or DS based on their definitions derived from the biological metaphor. The PAMP is extracted from the features that have high degree of abnormality which can certainly cause damage to the system. The SS is extracted from the features that have high degree of normality or steady-state system behaviour while the DS is extracted from the features that have high degree of abnormality associated with them when their values increase and indicate moderate degree of normality when their values decrease.

There are three common signal categorisation techniques used with the DCA in the literature, including the manual approach by relying on the expert knowledge of the problem domain [1], and the automatic methods based on PCA [2], fuzzy-rough set theory based-DCA (FRST–DCA) [4], or fuzzy inference systems [8,9]. After signal categorisation process, the DCA initialises a population of artificial DCs (often 100) in a sampling pool which are responsible for data items (i.e. antigens in biological term) sampling [1,10]. Then, a pre-defined number of DCs (often 10) are selected to perform context detection, as discussed in below.

Context Detection. Each DC in the sampling pool is assigned a migration threshold in order to determine the lifespan while sampling data items and simulate the function of csm as discussed in the Sect. 2. The migration threshold is determined by the characteristic behaviour of the dataset and the amount of data instances the DCs can collect, usually initialised in a Gaussian distribution [2]. In this phase, the selected DCs use a weight summation function for sampling based on three sets of pre-defined weights to process the input signals to obtain three output context values termed as csm, mDC and $smDC$:

$$Contex[csm, smDC, mDC] = \sum_{d=1}^{m} \frac{\sum_{i,j=1,1}^{3}(c_j * w_i^j)}{\sum_{i,j=1,1}^{3} w_i^j}, \tag{1}$$

where $c_j(j = 1, 2, 3)$, represent the PAMP, DS and SS signal values respectively; and $w_i^j(i, j = 1, 2, 3)$ represent the weights of csm, mDC and $smDC$ context, regarding PAMP, DS and SS, respectively. The weights are usually either pre-defined or derived empirically from the dataset.

Since the DCs sample multiple data items overtime, they accumulate the three output context values for the sampled data items to obtain three cumulative values in an incremental manner. Simultaneously, each DC continuously compares its cumulative csm value with the assigned migration thresholds; so, when the csm value exceeds the migration thresholds, it ceases sampling the data items, becomes either $smDC$ or mDC, and move to the context assignment phase.

Context Assignment. The cumulative context values of $smDC$ and mDC obtained from the detection phase are used to perform context assignment. If DC has a greater cumulative mDC than $smDC$, it is assigned a binary value of 1 and 0 otherwise. Then, the DC assigns this value to all the data items it has sampled. This information is then used in the classification phase to compute the number of anomalous data items presented in the dataset, given that those with a binary value of 1 are potentially anomalous, otherwise normal. Note that, once the DCs completed their life cycle in DCA, they are reset and returned to the sampling population in order to maintain the population size.

Label Assignment. The processed data items by DCs are analysed by deriving the Mature Context Antigen Value ($MCAV$) per data item. Firstly, the anomaly threshold of $MCAV$ is computed from the training dataset by taking the ratio of the total number of anomaly class's data samples to the total number of data samples present. Then, $MCAV$ value is determined from the ratio of the number of times a data item is presented in the mDC (i.e.; anomaly) context to the total number of presentation by multiple DCs. If a data item has the $MCAV$ value greater than the anomaly threshold, it is classified into the anomaly class, otherwise normal class.

3   Revised DCA Without Signal Categorisation

Features are conventionally combined into signals using arithmetic average functions, which are then used for the calculation of the context values of csm, $smDC$, and mDC using weighted summations function. Therefore, from mathematical point of view, these two level of linear weighted summation functions can be combined into one to directly calculate the context values from the selected features. Therefore, a revised DCA without the signal categorisation step is proposed in this section as demonstrated in Fig. 1. Of course, in this case, the challenge would be the determination of the weights of the revised weighted summation function. As demonstrated in Fig. 1, the general search algorithm, genetic algorithm (GA), is applied to search the optimal set of weights. After this, the last two phases of the revised DCA are exactly the same with their originals, and thus the rest of this section focuses only on the context detection phase and the determination of the weights using GA.

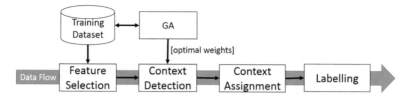

Fig. 1. The proposed DCA without signal categorisation

3.1 Context Detection Without Signal Categorisation

Suppose that u features have been selected through the feature selection pre-processing step, the migration threshold for an iDC is th, and m data instances have been sampled by an iDC overtime; its cumulative value of csm, denoted as $Context[csm]$, is determined using a generalised weighted function of Eq. 1 as expressed below:

$$Context[csm] = \sum_{d=1}^{m} \frac{\sum_{j=1}^{u}(x_j * w_{csm}^j)}{\sum_{j=1}^{u} w_{csm}^j}, \tag{2}$$

where x_j is the normalised value of selected features j, w_{csm}^j is the weight of attribute j regarding the csm. The weight w_{csm}^j is determined during the training process using any general search algorithm and the GA is used in this work as discussed in the next Subsect. 3.2.

As soon as the $context[csm]$ value of the iDC exceeds its assigned th, it ceases sampling data instances, develops to a full maturation (i.e., abnormal) or semi maturation (i.e.; normal), and then moves to the mature DC pool ready for the calculation of the cumulative values of the $smDC$ and mDC contexts. Once the mature DC pool reaches a pre-defined size, the cumulative $smDC$ and mDC context values of each DC are calculated. Given a mature or semi-mature DC, its cumulative $smDC$ and mDC context values regarding normal and abnormal classes can be calculated using weighted functions 3 and 4, respectively.

$$smDC = \sum_{d=1}^{m} \frac{\sum_{j=1}^{u}(x_j * w_{smDC}^j)}{\sum_{j=1}^{u} w_{smDC}^j}, \tag{3}$$

$$mDC = \sum_{d=1}^{m} \frac{\sum_{j=1}^{u}(x_j * w_{mDC}^j)}{\sum_{j=1}^{u} w_{mDC}^j}. \tag{4}$$

Although these equations share similar forms with Eq. 2, they take very different set of weights. All the weights are determined using a GA in this work as detailed below.

3.2 Weight Optimisation by GA

GA uses adaptive search heuristic techniques inspired by natural evolution to find the optimal or near-optimal solutions to optimisation problems [11]. GA has been successfully employed for weights optimisation in many tasks, such as neural networks [12], DCA [13] and rule base optimisation in fuzzy inference systems [14, 15] with significant performances. The main steps of weight determination for Eqs. 2, 3, and 4 in this work are summarised below.

(1) Individual representation: In this work, an individual (I) is designated as a possible solution that comprises of all the weights involved in Eqs. 2, 3 and 4. the individual can be represented as $I = \{w_{smDC}^1, w_{smDC}^2, .., w_{smDC}^u, w_{mDC}^1, w_{mDC}^2, .., w_{mDC}^u, w_{csm}^1, w_{csm}^2, .., w_{csm}^u\}$, where u is the total number of selected features during the pre-processing stage.

(2) Individual initialisation: The population $\mathbb{P} = \{I_1, I_2, ..., I_N\}$ is initialised by random numbers from a Gaussian distribution with a mean of 0 and a standard deviation of 5. Here, N is the size of population.

(3) Objective function: In this work, the objective function is taken as the classification accuracy of the DCA.

(4) Selection: The fitness proportionate selection method is adopted in this work for selecting individuals who reproduce. Therefore, the probability of an individual to become a parent is proportional to its fitness.

(5) Reproduction: In this work, the single point crossover and mutation operations were applied to reduce the likelihood of GA being trapped in local maxima and increase the probability of obtaining the global optimal solution.

(6) Iteration and termination: In this work, GA converges when the predefined maximum number of iterations is attained or the classification accuracy of the DCA exceeds the pre-specified threshold of the optimum accuracy. When GA terminates, the optimal weights are taken from the fittest individual in the current population. From this, the DCA can perform using this set of weights.

4   Comparative Study

This section details the experimental setup, results and analysis of a comparative study between the proposed approach and the original DCA. The datasets for evaluating the performance of the proposed technique were taken from the UCI machine learning repository [6]. The properties of these datasets are provided in Table 1. Note that, ten-fold cross-validation was applied to all datasets since no test datasets are provided in the repository.

4.1   Experimental Setup

Dataset Pre-processing: This work adopted the Information Gain (IG) method to select the most important features from the datasets [16]. Then, the selected features were normalized using the min-max normalisation.

DCs Initialisation and Sampling: A population of 100 DCs was initialised in the sampling pool, and the size of the mature pool was 10 DCs. The migration thresholds of DCs were initialised based on a Gaussian distribution with a mean of 7.5 and standard deviation of 1.

Table 1. Benchmark datasets

Dataset	#Samples	#Features
Mammographic Mass (MM)	961	6
Pima Indians Diabetes (PID)	768	8
Blood Transfusion Service Center (BTSC)	748	5
Wisconsin Breast Cancer (WBC)	699	9
Ionosphere (IONO)	351	34
Liver Disorders (LD)	345	7
Haberman's Survival (HS)	306	4
Statlog (Heart) (STAT)	270	13
Sonor (SN)	208	61
Spambase (SB)	4601	58

GA Parameters: In this study, the parameter values used for the GA are; mutation rate of 0.1, crossover rate of 0.95, 250 number of iterations and 50 individuals in a population.

4.2   Measurement Metrics

The performance of the proposed approach was measured in terms of sensitivity, specificity and accuracy. The results were compared with the conventional DCA based on PCA [2] and FRST-DCA [4] signal categorisation techniques. Sensitivity, specificity, and accuracy are defined as follows:

$$Sensitivity = \frac{TP}{TP + FN}$$
$$Specificity = \frac{TN}{TN + FP}$$
$$Accuracy = \frac{TP + TN}{TP + TN + FP + FN},$$

(5)

where TP, FP, TN, and FN refer respectively to true positive, false positive, true negative and false negative, respectively.

Furthermore, the performance of the proposed approach was evaluated via Precision, Recall and F1-Score metrics in order to understand how good is the model when there is an uneven class distribution in the dataset (i.e., class imbalance). Note that, high accuracy indicates that the model is doing better only when the dataset has balanced data samples between classes. F1-score is more efficient than accuracy when the datasets have an uneven class distributions.

The precision, recall and F1-score are computed as follows:

$$Precision = \frac{TP}{TP + FP}$$
$$Recall = \frac{TP}{TP + FN} \tag{6}$$
$$F1 - score = \frac{2 * Recall * Precision}{Recall + Precision}.$$

4.3 Results and Analysis

Sensitivity, Specificity and Accuracy Results: Table 2 presents the testing results on sensitivity, specificity and accuracy for the proposed approach, PCA-DCA and FRST-DCA. The best performing result among the three approaches for each dataset are marked in bold.

Table 2. Comparison of the Proposed approach with the RST-DCA and PCA

Dataset	Sensitivity (%)			Specificity (%)			Accuracy (%)		
	Proposed	FRST	PCA	Proposed	FRST	PCA	Proposed	FRST	PCA
MM	**100**	99.10	88.25	97.73	**99.03**	93.25	98.75	**99.06**	91.28
PID	**100**	99.40	94.66	99.21	**99.25**	94.95	**99.47**	99.34	94.80
BTSC	**100**	99.47	92.59	**99.65**	98.87	85.52	**99.73**	99.33	87.67
WBC	**100**	99.13	99.46	**98.35**	98.33	99.33	**99.43**	98.85	99.39
IONO	96.18	**97.77**	94.44	**100**	95.23	95.55	**98.57**	96.86	95.15
LD	**97.55**	89.50	93.82	**100**	90.34	90.10	**98.55**	89.85	91.82
HS	84.38	**88.88**	81.60	**100**	94.22	70.90	**95.10**	92.81	79.40
STAT	**100**	90.00	82.50	**100**	91.33	84.55	**100**	90.74	83.75
SN	**100**	97.29	93.82	97.00	**97.93**	90.10	**98.56**	97.59	91.82
SB	**100**	96.29	94.82	**99.71**	93.93	92.10	**99.83**	97.59	93.82

The sensitivity, specificity and accuracy results in this table indicate that, the proposed DCA overall performs better than the other two. In particular, the proposed DCA generated better performances compared to FRST-DCA and DCA-PCA except on three datasets where it was slightly outperformed by the FRST-DCA on accuracy for MM dataset, sensitivity for IONO and HS datasets and specificity for MM and SN datasets. The performance of the proposed DCA on all datasets in terms of accuracy range from 95% to 100%, so as sensitivity and specificity except the HS dataset which registered a sensitivity of 84.38%. This results demonstrate that, without mapping the features to signals, GA is able to generate the appropriate weights associated with the features.

The precision, recall and F1-score for the proposed DCA on all datasets are summarised in Fig. 2. Except of Haberman's Survival (HS) dataset where the proposed DCA has generated low recall, the rest datasets have all generated the best results with their precision ranging from almost 98% to 100%. The F1-score result for the majority of datasets range from 98% to 100%, indicating that, for imbalanced datasets, the proposed approach is notably a better choice with the DCA.

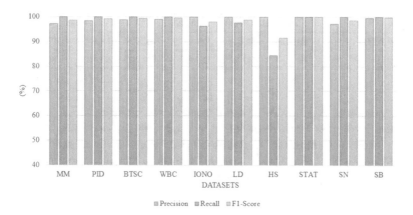

Fig. 2. Precision, Recall and F1-Score

4.4 Application of the Proposed Technique to Intrusion Detection Datasets

The proposed approach was also applied to two benchmark intrusion detection datasets which are more complex compared to the datasets present in Table 1. The testing datasets are provided with these datasets so the training dataset was used for training and the testing dataset was used for validation. The datasets are briefed below:

KDD99 Dataset contains 494,021 training records (97,278 normal and 396,743 anomalous) and 311,029 testing records (60,593 normal and 250,436 anomalous); each set with 41 attributes [17].

UNSW_NB15 Dataset contains 175,341 training records (56,000 normal and 119,341 anomalous) and 82,332 testing records (37,000 normal and 45,332 anomalous); each set with 49 attributes and more moderns attack types than the KDD99 [18].

Based on the sensitivity, specificity and accuracy, the proposed approach is only compared with the PCA-DCA since the FRST-DCA is practically not applicable to these datasets due to large number of samples. From Table 3, it is clear that the proposed approach is applicable to intrusion detection datasets with larger number of samples. The proposed approach has outperformed the PCA on the KDD99 dataset in terms of accuracy and sensitivity. Similarly, on

UNSW_NB15 dataset, the proposed version has performed better than the PCA on accuracy and specificity.

Table 3. Comparison of the Proposed approach and PCA on intrusion datasets

Dataset	Sensitivity (%)		Specificity (%)		Accuracy (%)	
	Proposed	PCA	Proposed	PCA	Proposed	PCA
KDD99	**99.98**	88.25	76.81	**93.25**	**94.05**	91.28
UNSW_NB15	87.48	**92.59**	**99.84**	85.52	**90.23**	89.67

The precision, recall and F1-score for the two intrusion detection datasets are shown in Fig. 3. Both datasets have produced better results on F1-score compared to the accuracy in Table 3, indicating that, the proposed technique is notably the better choice for DCA when applied to intrusion detection datasets which often contain uneven class distribution.

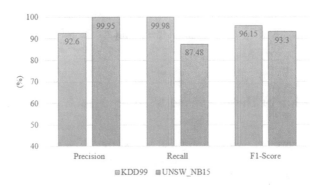

Fig. 3. Precision, Recall and F1-Score for IDS datasets

The experiments have proved that the signal categorisation is not a necessary step if the weights used in the context detection phase can be properly learned through a search algorithm. Actually, based on the experiments, better classification performance can be obtained for most of the situation if signal categorisation is not applied to the DCA, compared to that of the conventional DCAs with signal categorisation phase. Also, the DCA without signal categorisation is more convenient in handling the imprecision that may be caused by the lack of mathematical model for feature-to-signal mapping. Moreover, the DCA without the signal categorisation phase has a potential to eliminate the time, impression and complexity required to map the features to signal categories, which often leads to performance improvement.

5 Conclusions

This work analysed the necessity of the signal categorisation phase of DCA by proposing a DCA without the signal categorisation phase but with generalised context detection functions. The weights of the context detection functions are determined using the GA. The proposed approach was validated using ten binary classification datasets and two benchmark intrusion detection datasets. The experimental results show that, the DCA without signal categorisation produced comparable results to that of the conventional DCA while simultaneously eliminating time, impression and complexity required to map the features to their convenient signal categories. The precision, recall and F1-score results further prove that, the application of DCA without signal categorisation perform well on imbalance datasets as well as on large datasets without limitations. The possible future work is to extend the DCA to multiclass classification problems as all the existing DCA only targets binary classification problems.

References

1. Greensmith, J., Aickelin, U., Cayzer, S.: Introducing dendritic cells as a novel immune-inspired algorithm for anomaly detection. In: International Conference on Artificial Immune Systems, pp. 153–167. Springer (2005)
2. Gu, F.: Theoretical and empirical extensions of the dendritic cell algorithm. Ph.D. thesis, University of Nottingham (2011)
3. Chelly, Z., Elouedi, Z.: A survey of the dendritic cell algorithm. Knowl. Inf. Syst. **48**(3), 505–535 (2016)
4. Chelly, Z., Elouedi, Z.: Hybridization schemes of the fuzzy dendritic cell immune binary classifier based on different fuzzy clustering techniques. New Gener. Comput. **33**(1), 1–31 (2015)
5. Jensen, R., Shen, Q.: A rough set aided system for sorting www bookmarks. In: Asia-Pacific Conference on Web Intelligence, pp. 95–105. Springer (2001)
6. Dua, D., Graff, C.: UCI machine learning repository (1998)
7. Banchereau, J., Steinman, R.M.: Dendritic cells and the control of immunity. Nature **392**(6673), 245 (1998)
8. Yang, L., Chao, F., Shen, Q.: Generalised adaptive fuzzy rule interpolation. IEEE Trans. Fuzzy Syst. **25**(4), 839–853 (2017)
9. Elisa, N., Li, J., Zuo, Z., Yang, L.: Dendritic cell algorithm with fuzzy inference system for input signal generation. In: UK Workshop on Computational Intelligence, pp. 203–214. Springer (2018)
10. Elisa, N., Yang, L., Qu, Y., Chao, F.: A revised dendritic cell algorithm using k-means clustering. In: 2018 IEEE 20th International Conference on High Performance Computing and Communications, pp. 1547–1554. IEEE (2018)
11. Holland, J.H.: Genetic algorithms. Sci. Am. **267**(1), 66–73 (1992)
12. Juang, C.-F.: A hybrid of genetic algorithm and particle swarm optimization for recurrent network design. IEEE Trans. Syst. Man Cybern. Part B (Cybern.) **34**(2), 997–1006 (2004)
13. Elisa, N., Yang, L., Naik, N.: Dendritic cell algorithm with optimised parameters using genetic algorithm. In: 2018 IEEE Congress on Evolutionary Computation (CEC), pp. 1–8. IEEE (2018)

14. Naik, N., Diao, R., Shen, Q.: Dynamic fuzzy rule interpolation and its application to intrusion detection. IEEE Trans. Fuzzy Syst. **26**(4), 1878–1892 (2018)
15. Li, J., Yang, L., Yanpeng, Q., Sexton, G.: An extended Takagi-Sugeno-Kang inference system (TSK+) with fuzzy interpolation and its rule base generation. Soft Comput. **22**(10), 3155–3170 (2018)
16. Witten, I.H., Frank, E., Hall, M.A., Pal, C.J.: Data Mining: Practical Machine Learning Tools and Techniques. Morgan Kaufmann, San Francisco (2016)
17. KDD Cup 1999 Data. http://kdd.ics.uci.edu/databases/kddcup99/kddcup99.html/. Accessed 16 Dec 2018
18. Moustafa, N., Slay, J.: UNSW-NB15: a comprehensive data set for network intrusion detection systems (UNSW-NB15 network data set). In: Military Communications and Information Systems Conference (MilCIS), pp. 1–6. IEEE (2015)

Interval Construction and Optimization for Mechanical Property Forecasting with Improved Neural Networks

Tingyu Xie[1], Gongzhuang Peng[2], and Hongwei Wang[1(✉)]

[1] ZJU-UIUC Institute, Zhejiang University, Haining, China
hongweiwang@zju.edu.cn
[2] Engineering Research Institute, University of Science and Technology Beijing, Beijing, China

Abstract. Efficient and accurate predication of mechanical properties is the key to controlling the production process. In this paper, a novel Prediction Interval (PI) based method is proposed for forecasting strip steel properties. It specifically consists of a Lower Upper Bound Estimation (LUBE) technique for PI generation based on Particle Swarm Optimization (PSO) and a Coverage Width Symmetry-based Criterion (CWSC) for PI evaluation. To evaluate the proposed method, computational experiments are carried out on two numerical datasets and two real-world datasets from a strip steel production process. A comparison between the results obtained by this work and previous work shows that the proposed method is viable and achieves more advantages. Moreover, the PI constructed on the real-world datasets achieve better quality, demonstrating that the proposed method has good potential in real-world problems.

Keywords: Prediction Interval · Neural Network · Mechanical property forecasting

1 Introduction

As information technologies such as Internet of Things (IoT) develop rapidly, great innovations have been made in manufacturing industry to enhance sharing and collaboration. Cloud Manufacturing (CMfg) is the most widely applied one amongst these innovations. In CMfg, many products such as strip steels have complex non-linear production processes for which real-time parameter tuning and control of the production process is critical. In strip steel production, prediction of steel mechanical property is holds to key to real-time parameter tuning. It is not only useful for optimizing process parameters, but also helpful in the development of new steel products. The research on mechanical property prediction of hot rolled strip steels began in 1950s when Irvine and Pickering [1] proposed using mathematical models to predict mechanical properties.

There are mainly two kinds of prediction models for strip steel mechanical properties: metallurgical mechanism models and statistical models. The former, with a wide range of applications, essentially focuses on modeling using physical laws, but requires

© Springer Nature Switzerland AG 2020
Z. Ju et al. (Eds.): UKCI 2019, AISC 1043, pp. 223–234, 2020.
https://doi.org/10.1007/978-3-030-29933-0_19

many assumptions and simplifications to be made due to complexity and unde-tectability of most variables. The latter is much more accurate, although interpretability of result is not as good as the theoretical models.

Although some predication systems have been developed for mechanical property prediction, they generally have low accuracy and are considered unpractical to some degree [2]. The reasons are mainly two-fold. First, the measurement error of input variables lowers the accuracy to a large extent. Second, metallurgical mechanism models rely on special assumptions to address its complexity, which results in inevi-table errors. Meanwhile, statistical models can be unreliable when datasets are too small. As it is very difficult to systematically improve detection accuracy and the cost is unacceptably high, the effect is very limited by working on detection accuracy [2]. In addition, many errors in metallurgical mechanism models are unavoidable due to its definition. Thus, the development of advanced statistical models is an effective way of improving prediction performance. A difficulty in statistical models is the collection of data. For example, previous research shows that it takes 1000–10000 samples of data to establish a relatively stable statistical model for one kind of trip steel [3]. In the context of CMfg, data collection becomes more efficient, which paves the way for more research into better models.

Current prediction systems focus on generation of point forecasts. However, the reliability of point forecasts can drop significantly when the uncertainty of data increases [4]. The production system of strip steel is complex and has a high level of uncertainty, and thus point forecasts can hardly meet the prediction requirement in terms of accuracy and reliability. To address these problems, Prediction Interval (PI) has been proposed in recent literature. A PI is formed by a lower and an upper bound, which covers the target value in a probability of $((1 - \alpha)\%)$ called confidence level [5]. By providing prediction intervals of different width, PI can provide useful information about uncertainty level of data.

Traditional methods for PI construction, e.g. the Bayesian method, have low fea-sibility and high computational cost due to their complex definitions. They also rely on specific and complicated assumptions about data distribution, making the constructed PI less reliable. To address these problems, Khosravi et al. proposed a new method called Lower Upper Bound Estimation (LUBE) [6]. In LUBE, a Neural Network (NN) model with two outputs is adopted to directly output the two bounds of PI [6]. As LUBE method eliminates tedious assumptions about data distribution and heavy computational burden, it is more practical and helpful. The cost function of the LUBE method in [6] is a coverage width-based criterion called CWC, which measures PI quality by considering both coverage probability and PI width. Some new cost func-tions have also been introduced to improve the PI construction process, but they all based on coverage probability and PI width. An index called Coverage Width Symmetry-based Criterion (CWSC) proposed in [7] first takes the geometric structure of PI into consideration by adding PI symmetry into evaluation [1].

As the cost function in LUBE is nonlinear, discontinuous and nondifferentiable, traditional derivative-based algorithms cannot be applied to this problem [4]. Various evolution-based algorithms such as simulated annealing [6] and Particle Swarm Optimization (PSO) [4] have been applied in LUBE. This research proposes a novel PI-based solution for mechanical property prediction, with the purpose of improving both the accuracy and applicability. Specifically, the LUBE method is applied to generate PI, an improved PSO with mutation operator is used to minimize the cost function, and the comprehensive index CWSC is applied to measure PI quality.

The paper is organized as follows. Section 2 introduces PI assessment measures. Section 3 describes PSO-based NN for LUBE method. Data and experiments are presented in Sect. 4. Section 5 concludes this paper and introduces future work.

2 PI Assessment Measures

The three indicators called PI Coverage Probability (PICP), PI width and PI symmetry are introduced in this section, and a comprehensive measure combining these three indices is described.

2.1 PICP

PICP is the most fundamental feature of PI, which indicates the probability of the target value falling in the constructed PI. The definition of PICP is as follows:

$$PICP = \frac{1}{n} \sum\nolimits_{i=1}^{n} c_i \tag{1}$$

where n is the number of samples. c_i indicates whether the target value of sample i lies within the corresponding PI. If the PI covers the target, $c_i = 1$, otherwise, $c_i = 0$.

2.2 PINAW

If the width of the PI is wide enough, a high PICP can be easily reached. However, a PI with a very large width is much less informative and of no use. Thus, width is another important feature of PI. Normalized Average Width of PI (PINAW), which is also called Normalized Mean PI Width (NMPIW), was applied in [6], which has definition as follows:

$$PINAW = \frac{1}{nR} \sum\nolimits_{i=1}^{n} (U_i - L_i) \tag{2}$$

where U_i and L_i correspond to the upper and lower bounds of constructed PI for sample i; R is the range of target value. n is the same as in PICP.

2.3 PIS

Although PICP and PINAW has been worked well for construction of PI, the geometric structure of PI has never been discussed until PI Symmetry (PIS) is brought out in [7]. A PI with the target value fall on the bound is of little help, but still can get high PICP and low PINAW. PIS addresses this problem by considering the distance between target value and the midpoint of PI. The definition of PIS is as follows:

$$PIS = \frac{1}{n}\sum\nolimits_{i=1}^{n} \frac{|y_i - (U_i + L_i)/2|}{U_i - L_i} \tag{3}$$

where n, y_i, U_i and L_i have the same meanings as in last two indicators. According to the definition, PIS is the normalized mean distance between target value and the midpoint of the PI. If $PIS \leq 0.5$, y_i falls in the constructed PI, otherwise, it is not covered by the PI. Specifically, if $PIS = 0$, y_i falls right on the midpoint of the PI. Thus, a PI with $PIS \leq 0.5$ is valid, and a PI with smaller PIS has better quality.

2.4   Comprehensive Measure of PI

Large PICP, narrow PINAW and small PIS are features of a PI with high quality, among which PICP and PINAW are two conflicting indices because a higher PICP often leads to a wider width, and a narrower width is likely to result in a lower PICP. An integrated indicator called Coverage Width-based Criterion (CWC) is formed in [6] by combining PICP and PINAW into one goal. The definition as follows:

$$CWC = PINAW(1 + \gamma(PICP)e^{-\eta(PICP-\mu)}) \tag{4}$$

where μ represents the nominal confidential level of the constructed PI, which can be set equal to $(1 - \alpha)$. Thus, a PICP higher than μ is preferred. η magnifies the difference of PICP and μ. By using the exponent term of PICP, coverage probability can be sufficiently optimized to satisfy the assigned confidential level. $\gamma(PICP)$ is a step function defined as follows:

$$\gamma(PICP) = \begin{cases} 0, PICP \geq \mu \\ 1, PICP < \mu \end{cases} \tag{5}$$

During training, $\gamma(PICP)$ is constantly set to 1 to maximumly optimize PICP.

Despite CWC worked well in previous researches, it has disadvantages. If a zero width is found during training, CWC will be zero and stuck in this minimum value, but the PICP can be far from the optimal. In addition, CWC doesn't consider the geometric structure of PI. To improve the comprehensive indicator of PI, Coverage Width Symmetry-based Criterion (CWSC) was introduced in [7], which has the definition as follows:

$$CWSC = \gamma(PIS)e^{\eta_3(PIS-\mu_2)} + \eta_2 PIARW + \gamma(PICP)e^{-\eta_1(PICP-\mu_1)} \tag{6}$$

where η_1 and μ_1 are the same as in CWC. η_2 is the weight of PINAW. Same as PICP, η_3 and μ_2 are two hyperparameters controlling the PIS, but the difference is that a PIS lower than μ_2 is desired. $\gamma(PIS)$ is a step function just like $\gamma(PICP)$, and is set to 1 during training.

Since a PI with PIS larger than μ_2 or PICP lower than μ_1 is invalid, the exponents of PIS and PICP are applied in CWSC to efficiently optimize these two indices. Instead of using the product of PICP, PINAW and PIS, the sum of them is adopted. Because in this way, CWSC will not be trapped in minimum value of zero when a zero PI width is reached. And a sum is much easier to control.

At the beginning of training, PICP is always very small and far from satisfying μ_2. This leads to a heavy penalty given by the cost function. During training, the penalty term will decrease exponentially along with the increase of PICP. The penalty term of PIS works in the same. When PICP and PIS all reach the assigned level, the PINAW will play a more important role in optimization and the cost function begins to search for a narrower width. Finally, there will be a balance between the PICP, PIS and PINAW.

The reason why $\gamma(PICP)$ is set to a step function during testing is as follows: When $PICP < \mu$, by setting $\gamma(PICP) = 1$, the penalty of the undesired PICP is counted into evaluation. Once the PICP becomes not smaller than PINC, PIs are considered valid, and there is no need to consider the penalty term, so $\gamma(PICP)$ should be set to 0.

3 PSO-Based NN for LUBE Method

3.1 LUBE Method

LUBE method is first developed in [6]. It adopts an NN model with two output neurons to directly generate the PI. The first and second outputs of the NN correspond to the lower and upper bounds of PIs respectively. Figure 1 shows the general architecture of the NN applied in LUBE method.

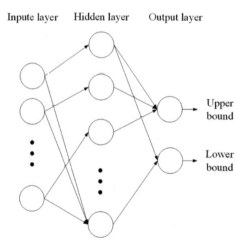

Fig. 1. NN model for LUBE method to generate upper and lower bounds of PIs.

In this paper, the activation of the hidden layer is sigmoid function, and the activation of the output layer is relu function. Different activations are applied in different cases. Sigmoid function is defined as follows:

$$\text{sigmoid}(x) = \frac{1}{1 + e^{-x}} \tag{7}$$

The relu function is as follows:

$$\text{relu}(x) = \max(x, 0) \tag{8}$$

The output of the ith hidden neuron is as below:

$$a_i = sigmoid(W_i \cdot x + b_i) \tag{9}$$

where $W_i = (w_{1i}, w_{2i}, \cdots, w_{ni})$ is the weights connecting the n inputs and the ith hidden neuron; b_i is the bias of ith hidden neuron.

The output of the NN is given by:

$$y_j = relu\left(\sum\nolimits_{i=1}^{m} h_{ij} \cdot a_i + b_j\right) \tag{10}$$

where m is the number of hidden units; $j = 1, 2$, representing the index of output neurons; h_{ij} is the weight connecting the ith hidden neuron and the jth output neuron; a_i is the output value of the ith hidden neuron; and b_j is the bias of the jth output neuron.

3.2   PSO-Based LUBE Method

The key idea of applying PSO in the LUBE method is encoding NN weights as the positions of particles. Meanwhile, velocities of particles represent the updating rates of NN weights. The rest of this section details the process of PSO-based LUBE method, which is illustrated in Fig. 2.

1. Parameter initialization

As NN weights are encoded as particles positions, the velocities and positions of particles are the parameters to be initialized. The quality of parameter initialization has a great impact on the quality of constructed PI and the convergency of the algorithm. Thus, an appropriate and effective approach of initialization should be applied.

2. Velocity and position update

The optimization of NN weights achieved through the update of articles' velocities and positions. The update rules are as follows [8, 9]:

$$v_n(t+1) = \omega v_n(t) + c_1 rand()\left(p_{best,n} - x_n(t)\right) + c_2 rand()(g_{best,n} - x_n(t))$$

$$x_n(t+1) = x_n(t) + v_n(t+1) \tag{11}$$

where v_n is the velocity on the nth dimension, $x_n(t)$ is the position on the nth dimension, ω is an inertia parameter controlling the extent of keeping the velocity still, $rand()$ is a random number between [0, 1], c_1 and c_2 are scaling parameters that control the influence of p_{best} and g_{best}. Besides, x_{max} and v_{max} are used to limit the absolute value of positions and velocities.

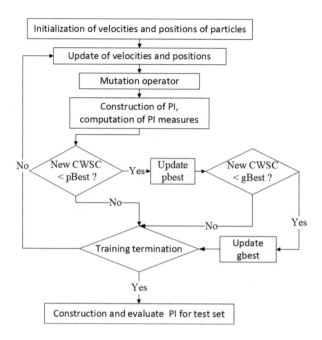

Fig. 2. The process of PSO-based NN for LUBE method

Through the update, a particle keeps moving towards the contemporary best position and its own historical best position to search for optimal solution.

3. Mutation of particles

Mutation has been widely applied in evolutionary algorithms to increase searching ability. Previous researches also show that the results of PSO can be improved by adopting mutation operator [10]. With mutation operator, a particle not only moves towards the contemporary best positions, but also flies around to expand the searching capacity.

4. PI construction

After parameter update and mutation complete, the model NN like the one showed in Fig. 1 carries out forward propagation and generates the upper lower bounds of PI. Next, the PI evaluation indices are calculated using equations described in Sect. 2.

5. p_{best} and g_{best} update

p_{best} and g_{best} guide particles to better positions during searching. g_{best} is the best particle in the whole swarm, which constructs PI with smallest CWSC, and p_{best} is the personal historical best solution of one particle. Thus, every particle has its corresponding p_{best}, while there is only one g_{best} in a whole group. Every time PI are constructed, particles get new CWSC value. If the new CWSC of a particle is smaller than its p_{best}, the p_{best} will be replaced by this particle. Similarly, if any new CWSC is smaller than g_{best}, g_{best} will be updated as the corresponding particle.

As shown in Fig. 2, Step 2 to 5 will be repeated until training termination.

4   Experiment

4.1   Datasets

We take the cold-Strip Steel of a steel mill as our study case in mechanical property prediction. As there are hundreds of procedures in the production of strip steel, sixteen important procedures are selected to offer the process parameters. These process parameters are the input of the prediction system, and the strength and tensile strength are the two predicted mechanical properties in this paper. Thus, we carry out the experiments of mechanical property prediction on two separate datasets: Strength (STR) and Tensile Strength (TS). For each dataset, the attributes are sixteen. To validate our method, we also generate two synthetic datasets which has also been studied in [4], and compare our results with the results in [4]. The two synthetic datasets are as follows:

1. The first one called Ding10 is a one-dimensional function: $f(x) = x^2 + \sin(x) + 2 + \varepsilon$, where x is a random number in $[-10, 10]$, and ε is the added noise which has a heterogeneous distribution.
2. The second one called HAS is a five dimensional function: $f(x_1, x_2, x_3, x_4, x_5) = 0.0647(12 + 3x_1 - 3.5x_2^2 + 7.2x_3^3)(1 + \cos(4\pi x_4))$, and it is added by a normally distributed noise whose variance is a constant value.

Table 1 summarizes all the four datasets studied in our experiments.

Table 1. Datasets of experiments

Datasets	Target	Samples	Attributes
1	A 1-D function with heterogeneous distributed noise (Ding10)	500	1
2	A 5-D function with normally distributed noise of constant variance (HAS)	300	5
3	The strength value of the produced strip steel (STR)	26571	16
4	The tensile strength value of the produced strip steel (TS)	26577	16

4.2   Methodologies

For each dataset, 70% of samples are taken as training set, the rest are set to be test set. As NN structure has a great influence on the training results, k-fold cross validation is applied on training set to select the optimal NN structure. In k-fold cross validation method, the training set is split into k subsets; every subset takes turns to be the validation set, with the rest being the training set to train a candidate NN structure [11]; each candidate NN is trained and validated for k times using k different training and validation sets; the mean or median value of evaluation index for k validation sets is taken as the measure of each candidate NN.

In this paper, 5-fold cross validation is adopted, and mean value of CWSC for five validation sets is taken to measure the quality of candidate NN structure. The structure with the smallest value of mean CWSC is taken as the optimal structure and adopted in the following training. All candidate NN structures are similar as the one showed in Fig. 1, except that they vary in the number of hidden neurons. For dataset Ding10 and dataset HAS, the numbers of hidden neuron are searched between [1, 20], while for dataset strength and tensile strength, as they are more complex, the numbers of hidden neuron are searched between [33, 50].

Dataset Ding10 is taken as an example here to illustrate the process of cross validation. Figure 3 shows the mean CWSC corresponding to different numbers of hidden neurons.

According to Fig. 3, NN structures with 8, 12 and 13 hidden neurons get the smallest CWSC. Take computational burden, model complexity and generalization into consideration, NN with 8 hidden neuron is chosen be the optimal structure.

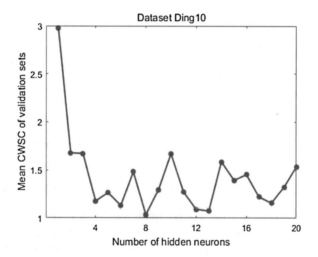

Fig. 3. CWSC value corresponding to different numbers of hidden neurons range in [1, 20].

For parameter initialization, previous researches show that Nguyen-Widrow (NW) method always obtains good results and is more stable among various initialization methods [4]. Thus, NW method is adopted here.

In mutation step, gaussian mutation is applied on particle positions in this paper, with the mean set as the parameter value and the standard deviation set as 10% of that value.

The rest of methodologies adopted in our experiments are showed in Fig. 2. Table 2 shows the values of CWSC parameters adopted in this paper.

4.3   Training Results

To illustrate the training process, Fig. 4 is drawn to show the change of cost function value also the CWSC of g_{best} during training.

Table 2. Values of CWSC parameters

Parameters	Values
η_1	5
η_2	80
η_3	5
μ_1	0.9
μ_2	0.5

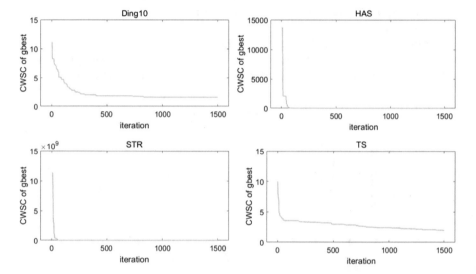

Fig. 4. The change of g_{best} CWSC of five datasets during iteration

As shown in Fig. 4, CWSC drops rapidly at the beginning of training for all datasets, and reaches a very small value in the end. For Ding10 and tensile strength, CWSC not only drops in the beginning, but also fitfully decreases during iteration. According to the training results, the adopted method is qualified with robust capacity for searching, high speed for convergency, and strong ability for of jumping out of local optima.

4.4   Testing Results and Discussion

Table 3 include the test results of four datasets. To validate the adopted method, the test results of Ding10 and HAS in [4] are also summarized in Table 3 for comparison. As the comprehensive measures of PI applied in [4] is CWC without PIS, only PICP and PINAW got in [4] are shown in the table.

For Ding10, the PINAW and PICP of adopted method are generally in the same level as they in [4], despite the PINAW in this paper is slightly larger than it of CWC-based method. For HAS, not only the PICP in this paper is smaller, but also the PINAW is much narrower. What's more, these two datasets both get a quite small PIS. Thus, the adopted method adopted is qualified of construction PI with high quality.

Table 3. Test results

Comparison	Datasets	PICP	PINAW	PIS	CWSC
CWSC with PSO	Ding10	0.9267	0.1191	0.2813	0.5954
	HAS	0.9333	0.1409	0.2614	0.7043
	STR	0.9379	0.6477	0.2541	3.238
	TS	0.9364	0.3326	0.2131	1.663
CWC with PSO	Ding10	0.9267	0.1149	/	/
	HAS	0.9111	0.6181	/	/

As shown in Table 3, each test set get a PICP and PIS superior to the assigned level, while the performance on PINAW varies from different datasets. The PINAW for Ding10 and HAS are very small, and for TS is in the middle, while for STR is largest. Through variant width, PI offers the information about the level of uncertainty in the data and provide the reliability of prediction. A wider PI always indicates higher level of uncertainty. The variance performance of PI width on STR and TS demonstrate the informativeness and effectiveness of PI, which can play a significant role in mechanical property prediction.

To visualize the test results, Fig. 5 was drawn to present the normalized PI constructed for test sets. Given space limitations, only the first 80 samples in each dataset was shown. According to Fig. 5, PI has variant performance for different mechanical properties. Compared with PI for TS, the PI for STR are much wider, indicating a higher level of uncertainty in strength. Meanwhile, for both mechanical properties, the coverage probabilities are very large, with a large percentage of target values falling on the middle area of PI as well. As for two synthetic datasets, PI width in both cases are sufficiently narrow, besides a high level of coverage rate and PI symmetry can been seen in both datasets. This intuitively demonstrate the effectiveness of the adopted method, and further shows the successful application of PI to mechanical property prediction.

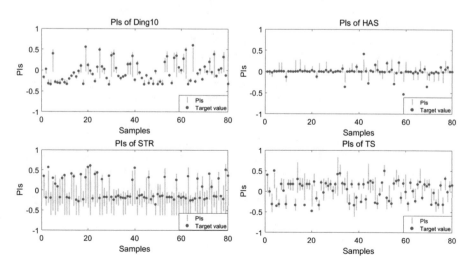

Fig. 5. The constructed PI of four test datasets

5   Conclusion

As existing prediction systems for mechanical property generally have low accuracy, PI is first applied to predict mechanical property in this paper. By applying PSO-based LUBE method, PI is generated effectively. To evaluate PI comprehensively, CWSC consisting of coverage probability, PI width and PI symmetry indices is adopted as the cost function. The prediction results on two synthetic datasets are superior compared with they are in previous work. Furthermore, PI with high quality are also obtained on two real datasets of two mechanical properties. The width of PI varies between datasets, which demonstrates PI is qualified to offer information about data uncertainty.

The experiments in this paper successfully applied PI on mechanical property prediction, which can be helpful in improving the control of production under CMfg.

To improve the quality of PI, NN with multiple hidden layers will be studied in the future. As there are plenty of advanced Machine Learning (ML) models proposed in literature, some effective models will also be studied to improve the performance of LUBE method.

References

1. Gan, Y.: Development of prediction technology for microstructure of thin slab continuous casting and rolling (TSCR) hot rolling process. Steel **38**(8), 10–15 (2003)
2. Guo, Z.H., Zhang, Q.L., Su, Y.C.: Thoughts on prediction technology of mechanical properties of hot rolled strip. Metall. Autom. **33**(2), 1–6 (2009)
3. Guo, Z.H., Su, Y.C., Zhang, Q.L.: Consideration on mechanical property prediction technology of hot rolled strip. Metall. Autom. **33**(2), 1–6 (2009)
4. Quan, H., Srinivasan, D., Khosravi, A.: Particle swarm optimization for construction of neural network-based prediction intervals. Neurocomputing **127**, 172–180 (2014)
5. Chatfield, C.: Calculating interval forecasts. J. Bus. Econ. Stat. **11**(2), 121–135 (1993)
6. Khosravi, A., Nahavandi, S., Creighton, D.: Lower upper bound estimation method for construction of neural network-based prediction intervals. IEEE Trans. Neural Netw. **22**(3), 337–346 (2011)
7. Zhang, H., Zhou, J., Ye, L.: Lower upper bound estimation method considering symmetry for construction of prediction intervals in flood forecasting. Water Resour. Manag. **29**(15), 5505–5519 (2015)
8. Panigrahi, B.K., Pandi, V.R., Das, S.: Adaptive particle swarm optimization approach for static and dynamic economic load dispatch. Energy Convers. Manag. **49**(6), 1407–1415 (2008)
9. Pulido, G.T., Coello, C.A.C.: A constraint-handling mechanism for particle swarm optimization. In: Congress on Evolutionary Computation. IEEE (2016)
10. Coello, C.A.C., Pulido, G.T., Lechuga, M.S.: Handling multiple objectives with particle swarm optimization. IEEE Trans. Evol. Comput. **8**(3), 256–279 (2004)
11. Wright, J.L., Manic, M.: Neural network architecture selection analysis with application to cryptography location. In: International Joint Conference on Neural Networks. IEEE (2010)

An Immunological Algorithm for Graph Modularity Optimization

A. G. Spampinato[1,2], R. A. Scollo[1], S. Cavallaro[2], M. Pavone[1(✉)],
and V. Cutello[1]

[1] Department of Mathematics and Computer Science, University of Catania,
V.le A. Doria 6, 95125 Catania, Italy
mpavone@dmi.unict.it, cutello@unict.it
[2] Institute for Biomedical Reasearch and Innovation,
Italian National Research Council, Via P. Gaifami 18, 95126 Catania, Italy
sebastiano.cavallaro@cnr.it

Abstract. Complex networks constitute the backbone of complex systems. They represent a powerful interpretation tool for describing and analyzing many different kinds of systems from biology, economics, engineering and social networks. Uncovering the community structure exhibited by real networks is a crucial step towards a better understanding of complex systems, revealing the internal organization of nodes. However, existing algorithms in the literature up-to-date present several crucial issues, and the question of how good an algorithm is, with respect to others, is still open. Recently, Newman [18] suggested modularity as a natural measure of the goodness of network community decompositions. Here we propose an implementation of an Immunological Algorithm, a population based computational systems inspired by the immune system and its features, to perform community detection on the methods of modularity maximization. The reliability and efficiency of the proposed algorithm has been validating by comparing it with Louvain algorithm one of the fastest and the popular algorithm based on a multiscale modularity optimization scheme.

Keywords: Immunological-inspired algorithms · Networks ·
Modularity optimization · Community structure · Opt-IA

1 Introduction

In modern interdisciplinary science, networks (graphs) are an extremely useful for the representation of a wide number of complex systems. A large variety of natural processes can be conveniently described and studied using graphs, where nodes are the elementary parts of the system and edges between them represent their mutual interactions [3,14]. Usually complex systems are organized in compartments, where each of them has a role and/or a function that satisfy a certain property of relative cohesiveness. In the context of the theory

© Springer Nature Switzerland AG 2020
Z. Ju et al. (Eds.): UKCI 2019, AISC 1043, pp. 235–247, 2020.
https://doi.org/10.1007/978-3-030-29933-0_20

of complex networks, compartments are represented by partitions of the set of nodes with a high density of internal links (whereas links between compartments have a comparatively lower density), called communities or modules [9, 12]. Finding compartments in a graph-theoretic background has become a fundamental problem in network science, since it may shed some light on the organization of complex systems and on their function. Different compartments often exhibit significantly different properties; therefore a global analysis of the network would be inappropriate and unfeasible. A detailed analysis of individual communities, instead, leads to more meaningful insights into the roles of individuals. Such an approach can also allow the visualization and the analysis of a large and complex network focusing on a new higher-level structure, where each identified communities can be compressed into a node belonging to the latter. Let us underline that classical algorithms for graph clustering are not suitable to reveal the properties of community structures. They are mostly based on optimal subdivisions of graphs in order to guarantee min-flow cut. Finding properties of community structures, instead, requires a complex analysis of link patterns and relations.

In the last few years, a very large amount of new computational techniques (often called network clustering), have been developed and are commonly used for community detection in graphs as well as to optimize a graph structure so to guarantee certain desired features, [5, 9, 17, 21]. Furthermore, many approaches have been proposed for finding such partitions and some different metrics for community structure evaluation have been introduced [16]. Whether or not to search for a hierarchical partition, where the communities are recursively subdivided into sub-communities, as well as the definition of the size of the communities (specified by the user or derived by the algorithm) along with other parameters, are the substantial differences between the proposed approaches.

In this work, we propose an implementation of an Immunological Algorithm, a population based computational systems inspired by the immune system and its features, to perform community detection on the methods of modularity maximization. To evaluate its reliability and efficiency, the proposed algorithm has been compared with Louvain's algorithm [4], one of the fastest and popular community detection methods in networks [2] that uses a multiscale modularity optimization scheme in order to maximize a modularity score for each community.

2   Community Detection and Modularity Maximization

Modularity is a benefit function that measures the quality of a particular partitioning of a graph into communities, and it was proposed by Newman [18]. Originally defined for undirected graphs, the definition of modularity has been subsequently extended to directed and weighted graphs [1, 13, 15]. The aim of community detection in graphs is to identify, by using only the information encoded in the graph topology, the modules and their hierarchical organization. Modularity maximization is the most popular and one of most widely used methods for community partition. It detects communities by searching over possible partitions of a graph, over which modularity is maximized. Given a subgraph, a.k.a. a cluster, the modularity function is defined as the difference between the actual density of edges inside the cluster and the expected density of such edges if the graph was random conditioned on its degree distribution [18]. This expected edge density depends on the chosen null model, a random copy of the original graph that maintains the structural properties but not those on the structure of the communities. The idea behind modularity is that a random graph does not have a clustering structure. The edge density of a cluster should be greater than the expected density of a subgraph whose nodes are randomly connected.

Given a graph $G = (V, E)$ with $|E| = m$, and given a partition of G with n_c clusters, the benefit function of modularity can be written as:

$$Q = \sum_{c=1}^{n_c} \left[\frac{l_c}{m} - \left(\frac{d_c}{2m} \right)^2 \right] \tag{1}$$

where, for each cluster, i.e. subgraphs c,

- l_c is the total number of edges and
- d_c is the sum of the degrees of its vertices,
- $\left(\frac{l_c}{m} \right)$ represents the fraction of edges inside a certain cluster and
- $\left(\frac{d_c}{2m} \right)^2$ the fraction of the expected edges if the graph was random (null model).

Although an important limit of resolution of the modularity measure has been underlined by Fortunato and Barthelemy [10], modularity seems to be a useful measure of the community structures. In fact, algorithms that search for graph partitions that offer optimal modularity are already proposed generally claim to be able to successfully find communities in very large and complex networks [15,16].

Unfortunately, in modularity optimization methods, overlapping between the communities is not allowed, i.e. each vertex of the graph can be inserted in a single community. Furthermore, it is possible to discover sub-communities by applying these algorithms iteratively, but it is not possible to discover partially overlapping communities. The modularity defined in a heuristic way, considers a good division one that places most of the edges of a network within groups and only some of them between groups. High values of modularity indicate good partitions. In particular, we desire a quality function Q which, given a network and a candidate division of that network into groups, assigns a score to each partition of a graph, so to rank partitions and evaluate when a partition is better than another (in a graph the partition corresponding to its maximum value should be the best, or at least a very good one). Maximization of modularity is therefore sought at all costs. Obviously, a brute force search to optimize Q is impossible, due to the enormous number of ways in which it is possible to partition a graph. Moreover, it has been proven that the optimization of modularity is an NP-complete problem [4], so it is highly unlikely to perform the optimization task and find an optimal solution in polynomial with respect to the graph dimension. Therefore, we need to turn to approximate methods of optimization, which can find fairly good approximations of maximum modularity in a reasonable time, through the use of appropriate algorithms particularly suitable when the solution space of a given problem is very large, and an exhaustive brute force search for the optimal solution is unfeasible.

3   The Immunological Algorithm

Immunological-inspired computation is nowadays a wide family of successful algorithms in searching and optimization that takes inspiration from the immune system (IS). What makes the IS interesting and source of inspiration is its defence dynamics and features that allow it to be able to protect living organisms against invaders and diseases. It is also a robust and efficient recognition system, able to detect and recognize the invaders, and distinguish between own cells, and foreign ones (*self/nonself discrimination*). Other really interesting and inspiring features that allow to design efficient solving methodologies are its high ability to learn, memory usage; self-regulation; associative retrieval; threshold mechanism, and the ability to perform parallel, and distributed cognitive tasks. In light of all the above, immunological heuristics are mainly focused on three general approaches: (1) clonal selection [6,19] (2) negative selection [11,20]; and (3) immune networks [22]. Such algorithms have been successfully employed in a variety of different application areas.

In this research work, we have developed an immunological algorithm inspired by the clonal selection theory, which belongs to a special class of the immunological heuristics family called *Clonal Selection Algorithms* (CSA) [7,8]. The developed algorithm is based on two main concepts/entities:

− antigen (Ag), i.e. the optimization problem to be solved, and
− B cell receptor, i.e. a point in the search space (solutions) for the problem Ag.

The major features and points of strength of this kind of heuristics are the operators: (i) cloning, (ii) inversely proportional hypermutation and (iii) aging. The first operator generates a new population of B cells centered on the highest affinity/fitness values; the second one explores the neighbourhood of each point of the search space, perturbing each solution with a probability which is inversely proportional to its fitness function; and the last one eliminates old solutions from the current population with the goal of introducing diversity and avoiding local minima during the evolutionary search process.

For simplicity, hereafter, we call the algorithm as OPT-IA. At each time step t OPT-IA maintains a population of B cells $P^{(t)}$ of size d (i.e., d candidate solutions). Each element of the population represents a possible subdivision of the vertices of the graph $G(V, E)$ in the community. More in details, if N is the cardinality of the set of vertices V, a B cell c will be a sequence (array) of n integers, between 1 and N, where $c[i] = j$ represents the fact that the vertex i is placed in the cluster j.

These elements are randomly initialized at the time step $t = 0$, using the uniform probability distribution. Once the population is initialized, the next step is to evaluate the fitness function for each B cell $\boldsymbol{x} \in P^{(t)}$ using the function *Compute_Fitness($P^{(t)}$)* which computes for each B cell c the value given by Eq. 1.

A description of OPT-IA is presented in the pseudocode shown in Algorithm 1. The rest of the section describes the main steps performed by the immunological algorithm (Algorithm 1).

Algorithm 1. Opt-IA (d, dup, ρ, τ)

1: $t \leftarrow 0$
2: $P^{(t)} \leftarrow$ Initialize_Population(d);
3: Compute_Fitness($P^{(t)}$)
4: **repeat**
5:      $P^{(clo)} \leftarrow$ Cloning($P^{(t)}, dup$);
6:      $P^{(hyp)} \leftarrow$ Hypermutation($P^{(clo)}, \rho$);
7:      Compute_Fitness($P^{(hyp)}$)
8:      $P_p^{(t)} \leftarrow$ Precompetition($P^{(t)}$);
9:      $P_a^{(t)} \leftarrow$ Aging_Random($P_p^{(t)}, \tau$);
10:     $P^{(t+1)} \leftarrow$ Selection($P_a^{(t)}, P^{(hyp)}$)
11:     $t \leftarrow t + 1$
12: **until** (termination criterion is satisfied)

Initialization phase: each element of the population, denoted by $P^{(t)}$ in the Algorithm 1, represents a possible subdivision of the vertices of the graph $G(V, E)$ in the community. In the initialization phase ($t = 0$), the elements of the population are randomly generated, assigning to the vertices a number included in the interval $[1, N]$, where $N = |V|$. The assigned number is the cluster number to which the vertex is assigned.

Cloning operator: the cloning operator has the simple purpose to duplicate dup times the elements of the population, creating an intermediate population $P^{(clo)}$ of dimensions $d \times dup$. To avoid a premature convergence of the algorithm, we made dup independent from the fitness of the element, If we had chosen to increase the number of clones for high fitness elements, we would have obtained quickly a very homogeneous population, causing in turn a poor exploration of the search space.

Hypermutation operators: the *hypermutation operator* acts on each element of the population $P^{(clo)}$ performing ρ mutations, where as for dup, ρ is a constant determined by the user. Again, we wanted to avoid a premature convergence of the algorithm, and so the mutation rate (ρ) does not depend upon the fitness value. The cloning operator, coupled with the hypermutation operator, performs a local search around the cloned solutions. The introduction of blind mutation produces individuals with higher affinity (i.e. higher fitness function values), which will be then selected forming the improved mature progenies.

We considered different types of mutation operators that can act on a single vertex of the sequence (*local operators*) or on a group of nodes (*global operators*). Among them, we distinguish:

- **TotalRandom:** randomly selects a vertex of the solution and randomly assigns it to a cluster among the N possible.
- **Equiprobality:** randomly selects a vertex of the solution and assigns it to a cluster among those existing at time t. Each cluster has the same probability of being selected.
- **Existing:** randomly selects a vertex of the solution and assigns it to a cluster among those existing at time t. The probability with which a cluster is selected is proportionate to its size: the larger the cluster, the greater the probability that the vertex will be assigned to that cluster.
- **Fuse:** randomly selects a solution cluster and assigns all its nodes to a randomly selected cluster among those existing at time t.
- **Destroy:** randomly selects a cluster of the solution and assigns a variable percentage of its nodes to clusters chosen at random from the N possible.

Precompetition: this function randomly chooses two individuals from the population and, if they have the same cluster number, it removes the lower fitness element with a 50% probability. This strategy makes it possible to maintain a more heterogeneous population during the evolutionary cycle, maintaining solutions with a different number of communities, so as to better explore the research space.

Random aging operator: to help the algorithm escape local maxima, we introduced a random aging operator. In details, at each iteration the elements of the population are removed with a given probability τ. The aging operator reduces premature convergences and keeps high diversity into the population.

Selection operator: at this point the new population $P^{(t+1)}$ is created for the next generation by using $(\mu + \lambda)$-*Selection operator*, which selects the best d survivors of the aging step from the populations $P^{(t)}$ and $P^{(hyp)}$.. Such operator, with $\mu = d$ and $\lambda = (d \times dup)$, reduces the offspring B cell population of size $\lambda \geq \mu$ – created by cloning and hypermutation operators – to a new parent population of size $\mu = d$. The selection operator identifies the d best elements from the offspring set and the old parent B cells, thus guaranteeing monotonicity in the evolution dynamics.

Termination: finally, the algorithm terminates its execution when the termination criterion is satisfied. In this research work, a maximum number of the fitness function evaluations FFE_{Max} has been considered for all experiments performed.

4   Results

In this section, we present the results obtained by OPT-IA, showing the competitiveness of the proposed approach with respect to the state-of the-art. In particular, to properly evaluate the performance of our proposed Clonal algorithm, we compared it with the deterministic Louvain algorithm, which is able to obtain good results in reasonable times. The analysis was conducted on a series of networks, most commonly used as a benchmark in community detection in graphs, whose set includes *dolphins*, *karate* and *ukfaculty*. Before running our algorithm, we need to set both the population size and the number of clones, i.e. the two fundamental parameters of the algorithm, d and *dup*. To determine the best values to assign to them, we studied the fitness trend as they varied. The points visible in the Fig. 1 represent the average values on *10 runs* relative to the graph *almost_lattice*, chosen for its particularly complex landscape.

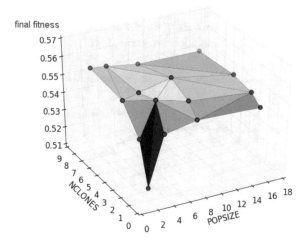

Fig. 1. Dependence of the algorithm on the parameters for *mut_rate* = *3*, on *almost_lattice* graph

The best combinations of *popsize* and *clone number* are respectively $d = 8$ and *dup* = 4. The comparison between them, carried out before the choice of the most advantageous combinations, is shown in Fig. 2, corresponding to three different values of the mutation rate ρ. It is clear that both *fuse* and *total_random* do not offer good results. The first one leads to premature convergence of fitness, while the second one, at each cycle, modifies the communities of a single element with few improvements in fitness. On the other hand, *equiprobality*, *existing* and *destroy* are very effective functions. The latter allows escaping from local optima due to the presence of isolated nodes. In light of the tests carried out, the *equiprobality* and *destroy* operators were used with the same probability, while the *fuse* operator was used with low probability in order to have a cluster aggregation tool that allows to compete if minimally with the *destroy* operator.

The choice of mutation rate was made on the *almost_lattice* instance by varying ρ (from 1 to 5). It was observed that the best results were obtained for

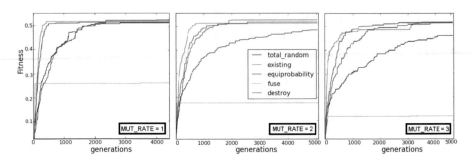

Fig. 2. Comparison of mutations for $MUT\_RATE = 1, 2, 3$ in the *dolphins* graph

$\rho = \{1, 2, 3\}$ (Fig. 3). Thus, at each t iteration, we chose to mutate the cloned elements a number of times included between $[1, \rho]$.

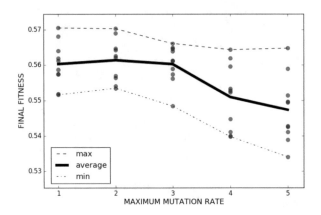

Fig. 3. Value of Fitness when MUT_RATE changes in *almost_lattice* graph

For each instance, the algorithm was executed *10 times*, and that's the reason why the best result obtained (max) is reported, with the relative number of communities, the worst (min) and the average with the standard deviation. A tuning was performed on the parameter τ used by the aging operator. The graph in the Fig. 4, shows the trend of the best individual for each iteration as the parameter τ varies: the values 1 and 0.001 represent the two extreme cases, that is the one in which all the elements or no element is discarded by the population; the intermediate values instead introduce a turnover among the elements of the population, producing diversity within the same population and avoiding that the algorithm converges in a premature way. The effectiveness of the operator is shown in the Fig. 5, in which the loss of the element with the best fitness corresponds, a few generations later, to the discovery of solutions with a higher fitness value (Fig. 6).

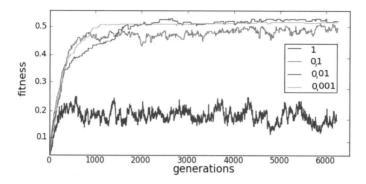

Fig. 4. Best fitness at every generation for various *RANDOM_DIE* action probabilities in the *dolphins* graph

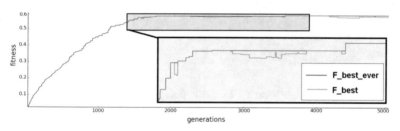

Fig. 5. Comparison between the fitness of the best individual at each cycle and that of the best up to that cycle

Each experiment was performed with population size $d = 8$, duplication parameter $dup = 2$, mutation rate $\rho = \{1, 2, 3\}$, probability of removal from the population $\tau = 0.01$ and maximum number of the fitness function evaluations $FFE_{Max} = 10^5$. Except for the *miserables* and *huckleberry* networks, for which probably more generations would be needed, in most cases, it succeeds in overcoming it. In fact the functions exploited have been chosen precisely to avoid the entrapment of dynamics in excellent premises, from which the deterministic it fails to escape instead. Furthermore, the number of communities found is not particularly influential in this disparity: there are cases in which the deterministic algorithm does not reach the global optimum despite having identified the correct number of clusters that maximizes modularity. Not even the number of nodes constituting N seems relevant in this regard. At the same time, from a simple graphical examination it is possible to note that the evolutionary (on the right) is a winner in cases where the number of links between one community and another is very large (below), therefore the network is rather complex, while for graphs with more evident clusters the two algorithms lead almost to the same result (above). In the latter case the landscape is in fact quite simple, so that the global optimum is easily reachable even from the deterministic algorithm, while in the former only the evolutionary algorithm manages to explore the entire research space well (Tables 1 and 2).

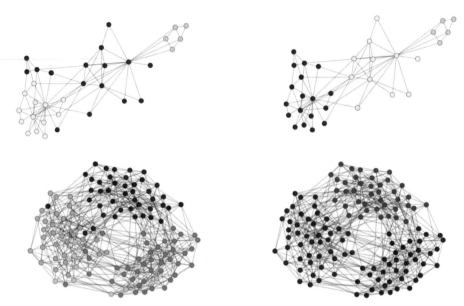

Fig. 6. Deterministic algorithm of Louvain (left) vs. Opt-IA (right) on the graphs *karate* (top) and *3mixed* (bottom)

Table 1. Experimental results of OPT-IA on the set of the networks benchmark. For each network is showed the number of vertices; min, max and average values; standard deviation; and number of communities found.

Instance	$|V|$	OPT-IA				
		Min	Max	Avg	Dev. St	Community
dolphins	62	0.5265	0.5285	0.5275	0.0008	5
karate	34	0.4198	0.4198	0.4198	0.0	4
ukfaculty	81	0.4488	0.4488	0.4488	0.0	4
miserables	77	0.5562	0.5600	0.5584	0.002	6
huckleberry	69	0.5308	0.5346	0.5334	0.0018	4
GN_benchmark2	128	0.4336	0.4336	0.4336	0.0	2
GN_benchmark4	128	0.5393	0.5393	0.5393	0.0	4
LFR_benchmark	128	0.1869	0.1980	0.1936	0.0037	5
almost_lattice	64	0.5436	0.5576	0.5576	0.0089	8
3mixed	128	0.4297	0.4297	0.4297	0.0	3

Table 2. OPT-IA vs. LOUVAIN, a deterministic algorithm.

Instance	$|V|$	LOUVAIN		OPT-IA	
		Q	Community	Q	Community
dolphins	62	0.5188	5	**0.5285**	5
karate	34	0.4156	4	**0.4198**	4
ukfaculty	81	**0.4488**	4	0.4488	4
miserables	77	0.5583	6	**0.5600**	6
huckleberry	69	**0.5346**	4	0.5346	4
GN_benchmark2	128	**0.4336**	2	0.4336	2
GN_benchmark4	128	**0.5393**	4	0.5393	4
LFR_benchmark	128	0.1560	6	**0.1980**	5
almost_lattice	64	0.5279	8	**0.5576**	8
3mixed	128	0.3682	5	**0.4297**	3

5 Conclusions

We have introduced a clonal algorithm, denoted Opt-IA, for the optimization
of modularity function. The discovery of community structures is essentially
based on the maximization of the quality function Q, proposed by Newman
[15]. The maximization of sought modularity has been achieved through the
proposed heuristic algorithm that, with its performance, gives a significantly
good guarantee on the solution quality. To drive the system to escape from local
optima, we use, at each cycle, two special operators, *equiprobality* and *destroy*,
that separate the community structures ensuring great improvements in fitness.
When the *equiprobality* and *destroy* procedures eventually finish, we apply a *fuse*
operator to have a cluster aggregation tool in order to refine the community
structure. The proposed method outperforms, in most cases, the deterministic
Louvain algorithm in terms of higher modularity values found and in less CPU
time for computer-generated graphs most commonly used as a benchmark in
community detection in networks. The number of communities found and the
number of nodes constituting the communities are not particularly influential in
this disparity. Furthermore, the evolutionary algorithm manages to well explore
the entire search space when the network is rather complex and the number of
links between one community and another is very large.

References

1. Bickel, P.J., Chen, A.: A nonparametric view of network models and newman
   girvan and other modularities. Proc. Natl. Acad. Sci. **106**(50), 21068–21073 (2009).
   https://doi.org/10.1103/PhysRevE.74.036104
2. Blondel, V.D., Guillaume, J.-L., Lambiotte, R., Lefebvre, E.: Fast unfolding of
   communities in large networks. J. Stat. Mech. Theory Exp. **2008**(10), P10008
   (2008). https://doi.org/10.1088/1742-5468/2008/10/P10008

3. Boccaletti, S., Latora, V., Moreno, Y., Chavez, M., Hwang, D.-U.: Complex networks: structure and dynamics. Phys. Rep. **424**(4–5), 175–308 (2006). https://doi.org/10.1016/j.physrep.2005.10.009

4. Brandes, U., Delling, D., Gaertler, M., Gorke, R., Hoefer, M., Nikoloski, Z., Wagner, D.: On modularity clustering. IEEE Trans. Knowl. Data Eng. **20**(2), 172–188 (2007). https://doi.org/10.1109/TKDE.2007.190689

5. Coscia, M., Giannotti, F., Pedreschi, D.: A classification for community discovery methods in complex networks. Stat. Anal. Data Min. ASA Data Sci. J. **4**(5), 512–546 (2011). https://doi.org/10.1002/sam.10133

6. Cutello, V., Lee, D., Nicosia, G., Pavone, M., Prizzi, I.: Aligning multiple protein sequences by hybrid clonal selection algorithm with insert-remove-gaps and block-shuffling operators. In: Proceedings of the 5th International Conference on Artificial Immune Systems (ICARIS). LNCS, vol. 4163, pp. 321–334 (2006). https://doi.org/10.1007/11823940_25

7. Cutello, V., Nicosia, G., Pavone, M., Prizzi, I.: Protein multiple sequence alignment by hybrid bio-inspired algorithms. Nucl. Acids Res. **39**(6), 1980–1992 (2011). https://doi.org/10.1093/nar/gkq1052

8. Cutello, V., Nicosia, G., Pavone, M., Timmis, J.: An immune algorithm for protein structure prediction on lattice models. IEEE Trans. Evol. Comput. **11**(1), 101–117 (2007). https://doi.org/10.1109/TEVC.2006.880328

9. Fortunato, S.: Community detection in graphs. Phys. Rep. **486**(3–5), 75–174 (2010). https://doi.org/10.1016/j.physrep.2009.11.002

10. Fortunato, S., Barthelemy, M.: Resolution limit in community detection. Proc. Natl. Acad. Sci. **104**(1), 36–41 (2007). https://doi.org/10.1073/pnas.0605965104

11. Fouladvand, S., Osareh, A., Shadgar, B., Pavone, M., Sharafia, S.: DENSA: an effective negative selection algorithm with flexible boundaries for selfspace and dynamic number of detectors. Eng. Appl. Artif. Intell. **62**, 359–372 (2016). https://doi.org/10.1016/j.engappai.2016.08.014

12. Girvan, M., Newman, M.E.: Community structure in social and biological networks. Proc. Natl. Acad. Sci. **99**(12), 7821–7826 (2002). https://doi.org/10.1073/pnas.122653799

13. Mucha, P.J., Onnela, J., Porter, M.: Communities in networks. Not. Am. Math. Soc. **56**, 1082–1097 (2009)

14. Newman, M.E.: The structure and function of complex networks. SIAM Rev. **45**(2), 167–256 (2003). https://doi.org/10.1137/S003614450342480

15. Newman, M.E.: Fast algorithm for detecting community structure in networks. Phys. Rev. E **69**(6), 066133 (2004). https://doi.org/10.1103/PhysRevE.69.066133

16. Newman, M.E.: Finding community structure in networks using the eigenvectors of matrices. Phys. Rev. E **74**(3), 036104 (2006). https://doi.org/10.1103/PhysRevE.74.036104

17. Newman, M.E.: Communities, modules and large-scale structure in networks. Nat. Phys. **8**(1), 25 (2012). https://doi.org/10.1038/nphys2162

18. Newman, M.E., Girvan, M.: Finding and evaluating community structure in networks. Phys. Rev. E **69**(2), 026113 (2004). https://doi.org/10.1103/PhysRevE.69.026113

19. Pavone, M., Narzisi, G., Nicosia, G.: Clonal selection - an immunological algorithm for global optimization over continuous spaces. J. Glob. Optim. **53**(4), 769–808 (2012). https://doi.org/10.1007/s10898-011-9736-8

20. Poggiolini, M., Engelbrecht, A.: Application of the feature-detection rule to the negative selection algorithm. Expert Syst. Appl. **40**(8), 3001–3014 (2013). https://doi.org/10.1016/j.eswa.2012.12.016
21. Porter, M.A., Onnela, J.-P., Mucha, P.J.: Communities in networks. Not. AMS **56**(9), 1082–1097 (2009)
22. Smith, S., Timmis, J.: Immune network inspired evolutionary algorithm for the diagnosis of Parkinsons disease. Biosystems **94**(1–2), 34–46 (2008)

Construction and Refinement
of Preference Ordered Decision Classes

Hoang Nhat Dau[1], Salem Chakhar[2,3]([✉]), Djamila Ouelhadj[1,3],
and Ahmed M. Abubahia[4]

[1] Department of Mathematics, Faculty of Technology, University of Portsmouth,
Portsmouth, UK
nhat.dau@myport.ac.uk, djamila.ouelhadj@port.ac.uk
[2] Portsmouth Business School, University of Portsmouth, Portsmouth, UK
[3] Centre for Operational Research and Logistics, University of Portsmouth,
Portsmouth, UK
salem.chakhar@port.ac.uk
[4] School of Computing, Faculty of Technology, University of Portsmouth,
Portsmouth, UK
ahmed.abubahia@port.ac.uk

Abstract. Preference learning methods are commonly used in multicriteria analysis. The working principle of these methods is similar to classical machine learning techniques. A common issue to both machine learning and preference learning methods is the difficulty of the definition of decision classes and the assignment of objects to these classes, especially for large datasets. This paper proposes two procedures permitting to automatize the construction of decision classes. It also proposes two simple refinement procedures, that rely on the 80-20 principle, permitting to map the output of the construction procedures into a manageable set of decision classes. The proposed construction procedures rely on the most elementary preference relation, namely dominance relation, which avoids the need for additional information or distance/(di)similarity functions, as with most of existing clustering methods. Furthermore, the simplicity of the 80-20 principle on which the refinement procedures are based, make them very adequate to large datasets. Proposed procedures are illustrated and validated using real-world datasets.

Keywords: Clustering · Preference learning · Classification ·
Classes construction

1 Introduction

Preference learning methods [3,12] are commonly used in multicriteria analysis. These methods are often used to build a preference model based on a sample of past decisions for further prescriptive decision purposes. Preference learning methods have been inspired by knowledge discovery techniques and preference modeling methods. The working principle of preference learning methods is similar to classical machine learning methods [10]: they use a subset of data, called

© Springer Nature Switzerland AG 2020
Z. Ju et al. (Eds.): UKCI 2019, AISC 1043, pp. 248–261, 2020.
https://doi.org/10.1007/978-3-030-29933-0_21

learning set, to extract some knowledge permitting to classify unseen objects. In contrary to machine learning approaches, preference learning methods assume that both attributes and decision classes are preference-ordered.

The definition of the learning set is a crucial step in the application of machine and preference learning methods [1,11]. It involves two operations: (i) the selection of a representative subset of objects, and (ii) their assignment into different pre-defined decision classes. A common issue to both to machine learning and preference learning methods is the difficulty of defining the decision classes. The situation is further complicated especially for large datasets.

One possible solution to define the learning set is to use one of existing multi-criteria ordered clustering methods e.g. [2,4–7,14]. However, these methods are very demanding in terms of additional information (such as preference parameters) and require the specification of a distance/(di)similarity functions.

In this paper, we propose two procedures automatizing the construction of decision classes. These procedures permit to reduce the cognitive effort required from the decision maker and then extend the application domain of preference learning methods. We also propose two simple 80-20 principle-based refinement procedures that permit to map the output of the construction procedures into a manageable set of decision classes.

The rest of the paper goes as follows. Section 2 provides the background. Sections 3 and 4 present the construction and refinement procedures, and Sect. 5 illustrates them using a real-world dataset. Section 6 concludes the paper.

2   Background

2.1   Problem Description

Let U be a non-empty finite set of objects and Q is a non-empty finite set of attributes such that $q : U \rightarrow V_q$ for every $q \in Q$. The V_q is the domain of attribute q, $V = \bigcup_{q \in Q} V_q$, and $f : U \times Q \rightarrow V$ is the information function defined such that $f(x, q) \in V_q$ for each attribute q and object $x \in U$. The value $f(x, q)$ corresponds to the evaluation of object x on attribute $q \in Q$. The domains of condition attributes are supposed to be ordered according to a decreasing or increasing preference. Such attributes are often called criteria.

The objective is to partition U into a finite number of preference-ordered decision classes $\mathbf{Cl} = \{Cl_t, t \in T\}$, $T = \{1, \cdots, n\}$, such that each $x \in U$ belongs to one and only one class.

2.2   Dominance Relation and Graph

The dominance relation Δ is defined for each pair of objects x and y as follows:

$$x \Delta y \Leftrightarrow f(x, q) \geq f(y, q), \forall q \in Q. \tag{1}$$

This definition applies for gain-type (i.e. benefit) criteria. For cost-type criteria, the symbol '$\geq$' should be replaced with '$\leq$'. Equation (1) implements the weak version of the dominance relation since all the inequality are large. Clearly, this version of dominance relation is reflexive (i.e. $x\Delta x$, $x \in U$) and transitive (i.e. if $x\Delta y$ and $y\Delta z$, then $x\Delta z$, $\forall x, y, z \in U$) (but in general not complete). Consequently, it defines a partial preorder on U. The strict version of the dominance relation Δ_s requires at least one strict inequality in Eq. (1). The strict dominance relation is no longer reflexive but still transitive. It is easy to see that $\Delta_s \subseteq \Delta$ (i.e. $x\Delta_s y \Rightarrow x\Delta y$, $\forall x, y \in U$).

The dominance relations on $U \times U$ can be summarized through a dominance matrix $M[x_{ij}]h \times h$ where $h = |U|$, i.e. the number of objects in U and

$$x_{ij} = \begin{cases} 1, \text{ If } x_i \Delta x_j, \\ 0, \text{ Otherwise.} \end{cases} \qquad (2)$$

The dominance relation defines a partial preorder on the objects set U. Any preorder can be represented by a directed graph, with elements of the set corresponding to vertices, and the order relation between pairs of elements corresponding to the directed edges between vertices. Therefore, the dominance relation can be represented as directed graph $G = (U, E)$ with elements U corresponding to vertices, and the dominance relation between pairs of elements corresponding to the directed edges E between vertices, defined as $E = \{(x, y) \in U \times U : x\Delta y\}$. The graph G is constructed using a top-to-bottom order. This means that if a node x dominates a node y, x appears above y in the graph.

2.3   Running Example

Table 1 is an extract from dataset used in Sect. 5. It provides the description of 10 objects with respect to a set $Q = \{A_1, A_2, A_3, A_4\}$ of four criteria, which are introduced later in Sect. 5. At this level, we just mention that all the criteria are to be maximized and that the first and fourth criteria are ordinal while the second and third are continuous.

Table 1. Information table

#	A_1	A_2	A_3	A_4
1	2	1312.5	26.25	2
2	2	1365	27.30	2
3	1	347.76	19.32	1
4	1	74.8	7.48	1
5	1	1117.98	62.11	2
6	1	1289.88	71.66	2
7	3	193.55	38.71	1
8	4	1313	13.13	2
9	3	326	3.26	1
10	3	2268	126.00	1

Table 2. The dominance matrix

	1	2	3	4	5	6	7	8	9	10
1	1	0	1	1	0	0	0	0	0	0
2	1	1	1	1	0	0	0	0	0	0
3	0	0	1	1	0	0	0	0	0	0
4	0	0	0	1	0	0	0	0	0	0
5	0	0	1	1	1	0	0	0	0	0
6	0	0	1	1	1	1	0	0	0	0
7	0	0	0	1	0	0	1	0	0	0
8	0	0	0	1	0	0	0	1	1	0
9	0	0	0	0	0	0	0	0	1	0
10	0	0	1	1	0	0	1	0	1	1

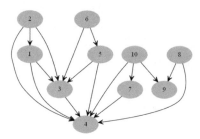

Fig. 1. Dominance graph

Table 2 is the dominance matrix associated with the dataset in Table 1. The corresponding dominance graph is shown in Fig. 1. We note that the self dominance relationships are omitted for simplicity.

3 Construction Procedures

The construction procedures are based on the dominance relation, which is the most elementary preference information. Let $Cl(x)$ be the target class of decision object $x \in U$. Then, the assignment of objects to decision classes relies on the following construction rules:

$$x\Delta y \Leftrightarrow Cl(x) \geq Cl(y), \forall x, y \in U. \tag{3}$$
$$x\Delta_s y \Leftrightarrow Cl(x) > Cl(y), \forall x, y \in U. \tag{4}$$

Based on these rules, we designed two basic procedures: Top-Down and Bottom-Up. These procedures are introduced in the rest of this section.

3.1   Top-Down Procedure

The set of decision classes can be induced from the directed graph $G = (U, E)$ using the following idea. First, we identify a minimal subset $N \subseteq U$ such that: (i) any decision object that is not in N is dominated by at least one decision object from N; and (ii) the objects in set N are incomparable (i.e. they do not dominate each other, except of the self dominance). The set N is called the *kernel* of graph G, the dominant subset or also the external stability. The elements of N are assigned to the most preferred decision class Cl_n. Then, the same procedure is used to identify the kernel N' of the sub-graph $G' = (U \setminus N, E')$ and the elements of set N' are then assigned to the second most preferred decision class Cl_{n-1}. The same procedure is repeated until all the objects are assigned. This idea is formalized in Algorithm 1. The procedure **Kernel** in this algorithm permits to compute the kernel of the graph given as parameter. Different procedures for computing a Kernel of graph are available in the literature [8,9]. Algorithm 1 runs in $O(\beta|U|)$ where β is the complexity of computing the kernel.

Algorithm 1. TopDown Procedure

Input : $S = \langle U, Q, V, f \rangle$, // information table.
Output: Cl, // equivalence classes.
1 $n \longleftarrow |U|$;
2 $Z \longleftarrow \emptyset$;
3 **while** $(Z \neq U)$ **do**
4 | $E \longleftarrow \{(x, y) : x, y \in U \setminus Z \text{ and } x \Delta y\}$;
5 | $G \longleftarrow (U \setminus Z, E)$;
6 | $Cl_n \longleftarrow \textbf{Kernel}(G)$;
7 | $Z \longleftarrow Z \cup Cl_n$;
8 |_ $n \longleftarrow n - 1$;
9 re-label decision classes $Cl_{|U|}, \cdots, Cl_{n+1}$ as $Cl_{|U|-n}, \cdots, Cl_1$;
10 $Cl \longleftarrow \{Cl_1, \cdots, Cl_n\}$;
11 **return** Cl;

Figure 2 illustrates the application of the Top-Down procedure on the dominance graph of Fig. 1. The nodes with bold boundary constitute the kernel. The procedure leads to four classes: $Cl_1 = \{4\}$, $Cl_2 = \{3\}$, $Cl_3 = \{1, 5, 7, 9\}$ and $Cl_4 = \{2, 6, 8, 10\}$, where Cl_4 is the most preferred class.

3.2   Bottom-Up Procedure

The set of decision classes can also be induced by examining the directed graph $G = (U, E)$ from bottom to up. First, we identify a minimal subset $M \subseteq U$ such that: (i) any decision object that is not in M dominates at least one decision object from M; and (ii) the objects in set M are incomparable. We call set M the *anti-kernel* of graph G. The elements of M are assigned to the less preferred decision class Cl_1. Then, the same procedure is used to identify the anti-kernel M' of the sub-graph $G' = (U \setminus M, E')$ and the elements of set M' are then assigned to the second less preferred decision class Cl_2. The same procedure is repeated until all the objects are assigned. This idea is formalized in Algorithm 2. The procedure **AntiKernel** in this algorithm permits to compute the anti-kernel of the graph given as parameter. The procedures for identifying the AntiKernel

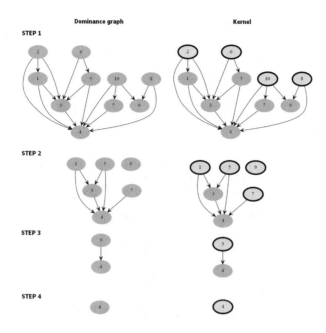

Fig. 2. Illustration of Top-Down procedure

of a graph can be obtained by adapting those used to compute graph Kernels [8,9]. Algorithm 2 runs in $O(\gamma|U|)$ where γ is the complexity of computing the anti-kernel.

Algorithm 2. BottomUp Procedure

Input : $S = \langle U, Q, V, f \rangle$, // information table.
Output: Cl, // equivalence classes.
1   $n \longleftarrow 1$;
2   $Z \longleftarrow \emptyset$;
3   **while** $(Z \neq U)$ **do**
4      $E \longleftarrow \{(x, y) : x, y \in U \setminus Z \text{ and } y\Delta x\}$;
5      $G \longleftarrow (U \setminus Z, E)$;
6      $Cl_n \longleftarrow$ **AntiKernel**(G);
7      $Z \longleftarrow Z \cup Cl_n$;
8      $n \longleftarrow n + 1$;
9   $Cl \longleftarrow \{Cl_1, \cdots, Cl_{n-1}\}$;
10   return Cl;

Figure 3 illustrates the application of the Bottom-Up procedure on the dominance graph of Fig. 1. The nodes with double circle boundary constitute the anti-kernel. The procedure leads to four classes: $Cl_1 = \{4, 9\}$, $Cl_2 = \{3, 7\}$, $Cl_3 = \{1, 5\}$ and $Cl_4 = \{2, 6, 8, 10\}$, where Cl_4 is the most preferred class.

4 Refinement of Decision Classes

The number of decision classes may be very high. Hence, there is a need to reduce the number of these classes to obtain a manageable set of decision classes. Two

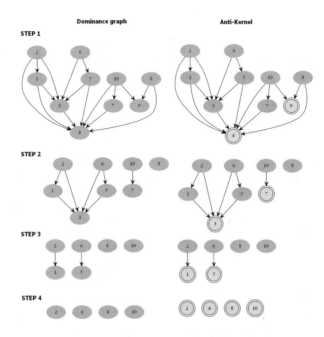

Fig. 3. Illustration of Bottom-Up procedure

simple refinement procedures that rely on the 80-20 rule are proposed in this paper. The 80-20 principle (also known as Pareto principle) relies on the fact that, for many events, roughly 80% of the effects come from 20% of the causes.

Let assume that there are n classes (with Cl_n is best class and Cl_1 is the worst) that should be mapped into $p < n$ classes. The first refinement procedure starts from the top and merge the 20% best classes into class Cl_p. The classes already assigned are removed and then the 20% of best classes of the remaining ones are merged to class Cl_{p-1}. The algorithm continues until the definition of $p - 1$ classes. At the end, the remaining classes are merged into class Cl_1.

This first solution may lead to large classes. A better solution consists in using the 80-20 rule but by considering objects instead of classes. Let assume that there are n classes (with Cl_n is best class and Cl_1 is the worst) that should be mapped into $p < n$ classes. Then, the working principle of the refinement procedure using 80-20 rule on decision objects is as follows. It starts from the top and merges a set of best classes that contain the best 20% objects into class Cl_p. The classes already assigned are removed and then the classes containing the top 20% of best remaining objects are merged into class Cl_{p-1}. The procedure continues until the definition of $p - 1$ classes. At the end, the remaining classes are merged into class Cl_1.

In both versions, a strict application of 80-20 rule may lead to large classes. One possible solution to handle this issue is to use a parameter ϑ to relax the merging conditions.

Algorithm 3. Refinement Using 80-20 Rule on Decision Objects

Input : $L = \{Cl_1, \cdots, Cl_n\}$, // initial classes.
         β, // integer.
Output: $O = \{K_1, \cdots, K_p\}$, // refined classes.

1   $i \longleftarrow n$;
2   **while** $(i \geq 2)$ **do**
3      $K_i \longleftarrow$ top classes in L containing the top $20\% \pm \vartheta$ best objects;
4      $L \longleftarrow L \setminus K_i$;
5      $i \longleftarrow i - 1$;
6   $K_1 \longleftarrow L$;
7   $O \longleftarrow \{K_1, \cdots, K_p\}$;
8   return O;

Algorithm 3 implements the second refinement procedure. In this algorithm, we assumed that there are n classes with Cl_n is best class and Cl_1 is the worst.

5 Application and Validation

5.1 Dataset

For the purpose of illustration, we consider a real-world dataset (reproduced from [13]) relative to spare parts management corresponding to a Chinese firm. The dataset is composed of 98 spare parts described with respect to four criteria, namely A_1 (Criticality), A_2 (Annual Dollar Usage), A_3 (Average Unit Cost), and A_4 (Lead Time) (see Table 3). The criteria A_2 and A_3 are continuous while A_1 and A_4 criteria are ordinal. The criterion A_1 can take one of four values 1, 2, 3 and 4 where 1 corresponds to the lowest criticality and 4 corresponds to highest critically. The possible values for A_4 are 1, 2 and 3 where 1 means a low lead

Table 3. Characteristics of considered criteria

Code	Name	Description	Preference	Data type
A_1	Criticality	It represents the influence of spare parts running out on the availability of equipment	Gain	Ordinal
A_2	Annual Dollar Usage	It is calculated by spare part cost multiply demand volume	Gain	Continuous
A_3	Average Unit Cost	It refers to spare part cost	Gain	Continuous
A_4	Lead Time	It refers to the time between the placement of an order and delivery of a new spare part from the firm's supplier	Gain	Ordinal

Table 4. Information table

#	A_1	A_2	A_3	A_4	#	A_1	A_2	A_3	A_4	#	A_1	A_2	A_3	A_4
1	2	1312.5	26.25	2	34	4	1260	12.6	2	67	3	2126.25	47.25	1
2	2	1365	27.3	2	35	4	2100	21	2	68	3	623.7	34.65	2
3	1	347.76	19.32	1	36	4	1050	10.5	2	69	3	420	4.2	2
4	1	74.8	7.48	1	37	4	1575	15.75	2	70	4	840	8.4	2
5	1	1117.98	62.11	2	38	4	578	5.78	2	71	4	3150	31.5	3
6	1	1289.88	71.66	2	39	3	2936	29.36	1	72	4	2625	26.25	3
7	3	193.55	38.71	1	40	4	19682.3	3936.46	2	73	4	13925.3	1392.53	3
8	4	1313	13.13	2	41	4	1444.2	24.07	1	74	3	199.5	5.25	2
9	3	326	3.26	1	42	3	463.05	13.23	1	75	3	472.5	9.45	2
10	3	2268	126	1	43	3	132.3	7.35	1	76	3	336	16.8	2
11	3	4134.6	91.88	1	44	1	2734.2	97.65	1	77	1	57.8	5.78	2
12	3	1587.6	88.2	1	45	1	3071.25	87.75	1	78	1	161.84	5.78	2
13	3	2063.4	54.3	1	46	4	785.7	17.46	1	79	3	840	8.4	2
14	3	1786.4	44.66	1	47	4	955.2	11.94	1	80	4	840	8.4	3
15	3	10365.75	121.95	1	48	1	28.44	1.58	1	81	4	2625	26.25	3
16	3	770.26	20.27	1	49	4	851	8.51	2	82	4	2100	21	3
17	3	2646	52.92	1	50	1	352.8	8.82	1	83	4	25725	257.25	3
18	3	113.4	5.67	1	51	3	105.84	2.94	1	84	4	40056	400.56	3
19	1	650	65	1	52	1	1304.48	21.04	1	85	4	3780	126	2
20	1	418.88	14.96	1	53	4	3580.5	65.1	2	86	3	882	29.4	2
21	1	948.3	31.61	1	54	2	1325.52	73.64	1	87	1	1470	36.75	2
22	3	410.7	13.69	1	55	4	18375	1837.5	3	88	3	126	12.6	2
23	3	26995.6	2699.56	2	56	2	236.7	2.63	1	89	4	1071	17.85	2
24	4	746	7.46	2	57	2	862	8.62	1	90	3	121.1	3.46	2
25	4	3150	31.5	2	58	3	735	7.35	1	91	3	43.56	2.42	2
26	4	3675	36.75	2	59	1	2315.34	128.63	1	92	1	823.2	29.4	2
27	3	27562.5	1837.5	2	60	1	1984.5	110.25	1	93	1	1029	5.78	2
28	4	840	8.4	3	61	1	157.5	15.75	1	94	4	1025.55	22.79	2
29	4	1670	16.7	2	62	1	340.2	18.9	2	95	4	2688	33.6	2
30	4	1754	17.54	2	63	1	642.6	35.7	2	96	3	1470	29.4	2
31	4	437	4.37	2	64	4	346.5	34.65	2	97	2	264.6	14.7	2
32	4	2625	26.25	2	65	3	1890	189	1	98	4	11025	5512.5	3
33	4	462	4.62	2	66	3	567	31.5	1					

time and 3 means a high lead time. All the criteria are benefit-type (i.e. the higher their values, the more important spare part is). The evaluation of the spare parts with respect to criteria is given in Table 4.

5.2   Application and Results

We applied the Top-Down and Bottom-Up procedures using the dataset given in Table 5. The results of these procedures are given in Table 5. As shown in this table, the use of the Top-down procedure leads to 22 classes with class #1 is the most preferred and class #22 is the worst while the application of the Bottom-Up procedure leads to 23 classes with class #23 is most preferred and class #1 is the worst. As shown in Table 5, both the Top-Down and Bottom-Up procedures lead to a high number of classes and should be mapped into a reduced set of classes. Let assume that the obtained decision classes should be mapped into $p = 3$ ordered decision classes, labelled A, B and C, respectively.

In this paper, we used the second refinement procedure. The application of this procedure on the output of Top-Down and Bottom-Up procedures is summarized in Table 6. In both cases, we used a relaxation parameter of $\vartheta = 5$. Table 6 shows that the refinement of the Top-Down procedure's output leads to the assignment of 21 spare parts to class A, 13 to class B and 64 to class C, which makes a percentage of 21.43%, 13.27% and 65.30%, respectively. It also shows that the refinement of the Bottom-Up procedure's output leads to relatively the same rate with the assignment of 19 spare parts to class A, 16 to class B and 63 to class C, which makes a percentage of 19.39%, 16.33% and 64.28%, respectively.

Table 5. Results of Top-Down and Bottom-Up procedures

Preference order	Top-Down		Bottom-Up	
	Class ID	Content	Class ID	Content
Best	1	23, 27, 40, 55, 84, 98	1	55, 84
	2	73, 83,	2	40, 73, 83, 98
	3	15, 59, 65, 71, 85	3	85
	4	6, 10, 11, 26, 44, 53, 72, 81	4	26, 53, 71
	5	5, 12, 13, 17, 25, 39, 45, 60, 64, 68, 82, 87, 95	5	25, 72, 81, 95
	6	32, 54, 63, 67, 96, 28, 80	6	32, 82
	7	2, 14, 19, 35, 41, 86, 94	7	35
	8	1, 7, 16, 21, 30, 66, 89, 92	8	30
	9	29, 46, 52, 62, 76	9	29
	10	3, 37	10	37
	11	8, 20, 22, 42, 61, 97	11	8
	12	34, 49, 88	12	23, 27, 34, 89
	13	36, 47,	13	15, 36, 94
	14	57, 70, 75, 93	14	11, 28, 49, 80
	15	24, 79	15	10, 17, 70
	16	38, 50, 58	16	13, 24, 65, 67, 96
	17	4, 33, 43, 74, 78	17	2, 6, 12, 14, 38, 39, 41, 59, 68, 86
	18	18, 31, 77	18	1, 5, 16, 33, 44, 45, 46, 47, 54, 60, 64, 66, 79, 87
	19	69	19	7, 19, 21, 22, 31, 42, 52, 58, 63, 75, 76, 92
	20	9, 90	20	3, 20, 43, 62, 69, 74, 88, 93, 97
	21	51, 56, 91	21	9, 18, 50, 57, 61, 78, 90
	22	48	22	4, 51, 56, 77, 91
Worst			23	48

Table 6. Results of refinement

Preference order	Top-Down				Bottom-Up			
	Class ID	Content	Nb of objects	%	Class ID	Content	Nb of objects	%
Best	A	23, 27, 40, 55, 84, 98, 73, 83, 15, 59, 65, 71, 85, 6, 10, 11, 26, 44, 53, 72, 81	21	21.43	A	55, 84, 40, 73, 83, 98, 85, 26, 53, 71, 25, 72, 81, 95, 32, 82, 35, 30, 29	19	19.39
	B	5, 12, 13, 17, 25, 39, 45, 60, 64, 68, 82, 87, 95	13	13.27	B	37, 8, 23, 27, 34, 89, 15, 36, 94, 11, 28, 49, 80, 10, 17, 70	16	16.33
	C	32, 54, 63, 67, 96, 28, 80, 2, 14, 19, 35, 41, 86, 94, 1, 7, 16, 21, 30, 66, 89, 92, 29, 46, 52, 62, 76, 3, 37, 8, 20, 22, 42, 61, 97, 34, 49, 88, 36, 47, 57, 70, 75, 93, 24, 79, 38, 50, 58, 4, 33, 43, 74, 78, 18, 31, 77, 69, 9, 90, 51, 56, 91, 48	64	65.30	C	13, 24, 65, 67, 96, 2, 6, 12, 14, 38, 39, 41, 59, 68, 86, 1, 5, 16, 33, 44, 45, 46, 47, 54, 60, 64, 66, 79, 87, 7, 19, 21, 22, 31, 42, 52, 58, 63, 75, 76, 92, 3, 20, 43, 62, 69, 74, 88, 93, 97, 9, 18, 50, 57, 61, 78, 90, 4, 51, 56, 77, 91, 48	63	64.28

5.3 Validation

To validate the results, we applied the preference learning method Dominance-based Rough Set Approach (DRSA) [12] on the output of refinement procedures. The input of DRSA is a learning dataset representing the description of a set of objects with respect to a set of criteria. The main output of DRSA is a collection of decision rules. A decision rule is a consequence relation $E \rightarrow H$ (read as If E, then H) where E is a condition (evidence or premise) and H is a conclusion (decision). Each elementary condition is built upon a single criterion while a consequence is defined based on a decision class. The obtained decision rules can then be applied to classify unseen objects.

We applied the DRSA using the outputs (after refinement) of the Top-Down and Bottom-Up procedures as learning sets. This led to a collection of decision rules, which are then validated through the reclassification of the spare parts in the learning sets using the inferred decision rules. The results of reclassification are summarized in the confusion matrices given in Table 7. As shown in this table, the reclassification using the results obtained from the Bottom-Up procedure shows a perfect match since all the spare parts have been re-classified to the same class. In turn, the reclassification using the results obtained from the Top-Down procedure shows a rate of 92% of correct classifications and a rate of 8% of misclassifications. This holds because some spare parts have been assigned to more than one class.

Table 7. Confusion matrices

Original	Possible			Original	Possible		
	C	B	A		C	B	A
C	64/64	0/1	0/1	C	63/63	0/0	0/0
B	0/5	13/13	0/2	B	0/0	16/16	0/0
A	0/0	0/1	21/21	A	0/0	0/0	19/19
	Top-Down Procedure				Bottom-Up Procedure		

To further evaluate the results of the DRSA, we used a series of well-known non-parametric statistics to measure the correlation between the assignments provided by the Top-Down and Bottom-Up procedures and those generated based on the corresponding rules. The result of the statistical analysis is summarized in Table 8. Let us first note that for the Top-Down procedure, we distinguished two cases concerning the 9 ambiguous assignments: (i) case of best choice in which the assignment intervals have been reduced into a single assignment equal to the one generated by Top-Down procedure, and (ii) worst choice in which the assignment intervals have been reduced into a single assignment different from the one generated by Top-Down procedure. Based on Table 8, we can conclude that all the statistics show a perfect agreement for the case of Bottom-Up procedure and Top-Down with best choices procedure. For the Top-Down procedure with worst choices, all the statistics show a relatively high to very high agreement.

Table 8. Statistical analysis

Statistics	Kendall's tau	Spearman's rho	Unweighted kappa	Weighted kappa
Top-Down with best choices vs rules	1	1	1	1
Top-Down with worst choices vs rules	0.8698	0.8893	0.8257	0.8714
Bottom-Up vs rules	1	1	1	1

6 Conclusion

The paper addresses the problem of ordered decision classes construction and refinement. It has several theoretical and practical contributions. Firstly, it proposes two procedures permitting to 'automatise' the construction of decision classes. These procedures rely on the most elementary preference relation, namely dominance relation. Thus and in contrary to most of existing clustering methods, the proposed procedures avoid the need for additional information or distance/(di)similarity functions. Secondly, it extends the application domain of preference learning methods, especially the DRSA, to decision problems involving large datasets. Thirdly, it introduces two refinement procedures that rely on the 80-20 principle. The refinement procedures permit to map the output of the construction procedures into a manageable set of decision classes. The simplicity of the 80-20 principle on which the refinement procedures are based, make them very adequate to large datasets.

Several topics need to be investigated in the future. The first topic concerns the use of a mixed construction procedure by combining the results of Top-Down and Bottom-Up procedures. The second topic is related to the application and performance evaluation of the procedures with very large data sets. The last topic concerns the use of other refinement techniques.

References

1. Albatineh, A., Niewiadomska-Bugaj, M.: MCS: a method for finding the number of clusters. J. Classification **28**(2), 184–209 (2011)
2. Baroudi, R., Bahloul, S.: A multicriteria clustering approach based on similarity indices and clustering ensemble techniques. Int. J. Inf. Technol. Decis. Mak. **13**(04), 811–837 (2014)
3. Bregar, A., Györkös, J., Jurič, M.: Interactive aggregation/disaggregation dichotomic sorting procedure for group decision analysis based on the threshold model. Informatica **19**(2), 161–190 (2008)
4. de la Paz-Marín, M., Gutiérrez, P., Hervás-Martínez, C.: Classification of countries' progress toward a knowledge economy based on machine learning classification techniques. Expert Syst. Appl. **42**(1), 562–572 (2015)
5. De Smet, Y.: P2CLUST: an extension of PROMETHEE II for multicriteria ordered clustering. In: 2013 IEEE International Conference on Industrial Engineering and Engineering Management (IEEM), pp. 848–851, December 2013
6. De Smet, Y.: An extension of PROMETHEE to divisive hierarchical multicriteria clustering. In: 2014 IEEE International Conference on Industrial Engineering and Engineering Management (IEEM), pp. 555–558, December 2014
7. De Smet, Y., Nemery, P., Selvaraj, R.: An exact algorithm for the multicriteria ordered clustering problem. Omega **40**(6), 861–869 (2012)
8. Emmert-Streib, F., Dehmer, M., Shi, Y.: Fifty years of graph matching, network alignment and network comparison. Inf. Sci. **346–347**, 180–197 (2016)
9. Ghosh, S., Das, N., Calves, T.G., Quaresma, P., Kundu, M.: The journey of graph kernels through two decades. Comput. Sci. Rev. **27**, 88–111 (2018)
10. Gilboa, I., Schmeidler, D.: Case-based knowledge and induction. IEEE Trans. Syst. Man Cybern. Part A **30**(2), 85–95 (2000)
11. Gionis, A., Mannila, H., Tsaparas, P.: Clustering aggregation. ACM Trans. Knowl. Discov. Data **1**(1) (2007). Article 4

12. Greco, S., Matarazzo, B., Słowiński, R.: Rough sets theory for multicriteria decision analysis. Eur. J. Oper. Res. **129**(1), 1–47 (2001)
13. Hu, Q., Chakhar, S., Siraj, S., Labib, A.: Spare parts classification in industrial manufacturing using the dominance-based rough set approach. Eur. J. Oper. Res. **262**(3), 1136–1163 (2017)
14. Rocha, C., Dias, L.: MPOC: an agglomerative algorithm for multicriteria partially ordered clustering. 4OR **11**(3), 253–273 (2013)

Improving Imbalanced Students' Text Feedback Classification Using Re-sampling Based Approach

Zainab Mutlaq Ibrahim$^{(\boxtimes)}$, Mohamed Bader-El-Den, and Mihaela Cocea

University of Portsmouth, Lion Terrace, Portsmouth PO1 3HE, UK
{zainab.mutlaq-ibrahim,mohamed.bader,mihaela.cocea}@port.ac.uk

Abstract. Class imbalance is a major problem in text classification, the problem happens when the used machine learning algorithm biases towards the majority class, so this makes it incorrectly classifies minority class instances. To get over this problem, investigators use the Synthetic Minority Oversampling Technique (SMOTE), it is pre-processing algorithm which was proven as a very good solution for handling imbalanced data sets. In this paper an empirical study have been executed to handle three imbalanced data sets in text format using SMOTE, the recall of all minority classes significantly improved in addition of significant improvement in all models overall performance.

Average classes' recall was improved significantly, by 0.15, 0.09, 0.10 in classification of ASS, FDS, NASS data sets respectively. While the recall for the minority class has significantly increased, ASS (0.23), FDS (0.08), and NASS (0.15).

Keywords: Imbalanced data set · SMOTE · Text feedback · Text classification

1 Introduction

Class imbalance is considered to be a big problem in text classification, the problem appears when classes do not make up an equal portion of a data-set. For example, in a simple two classes case, a balanced state would have the class priority of both classes approximately equal to each other.

In case of an imbalanced problem, the majority class has much more priority than the minority class.

It is very important to correctly classify all instances in a data set whether they belong to majority or minority class, in some cases, it is costly to dis-classify minority instances such as in cancerous cells [1], breast and colon cancer [2], this miss-classification can lead to wrongly diagnosis of the illnesses, mess up and delay efficient and quick treatment in both cases.

Wrong classification in fraud detection [3,4], information retrieval [5], marketing [6], and keyword extraction [7] can lead to frauds success, wrong decision and results or marketing the wrong product. In addition, detecting oil-spill wrongly can lead to environment pollution and harming the nature [8].

© Springer Nature Switzerland AG 2020
Z. Ju et al. (Eds.): UKCI 2019, AISC 1043, pp. 262–267, 2020.
https://doi.org/10.1007/978-3-030-29933-0_22

Without considering imbalanced priorities, a classifier may learn to always predict the majority class. The cost of incorrectly classifying the minority class may be extremely high and not acceptable [9,10].

Approaches have been proposed to deal with class imbalance problem that include re-sampling techniques which changes the priors in the training set by either generating more instances of minority class or omitting instances from majority class.

More techniques for dealing with class imbalanced include appropriate feature selection [11], cost-sensitive learners that consider miss-classification cost in the learning phase [12], one class learner [10,13], and hybrid of the above techniques.

In this paper we are going to use re-sampling method because it showed good results [14], researches have shown a strong relation between re-sampling methods and cost-sensitive techniques [12], and also re-sampling method is easy to implement.

The rest of the paper includes: Sect. 2 which outlines some related work that deal with class imbalance, Sect. 3 describes and illustrates the framework, Sect. 4 details the empirical study, and Sect. 5 concludes the paper.

2 Related Work

Ferdenands et al. explained different techniques to deal with class imbalance problem, two of them refer to oversampling and down sampling the data set [15]. Oversampling methods have been proposed and adopted in different studies, Synthetic Minority Oversampling Technique (SMOTE) is an oversampling technique, in this technique more instances of minority class are generated.

Mohasseb et al. improved Naive bays classifier performance significantly [16] by implementing such a method, while Sisovic et al. slightly improved their clustering model performance [17]. Awad et al. enhanced the performance of all models using SMOTE method [18]. LV et al. used SMOTE to improve the recognition of their model which is to distinguish and recognize users who steal electricity [19].

Down sampling technique is to take out a set of the majority class to balance the data set, this technique is used more often in image processing, Wang and ec al used this technique to get image super-resolution [20], and Lin et al. used it to improve image compression at low bit rates [21].

3 Data Distribution

The used data set in this study was collected from school of computing/university of Portsmouth between 2012–2016 as end of unit feedback, for more details about the data set please see [22].

Figure 1 shows the imbalanced class distribution (positive, negative), in the full data set (FDS), assessment related (ASS), and not assessment related (NASS).

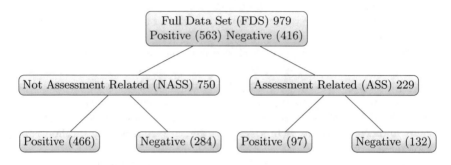

Fig. 1. Data set distribution [22].

4   Proposed Framework

Figure 2 illustrates the proposed framework of dealing with imbalanced data set using SMOTE method.

The proposed framework first transforms the text documents into structured data, there is an imbalanced problem regarding sentiment classes (positive, negative) as illustrated in Fig. 1 Sect. 2, so the framework has two routes, firstly it follows the 'yes' route after binary question whether the data set is balanced or not (the black arrows) to conclude result 1, and secondly it follows the grey arrows to conclude result 2.

Text pre-processing can include, remove numbers, punctuation marks, stop words and words that are less than three letters in addition to apply the SMOTE algorithm. SMOTE method works better with binary labels [16], this make it perfect for our data sets, it runs an oversampling approach to re-balance the original training set. It applies a simple reproduction of the minority class instances. The main task of SMOTE is to introduce artificial and unreal examples. This new data is created by insertion between several minority class instances that are within a defined neighborhood.

In this paper, Support vector machine algorithm was used to build the classification model, it is a powerful tool for text classification, it has the power to determine an optimal separating point that labels records into different categories.

The final Phase is to evaluate models in both routes to see and compare their performance using recall precision, accuracy, and F-Measure.

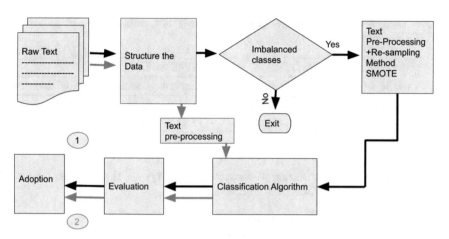

Fig. 2. Proposed framework for handling imbalanced data set

5 Empirical Study

In all experiments to build the needed models, a computer desktop was used with quad 2.33 GHZ CPU, 4GB RAM, and windows 7 operating system.

WEKA a graphic user interface (GUI) was used to pre-process, classify, and re-sample our data sets.

Support Vector Machine (SVM) was used as a machine learning algorithm for automatic text classification, it was applied using data set and its subsets that used in [22] to build three models, the performance of these models evaluated using Accuracy, Precision, Recall, and F-Measure measurements, the study used 10-fold cross validation.

This study results show the success of handling imbalanced data of classification models' performance using SMOTE technique, first this paper shows the results of applying SVM algorithm in classification process without using SMOTE, results were listed in [22] and second it shows the results of rerunning the experiment using smote technique, see Table 1.

5.1 Results

Table 1 presents overall classification performance details of SVM classifier using both the SMOTE algorithm and without using it. The results show better classifier performance when handling the class imbalance.

To shed the light in how did the classifier classify the minority class in the above data sets we need to have a look at the recall results which are illustrated in Table 2. The results show that classification of the minority class in all data set has improved, in FDS data set, the recall increased by 0.08, in the NASS data set case the recall increased by 0.15 while in ASS data set the recall significantly increased by 0.23 in total.

Table 1. Overall SVM classifier performance without/with the implementation of SMOTE algorithm

	SVM without SMOTE				SVM with (SMOTE)			
Data set	Accuracy	Precision	Recall	F-Measure	Accuracy	Precision	Recall	F-Measure
ASS	0.69	0.70	**0.70**	0.69	0.85	0.88	**0.85**	0.84
FDS	0.76	0.76	**0.74**	0.74	0.83	0.83	**0.83**	0.83
NASS	0.76	0.75	**0.73**	0.74	0.85	0.88	**0.85**	0.84

Table 2. Recall SVM classifier performance without/with the implementation of SMOTE algorithm P = Positive, N = negative

	SVM without SMOTE			SVM with (SMOTE)		
	FDS	NASS	ASS	FDS	NASS	ASS
P/N	563/416	466/284	97/132	563/416	466/284	97/132
P	0.85	0.84	**0.64**	0.92	0.75	**0.87**
N	**0.63**	**0.70**	1	**0.71**	**0.85**	0.79

6 Conclusion and Future Work

This study proposed a framework for handling class imbalance issue using SMOTE algorithm and utilizing Uni gram feature.

Empirically, results have shown that the proposed framework worked well for the used data set classification, and improved all models in terms of accuracy, precision, recall, and F-measure.

Average classes' recall was improved significantly, by 0.15, 0.09, 0.10 in classification of ASS, FDS, NASS data sets respectively.

The recall for the minority class has significantly increased using SMOTE for all data sets, the best recall improvement was in ASS data set (positive class), it increased by 0.23, the second was in NASS (negative class) data set which increased by 0.15, and finally FDS (negative class) by 0.08.

Future work will include applying more techniques for dealing with class imbalanced such as appropriate feature selection, cost-sensitive learners that consider mis classification cost in the learning, and the SMOTE algorithm with different feature representations such as bi-gram and Part of Speech (PoS).

References

1. Rushi, L., Snehalata, D.: Class imbalance problem in data mining review. arXiv preprint arXiv:1305.1707 (2013)
2. Majid, A., Ali, S., Iqbal, M., Kausar, N.: Prediction of human breast and colon cancers from imbalanced data using nearest neighbor and support vector machines. Comput. Methods Progr. Biomed. **113**(3), 792–808 (2014)

3. Phua, C., Alahakoon, D., Lee, V.: Minority report in fraud detection: classification of skewed data. ACM SIGKDD Explor. Newsl. **6**(1), 50–59 (2004)
4. Chan, P.K., Stolfo, S.J.: Toward scalable learning with non-uniform class and cost distributions: a case study in credit card fraud detection. In: KDD, vol. 1998, pp. 164–168 (1998)
5. Turney, P.D.: Learning algorithms for keyphrase extraction. Inf. Retr. **2**(4), 303–336 (2000)
6. Ling, C.X., Li, C.: Data mining for direct marketing: problems and solutions. In: KDD, vol. 98, pp. 73–79 (1998)
7. Lewis, D.D., Gale, W.A.: A sequential algorithm for training text classifiers. In: SIGIR1994, pp. 3–12. Springer (1994)
8. Kubat, M., Holte, R.C., Matwin, S.: Machine learning for the detection of oil spills in satellite radar images. Mach. Learn. **30**, 195–215 (1998)
9. Liu, A., Ghosh, J., Martin, C.E.: Generative oversampling for mining imbalanced datasets. In: DMIN, pp. 66–72 (2007)
10. Sharma, S., Bellinger, C., Krawczyk, B., Zaiane, O., Japkowicz, N.: Synthetic oversampling with the majority class: a new perspective on handling extreme imbalance. In: 2018 IEEE International Conference on Data Mining (ICDM), pp. 447–456. IEEE (2018)
11. Zheng, Z., Xiaoyun, W., Srihari, R.: Feature selection for text categorization on imbalanced data. ACM SIGKDD Explor. Newsl. **6**(1), 80–89 (2004)
12. Zadrozny, B., Langford, J., Abe, N.: Cost-sensitive learning by cost-proportionate example weighting. In: ICDM, vol. 3, pp. 435 (2003)
13. Raskutti, B., Kowalczyk, A.: Extreme re-balancing for svms: a case study. ACM SIGKDD Explor. Newsl. **6**(1), 60–69 (2004)
14. Collell, G., Prelec, D., Patil, K.R.: A simple plug-in bagging ensemble based on threshold-moving for classifying binary and multiclass imbalanced data. Neurocomputing **275**, 330–340 (2018)
15. Fernández, A., García, S., Galar, M., Prati, R.C., Krawczyk, B., Herrera, F.: Learning from imbalanced data sets. Springer (2018)
16. Mohasseb, A., Bader-El-Den, M., Cocea, M., Liu, H.: Improving imbalanced question classification using structured smote based approach. In: 2018 International Conference on Machine Learning and Cybernetics (ICMLC), vol. 2, pp. 593–597. IEEE (2018)
17. Šišović, S., Matetic, M., Bakaric, M.B.: Clustering of imbalanced moodle data for early alert of student failure, pp. 165–170, January 2016
18. Awad, A., Bader-El-Den, M., McNicholas, J., Briggs, J.: Early hospital mortality prediction of intensive care unit patients using an ensemble learning approach. Int. J. Med. Inf. **108**, 185–195 (2017)
19. Lv, D., Ma, Z., Yang, S., Li, X., Ma, Z., Jiang, F.: The application of smote algorithm for unbalanced data. In: Proceedings of the 2018 International Conference on Artificial Intelligence and Virtual Reality, pp. 10–13. ACM (2018)
20. Wang, Y., Wang, L., Wang, H., Li, P.: Information-compensated downsampling for image super-resolution. IEEE Signal Process. Lett. **25**(5), 685–689 (2018)
21. Lin, W., Dong, L.: Adaptive downsampling to improve image compression at low bit rates. IEEE Trans. Image Process. **15**(9), 2513–2521 (2006)
22. Ibrahim, Z.M., Bader-El-Den, M., Cocea, M.: Mining unit feedback to explore students' learning experiences. In: UK Workshop on Computational Intelligence, pp. 339–350. Springer (2018)

Detection, Inference and Prediction

Phoneme Aware Speech Synthesis via Fine Tune Transfer Learning with a Tacotron Spectrogram Prediction Network

Jordan J. Bird$^{(\boxtimes)}$, Anikó Ekárt, and Diego R. Faria

School of Engineering and Applied Science, Aston University, Birmingham, UK
{birdj1,a.ekart,d.faria}@aston.ac.uk

Abstract. The implications of realistic human speech imitation are both promising but potentially dangerous. In this work, a pre-trained Tacotron Spectrogram Feature Prediction Network is fine tuned with two 1.6 h speech datasets for 100,000 learning iterations, producing two individual models. The two Speech datasets are completely identical in content other than their textual representation, one follows the standard English language, whereas the second is an English phonetic representation in order to study the effects on the learning processes. To test imitative abilities post-training, thirty lines of speech are recorded from a human to be imitated. The models then attempt to produce these voice lines themselves, and the acoustic fingerprint of the outputs are compared to the real human speech. On average, English notation achieves 27.36%, whereas Phonetic English notation achieves 35.31% similarity to a human being. This suggests that representation of English through the International Phonetic Alphabet serves as more useful data than written English language. Thus, it is suggested from these experiments that a phonetic-aware paradigm would improve the abilities of speech synthesis similarly to its effects in the field of speech recognition.

Keywords: Speech synthesis · Fine tune learning · Phonetic awareness · Fingerprint analysis · Tacotron

1 Introduction

> *Are there imaginable digital computers which would do well in the imitation game?*
>
> *Alan M. Turing*
> *1950*

Artificial Intelligence researchers often seek the goal of intelligent human imitation. In 1950, Alan Turing proposed the Turing Test, or as he famously called it, the *'Imitation Game'* [1]. In the seven decades since this paradigm-altering

© Springer Nature Switzerland AG 2020
Z. Ju et al. (Eds.): UKCI 2019, AISC 1043, pp. 271–282, 2020.
https://doi.org/10.1007/978-3-030-29933-0_23

query, Computer Scientists continue to seek improved methods of true imitation of the multi-faceted human nature. In this paper, work explores a new method towards the imitation of human speech in an audial sense. In this competition of two differing data representation methods, rather than a human judge, statistical analyses work to distinguish the differences between real and artificial voices. The ultimate goal of such thinking is to discover new methods of artificial speech synthesis in order to fool a judge when discerning between it and a real human being, and thus, explore new strategies of winning an Imitation Game.[1]

Speech Synthesis is a rapidly growing field of artificial data generation not only for its usefulness in modern society, but for its forefront in computational complexity. The algorithm resource usage for training and synthesising human-like speech is taxing for even the most powerful hardware available to the consumer today. When hyper-realistic human speech synthesis technologies are reached, implications when current security standards are considered are somewhat grave and dangerous. In a social age where careers and lives could be dramatically changed, or even ruined by public perception, the ability to synthesise realistic speech could carry world-altering consequences. This report serves not only as an exploration into the effects of phonetic awareness in speech synthesis as an original scientific contribution, but also as a warning and suggestion of path of thought for the information security community. To give a far less grave example of the implications of speaker-imitative speech synthesis, there are many examples of disease or accident that result in a person losing their voice. For example, Motor Neurone Disease causes this through weakness in the tongue, lips, and vocal chords [2,3]. In this study, only 1.6 h of data are used for fine tune transfer learning in order to derive realistic speech synthesis, and of course, would likely show better performance with more data. Should enough data be collected before a person loses their ability to speak, a Text-To-Speech (TTS) System developed following the pipeline in this study could potentially offer a second chance by artificially augmenting a digital voice which closely sounded to the voice that was unfortunately lost.

This project presents a preliminary state-of-the-art contribution in the field of speech synthesis for human-machine interaction through imitation. In this paper, two differing methods are presented for data preprocessing before a deep neural network in the form of Tacotron learns to synthesise speech from the data. Firstly, the standard English text format is benchmarked, and then compared to a method of representation via the International Phonetic Alphabet (IPA) in order to explore the effects on the overall data. State-of-the-art implementations of Speech Synthesis often base learning on datasets of raw text via speech dictation, this study presents preliminary explorations into the new suggested paradigm of phonetic translations of the original English text, rather than raw text.

---

[1] Demonstrations are available at http://jordanjamesbird.com/tacotron/tacotrontest. html.

The remainder of this paper is as follows. Firstly, a background of Phonetic English, the Tacotron model and statistical acoustic fingerprinting are explored in Sect. 2. Secondly, the method outlining this experiment is described in Sect. 3; the method is comprised of preprocessing, training, and statistical validation. Though secondary scientific research is reviewed within the background section, due to the practices involved, citations are also given within the Method section where appropriate. Finally, Sect. 5 presents future research projects based on the findings of this work, implications of said findings, and a final conclusion.

2   Background

2.1   English Language and Its Phonetics

The English language in its modern form is an amalgamation of largely Old English and Old French which were mostly spoken in their respective countries until the Anglo-Saxon invasion and subsequent conquering of England in 1066. The language of the upper classes of Anglo-Saxon Britain, and thus the most of the literary works, were a form of Anglo-Norman. The Anglo-Norman language, with influence from surrounding European Nations, then began to form into the English language that is spoken today [4]. Research suggests that although the language has changed greatly in the following 953 years, the phonetic structure has undergone relatively slight changes [5], thus showing that phonetics are a better representation of language than the written word. This is known as the study of phonology [6]. All of the spoken phonemes were formally presented in 1888 through the IPA, or International Phonetic Alphabet.

The biological limitations of the human voice thus limit the number of sounds that can be produced. These are the methods of *Labial, Dental, Alveolar, Postulveolar, Palatal, Velar, or Glottal*. The methods can then in turn be affected as *Nasal, Plosive, Fricative, or Approximant* [7]. Such categories contain all sounds that make up all human languages since the discovery of the method of spoken language [8]. In the International and English Phonetic Alphabets, sounds are denoted by 44 unique symbols representing each of the different sounds in the British dialect.

Previous work found success in replacing the nominal outputs of a speech recognition system with phonetic representation [9]. This research project explores the effects of phonetic representation but for the synthesis of speech rather than its recognition.

2.2   Tacotron

Tacotron is a Spectrogram Feature Prediction Deep Learning Network [10, 11] inspired by the architectures of Recurrent Neural Networks (RNN) in the form of Long Short-term Memory (LSTM). The Tacotron model uses character embedding to represent a text, as well as the spectrogram of the audio wave. Recurrent architectures are utilised due to their ability of temporal awareness, since speech is a temporal activity [12,13]. That is, where frame n does not occur at the start or end of the wave, it is directly influenced and thus has predictive ability both to and for frames $n-1$ and $n+1$. Since audio may possibly be lengthy, a nature in which recurrence tends to fail, 'attention' is modelled in order to allow for long sequences in temporal learning and as such its representation [14].

Actual speech synthesis, the translation of spectrogram to audio data, is performed via the Griffin-Lim algorithm [15]. This algorithm performs the task of signal estimation via its Short-time Fourier Transform (STFT) by iteratively minimising the Mean Squared Error (MSE) between estimated STFT and modified STFT. STFT is a Fourier-transform in which the sinusoidal frequency of content of local sections of a signal are determined [16].

Alternate notations of English through encoding and flagging have been shown to provide more understanding of various speech artefacts. A recent work by researchers at Google found that spoken prosody could be produced [17]. The work's notation allowed for the patterns of stress and intonation in a language. The implementation of a Wave Network [18] has shown to produce similarity gains of 50% when use in addition to the Tacotron architecture.

2.3 Acoustic Fingerprint

Acoustic Fingerprinting is the process of producing a summary of an audio signal in order to identify or locate similar samples of audio data [19]. To produce similarity, alignment of audio is performed and subsequently the two time-frequency graphs (spectrograms) have the distance between their statistical properties such as peaks measured. This process is performed in order to produce a percentage similarity between a pair of audio clips.

Fingerprint similarity measures allow for the identification of data from a large library, the algorithm operated by music search engine Shazam allows for the detection of a song from a database of many millions [20]. Detection in many cases was succesfully performed with only a few milliseconds of search data. Though this algorithm is often used for plagiarism detection and search engines within the entertainment industries, to spoof a similarity would argue that the artificial data closely matches that of real data. This is performed in this experiment by comparing the fingerprint similarities of audio produced by a human versus the audio produced by the Griffin-Lim algorithm on the spectrographic prediction of the Tacotron networks.

3 Method

3.1 Data Collection and Preprocessing

An original dataset of 950 MB (1.6 h, 902 .wav clips) of audio was collected and preprocessed for the following experiments. This subsection describes the processes involved. Due to security concerns, the dataset is not available and is thus described in greater detail within this section.

The 'Harvard Sentences'[2] were suggested within the *IEEE Recommended Practices for Speech Quality Measurements* in 1969 [21]. The set of 720 sentences and their important phonetic structures are derived from the IEEE Recommended Practices and are often used as a measurement of quality for Voice over Internet Protocol (VoIP) services [22,23]. All 720 sentences are recorded by the subject, as well as tense or subject alternatives where available ie. sentence 9 *"Four hours of steady work faced us"* was also recorded as *"We were faced with four hours of steady work"*.

The aforementioned IEEE best practices were based on ranges of phonetic pangrams. A sentence or phrase that contains all of the letters of the alphabet is known as a pangram. For example, *"The quick brown fox jumps over the lazy dog"* contains all of the English alphabetical characters at least once. A phonetic pangram, on the other hand, is a sentence or phrase which contains examples of all of the phonetic sounds of the language. For example, the phrase *"that quick beige fox jumped in the air over each thin dog. Look out, I shout, for he's foiled you again, creating chaos"* requires the pronunciation of every one of the 45 phonetic sounds that make up British English. 100 British-English phonetic pangrams are recorded.

The final step of data collection was performed in order to extend the approximately 500 MB of data closer to the 1 GB mark, random articles are chosen from Wikipedia, and random sentences from said articles are recorded. Ultimately, all of the data was finally transcribed into either raw English text or phonetic structure (where lingual sounds are replaced by IPA symbols), in order to provide a text input for every audio data. From this the two datasets are produced, in order to compare the two pre-processing approaches. All of the training takes place via the 2816 CUDA cores of an Nvidia GTX 980Ti Graphics Processing Unit, with the exception of the Griffin-Lim algorithm which is executed on an AMD FX8320 8-Core Central Processing Unit at a clock speed of 3500 MHz.

3.2   Fine Tune Training and Statistical Validation

The initial network is trained on the LJ Speech Dataset[3] for 700,000 iterations. The dataset contains 13,100 clips of a speaker reading from non-fiction books along with a transcription. The longest clip is 10.1 s, the shortest is 1.1 s, and the average duration of clips are 6.5 s. The speech is made up of 13,821 unique words at which there are an average of 17 per clip. Following this, the two datasets of English language and English phonetics are introduced and fine tune training occurs for two different models for 100,000 iterations each. Thus, in total, 800,000 learning iterations have been performed where the final 12.5% of the learning has been with the two differing representations of English.

---

[2] https://www.cs.columbia.edu/~hgs/audio/harvard.html.

[3] https://keithito.com/LJ-Speech-Dataset/.

Table 1. Ten strings for benchmark testing which are comprised of all English sounds

ID	String
1	"Hello, how are you?"
2	"John bought three apples with his own money."
3	"Working at a University is an enlightening experience."
4	"My favourite colour is beige, what's yours?"
5	"The population of Birmingham is over a million people."
6	"Dinosaurs first appeared during the Triassic period."
7	"The sea shore is a relaxing place to spend One's time."
8	"The waters of the Loch impressed the French Queen"
9	"Arthur noticed the bright blue hue of the sky."
10	"Thank you for listening!"

For comparison of the two models, statistical fingerprint similarity is performed. This is due to model outputs being of an opinionated quality, ie. how realistic the speech sounds from a human point of view. This is not presented in the benchmarking of models, and thus comparing the loss of the two model training processes would yield no opiniative measurement. To perform this, natural human speech is recorded by the subject that the model is trained to imitate. The two models both also produce these phrases, and the fingerprint similarity of the models and real human are compared. A higher similarity suggests a better ability of imitation, and thus better quality speech produced by the model.

A set of 10 strings are presented in Table 1. Overall, this data includes all sounds within the English language at least once. This validation data is recorded by the human subject to be imitated, as well as the speech synthesis models. Each of the phrases are recorded three times by the subject, and comparisons are given between the model and each of the three tests, comprising thirty tests per model.

Figures 1, 2 and 3 show examples of spectrographic representations when both a human being and a Tacotron network speak the sentence *"Working at a University is an Enlightening Experience"*. Though frequencies are slightly mismatched in that the network seems to be predicting higher frequencies than those in the human speak, the peaks within the data discerning individually-spoken words are closely matched by the Tacotron prediction. Though the two predictions look similar, the fingerprint similarity of the phonetically aware prediction is far closer to a human than otherwise, this is due to the fingerprint consideration of most important features rather than simply the distance between two matrices of values. Additionally, the timings of values are not considered, the algorithm produces a best alignment of the pair of waves before analysing their similarity. For example, the largest peak is the first syllable of the word "University", and thus those two peaks would be compared, rather than the differing

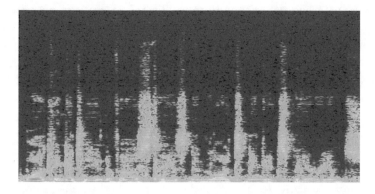

Fig. 1. Spectrogram of *"Working at a University is an Enlightening Experience"* when spoken by a human being.

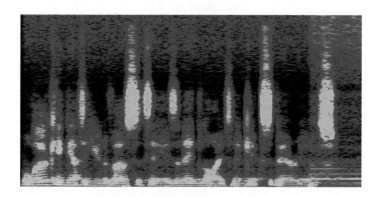

Fig. 2. Spectrogram of *"Working at a University is an Enlightening Experience"* when predicted by the English written text Tacotron network.

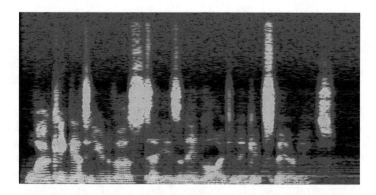

Fig. 3. Spectrogram of *"Working at a University is an Enlightening Experience"* when predicted by the phonetically aware Tacotron network.

data if alignment had not been performed. Therefore, silence before and after a spoken phrase is not considered, rather, only the phrase from its initial inception to final termination.

4   Preliminary Results

Within this section, the preliminary results are presented. Firstly, the acoustic fingerprint similarities of the models and human voices are compared. Finally, the average results for the two models are compared with one another.

Table 2. Thirty similarity tests performed on the raw English speech synthesis model with averages of sentences and overall average scores. Failures are denoted by **F**. Overall average is given as the average of experiments 1, 2 and 3.

Phrase	Experiment			
	1	2	3	Avg.
1	F	F	F	0 (3F)
2	22.22	22.22	66.67	37.02
3	56.6	56.7	75.4	62.9
4	0	51.28	0	17.09 (2F)
5	6	2	4	4
6	20	41.67	62.5	41.39
7	55.56	18.52	22.43	32.17
8	24.39	24.39	48.78	32.55
9	22.72	22.72	22.72	22.72
10	F	71.4	F	23.8 (2F)
Avg.	20.74	31.09	30.25	**27.36**

Table 2 shows the results for the tests on the raw English speech dataset. Of the thirty experiments, 23% (7/30) were failures and had no semblance of similarity to the natural speech. One test, phrase 1, was a total failure with all three experiments scoring zero. Overall, the generated data resembled the human data by an average of 21.07%.

Table 3 shows the results for the tests on the phonetic English speech dataset. Of the thirty experiments, 13% (4/30) were failures and had no semblance of similarity to the natural speech, this was slightly was lower than the raw English dataset. This said, there did not occur an experiment with complete catastrophic failure in which all three tests scored zero. On an average of the three experiments, the human data and the generated data were 35.31% similar. Figure 4 shows the average differences between the acoustic fingerprints of human and artificial data in each of the ten sets of three experiments.

Table 3. Thirty similarity tests performed on the phonetic English speech synthesis model with averages of sentences and overall average scores. Failures are denoted by **F**. Overall average is given as the average of experiments 1, 2 and 3.

Phrase	Experiment			
	1	2	3	Avg.
1	58.8	0	0	19.6 (2F)
2	85.7	28.57	57.14	57.14
3	93.7	78.12	46.88	72.9
4	51.28	25.64	25.64	34.19
5	38.46	38.46	39	38.64
6	35.71	35.71	17.8	29.74
7	34.5	34	17.2	28.57
8	43.4	21.7	43.48	36.19
9	20.4	20	26.4	22.27
10	0	41.6	0	13.89 (2F)
Avg.	46.19	32.38	27.35	**35.31**

In comparing head-to-head results, the phonetics dataset produced experiments that on average outperformed the written language dataset in six out of ten cases. This said, experiment nine was extremely close with the two models achieving 22.27% and 22.72% with a negligible difference of only 0.45%. In the cases where the language set outperformed the phonetics set, the difference between the two were much smaller than the vice versa outcomes. In terms of preliminary results, the phonetic representation of language has gained the best results in human speech imitation when comparing the acoustic fingerprint metrics. Often, inconsistencies occur in the similarity of human and robotics speech (in both approaches); this is likely due to either a lack of enough data within the training and validation sets, or an issue of there not being enough training time in order to form a stable model that produces consistent output - or, of course, a combination of the two. Further exploration via future experimentation could pinpoint the cause of inconsistency.

5   Future Work and Concluding Implications

Although results suggested the phoneme aware approaches were preliminarily more promising than raw English notation, the phonetic awareness approach was faced with a disadvantage from the fine-tuning process. The pre-existing model was trained with raw English language on a US-dialect, and fine tuned for raw English language in GB-dialect as well as English phonetics in GB-dialect. Thus, the phonetic model would require more training in order to overcome the disadvantaged starting point it faced. For a more succinct comparison, future models should be trained from an initial random distribution of network weights for their

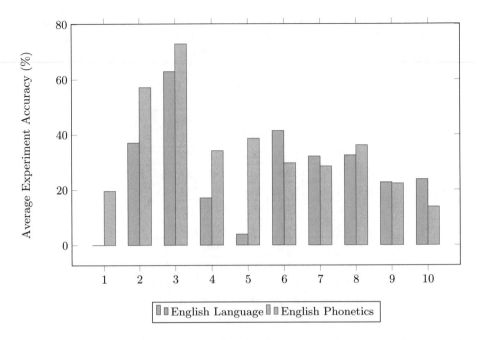

Fig. 4. Comparison of the two approaches for the average of ten sets of three experiments.

respective datasets. In addition to this, it must be pointed out that input data from the English written text dataset had 26 unique alphabetic values whereas this is extended in the second dataset since there are 44 unique phonemes that make up the spoken English language in a British dialect. Statistical validation through the comparison of acoustic fingerprints are considered, with similarities to real speech compared on the same input sentence or phrase. Though an acoustic fingerprint does give a concrete comparison between pairs of output data, human opinion is still not properly reflected. For this, as the Tacotron paper did, Mean Opinion Score (MOS) should also be performed. MOS is given as $MOS = \frac{\sum_{n=1}^{N} R_n}{N}$, where R are rating scores given by a subject group of N participants. Thus, this is simply the average rating given by the audience. MOS requires a large audience to give their opinions, denoted by a nominal score, in order to rate the networks in terms of human hearing; human hearing and audial understanding is an ability a Turing Machine does not yet have. Such MOS would allow for a second metric, real opinion, to also provide a score. A multi-objective problem is then presented through the maximisation of acoustic fingerprint similarities as well as the opinion of the audience. Additionally, other spectrogram prediction paradigms such as Tacotron2 and DCTTS should be studied in terms of the effects of English vs. Phonetic English.

As mentioned in the previous section, further work should also be performed into pinpointing the cause of inconsistent output from the models. Explorations into the effects of there being a larger dataset as well as more training time for the model could discover the cause of inconsistency and help to produce a stronger training paradigm for speech synthesis.

To conclude, 100,000 extra iterations of training on top of a publicly available dataset, then fine-tuned on a human dataset of only 1.6 h worth of speech translated to phonetic structure, has produced a network with the ability to reproduce new speech at 35.31% accuracy. It is not out of the question whatsoever for post-processing to enable the data to be completely realistic, which could then be 'leaked' to the media, the law, or otherwise. Such findings present a dangerous situation, in which a person's speech could be imitated to create the illusion of evidence that they have said such things that in reality they have not. Though this paper serves primarily as a method of maximising artificial imitative abilities, it should also serve as a grave warning in order to minimise the potential implications on an individual's life. Future information security research should, and arguably must, discover competing methods of detection of spoof speech in order to prevent such cases. On the other hand, realistic speech synthesis could be used in realtime for more positive means, such as an augmented voice for those suffering illness that could result in the loss of the ability of speech.

References

1. Turing, A.M.: Computing Machinery and Intelligence (1950)
2. Locock, L., Ziebland, S., Dumelow, C.: Biographical disruption, abruption and repair in the context of motor neurone disease. Sociol. Health Illn. **31**(7), 1043–1058 (2009)
3. Yamagishi, J., Veaux, C., King, S., Renals, S.: Speech synthesis technologies for individuals with vocal disabilities: voice banking and reconstruction. Acoust. Sci. Technol. **33**(1), 1–5 (2012)
4. Baugh, A.C., Cable, T.: A History of the English Language. Routledge, Abingdon (1993)
5. Loyn, H.R.: Anglo Saxon England and the Norman Conquest. Routledge, London (2014)
6. Fromkin, V., Rodman, R., Hyams, N.: An Introduction to Language. Cengage, Boston (2006)
7. Titze, I.R., Martin, D.W.: Principles of Voice Production. Prentice-Hall, Englewood Cliffs (1994)
8. Menzel, W., Atwell, E., Bonaventura, P., Herron, D., Howarth, P., Morton, R., Souter, C.: The ISLE corpus of non-native spoken English. In: Proceedings of LREC 2000: Language Resources and Evaluation Conference, vol. 2, pp. 957–964. European Language Resources Association (2000)
9. Bird, J.J., Wanner, E., Ekart, A., Faria, D.R.: Phoneme aware speech recognition through evolutionary optimisation. In: The Genetic and Evolutionary Computation Conference, GECCO (2019)
10. Wang, Y., Skerry-Ryan, R., Stanton, D., Wu, Y., Weiss, R.J., Jaitly, N., Yang, Z., Xiao, Y., Chen, Z., Bengio, S., et al.: Tacotron: Towards end-to-end speech synthesis. arXiv preprint arXiv:1703.10135 (2017)

11. Tachibana, H., Uenoyama, K., Aihara, S.: Efficiently trainable text-to-speech system based on deep convolutional networks with guided attention. In: 2018 IEEE International Conference on Acoustics, Speech and Signal Processing (ICASSP), pp. 4784–4788. IEEE (2018)
12. Sak, H., Senior, A., Beaufays, F.: Long short-term memory recurrent neural network architectures for large scale acoustic modeling. In: Fifteenth Annual Conference of the International Speech Communication Association (2014)
13. Li, X., Wu, X.: Constructing long short-term memory based deep recurrent neural networks for large vocabulary speech recognition. In: 2015 IEEE International Conference on Acoustics, Speech and Signal Processing (ICASSP), pp. 4520–4524. IEEE (2015)
14. Bahdanau, D., Cho, K., Bengio, Y.: Neural machine translation by jointly learning to align and translate. arXiv preprint arXiv:1409.0473 (2014)
15. Griffin, D., Lim, J.: Signal estimation from modified short-time fourier transform. IEEE Trans. Acoust. Speech Signal Process. **32**(2), 236–243 (1984)
16. Sejdić, E., Djurović, I., Jiang, J.: Time-frequency feature representation using energy concentration: an overview of recent advances. Digit. Signal Process. **19**(1), 153–183 (2009)
17. Skerry-Ryan, R., Battenberg, E., Xiao, Y., Wang, Y., Stanton, D., Shor, J., Weiss, R. J., Clark, R., Saurous, R.A.: Towards end-to-end prosody transfer for expressive speech synthesis with tacotron. In: International Conference on Machine Learning, pp. 4693–4702 (2018)
18. Zhang, M., Wang, X., Fang, F., Li, H., Yamagishi, J.: Joint training framework for text-to-speech and voice conversion using multi-source tacotron and wavenet. arXiv preprint arXiv:1903.12389 (2019)
19. Bormans, J., Gelissen, J., Perkis, A.: MPEG-21: the 21st century multimedia framework. IEEE Signal Process. Mag. **20**(2), 53–62 (2003)
20. Wang, A., et al.: An industrial strength audio search algorithm. In: ISMIR, vol. 2003, pp. 7–13, Washington, DC (2003)
21. IEEE.: IEEE Transactions on Audio and Electroacoustics, vol. 21. IEEE (1973)
22. Yochanang, K., Daengsi, T., Triyason, T., Wuttidittachotti, P.: A comparative study of VoIP quality measurement from G. 711 and G. 729 using PESQ and thai speech. In: International Conference on Advances in Information Technology, pp. 242–255. Springer (2013)
23. Yankelovich, N., Kaplan, J., Provino, J., Wessler, M., DiMicco, J.M.: Improving audio conferencing: are two ears better than one? In: Proceedings of the 2006 20th Anniversary Conference on Computer Supported Cooperative Work, pp. 333–342. ACM (2006)

Predicting Hospital Length of Stay for Accident and Emergency Admissions

Kieran Stone[1], Reyer Zwiggelaar[1], Phil Jones[2], and Neil Mac Parthaláin[1(✉)]

[1] Department of Computer Science, Aberystwyth University, Ceredigion, Wales
{kis12,rzz,ncm}@aber.ac.uk
[2] Hywel Dda University Health Board, Bronglais District General Hospital,
Aberystwyth, Ceredigion, Wales
phil.jones@wales.nhs.uk

Abstract. The primary objective of hospital managers is to establish appropriate healthcare planning and organisation by allocating facilities, equipment and manpower resources necessary for hospital operation in accordance with a patients needs while minimising the cost of healthcare. Length of stay (LoS) prediction is generally regarded as an important measure of inpatient hospitalisation costs and resource utilisation. LoS prediction is critical to ensuring that patients receive the best possible level of care during their stay in hospital. A novel approach for the prediction of LoS is investigated in this paper using only data based upon generic patient diagnoses. This data has been collected during a patients stay in hospital along with other general personal information such as age, sex, etc. A number of different classifiers are employed in order to gain an understanding of the ability to perform knowledge discovery on this limited dataset. They demonstrate a classification accuracy of around 75%. In addition, a further set of perspectives are explored that offer a unique insight into the contribution of the individual features and how the conclusions might be used to influence decision-making, staff and resource scheduling and management.

Keywords: Data mining · Classification · Medical informatics · Hospital length of stay · Decision support

1 Introduction

In order to ensure an optimum level of patient care and management, healthcare systems are placing increasing emphasis on effective resource management and forecasting. Primarily, there is a need to improve the quality of patient care whilst simultaneously prioritising patient needs and reducing the associated cost of care [1]. Several techniques have been developed to predict admissions [2], patient bed needs and overall bed utilisation [3]. The overarching aim of these techniques is to provide reliable prediction for the LoS of individual patients as well as an understanding of the factors which have an influence on it.

© Springer Nature Switzerland AG 2020
Z. Ju et al. (Eds.): UKCI 2019, AISC 1043, pp. 283–295, 2020.
https://doi.org/10.1007/978-3-030-29933-0_24

Patient hospital length of stay can be defined as the number of days that an in-patient will remain in hospital during a single admission event [4]. LoS provides an enhanced understanding of the flow of patients through hospital care units which is a significant factor in the evaluation of operational functions in various healthcare systems. LoS is regularly considered to be a metric which can be applied to identify resource utilisation, cost and severity of illness(es) [5]. Previous work has sought to group patients by their respective medical condition(s). This approach assumes that each disease, condition or procedure is associated with a predefined, recommended LoS. However, LoS can be affected by a great number of different factors which tend to extend the original target LoS including (but not limited to): a patient's level of fitness, medical complications, social circumstances, discharge planning, treatment complexity, etc.

The work in the literature that utilises data mining approaches to perform knowledge discovery, relies on datasets (and methodologies) that are often disease, condition or patient-group specific [6]. There is no work which utilises general admissions data in order to gain an understanding of the more generic factors that influence length of stay. Furthermore, very limited work focuses solely upon sparse diagnostic information, relying instead on specialised condition-relative data only.

In this paper, a novel approach for the prediction of LoS is investigated by employing a limited dataset which contains only diagnostic code information, along with generic patient information such as *age*, *sex*, and *postcode* (to give an indication of geographical catchment/demographic). This data however, can be very sparse, as some patients tend to have more recorded data in the form of diagnostic codes. Additionally, several confounding factors can be present which mean that vagueness and uncertainty plays a large part in determining the outcome. The data is analysed using a selection of both traditional learning approaches and those based upon fuzzy and fuzzy-rough sets.

The remainder of the paper is structured as follows: In Sect. 2, a broad overview of the current approaches to predicting LoS is presented along with an appraisal of the current state-of-the-art. In Sect. 3, the data is presented and the approach to dealing with that data is discussed. In Sect. 4, the experimental evaluation is presented along with the results. Finally, some conclusions are drawn and topics for further exploration are identified and discussed.

2   Background

The earliest studies of LoS in the 1970's adopted operations research-based approaches; employing techniques such as Compartmental Modelling, Hidden Markov Models and Phase-type Distributions [7]. These methods whilst providing a good level of predictive accuracy, often suffer from a number of shortcomings which mean that they are less suited to modern and dynamic hospital environments where uncertainty and rapidly changing care needs can affect the day-to-day running of a hospital. One of the other major disadvantages of such methods is the long simulation times required in order to generate reliable results.

Allied to this is the need for very large amounts of subjective expert input and data in order to correctly 'tune' the approach such that the output or prediction is useful. As a consequence of these factors, the process of actually training such approaches, means that these methods are also very resource and effort intensive [8]. More specifically, compartmental modelling is a well established and mathematically sound methodology, however it is based on a single day census of beds and as such can be highly dependent on the day the census was carried out. These models also do not allow for continuous time frames [9]. Phase-type distributions provide detailed insights into the causality of LoS but make use of more complex methodologies than traditional methods and thus require considerably more effort to ensure that model output is clinically meaningful [10].

Another set of approaches to predicting LoS are those based upon statistical and arithmetic foundations. Such approaches include: [11]. These have proven to be useful for estimating LoS, but focus on the problem of *average* LoS, which is a much less informative measure than the more general problem. Such methods tend to be easy to use and are rooted in statistical theory and are therefore easy to reproduce. However, they also suffer from a number of weaknesses; many of the approaches adopt foundations that are not transparent to human scrutiny e.g. Principal Component Analysis (PCA). In addition, many of the methods do not account for the inherent complexity and uncertainty that is typically found in LoS data [7].

More recently, machine learning and data mining approaches have been applied to the LoS problem in an attempt to improve modelling, but also to aid in dealing with more complex data that is associated with LoS prediction. Much of the work in this area produces very accurate models with performance in the high 80% - low 90% [12] but using sub-symbolic learning methods. Transparency is central to good clinical practice and clinicians may be reluctant to accept the output of such methods when there is no explicit explanation for the derivation of the results and rules [13]. The work in this paper is based solely upon methods that are transparent to human scrutiny.

3 Data

The data for this work is drawn from the Hywel Dda Health Board, and collected at *Bronglais General Hospital, Aberystwyth, Ceredigion, Wales* between March 2016 and March 2017. It consists of data relating to 6,543 admissions that were admitted via the *Accident and Emergency* department. It is balanced in terms of the number of male vs. female patients. The data was anonymised to remove any personally identifiable attributes. The general process of data collection and patient admission is shown in Fig. 1.

The data has 16 conditional features and these include attributes such as: a patient's age and sex, the first four letters of their postcode, their primary diagnosis, followed by up to 12 secondary diagnoses, (depending on which conditions are diagnosed for the patient). There are no missing values in the dataset. However, 12 data objects which did not have an associated discharge date were

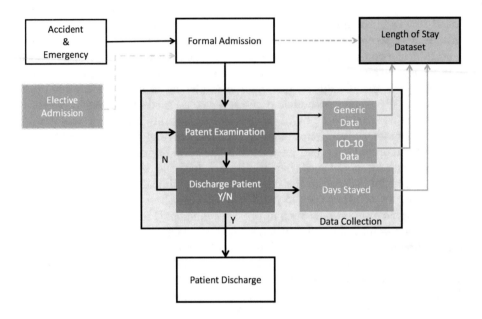

Fig. 1. Patient data collection process

removed along with records which did not have any diagnoses, leaving 6,531 records. A summary of the data characteristics can be found in Table 1 and the numbers of patients and days stayed are shown in Fig. 2.

The diagnoses features are recorded using the ICD-10 classification (International Statistical Classification of Diseases and Related Health Problems). This is a World Health Organisation (WHO) system of comprehensive medical classification. It contains codes for: diseases, signs and symptoms, abnormal findings, complaints, external causes of injury or diseases and even social circumstances. For the data under consideration, the features representing each of the ICD-10 diagnostic codes has a potential domain of up to 70,000 different values. However, the actual number of unique ICD-10 codes for the data used in this work is 1,393. These individual diagnostic codes are then grouped by ICD block e.g. I25.5 - *Ischaemic cardiomyopathy* and I51.7 - *Cardiomegaly* are drawn from the same block and so are placed in the same group relating to heart conditions in the data: I00–99. This reduces the possible number of feature values for each of the diagnoses to a maximum of 22. The rationale for performing this step is that feature values may become too sparse for some individual ICD-10 codes. In addition, from a clinical standpoint, many closely related codes are often overlapping (or are neighbouring) or may only differ slightly depending on the examining clinician's interpretation - see discussion. In terms of data acquisition, the data is coded at the time of discharge by the information department. The data is thus a combination of electronically entered data from time of admission and time of discharge as well as data entered by experienced clinical coders who review the notes and enter ICD-10 codes for primary diagnosis (i.e reason for

Table 1. LoS dataset

Feat. name	Type	Range of vals.
age at admission	integer	0–105
four character postcode	discrete	AA00 - ZZ99
sex	binary	m/f
Primary diagnosis	alpha-numeric	22 discrete
secondary diagnosis 1	alpha-numeric	22 discrete
secondary diagnosis 2	alpha-numeric	22 discrete
secondary diagnosis 3	alpha-numeric	22 discrete
secondary diagnosis 4	alpha-numeric	22 discrete
secondary diagnosis 5	alpha-numeric	22 discrete
secondary diagnosis 6	alpha-numeric	22 discrete
secondary diagnosis 7	alpha-numeric	22 discrete
secondary diagnosis 8	alpha-numeric	22 discrete
secondary diagnosis 9	alpha-numeric	22 discrete
secondary diagnosis 10	alpha-numeric	22 discrete
secondary diagnosis 11	alpha-numeric	22 discrete
secondary diagnosis 12	alpha-numeric	22 discrete
days stayed	numeric	0–200

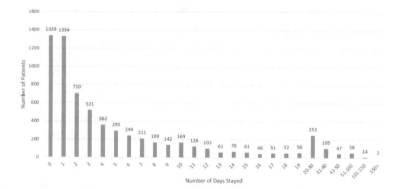

Fig. 2. The number of patients and days stayed

admission) and secondary diagnoses. Demographic data is stored digitally and is updated as and when required (e.g address, registered GP).

It is important to note that quite often patients admitted with the same primary diagnosis can have differing LoS particularly if any given patient is discharged on the same day as admission. The number and types of diagnoses therefore are not related to length of stay. Indeed, it is possible to have a high number of secondary diagnostic codes attached to a patient who only had a short stay. The converse is also true; a patient may have a long LoS but only a single diagnostic code recorded in the data.

4   Application

In this section, the methodology that is used to explore the data is presented. The first task was to use the data to determine what the predictive accuracy might be when framed as a regression problem. This provides some important indications as to the ability to model LoS in this way. However, from a hospital resource management standpoint, this may not be as useful as determining lengths of stay that are longer than a single day - since each 24 h stay (or part thereof) has an associated cost. With this in mind the problem is re-framed as a classification problem. The different experimental configurations for these two problems are described in the following section.

4.1   Experimental Setup

The 10 learning techniques employed here for building models for the task of classification and regression are drawn from both traditional and fuzzy/fuzzy-rough approaches. A preprocessing step in the form of feature ranking and selection is performed to gain a understanding of the influence of the features.

Each of the models were evaluated using stratified 10×10 fold cross-validation and a paired t-test (significance $= 0.05$) is used to assess statistical significance. The traditional approaches employed are J48 (a C4.5 decision tree based learner) [14], NB (NaiveBayes), JRIP (a rule-based learner) [15] and PART (also a rule-based learner) [16]. The fuzzy and fuzzy-rough approaches are FNN (fuzzy nearest neighbour [17]), FURIA [18], QSBA [19], FRNN, VQNN, OWANN (all fuzzy-rough learners) [20]. The Zero-Rule (majority class) was also employed as a benchmark.

The experimentation is divided into three parts as shown in Fig. 3: (1) a regression analysis, (2) as a binary classification problem and finally (3) an investigation of the influence of the features for part (2).

In order to frame the prediction of LoS as a regression problem, the exact number of days that each patient stayed was used as a decision. The number of nearest-neighbours to consider during learning for the nearest-neighbour based classifiers were incremented by two up to 5 and evaluated as part of a stratified 10×10 fold cross validation. The exception to this was FRNN which used the default value for NN as this does not affect performance [20].

For classification, *days stayed* was divided into two intervals *(0–1 days stayed)* which represented *short stays* of up to one day and *(2–200 days stayed)* which represented *long stays*. The problem that is addressed as a result, is that of: "Can a patient stay of 0–1 days or greater than one day be classified?" Once again, the number of nearest-neighbours is incremented in the same way as the previous approach. The remainder of the learners employed their respective default parameters.

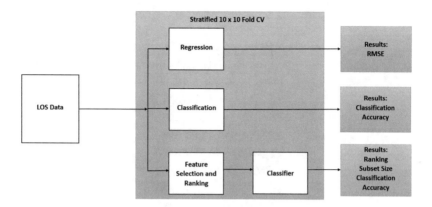

Fig. 3. Experimentation

The investigation of the contribution of the features and their respective influences was performed by employing two different approaches (1) a fuzzy-rough feature subset selection (FRFS) [21] with a greedy hill-climbing search as part of a 10×10 fold cross validation. (2) fuzzy-rough feature ranking (FRFR) - a ranking of the features using the same evaluator as FRFS but using only single feature candidate subsets in order to calculate the contribution of individual features. Again this was performed as part of a 10×10 fold stratified cross validation to ensure that the results were robust and the average rank is reported.

LoS as a Regression Problem. The initial investigation of the data involved attempting to determine if it could be used to predict the LoS when framed as a regression problem. Each of the regression models were created with the intent of predicting the LoS with a specific set of feature values. However, the data in this state is quite sparse with a large percentage of very specific diagnoses. Also, each patient had large variation in the number of days stayed. The fit of the models was estimated using root-means squared error (RMSE). It can be seen from Table 2 that each of the learners yield large values of RMSE. Note that not all of the learners presented here can be used for regression problems, i.e. where the decision is real-valued. Hence, no results are presented for such methods (QSBA, FURIA, J48, JRip, PART and NaiveBayes).

Although a small decrease in RMSE is apparent most noticeably in the performance of FNN, from 18.58 to 13.35 by varying the number of nearest-neighbours. The overall performance of each of the algorithms remains poor. It's also important to note that all of the algorithms perform worse than ZeroR with 11.34.

LoS as a Classification Problem. As mentioned previously, the overall sparsity of the data is likely to have an effect on the prediction of LoS using a regression approach. Additionally, from a hospital resource administration point of view, it is more clinically meaningful to reliably predict short stays as the

Table 2. Summary of regression results

Learner	RMSE (s.d)
ZeroR	11.34(1.50)
FNN(1)	18.58(2.49)*
FNN(3)	13.62(0.78)*
FNN(5)	12.35(0.68)*
FRNN	12.00(0.88)
VQNN(1)	11.99(0.86)
VQNN(3)	11.99(0.87)
VQNN(5)	11.98(0.88)
OWANN(1)	11.99(0.86)
OWANN(3)	11.99(0.87)
OWANN(5)	11.98(0.88)

required level of cost and care is negligible in comparison to patients that stay longer than a single 24 h period. Consequently, rather than predicting the exact LoS with data that has an inherent level of sparsity, it may be more useful to group patient LoS in terms of short stays and longer stays where the required level of care and cost associated with LoS is clinically significant.

Given that it is more clinically meaningful to reliably estimate 0–1 days stayed versus longer periods, the problem was re-framed as a classification task. The learners in Sect. 4.1 were once again employed using the same experimental setup. 10 × 10 fold cross validation was used to classify *short stay* or *long stay* as a binary classification problem. This had the effect of a small imbalance in the decision class - 59% *long stay*: 41% *short stay*. The classification results are shown in Table 3.

As can be seen in Table 3 the ZeroR metric simply returns the *long stay* class for all instances as it is the majority class: 59.07%. These results demonstrate that all of the fuzzy/fuzzy-rough NN algorithms perform statistically worse than the ZeroR benchmark indicating that they are only comparable with chance. This is possibly due to the crisp or discrete-valued nature of the data. Increasing the number of nearest-neighbours appears only to decrease the performance of each learner further. Of the fuzzy/fuzzy-rough approaches, only QSBA and FURIA were shown to statistically outperform the baseline performance with accuracies of 73.32% and 75.07% respectively. The highest accuracies were achieved using J48, JRIP, and NaiveBayes. FURIA and QSBA offer at least comparable performance with the other standard rule based algorithms without any parameter tuning. Note that there is no statistical significance between the results of these five learners. NaiveBayes, QSBA, FURIA, PART, JRip and PART all yield the highest values for ROC with scores over 0.7. It is interesting to note that the nearest-neighbour learners perform well for short stay and poorly for long stay however, the converse is true with QSBA, FURIA and the other tra-

Table 3. Summary of classification results

Learner	Short Stay Accy.(%)	Long Stay Accy.(%)	Overall Accy.(%) (s.d.)	ROC
ZeroR	0	100	59.07(0.06)	0.49
FNN(1)	70.52	45.56	55.78(1.87)*	0.58
FNN(3)	74.44	40.61	54.46(1.76)*	0.57
FNN(5)	75.83	38.54	53.8 (1.88)*	0.57
FRNN	65.13	48.18	55.12(1.89)*	0.61
VQNN(1)	70.44	45.74	55.85(1.89)*	0.58
VQNN(3)	74.37	39.39	53.71(1.77)*	0.62
VQNN(5)	76.35	37.24	53.25(1.64)*	0.64
OWANN(1)	70.44	45.74	55.85(1.89)*	0.58
OWANN(3)	74.37	39.39	53.71(1.77)*	0.64
OWANN(5)	76.35	37.24	53.25(1.64)*	0.65
QSBA	47.10	91.49	73.32(1.52)v	0.73
FURIA	59.29	86.00	75.07(1.88)v	0.73
J48	60.19	86.44	75.70(1.62)v	0.77
JRip	58.99	86.80	75.42(1.66)v	0.74
PART	45.30	89.76	71.56(1.65)v	0.76
NaiveBayes	65.91	79.62	74.01(1.72)v	0.81

* indicates statistically worse than ZeroR, 'v' indicates statistically better

ditional learners with each achieving >79%. This could be due to the tendency of NN algorithms to overfit the minority class because they only consider local neighbourhoods. The overall poor performance of the NN learners may be due to the discrete nature of the data itself and fuzzifying either the domain of the decision feature or indeed the conditional features may help in this respect - see conclusion.

Feature Selection and Ranking. The results of the feature analysis are shown in Table 4. The second column of this table is the frequency of occurrence of that feature in the selected subset for each of the 10 randomisations of the data across the 10 folds of classification; totalling 100 results. The third column shows the average individual rank for each feature, again as part of a 10-fold cross-validation. These results demonstrate that there is considerable variation between the relative importance of each of the features individually and in combination with one another. It is clear for FRFS that eight of the features always appear in the selected subset; *age at admission, four characters from postcode, sex, primary diagnosis* and *secondary diagnosis 1-4*. Quite often, the contribution of individual features in a dataset in isolation is not very useful, as it doesn't model the dependencies between attributes [21]. This means that several low ranked features when combined as a subset could have higher dependency than higher ranked (but somewhat redundant) features. This is reflected when observing *secondary diagnosis 8-12* as these never appear in any of the FRFS selected feature subsets. However, *secondary diagnosis 11 & 12* were ranked

6.36 and 6.3 respectively when considered in isolation by FRFR. Interestingly, *age at admission* had an average rank of 11.17 out of 16 using FRFR, despite appearing in 100% of the FRFS subsets. Also, the average ranks for each of the features revealed the *four character postcode* to be the most important feature along with *primary diagnosis, Secondary diagnoses 1 & 2* and *Sex* as the next most important.

Table 4. Feature analysis

Feature	FRFS freq.	FRFR avg. rank
age at admission	100%	11.17
four character postcode	100%	1
sex	100%	5.92
primary diagnosis	100%	2.3
secondary diagnosis 1	100%	3.99
secondary diagnosis 2	100%	4.36
secondary diagnosis 3	100%	8
secondary diagnosis 4	100%	11.59
secondary diagnosis 5	70%	10
secondary diagnosis 6	52%	14
secondary diagnosis 7	80%	15
secondary diagnosis 8	0%	13
secondary diagnosis 9	0%	11
secondary diagnosis 10	0%	10
secondary diagnosis 11	0%	6.36
secondary diagnosis 12	0%	6.3

Attribute Selected Classifiers. Using the best performing learners from Table 3, a feature selection preprocessing step was carried out. The results are shown in Table 5, which offers a comparison of classification accuracy for FRFS and FRFR. The overall performance of each of the learners is comparable to the performance shown in Table 3 with J48 and JRip both scoring slightly higher than before with accuracies of 76.39% and 75.74% respectively. Notably for FRFS, regardless of the number of features that have been selected the performance of the learners is almost identical to that of the unreduced dataset. This is an indication that there is some redundancy in the features as demonstrated in Table 4. Note that the top 11 ranked features are used here for FRFR to build classification models as this is the subset size that is typically returned by FRFS. As a result, a notable decrease in performance in terms of FRFR is evident for all of the learners.

Table 5. Attribute selected classifier results

Selection	Learner	Subset size	CA (s.d)	ROC Area
FRFS	J48	10.79	76.39%(1.54)	0.8
FRFS	JRip	10.79	75.74%(1.60)	0.74
FRFS	FURIA	10.79	74.93%(2.04)	0.73
FRFS	QSBA	10.79	73.28%(1.52)	0.74
FRFR	J48	11*	70.75%(4.60)	0.74
FRFR	JRIP	11*	69.71%(4.81)	0.68
FRFR	FURIA	11*	68.63%(5.08)	0.70
FRFR	QSBA	11*	64.31%(6.96)	0.59

*indicates that the 11 top-ranked features were used to build classifier learners

4.2   Discussion

The primary aim of this work is to provide a reliable estimation of LoS. When this was framed as a regression problem, the results were poor regardless of the learner employed. By changing the task to that of classification, where the two classes represent a more meaningful outcome (from a clinical and administrative standpoint) as *short stay* and *long stay*, it is possible to achieve ∼76.4% correct classification accuracy. It is interesting to note that there are very small differences between the standard deviations in Table 3 for all of the best performing learners, possibly indicating that there is not much more to be discovered from this data without further expert input. To provide a more accurate and clinically meaningful prediction of LoS, clearly a much more targeted approach is required which would encompass a patient's social factors, demographics and medical history etc.

The investigation of the features and their contribution to the classification outcomes provides some useful insights. It is clear that regardless of the approach involved that the first four secondary diagnosis features are important in determining LoS. The feature *four character postcode* also has the highest rank for FRFR indicating that geographical addresses play an important part in determining length of stay.

One of the potential limitations of this work is the use of ICD-10 codes in the data. These diagnostic codes represent diseases, signs and symptoms as well as other abnormal findings. Whilst they are used on a daily basis in a health care setting, in the context of LoS, the large groupings may not be useful in their current format. The reason for this is that codes can sometimes encompass disparate sets of disease classifications. This is an area that requires further investigation and expert input in order to address these shortcomings.

5 Conclusion

Determining LoS with high accuracy can be difficult as the task is confounded by multiple competing factors such as patient care and characteristics, social factors, and morbidity. All of these have the potential to extend patient LoS. Nevertheless, in this work the best performing learners were able to predict short stays and long stays with an average accuracy of 75%. This is clinically significant as there is a considerable difference in cost and care associated hospital LoS. The ability to predict stay for a longer or shorter periods enables hospital managers to allocate resources, improve patient care and deliver increased hospital function. This work determined that the *first four characters of postcode* was the most useful factor in classifying LoS. This is particularly important given that the data was collected from a hospital in a rural setting, which is significantly larger and more demographically varied than that of urban hospitals. It is also apparent from the feature analysis that the increasing number of secondary diagnoses that exist within the data become redundant regardless of LoS.

The novel approach to assessing LoS presented uses only generic patient diagnosis data. Although this is a preliminary study, it has highlighted the fact that there is limited knowledge that can be discovered from the data. In this work, patient diagnoses were grouped according to their ICD-10 classifications which may not be ideal - further expert input would help to improve this. However, further division into more granular or more related classifications may provide some extra leverage. As an alternative to this, the modelling could be improved by fuzzifying the domain of the decision feature with clinically informed categories of stay such as: *very short, short, medium, long,* etc.

Acknowledgements. Kieran Stone would like to acknowledge the financial support for this research through Knowledge Economy Skills Scholarship (KESS 2). It is part funded by the Welsh Government's European Social Fund (ESF) convergence programme for West Wales and the Valleys. WEFO (Welsh European Funding Office) contract number: C80815.

References

1. Garg, L., McClean, S.I., Barton, M., Meenan, B.J., Fullerton, K.: Intelligent patient management and resource planning for complex, heterogeneous, and stochastic healthcare systems. IEEE Trans. Syst. Man Cybern. - Part A: Syst. Hum. **42**, 1332–1345 (2012). https://doi.org/10.1109/TSMCA.2012.2210211
2. Kelly, M., Sharp, L., Dwane, F., Kelleher, T., Comber, H.: Factors predicting hospital length-of-stay and readmission after colorectal resection: a population-based study of elective and emergency admissions. BMC Health Serv. Res. **12**(1), 77 (2012)
3. Harper, P.R., Shahani, A.K.: Modelling for the planning and management of bed capacities in hospitals. J. Oper. Res. Soc. **53**(1), 11–18 (2002)
4. Huntley, D.A., Cho, D.W., Christman, J., Csernansky, J.G.: Predicting length of stay in an acute psychiatric hospital. Psychiatr. Serv. **49**(8), 1049–1053 (1998)

5. Chang, K.C., Tseng, M.C., Weng, H.H., Lin, Y.H., Liou, C.W., Tan, T.Y.: Prediction of length of stay of first-ever ischemic stroke. Stroke **33**(11), 2670–2674 (2002)
6. Awad, A., Bader-El-Den, M., McNicholas, J.: Patient length of stay and mortality prediction: a survey. Health Serv. Manag. Res. **30**(2), 105–120 (2017)
7. Marshall, A., Vasilakis, C., El-Darzi, E.: Length of stay-based patient flow models: recent developments and future directions. Health Care Manag. Sci. **8**(3), 213–220 (2005)
8. Altinel, I.K., Ulaş, E.: Simulation modeling for emergency bed requirement planning. Ann. Oper. Res. **67**(1), 183–210 (1996)
9. Vasilakis, C., Marshall, A.H.: Modelling nationwide hospital length of stay: opening the black box. J. Oper. Res. Soc. **56**(7), 862–869 (2005)
10. Faddy, M.J., McClean, S.I.: Analysing data on lengths of stay of hospital patients using phase-type distributions. Applied Stochastic Models in Business and Industry **15**(4), 311–317 (1999)
11. Wey, S.B., Mori, M., Pfaller, M.A., Woolson, R.F., Wenzel, R.P.: Hospital-acquired candidemia: the attributable mortality and excess length of stay. Arch. Intern. Med. **148**(12), 2642–2645 (1988)
12. Hachesu, P.R., Ahmadi, M., Alizadeh, S., Sadoughi, F.: Use of data mining techniques to determine and predict length of stay of cardiac patients. Healthcare informatics research **19**(2), 121–129 (2013)
13. Tu, J.V.: Advantages and disadvantages of using artificial neural networks versus logistic regression for predicting medical outcomes. J. Clin. Epidemiol. **49**(11), 1225–1231 (1996)
14. Quinlan, J.R.: C4. 5: Programs for Machine Learning. Morgan Kaufmann (1993)
15. Cohen, W.W.: Fast effective rule induction. In Machine Learning Proceedings 1995 (pp. 115-123). Morgan Kaufmann (1995)
16. Frank, E., Witten, I.H.: Generating accurate rule sets without global optimization (1998)
17. Keller, J.M., Gray, M.R., Givens, J.A.: A fuzzy k-nearest neighbor algorithm. IEEE transactions on systems, man, and cybernetics **4**, 580 585 (1985)
18. Hühn, J., Hüllermeier, E.: FURIA: an algorithm for unordered fuzzy rule induction. Data Min. Knowl. Disc. **19**(3), 293–319 (2009)
19. Rasmani, K.A., Shen, Q.: Modifying weighted fuzzy subsethood-based rule models with fuzzy quantifiers. In 2004 IEEE International Conference on Fuzzy Systems (IEEE Cat. No. 04CH37542) (Vol. 3, pp. 1679-1684). IEEE (2004, July)
20. Cornelis, C., Verbiest, N., Jensen, R.: Ordered weighted average based fuzzy rough sets. In International Conference on Rough Sets and Knowledge Technology (pp. 78-85). Springer, Berlin, Heidelberg (2010, October)
21. Jensen, R., Shen, Q.: New approaches to fuzzy-rough feature selection. IEEE Trans. Fuzzy Syst. **17**(4), 824–838 (2009)

KOSI- Key Object Detection for Sentiment Insights

Daniel Dimanov$^{(\boxtimes)}$ and Shahin Rostami$^{(\boxtimes)}$

Faculty of Science and Technology, Bournemouth University,
Bournemouth BH12 5BB, UK
{i7461730,srostami}@bournemouth.ac.uk
https://research.bournemouth.ac.uk/project/ciri/

Abstract. This paper explores an original approach of using computer vision, data mining and an expert system to facilitate marketers and other interested parties to take automated data-driven decisions with the use of actionable insights. The system uses a state-of-the-art algorithm to retrieves all the images of a desired Instagram user profile. Then, the data is passed through a combination of different convolutional neural networks for object detection and a rule-based translation system to determine the interests of this profile user. Further, using a separately trained convolutional neural network with an original dataset developed as part of this study, personality insights are derived. The results from the conducted experiments yield a satisfactory prediction of interests and not very promising results for the personality prediction.

Keywords: Image analytics · Data mining ·
Convolutional Neural Networks

1 Introduction

The rising interest in social media marketing [23] has become the reason many companies are exploring new audience engagement tools in aid of gaining exposure and achieving better customer reach [8]. Many marketing companies struggle to broadcast their advertisements and messages to their desired target group, determined by their common interests or other common denominators. Currently, this is done by looking through fan pages, searching keywords in shares, or even looking at other key influencers in the given industry and targeting their followers in pursuit driven by the prejudice that these people are interested in a certain type of merchandise or product [19]. Some other strategies focus more on location, gender and hashtags, which, however, are very broad and inefficient for narrowing down the specific target audience. As a consequence, more software systems with a deeper level of analytics and insights are starting to emerge, like IBM's Watson Analytics[1]. Marketing agencies, as well as any other businesses

---

[1] https://www.ibm.com/watson-analytics.

© Springer Nature Switzerland AG 2020
Z. Ju et al. (Eds.): UKCI 2019, AISC 1043, pp. 296–306, 2020.
https://doi.org/10.1007/978-3-030-29933-0_25

interested in marketing, need a system that can aid narrowing down the target group, utilising in an optimal way most of the data already available on social media. Data mining, even though dating back to more than two decades ago, has become one of the most popular tools for gathering and further analysing data in many different sectors [7]. The amounts of digital pictures taken and shared are drastically increasing [24]. Nearly five trillion photos are stored online at a rising rate of more than a trillion a year [18]. A huge part of the available data is left out when only text is considered for information extraction [28]. Since these valuable pieces of information are usually left out and there are a rising number of social media platforms focused purely on pictures and videos, this paper combines the concepts of data extraction, image processing and analysis to provide the end user with valuable insights for a chosen social media profile. The aim of this paper is to explore an algorithm that can predict interests and personality traits insights for Instagram users based purely on their posted images. Such a system would be able to facilitate the conversion of people's social media data into highly valuable information to enable marketers and other interested parties to make data-driven decisions in two aspects. First, to have quantitative measures to select potential influencers and have clear facts for sponsorship negotiation. Secondly, to quantify the effectiveness of their campaigns.

The remainder of the paper is organised in the following way. Section 2 introduces the concept of computer vision and machine learning approaches used in the algorithm. Section 3 discusses the overall structure of the algorithm and presents insights about how the algorithm works. Section 4 examines the performance of the algorithm in terms of accuracy, speed and fit for purpose. Section 5 is a discussion of what the algorithm capabilities are, what can it achieve, future possible applications and significance of the developed algorithm. Finally, Sect. 6 explores what future work can be done to build on the algorithm and surpass the state-of-art.

2   Computer Vision

Computer vision is an increasingly popular field of machine learning [25]. Computer vision is the translation of raw data input (still picture or video) into valuable information. The value of this information depends on the context it is used in and on the requirements [2]. Computer vision is the term used to refer to the ability of the machines (computers) to perceive entities like a human being. This term is often used to denominate the combination between concepts like image processing, image analysis, object detection and the deep learning regularly used to achieve the given goals of a project [26]. Convolutional Neural Networks (CNNs) are crucial for computer vision since these algorithms imitate the processes occurring in the visual cortex in human's brain [15]. CNNs are a type of artificial multi-layer neural network that applies convolutions on the input layer of the network [29]. It shares the concepts of the neuron and the adjustable weights and overall are similar to Artificial Neural Networks (ANNs), but also different, because of the additional convolutional, pooling and flattening layers [11,29]. CNNs consist of five or three layers of operations depending on the level of detail and complexity of each layer. In this paper, they will be split into five step-wise operations from the raw input to the output, which will not be included as separate layers. The first layer is the convolutional layer [5], followed by the hidden layer (ReLU activated usually [14]), Pooling [21], Flattening [12] and fully connected layers (ANN) [27] where the input for every layer is the output of the previous one [29]. These powerful algorithms rely on classification to recognise complex patterns or in the case of computer vision - objects.

Machine learning algorithms are used in a wide range of areas, such as medical science [1], concealed weapon detection [20], and retail forecasting [17].

An example of an implementation of this concept is the Single Shot MultiBox Detection (SSD) algorithm[2] [16]. SSD uses a single deep neural network trained to recognise objects in images and their position in the image. The power of SSD comes from convolving the images many times before detecting features with filters. Thus, the algorithm can find abstract features, using which it recognises what the object in the corresponding box is [4]. In the training process, it is fed with images, which have labelled boxes defining the ground truth on what objects are present in the image. In the paper of Liu et al. [16] the training process is based on Pascal VOC and COCO experiments. Moreover, in the paper a comparison is drawn with the relative state-of-the-art systems and SSD, even though simpler, proves to be superior and faster (Example of usage: Fig. 1).

Fig. 1. SSD algorithm used to recognise a cat in video frame

---

[2] https://github.com/amdegroot/ssd.pytorch.

3   KOSI Algorithm

Result: Target user interests and personality predictions
while *authentication is rejected* **do**
   if *authentication is successful* **then**
      Create user directory;
      Open sub-process executing API calls;
      Collect all user photos;
      Store only salient data (no descriptions or tags);
   else
      prompt for new credentials;
   end
end
Initialise counters and start timer;
Pass all images to the SSD algorithm;
load SSD model $\leftarrow$ pre-trained_weights;
   VGG16 $\leftarrow$ pretrained VGG16 on ImageNET;
   dir $\leftarrow$ user's directory
   for each file in [dir] **do**
      if video **then**
         array $\leftarrow$ split_in_frames(video);
         for each image in [array] **do**
            image $\leftarrow$ SSD.add_bounding_boxes(image);
            dir $\leftarrow$ dir $\cup$ $\{image\}$;
         end for
      end if
      image $\leftarrow$ base_transformations(image);
      image $\leftarrow$ rescale_and_reshape(image);
      initialise $objects\_found[]$
      for each bounding_box **do**
         $confidence \leftarrow$ bounding_box.get_confidence();
         if $confidence > 0.6$ **then**
            $objects\_found[object] + = 1$;
         end if
      end for
      predictions$[] \leftarrow VGG16(file)$;
      object $\leftarrow$ decode_object() $\leftarrow$ where **predictions[id]** $> 60\%$;
      $objects\_found[id] + = 1$;
      $predictions\_ranked \leftarrow$ rank(predictions);
      if $\sum_{i=1}^{2} predictions\_ranked[object \leftarrow decode\_object(i)] > 75\%$ **then**
         $\sum_{i=1}^{2} objects\_found[predictions\_ranked[object \leftarrow decode\_object(i)]] + = 1$;
      end if
   end for
interests $\leftarrow$ Rule-based_system(objects_found);
personalities $\leftarrow$ personalities_module(dir);

Algorithm 1: KOSI Algorithm

Result: Key objects are translated to interests

```
ic ← image_count ← dir
objects[] ← KOSI;
objects[] ← decode(objects[])
exceptions[] ← Special objects derived from user study including "Person",
"dog"(and other animals) , "volleyball"(and other sports),
"confectionery","toyshop", "bookshop", "self_propeller_vehicle" ...;
interests[] ;
for each object in objects[] do
    synset ← object;
    if topic_domains ∈ synset then
        td ← topic_domains;
        for each t ∈ td do
            for each l ∈ t.lemmas[] do
                interests[l.name] += objects[object];
            end for
        end for
    else
        while synset.min_depth > 8 do
            if synset ∉ exceptions[] then
                hyp_n ← syn.hypernyms[0];
                interests[hyn_n.lemmas[0]]+= object[object];
            end if
        end while
    end if
end for
categories[] ← 140 object categories(manually derived)
for each interest in interests[] do
    interests[] ← categorise(interests[], categories[]);
end for
interests_counter ← occurrence, interest ← interests[];
all_int_count ← ∑ᵢⁱⁿᵗᵉʳᵉˢᵗˢ⁽ⁱⁿᵗᵉʳᵉˢᵗˢ⁻ᶜᵒᵘⁿᵗᵉʳ⁾ interests_counter[i];
final_interests[]
for o, i ← interests_counter do
    if o >= all_int_count * 0.05 then
        if i = "Person" then
            if o > 2.5ic then
                i = "Group activities";
            else if 2.5ic >= o > 1.5ic then
                i = "Relationship"
            else if o > 0.6ic then
                i = "Appearance"
            else
                i = "People"
            end if
        end if
        final_interests[] ← i;
    end if
end for
interests ← Rule-based_system(objects_found);
```

Algorithm 2: Rule based system

The algorithm (Algorithm 1) can be explored as having four separate layers. The first one is data extraction, which is responsible for collecting all pictures of an Instagram user, specified by the user of the system. The second one is object detection, which uses computer vision and a combination of two CNNs to discover all objects in all collected photos. The third layer is sentiment analysis, where the found objects are categories using WordNET[3] and an expert system is created and used to determine the potential interests of the profile's user based on the key objects found in the pictures and the number of their occurrences (Algorithm 2). The fourth one is the personality insights detection.

The first layer consist of a data mining algorithm, which collects all images that will be analysed next. In the context of the project these images are collected from the specified user's profile. For the second layer, CNNs are used in two different ways. The data first goes through an SSD neural network (Please refer to Sect. 2), which was developed to recognise over 30 objects using Pascal VOC and COCO classes [9]. Pre-trained model on these classes is implemented together with an official pytorch implementation of the SSD model described in Liu's et al. paper [16] from De Groot [6]. Firstly, all images go through a series of transformations, values of which are the suggested ones from the study of Liu [16] and are passed through the SSD model. SSD uses boxes to determine the objects in the images, this is useful, since it can detect all objects present in the image and their placement. It can also be used to detect objects in video frames, but that function was purposefully left out in the development because it's a trade-off, which is performance. All objects and their count are stored in Counter variable and the same variable is further used for the next step in the object detection. Because of the relatively few classes the SSD is trained and developed for, a different approach was implemented as well to combine the quality of the SSD detections and to fill the gap in the left-out objects. "A high-level neural networks API, written in python" called Keras[4] [10] was used to develop a second sieve. In this step, VGG16 [22] model is used. It is used together with ImageNet[5], which is an enormous dataset, containing millions of images in over a thousand classes. The drawback of this approach is that the images are not box-labelled, thus a prediction of the whole image is done and in the case of Instagram there is rarely just one object that the image features. Thus a threshold for certainty of the predictions is introduced, below which predicted objects are ignored. All other predictions or a combination of the them (based on predefined rules) are added to the Counter and further analysed using a rule-based expert system. The last part of the algorithm is the personalities insights extraction, which uses the raw input of images from the first layer and since parallelism is not yet implemented it runs last. A novel CNN is used to determine the possible profile user's personality according to 16personalities[6].

---

[3] https://wordnet.princeton.edu.
[4] https://keras.io.
[5] http://www.image-net.org.
[6] https://www.16personalities.com.

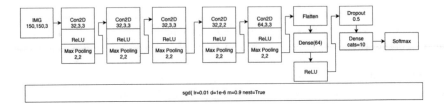

Fig. 2. CNN used for personalities detection

The personalities CNN (Fig. 2) predicts the profile user's personality based on the training set provided by a conducted study. The study is psychology based qualitative research in the form of personalised interviews with a group of 30 people, who were invited to participate. During the interviews, the participants were asked to take a personality test based on [3]. Since experience and expertise of psychology was required, an expert in the field of psychology was invited to supervise the interviews. The personality traits are motivated from the Big Five [13] study and 16 personalities test, used to determine the participant's personalities.

4 Results

This section presents and interprets some key findings of the paper as well as illustrate the performances of different modules.

Based on the user study and further analysis of all the information the objects collected could be categorised into 584 categories, referring to 71 interests. The object detection algorithm was tested with a manually crafted test set, which contains 244 different images with a total of 708 objects. 636 objects were correctly recognised, resulting in 89.83% performance. Out of the 244 images 214 were labelled correctly and the other 30 contained either false positive results or failed to recognise some of the objects, which is 87.70% accurate. Since the test

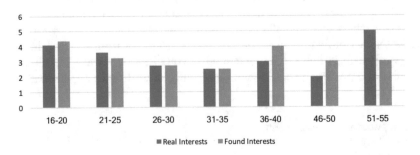

Fig. 3. Predictions in different age groups predicted vs. actual

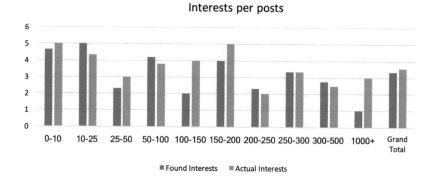

Fig. 4. Interests found per posts predicted vs. actual

data is insufficient and with relatively low diversity to fully determine the algorithm's accuracy with certainty, the accuracy of the object detection is yet to be finally determined. It can be argued that the findings in Fig. 3 do not necessarily explore the true performance for the corresponding age groups, because of the relatively small sample size of participants. The number of participants from each group has been attempted to be as evenly distributed and minor deviations do exist in the 30–50 groups, where results can be more easily swayed by outlier data, but overall the sample for the test set was carefully studied and is considered to be fairly representative for the problem domain. The rule-based system, although developed manually, may be regarded as a basis for improvement and automation using results from this exact algorithm. The categories devised from the system are based on human-logic, but the logic is transferable since a data-set could be developed and serve for training a semi-supervised algorithm. Nevertheless, an overall 93,48% accuracy is achieved in predicting interests, which is reduced to 81,13% after removal of false-positive results (Please refer to Figs. 3 and 4). A performance evaluation was done comparing the posts of an individual with the elapsed time of the system illustrated by Fig. 5.

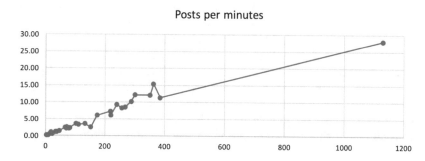

Fig. 5. Performance measured with posts successfully analysed compared to the time elapsed (in minutes)

5   Conclusions and Discussion

This paper explores an original approach of using computer vision, data mining and an expert system to take automated data-driven decisions with the use of actionable insights. The paper first introduces the problem and the proposed solution. It sets the aim, which is to convert people's social media data into actionable insights. The algorithm this paper discusses is constructed using a mining algorithm to gather the user images data, followed by two state-of-art CNNs to detect objects in the images and a rule-based system to further narrow down, categories and interpret the interests of this user. The rule-based system, even though manual can act as a basis for dataset creation, which can serve as a training and testing set for supervised or semi-supervised algorithm. As part of this study, two original datasets are created and original actionable insights are derived from the data.

6   Future Work

Based on the devised findings and analysis, as well as the whole methodology some limitations, recommendations and suggestions for future work are produced. Starting with one of the most imminent ones, in order for this project to comply with the new GDPR agreement and the depreciated Instagram API changes to the system seem to be required. Another gap in the system is that the rule-based system translates categories to interest using expert knowledge. Not only it misses some of the important objects and factors, but more data is needed for testing and understanding how objects and images relate to interests. However, the system's rule-based approach may be used for data-set creation for the training of a neural network. Therefore, a recommendation is that the manual decisions in the algorithm are substituted by a semi-supervised system in the future. Another recommendation is that bigger and much more diverse data-set is used when decisions are taken since all participants in the study are mainly aged between 18 and 29. The data-set created by the author is, although for future work more expertise is needed for deriving such conclusions, reliable, however, it is relatively small. For a convolutional neural network to be able to precisely determine such intended information a much larger data-set should be considered. Performance wise, all working algorithms and systems may be used asynchronously on different threads to conserve time and to utilise computing resources. Moreover, all results are achieved on MacBook Pro 2016 with Intel i7 and all training, as well as processing, is using only the CPU, because of inability to use the graphics card with CUDA and make it easy for integration. Thus, further optimisation may be done using GPU-acceleration, which will significantly speed up all computing processes and yield much better performance.

References

1. Bellegdi, S.A., Mohandes, S., Soufan, O.M., Arafat, S.: Computational intelligence for cardiac arrhythmia classification. UKCI **2011**, 93–97 (2011)
2. Bradski, G., Kaehler, A.: Learning opencv:[computer vision with the opencv library], [nachdr.] (2011)
3. Briggs, K.C.: Myers-Briggs Type Indicator. Consulting Psychologists Press, Palo Alto (1976)
4. Cai, Z., Vasconcelos, N.: Cascade R-CNN: delving into high quality object detection. In: Proceedings of the IEEE Conference on Computer Vision and Pattern Recognition, pp. 6154–6162 (2018)
5. Ciresan, D.C., Meier, U., Masci, J., Gambardella, L.M., Schmidhuber, J.: Flexible, high performance convolutional neural networks for image classification. In: Twenty-Second International Joint Conference on Artificial Intelligence (2011)
6. DeGroot, A., Brown, E.: Ssd pytorch. Github (2018)
7. Delen, D., Eryarsoy, E., Seker, S.: Introduction to data, text and web mining for business analytics minitrack. In: Proceedings of the 50th Hawaii International Conference on System Sciences (2017)
8. Erevelles, S., Fukawa, N., Swayne, L.: Big data consumer analytics and the transformation of marketing. J. Bus. Res. **69**(2), 897–904 (2016)
9. Everingham, M., Van Gool, L., Williams, C.K., Winn, J., Zisserman, A.: The pascal visual object classes (voc) challenge. Int. J. Comput. Vision **88**(2), 303–338 (2010)
10. Gulli, A., Pal, S.: Deep Learning with Keras. Packt Publishing Ltd., Birmingham (2017)
11. Hussain, M., Bird, J.J., Faria, D.R.: A study on CNN transfer learning for image classification. In: UK Workshop on Computational Intelligence, pp. 191–202. Springer (2018)
12. Jin, J., Dundar, A., Culurciello, E.: Flattened convolutional neural networks for feedforward acceleration. arXiv preprint arXiv:1412.5474 (2014)
13. Judge, T.A., Higgins, C.A., Thoresen, C.J., Barrick, M.R.: The big five personality traits, general mental ability, and career success across the life span. Pers. Psychol. **52**(3), 621–652 (1999)
14. Kuo, C.C.J.: Understanding convolutional neural networks with a mathematical model. J. Vis. Commun. Image Represent. **41**, 406–413 (2016)
15. Lawrence, S., Giles, C.L., Tsoi, A.C., Back, A.D.: Face recognition: a convolutional neural-network approach. IEEE Trans. Neural Networks **8**(1), 98–113 (1997)
16. Liu, W., Wen, Y., Yu, Z., Yang, M.: Large-margin softmax loss for convolutional neural networks. In: ICML, vol. 2, p. 7 (2016)
17. Miller, S.M., Popova, V., John, R., Gongora, M.: Improving resource planning with soft computing techniques. In: Proceedings of UKCI 2008, De Montfort University, Leicester, UK, pp. 37–42 (2008)
18. Paglen, T.: Invisible images: your pictures are looking at you. Architectural Des. **89**(1), 22–27 (2019)
19. Pradiptarini, C.: Social media marketing: measuring its effectiveness and identifying the target market. UW-L. Undergraduate Res. **XIV**, 1–11 (2011)
20. Rostami, S., O'Reilly, D., Shenfield, A., Bowring, N.: A novel preference articulation operator for the evolutionary multi-objective optimisation of classifiers in concealed weapons detection. Inf. Sci. **295**, 494–520 (2015)
21. Scherer, D., Müller, A., Behnke, S.: Evaluation of pooling operations in convolutional architectures for object recognition. In: International Conference on Artificial Neural Networks, pp. 92–101. Springer (2010)

22. Simonyan, K., Zisserman, A.: Very deep convolutional networks for large-scale image recognition. arXiv preprint arXiv:1409.1556 (2014)
23. Stelzner, M.: Social media marketing industry report (2017)
24. Stenroos, O., et al.: Object detection from images using convolutional neural networks (2017)
25. Szegedy, C., Vanhoucke, V., Ioffe, S., Shlens, J., Wojna, Z.: Rethinking the inception architecture for computer vision. In: Proceedings of the IEEE Conference on Computer Vision and Pattern Recognition, pp. 2818–2826 (2016)
26. Szeliski, R.: Computer vision, texts in computer science (2011)
27. Vo, D.M., Lee, S.W.: . Semantic image segmentation using fully convolutional neural networks with multi-scale images and multi-scale dilated convolutionsMultimedia Tools and Applications, pp. 1–19 (2018)
28. Welz, B., Rosenberg, A.: Exponential growth. In: SAP Next-Gen, pp. 31–35. Springer (2018)
29. Wu, J.: Introduction to convolutional neural networks. National Key Lab for Novel Software Technology. Nanjing University. China pp. 5–23 (2017)

Detection of Suicidal Twitter Posts

Fatima Chiroma[1(✉)], Mihaela Cocea[1], and Han Liu[2]

[1] University of Portsmouth, Portsmouth, UK
{fatima.chiroma,mihaela.cocea}@port.ac.uk
[2] Cardiff University, Cardiff, UK
LiuH48@cardiff.ac.uk

Abstract. As web data evolves, new technological challenges arise and one of the contributing factors to these challenges is the online social networks. Although they have some benefits, their negative impact on vulnerable users such as the spread of suicidal ideation is concerning. As such, it is vital to fine tune the approaches and techniques in order to understand the users and their context for early intervention. Therefore, in this study, we measured the impact of data manipulation and feature extraction, specifically using N-grams, on suicide-related social network text (tweets). We propose a diversified ensemble approach (multi-classifier fusion) to improve the detection of suicide-related text classification. Four machine classifiers were used for the fusion: Support Vector Machine, Random Forest, Naïve Bayes and Decision Tree. The results of our proposed approach have shown that the multi-classifier fusion has improved the detection of suicide-related text and, also, that Support Vector Machine has shown some promising results when dealing with multi-class datasets.

Keywords: Ensemble learning · Suicide-related tweets · Text classification

1 Introduction

The ability of machine learning to automatically detect and uncover patterns in data for the purpose of prediction as well as to enhance decision making has, in the last decade, led to the rapid increase in its application across diverse areas including Healthcare [15], Finance [31] and Law Enforcement [11], to name a few. Law enforcement and other intelligence organizations have used machine learning approaches and techniques to explore large databases efficiently [11]. Unfortunately, the emergence of the online social networks (often referred to as social media) has created a new source for immense and uncontrollable data generation, in addition to transforming the way crime and victimisation is understood, experienced and committed [19].

This type of crimes, such as hate crime which is a criminal offence that is prejudicially targeted towards someone based on a personal characteristic [6,14], make up around two percent of crimes based on the notifiable reports recorded

© Springer Nature Switzerland AG 2020
Z. Ju et al. (Eds.): UKCI 2019, AISC 1043, pp. 307–318, 2020.
https://doi.org/10.1007/978-3-030-29933-0_26

by the UK Home Office [14]. However, the percentage is believed to be higher as this type of crime is largely unreported to the police [8]. Additionally, hate crimes often follow hate speech and over the last decade hate speech online has significantly increased [4,24]. Online hate speech such as misogyny and the spread of suicidal ideation may impact vulnerable social network users as they are at potential risk of harming themselves due to the information they receive [9,12].

Furthermore, several studies have shown the correlation between social media and suicidal behaviour [17,27,30]. Hence, there is a need to understand social network users and the contents they post for the enhancement of existing approaches and techniques, for possible interventions, as well as keeping up with the evolving web. Therefore, in this study, we use Twitter posts that were extracted using suicide-related search terms to propose an approach based on text classification, a machine learning task, to investigate whether employing ensemble learning would lead to an improved detection of suicidal risk from social media text. According to [7], Twitter is a logical source for suicide-related communications as users are more likely to deindividualize and express themselves emotionally while other studies [9,10] have shown that people are more likely to seek for support through social networks, such as Twitter, than professional help due to anonymity and concerns of social stigmatization.

Detection of suicidal risk from social media text using automatic techniques, such as text classification, has only recently started to be explored, with only few studies [9,12,13,20] reported. In this paper, we further investigate the performance of classifiers, while exploring ensemble learning. In particular, we are investigating the influence of ensemble learning, i.e. the use of several classifiers, on classification performance in comparison with using individual classifiers.

The rest of the paper is organized as follows: Sect. 2 describes the background and related work on social network suicide-related communications and machine learning, focusing specifically on text classification; Sect. 3 provides details of the proposed approach including the experimental process; in Sect. 4 the results obtained are presented and discussed; and Sect. 5 draws the conclusions which include a summary of contributions of the paper and future directions.

2   Background and Related Work

In this section, background related to text classification is covered, as well as related work on suicidal ideation on social media.

2.1   Text Classification

Classification is one of the most prominent machine learning tasks where the category of an unseen instance is judged [12] and it typically involves employing an algorithm to build a model to identify an instance's category. Furthermore, classification relating to text is referred to as text classification and it can also be defined as the assigning of a pre-defined class to a textual instance in a dataset [3]. Although the concept of classification, regardless of type, seems quite straight forward, it is complex and cannot guarantee an accurate classification for unseen instances, especially when dealing with real world data which contains many irrelevant, noisy and redundant features [18].

Furthermore, several studies have identified some problems relating to the mis-classification of data which includes class imbalance [2,21,23]. Class imbalance is the insufficient representation of the target or minority class in a dataset [23], and most machine learning algorithms do not consider the underlying distribution of a dataset, generally leading to good performance on the detection of majority classes (as the algorithm has more instances to learn from) and poor performance on the minority classes (as the algorithm may not have been exposed to sufficient information to learn reliable patterns).

2.2   Suicide and Online Social Networks

Text classification relating to suicide-related communications is still in its infancy stage; as such, the research in this area is limited. Some studies have been carried out using text classification to try and detect social network users that are at risk of suicide. An example of such a study is [1], where they used machine learning to identify risk factors relating to suicide from Twitter conversations and they found a strong correlation between geographical suicide rates and Twitter data, with an accuracy of approximately 63%.

Additionally, in their study, [9] used machine classifiers to classify suicide-related Twitter communications. Their baseline experiment achieved an F-measure of 0.702 for all their (seven) classes, however, they further improved the results to 0.728 by applying an ensemble learning approach. Another study was carried out by [12], where they conducted a baseline experiment to measure the performance of popular machine classifiers in distinguishing suicide-related communications from Twitter. They acquired these dataset from [9] and used Decision Tree, Naive Bayes, Random Forest and Support Vector Machine for the classification. Their result showed an F-measure of up to 0.778 was achieved by the Decision Tree classifier.

3   Experimental Approach

We propose an approach based on ensemble learning, to investigate whether it would lead to an improved detection of suicide-related communications. Figure 1 provides an overview of the experimental approach which consists of four stages: Data Preparation, Feature Preparation, Individual Classification and Ensemble Classification. Furthermore, this experiment was carried using Knime[1], an open-source data analytics tool.

---

[1]   https://www.knime.com.

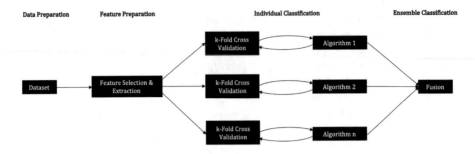

Fig. 1. The experimental approach

3.1   Data Preparation

This is the initial stage of the approach where a dataset is cleaned and partly transformed for modelling. It comprises of the data collection, data manipulation and pre-processing.

Data Collection: Twitter is a good source for suicide-related communications as stated in Sect. 1. Therefore, in our studies, we used 2,000 suicide-related communication tweets from [9], which were collected from Twitter using the Twitter Streaming Application Programming Interface (API). They used search keywords from reported news and lexicon of terms such as *don't want to exist* and *Kill myself* which were derived from known suicide websites for the collection. These were further annotated by four human annotators from CrowdFlower[2], a crowd sourcing online service, into seven suicide categories as shown in Table 1. These categories were developed by [9] with expert researchers in the area of suicide to best capture people's general representation when communicating on suicide topics. Additionally, following established methods [9,27], we also discarded tweets that have less than 75% annotator agreement leaving a total of 1064 tweets. The tweets are organised into several datasets for experimentation, as outlined below.

Data Manipulation: The data can be organised in different datasets reflecting different labeling schemes derived from the initial seven labels, which can be manipulated and categorized based on the level of similarity or dissimilarity. For example, class 1 is suicide and class 2 is the flippant reference to suicide, however classes 3 to 7 are about suicide in other contexts, i.e. not in the context of a person considering the possibility of committing suicide. Table 2 describes the data manipulation distributions – although all the datasets are from the same original data, the classes, size and complexity have changed for the resulting datasets.

---

[2] http://www.crowdflower.com.

Table 1. Instances per class (Adapted from [9])

	Class	Type	Description	Instances
	1	Suicide	Possible suicidal intent	159
	2	Flippant	Un-serious reference to suicide	133
	3	Campaign	Suicide petitions	158
	4	Support	support or information	178
	5	Memorial	Condolences or memorial	142
	6	Reports	Suicide reports excluding bombing	165
	7	Other	None of the above	129
Total:	**7**	-	-	**1064**

Table 2. Data manipulation distribution (Adapted from [12])

Class	Dataset	Class name	Raw	Processed	Total
1	Binary-class	Suicide	159	156	289
2		Flippant	133	133	
1	Three-class	Suicide	159	156	1060
2		Flippant	133	133	
3, 4, 5, 6, 7		Non-suicide	772	771	
1	Seven-class	Suicide	159	156	1060
2		Flippant	133	133	
3		Campaign	158	158	
4		Support	178	178	
5		Memorial	142	142	
6		Reports	165	165	
7		Other	129	128	

Pre-processing: When dealing with real world textual datasets, especially social network user generated datasets, pre-processing is vital. These datasets contain noise and redundant information; therefore, the use of pre-processing techniques lead to the removal of irrelevant features and reduces the vector space, thereby improving classification performance [5,16]. For this study, standard and established pre-processing methods [9,16,25] were applied, which include the removal of stop words, words containing numbers, punctuations, URLs and non-ASCII characters, Part-of-speech tagging and reducing terms to their stem for redundancy reduction (these reduced the number of instances from 1064 to 1060).

3.2   Feature Preparation

Subsequent to pre-processing, the Bag-of-words representation, which ignores the syntactic and semantic information and treats the text as a collection of words [25], was used to extract relevant unigrams (i.e. individual words) based on their term frequencies, while the n-gram approach (i.e. each feature is a term including n words) was used to extract terms between 2 and 5 words, also based on term frequencies. However, we found that the use of 4-grams and 5-grams degrades the classification performance due to almost non-existent representation of the target classes (i.e. suicide as well as flippant). Consequently, we report only the experiments using 1 to 3-grams and the number of features for each of the three datatsets are displayed in Table 3. Additionally, dimensionality reduction is typically applied to textual data, as it is known that high dimensional text data hinders classifiers' performance [22,29]. However, in this study, no dimensionality reduction technique has been applied as the number of features are small and dimensionality reduction is usually applied to datasets with hundreds of thousands of features [28].

Table 3. NGrams per dataset

Dataset	1 Gram	1–3 Gram
Binary-class	599	2,075
Three-class	2,223	10,990
Seven-class	2,223	10,990

3.3   Individual Classification

This phase consists of the training and evaluation of the individual classifiers, where we use two sets of features: 1 Gram and 1–3 Gram. The machine learning algorithms that are used to train both sets of features were chosen based on their performance from previous studies [12]. These classifiers are Decision Tree (DT), Naïve Bayes(NB), Random Forest (RF) and Support Vector Machine (SVM). This was done for each of the datasets, i.e. the binary-class, three-class and seven-class datasets. Additionally, this (individual classification) phase will be used as the baseline for comparison with the ensemble classification (Multi-classifier fusion) results – please refer to Fig. 1 for these phases.

Training Setup: For training, stratified sampling was used, which preserves the original class distribution for the training and test data. 10-fold cross-validation was used given the relatively small size of the data, especially for the binary-class dataset.

Evaluation: Evaluation as a process allows us to determine the extent to which an objective has been attained [26]. For text classification purposes, the standard classification metrics, such as Precision, Recall and F-measure are used to measure the performance. Accuracy is not typically used for measuring the performance of text classification, especially for data with class imbalance, as the higher results of the majority class (which is often not the one of interest) gives the false impression of a good performance. Consequently, the F-measure is preferred to accuracy as it can reflect how the overall performance is affected due to the low performance for some classes.

3.4   Ensemble Classification

The ensemble classification stage uses the ensemble learning approach, which involves combining (fusing) the outputs of each classifier through techniques such as majority voting or algebraic formulas (e.g. weighted sum). In our approach, we use majority voting but excluded Random Forest as part of the ensemble, as it is already an ensemble method and will provide another point of comparison i.e. between the two ensembles.

4   Results and Discussion

This study investigated whether the use of ensemble learning will improve the detection of suicide-related communications on social media, and more specifically, from Twitter. The experimental investigation in this paper builds on previous work by [9], as well as our own previous studies [12] by using the same dataset, pre-processing techniques and machine classifiers for this study. Additionally, the results from this study are presented in two categories: (a) the individual classification results which comprises of the results for the individual classifiers and (b) the ensemble classification results, the results for the individual classifiers by applying the ensemble learning approach.

4.1   Individual Classification

We report the results for each of the datasets, i.e. binary-class, three-class and seven-class; we present the overall results (see Fig. 2), as well as per class, as we are interested in the performance of the *Suicide* and *Flippant* classes in particular, as these are the ones reflecting suicide risk. The results from the binary-class dataset indicate an F-measure between 0.411 to 0.776 was achieved for the 1 Gram whereas a similar but lower result of 0.380 to 0.771 (see Fig. 2) was achieved for the 1–3 Gram. The suicide class has a higher F-measure of up to 0.80 than the flippant class (see Table 4) in both 1 Gram and 1–3 Gram, which was achieved by SVM. The lowest performing classifiers for the binary-class dataset is NB with an F-measure of 0.38 and 0.41, respectively; Furthermore, both NB and RF seem to have performed well on the suicide class, but performed (very) poorly on the flippant class.

In addition, the performance of the classifiers for the three-class dataset (see Fig. 2 and Table 5) and seven-class dataset (see Fig. 2, Tables 6 and 7) has varying performance ranging from 0.00 to 0.90 for the three-class and 0.04 to 0.74 for the seven-class. NB had the worst performance for all the dataset categories i.e. binary-class, three-class and seven-class. Furthermore, DT and SVM are the two highest performing classifiers in this phase, however their performance varies depending on the dataset. For instance, DT had the highest F-measure for the binary-class, whereas SVM had the best performance for the three-class and seven-class dataset; which may imply that SVM performs better with larger and/or multi-class datasets.

Table 4. Individual classification results: binary-class

Classifier	Measure	1 Gram		1–3 Gram	
		Suicide	*Flippant*	*Suicide*	*Flippant*
DT	Recall	0.731	0.820	0.757	0.789
	Precision	0.826	0.722	0.821	0.719
	F-measure	0.776	0.768	0.788	0.753
NB	Recall	0.942	0.075	0.988	0.023
	Precision	0.544	0.526	0.562	0.600
	F-measure	0.690	0.132	0.717	0.043
RF	Recall	0.917	0.444	0.982	0.203
	Precision	0.659	0.819	0.610	0.900
	F-measure	0.767	0.576	0.753	0.331
SVM	Recall	0.821	0.729	0.822	0.707
	Precision	0.780	0.776	0.781	0.758
	F-measure	0.800	0.752	0.801	0.732

Table 5. Individual classification results: three-class

Classifier	Measure	1 Gram			1–3 Gram		
		Suicide	*Flippant*	*Non-suicide*	*Suicide*	*Flippant*	*Non-suicide*
DT	Recall	0.603	0.316	0.883	0.604	0.293	0.901
	Precision	0.537	0.512	0.848	0.551	0.506	0.865
	F-measure	0.568	0.391	0.865	0.576	0.371	0.882
NB	Recall	0.506	0.053	0.970	0.485	0.023	0.983
	Precision	0.840	0.280	0.794	0.891	0.250	0.816
	F-measure	0.632	0.089	0.874	0.628	0.041	0.892
RF	Recall	0.615	0.015	0.914	0.538	0.000	0.948
	Precision	0.568	0.182	0.801	0.615	0.000	0.819
	F-measure	0.591	0.028	0.854	0.574	0.000	0.879
SVM	Recall	0.641	0.226	0.921	0.675	0.278	0.929
	Precision	0.637	0.536	0.838	0.640	0.544	0.878
	F-measure	0.639	0.317	0.877	0.657	0.368	0.903

Table 6. Individual classification (1 Gram) results: seven-class

Datasets	DT			NB			RF			SVM		
	P	R	F	P	R	F	P	R	F	P	R	F
Suicide	0.67	0.54	0.60	0.50	0.87	0.63	0.88	0.43	0.57	0.74	0.58	0.65
Campaign	0.63	0.59	0.61	0.24	0.91	0.38	0.67	0.68	0.67	0.64	0.67	0.66
Flippant	0.42	0.35	0.38	0.05	0.54	0.10	0.17	0.48	0.25	0.33	0.41	0.37
Support	0.55	0.68	0.60	0.98	0.21	0.35	0.77	0.47	0.58	0.70	0.71	0.71
Memorial	0.36	0.30	0.33	0.08	0.41	0.13	0.23	0.39	0.29	0.47	0.36	0.41
Reports	0.35	0.45	0.39	0.10	0.55	0.17	0.26	0.60	0.37	0.42	0.54	0.47
Other	0.28	0.41	0.33	0.16	0.69	0.25	0.44	0.66	0.53	0.49	0.56	0.52

Table 7. Individual classification (1–3 Gram) results: seven-class

Datasets	DT			NB			RF			SVM		
	R	P	F	R	P	F	R	P	F	R	P	F
Suicide	0.71	0.46	0.56	0.48	0.91	0.63	0.76	0.50	0.61	0.79	0.54	0.64
Campaign	0.73	0.68	0.70	0.41	0.99	0.58	0.57	0.79	0.66	0.72	0.77	0.74
Flippant	0.35	0.39	0.37	0.02	0.60	0.04	0.12	0.52	0.20	0.29	0.41	0.34
Support	0.66	0.84	0.74	0.99	0.27	0.42	0.95	0.38	0.54	0.74	0.75	0.74
Memorial	0.46	0.32	0.38	0.14	0.88	0.24	0.12	0.83	0.22	0.52	0.37	0.44
Reports	0.31	0.43	0.36	0.06	0.83	0.11	0.21	0.68	0.32	0.33	0.55	0.41
Other	0.36	0.55	0.44	0.26	0.95	0.41	0.35	0.86	0.49	0.50	0.55	0.53

4.2 Ensemble Classification

The results show that each combination performed differently on each dataset. Interestingly, combining DT, NB and SVM gave a higher performance than combining only DT and SVM even though NB has the lowest performance amongst all the classifiers. Additionally, there is an improved performance when the ensemble learning is applied except in two cases (1 Gram for seven-class and 1–3 Gram for three-class) where SVM has the higher performance. Also, the multi-classifier fusion has outperformed RF for all the datasets (Fig. 3).

Additionally, in previous work [12], the feature extraction methods used were the bag-of-words and document frequency to generate only unigrams however in this study we further explored the use of bigrams and trigrams. Furthermore, applying the ensemble approach gave an improved performance compared with the individual classifiers for all the datasets, except in two cases mentioned earlier. The combination of DT, NB and SVM gave the best performance for all the datasets while removing NB from the ensemble combination marginally deteriorates the performance of all the datatsets except the binary-class dataset.

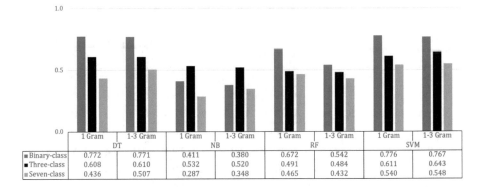

Fig. 2. Training results (F-measure) per dataset

	1 Gram	1-3 Gram	1 Gram	1-3 Gram	1 Gram	1-3 Gram	1 Gram	1-3 Gram
	DT		NB		RF		SVM	
Binary-class	0.772	0.771	0.411	0.380	0.672	0.542	0.776	0.767
Three-class	0.608	0.610	0.532	0.520	0.491	0.484	0.611	0.643
Seven-class	0.436	0.507	0.287	0.348	0.465	0.432	0.540	0.548

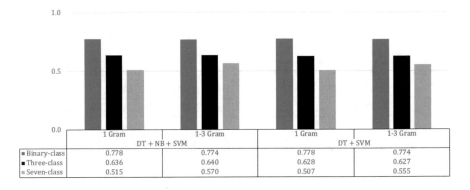

	1 Gram	1-3 Gram	1 Gram	1-3 Gram
	DT + NB + SVM		DT + SVM	
Binary-class	0.778	0.774	0.778	0.774
Three-class	0.636	0.640	0.628	0.627
Seven-class	0.515	0.570	0.507	0.555

Fig. 3. Ensemble classification: F-measure

5   Conclusion and Future Direction

In this paper, we investigated whether employing ensemble learning would lead to an improved detection of suicidal risk from social media text. Although there is evidence of classification improved, the improvement when compared to the worst performing classifier is significant but it is not significant when compared to the best performing classifiers. Consequently, the use of ensemble learning with majority voting improves the prediction of suicide-related text, but only marginally. In future work, we will investigate alternative ways of fusing the classifiers and their influence on performance.

Acknowledgement. This is an independent research that is supported by the Petroleum Development Technology Fund (PTDF) and the Department of Health Policy Research Programme (Understanding the Role of Social Media in the Aftermath of Youth Suicides, Project Number 023/0165). The views expressed in this publication are those of the authors and not necessarily those of PTDF or Department of Health.

References

1. Abboute, A., Boudjeriou, Y., Entringer, G., Azé, J., Bringay, S., Poncelet, P.: Mining twitter for suicide prevention. In: International Conference on Applications of Natural Language to Data Bases/Information Systems, pp 250–253. Springer (2014)
2. Ali, A., Shamsuddin, S.M., Ralescu, A.L.: Classification with class imbalance problem: a review. Int. J. Adv. Soft Comput. Appl. **7**(3), 176–204 (2015)
3. Allahyari, M., Pouriyeh, S., Assefi, M., Safaei, S., Trippe, E.D., Gutierrez, J.B., Kochut, K.: A brief survey of text mining: classification, clustering and extraction techniques. In: Proceedings of KDD Bigdas, Halifax, Canada, August 2017, p. 13 (2017)
4. Banks, J.: Regulating hate speech online. Int. Rev. Law Comput. Technol. **24**(3), 233–239 (2010)
5. Barbosa, L., Feng, J.: Robust sentiment detection on twitter from biased and noisy data. In: Proceedings of the 23rd International Conference on Computational Linguistics: Posters, pp. 36–44 (2010)
6. Burnap, P., Williams, M.L.: Hate speech, machine classification and statistical modelling of information flows on twitter: interpretation and communication for policy decision making. Proc. IPP **2014**, 1–18 (2014)
7. Burnap, P., Williams, M.L.: Cyber hate speech on twitter: An application of machine classification and statistical modeling for policy and decision making. Policy Internet **7**(2), 223–242 (2015)
8. Burnap, P., Williams, M.L.: Us and them: identifying cyber hate on twitter across multiple protected characteristics. EPJ Data Sci. **5**(1), 11 (2016)
9. Burnap, P., Colombo, W., Scourfield, J.: Machine classification and analysis of suicide-related communication on twitter. In: Proceedings of the 26th ACM Conference on Hypertext & Social Media, pp. 75–84. ACM (2015)
10. Cavazos-Rehg, P.A., Krauss, M.J., Sowles, S., Connolly, S., Rosas, C., Bharadwaj, M., Bierut, L.J.: A content analysis of depression-related tweets. Comput. Hum. Behav. **54**, 351–357 (2016)
11. Chen, H., Chung, W., Xu, J.J., Wang, G., Qin, Y., Chau, M.: Crime data mining: a general framework and some examples. Computer **37**(4), 50–56 (2004)
12. Chiroma, F., Liu, H., Cocea, M.: Text classification for suicide related tweets. In: 2018 International Conference on Machine Learning and Cybernetics (ICMLC), vol. 2, pp. 587–592. IEEE (2018)
13. Colombo, G.B., Burnap, P., Hodorog, A., Scourfield, J.: Analysing the connectivity and communication of suicidal users on twitter. Comput. Commun. **73**, 291–300 (2016)
14. Corcoran, H., Smith, K.: Hate crime, England and Wales, 2015/16 (2016). https://assets.publishing.service.gov.uk/government/uploads/system/uploads/attachment_data/file/559319/hate-crime-1516-hosb1116.pdf
15. Dipnall, J.F., Pasco, J.A., Berk, M., Williams, L.J., Dodd, S., Jacka, F.N., Meyer, D.: Fusing data mining, machine learning and traditional statistics to detect biomarkers associated with depression. PLoS ONE **11**(2), 1–23 (2016)
16. Haddi, E., Liu, X., Shi, Y.: The role of text pre-processing in sentiment analysis. Procedia Comput. Sci. **17**, 26–32 (2013)
17. Jashinsky, J., Burton, S.H., Hanson, C.L., West, J., Giraud-Carrier, C., Barnes, M.D., Argyle, T.: Tracking suicide risk factors through Twitter in the US. Crisis **35**(1), 51–59 (2014)

18. Li, J., Cheng, K., Wang, S., Morstatter, F., Trevino, R.P., Tang, J., Liu, H.: Feature selection: a data perspective. ACM Comput. Surv. (CSUR) **50**(6), 94 (2017)
19. McGovern, A., Milivojevic, S.: Social media and crime: the good, the bad and the ugly (2016). https://theconversation.com/social-media-and-crime-the-good-the-bad-and-the-ugly-66397
20. O'Dea, B., Wan, S., Batterham, P.J., Calear, A.L., Paris, C., Christensen, H.: Detecting suicidality on twitter. Internet Interv. **2**(2), 183–188 (2015)
21. Picek, S., Heuser, A., Jović, A., Bhasin, S., Regazzoni, F.: The curse of class imbalance and conflicting metrics with machine learning for side-channel evaluations. IACR Trans. Cryptographic Hardware Embed. Syst. **2019**(1), 209–237 (2018)
22. Rehman, A., Javed, K., Babri, H.A., Saeed, M.: Relative discrimination criterion-a novel feature ranking method for text data. Expert Syst. Appl. **42**(7), 3670–3681 (2015)
23. Sagi, O., Rokach, L.: Ensemble learning: a survey. Wiley Interdiscip. Rev. Data Mining Knowl. Discov. **8**(4), e1249 (2018)
24. Schmidt, P.: Human rights online (2018). http://www.inach.net/wp-content/uploads/2018/05/INACH_HumanRightsOnline.pdf
25. Schütze, H., Manning, C.D., Raghavan, P.: Introduction to Information Retrieval, vol. 39. Cambridge University Press, Cambridge (2008)
26. Steele, S.M.: Program evaluation-a broader definition. J. Extension **8**(2), 5–17 (1970)
27. Sueki, H.: The association of suicide-related Twitter use with suicidal behaviour: a cross-sectional study of young internet users in Japan. J. Affect. Disord. **170**(September 2014), 155–160 (2015)
28. Tang, B., He, H., Baggenstoss, P.M., Kay, S.: A bayesian classification approach using class-specific features for text categorization. IEEE Trans. Knowl. Data Eng. **28**(6), 1602–1606 (2016)
29. Uğuz, H.: A two-stage feature selection method for text categorization by using information gain, principal component analysis and genetic algorithm. Knowl. Based Syst. **24**(7), 1024–1032 (2011)
30. Won, H.H., Myung, W., Song, G.Y., Lee, W.H., Kim, J.W., Carroll, B.J., Kim, D.K.: Predicting national suicide numbers with social media data. PLoS ONE **8**(4) (2013). https://doi.org/10.1371/journal.pone.0061809
31. Yao, J., Zhang, J., Wang, L.: A financial statement fraud detection model based on hybrid data mining methods. In: 2018 International Conference on Artificial Intelligence and Big Data (ICAIBD), pp. 57–61. IEEE (2018)

Improving Session Based Recommendation by Diversity Awareness

Ramazan Esmeli[(✉)], Mohamed Bader-El-Den, and Hassana Abdullahi

University of Portsmouth, Portsmouth PO1 2UP, UK
{ramazan.esmeli,mohamed.bader,hassana.abdullahi}@port.ac.uk

Abstract. Recommender systems help users to discover and filter new and interesting products based on their preferences. Session-Based Recommender systems are powerful tools for anonymous e-commerce visitors to understand their behaviours and recommend useful products. Diversity in the recommendations is an important parameter due to increasing the opportunity of recommending new and less similar items that users interacted. Effect of diversity has been investigated in many works for the collaborative filtering-based Recommender systems. However, for session-based Recommender systems, exploring the effect of diversity is still an open area. In this paper, we propose an approach to calculate the diversity level of the items in the session logs and analyse the effect of diversity level on the session-based recommendation. In order to test the impact of diversity awareness, we propose a sequential Item-KNN recommendation model. The final recommendation list is created as a contribution of the interacted items in the session that depends on the diversity level between last interacted item of the session. We conduct several experiments to validate our diversity aware model on a real-world dataset. The results show that diversity awareness in the sessions helps to improve the performance of Recommender system in terms of recall and precision evaluation metrics. Also, the proposed method can be applied to other sequential Recommender system methods, including deep-learning based Recommender systems.

Keywords: Session based recommender systems · Diversity · Context awareness

1 Introduction

Nowadays, due to the increase of available online data, users have many options in digital platforms, and they need to spend the effort to find the most relevant information for them. Information overloading, especially in the e-commerce domain, attracts researchers' and businesses' attention to develop approaches to filter most relevant products to their customers. Recommender Systems (RS) have been used more than a decade on digital platforms in different domains including movie, music and e-commerce [1] to help users to filter most relevant and preferred items for them.

© Springer Nature Switzerland AG 2020
Z. Ju et al. (Eds.): UKCI 2019, AISC 1043, pp. 319–330, 2020.
https://doi.org/10.1007/978-3-030-29933-0_27

RS are mainly divided into three categories based on the method used. These methods are Collaborative Filtering (CF) [18],Content-Based recommendation (CB) [4] and Hybrid RS [1]. In CF, user histories are used to train RS models, while in CB RS user and item attributes are utilised to create the models. However, Hybrid RS are built as a combination of CF and CB RS. There are drawbacks and advantages of these different recommendation methods. For example, for CF methods, user past behaviours are needed, for CB RS, user and item attributes need to be known, and Hybrid RS can take longer time and be computationally expensive.

Recently, new methods are emerging where there are missing user-interaction history and user-item attributes. For example, Session-Based Recommendation Systems (SBRS) emerge. In SBRS recommendations need to be given based on the only on-going session since users are anonymous and there is no other user-interaction history except the current session [13,15,27]. Many different methods have been developed to adapt to RS algorithms for anonymous users and short user-item interaction history including Item-Item KNN [22], Recurrent Neural Network (RNN) [13], Markov Chain (MC) [29], and various type of Matrix Factorisation (MF) approaches [11,17]. Recent studies show that different version of RNN and modified Item-Item KNN methods [24] give promising results on SBRS domain in terms of recommendation quality by adapting to user intentions in the sessions and personalised recommendations.

The main aim of the developed methods in the SBRS domain is to improve the personalisation and user satisfaction by giving precise recommendations. Diversity in the recommendation area is a factor which can affect user satisfaction and recommendation quality [10]. In the literature, there are many works on diversity to analyse its effect on RS' performance [5,10,19]. However, in the SBRS domain, this factor is still an open area that needs to be investigated.

In this work, we propose Context and Diversity Aware (CDA) framework that measures the diversity on the on-going user session and adjust the next product recommendations based on the diversity level in the session. We evaluate the proposed framework on a real-world session based dataset. Computational results show that considering the interacted item diversity in the session improves the recommendation accuracy.

The main contributions of this paper are:

1. develop an approach to calculate diversity level of the items in the e-commerce sessions.
2. diversity level and session context integration into the SBRS framework.
3. validation of the proposed framework on a real-world dataset.

Rest of the paper is organized as follows. Section 2 discusses existing models in brief, which will be helpful in understanding SBRS, diversity and context awareness. Section 3 explains the proposed framework that identify the process through diversity measurement and context awareness. Also, this section will include the explanation of various steps used in the framework. Section 4 presents the data analysis for the dataset and the experiment results of the proposed models. Section 5 presents the conclusion, recommendation and future scope for this work.

2   Related Works

This section gives an overview of SBRS, diversity and context awareness in the users' session logs.

2.1   Session Based Recommendation Systems

Traditional RS methods, for example, CF, CB and Hybrid heavily depend on user information and user history [1]. However, customers visiting the e-commerce platforms mostly prefer to browse anonymously [16]. In this case, traditional RS approaches suffer from cold start problem [21]. Therefore, traditional RS models cannot create recommendations since these models need to learn latent factors of the users as a vector to represent them [21].

Item-Item RS methods generally have been used when users are anonymous [14,16] since item correlations are trained by looking at their co-occurrence in sessions. Amazon is one of the well-known examples of using item-item RS for SBRS [22]. Recently, Deep learning has been applied to SBRS domain in several works [13,14,16]. Their results show promising results in comparison to item-item RS. Also, in [16,24], authors modified the item-item KNN RS to have more personalised and sequential aware model. Their experiment results on using same dataset showed that modified KNN models had better accuracy performance than RNN models.

MC based and combination of MF and MC based approaches [11,17,24] are also used in the case of missing user information. However, these models are computationally expensive compared to item-item KNN approaches [24].

2.2   Diversity in RS

Diversity is defined as the opposite of similarity. Moreover, diversification is introduced as one of the option to deal with over-fitting in recommendation [19]. Many diversity aware algorithms are proposed, and several workshops are organised to measure the effect of diversification on RS quality [19]. [9] examined the effect of the level of diversity on RS accuracy. Their results showed that most of the RS algorithms recommend fewer diversity products to increase RS accuracy.

Furthermore, [28] investigated novelty and diversity in the RS domain in terms of their effect on RS accuracy. The author calculated the diversity by taking into account the product position in the recommendation list and its similarity with other recommended items. Also, [5] created a measurement formulation of diversity in CB-RS by measuring the similarity of the given recommendations. They had a different recommendation list which has various level of diversity, and they conducted a user study. They found that when the diversity increases for a certain level, the user satisfaction improves. However, above the identified level, users complained about seeing irrelevant recommendations to their preferences. Also, in [20], authors analysed the temporal diversity in CF, using user-user KNN and Singular Value Decomposition (SVD) methods. In their method, they trained the models using records from different time frames then they re-ranked the recommendation list based on the diversity rate between the most popular items. Also, [6] analysed diversity in the top-N RS. They re-ranked the recommendations by selecting the most diverse items to each other. Finally, [10] for the long tail problem (less popular items appears in the tail of list) in the recommendation system, they proposed an optimisation method that increases the diversity of RS. Their results show that the accuracy of RS is not affected; however, less popular items appeared in the top of the list.

Previous works [6,10,20] focused diversity in CF based Recommender systems. They mostly focused to diversity analysing on recommended items. Also, they re-ranked final recommendation list. However, in our work, we design a method that takes into diversity consideration before creating last recommendation list. Therefore, we do not apply re-ranking in final recommendation list.

2.3 Context Awareness in Session-Based Recommender Systems

User sessions are generally interpreted as a context in SBRS domain [27]. Users' interest shifting and browsed items in the session, the time difference between the clicked item, discounts on the item can be considered as the factors that build the session context [16,27]. RNN based models can take into consideration the clicked items in the session and can give the next recommendation based on the clicked items. Also, RNN models can handle the popularity of the viewed items in the session as context [26]. Moreover, [15] adopted short-term user's intention to filter recommendations from various RS algorithms, including Bayesian Personalised Ranking (BPR) and Session-KNN. Also, paper [23] adopted CF recommendation for news broadcasting websites by using users' last intention and they re-ranked final recommendation list in terms of similarity measured by short term content features and recommended item features.

In our work, we use the users' short term intention (last viewed item in the session) to find the diversity level with previously seen products. So, in our work, we change the context contribution to final recommendation list based on diversity levels between last interacted item and the other item in the session. However, we do not apply re-ranking to recommendation list since created recommendation list will have already diversity awareness and ranked.

3   Proposed Approach

The proposed framework consists of 4 phases (Fig. 1). In this section, the phases are described.

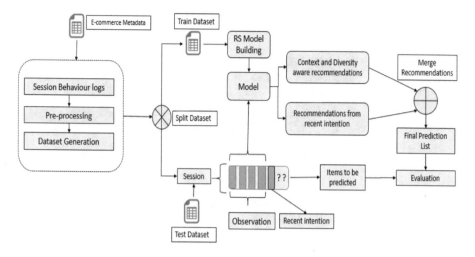

Fig. 1. Proposed context and diversity aware SBRS framework

3.1   Phase 1

In this phase, we prepare the session logs to fit RS algorithms. First, we select the related attributes from datasets, which are timestamp that shows when the session started, session id and item id, which are stored as string character in the dataset. We apply label encoding to make inputs suitable for RS algorithms. Also, in this phase, we clean the sessions which have less than 2 item interactions.

3.2   Phase 2

In this phase, we split the dataset as train and test. We use the last 2 days of the sessions as test dataset, and the others for train dataset. Also, in this phase, we train the Session-Item-Item KNN model [13].

3.3   Phase 3

One of the key contributions of this work is highlighted in this phase. In this phase, we build a diversity-aware SBRS, in which we use the diversity level of items between last interacted item to create a recommendation list (Fig. 2).

Diversity Level Calculation. Diversity means in our context, the dissimilarity level between two products in a session. We follow a novel method to find the similarity of two items in the context of the given session to calculate their dissimilarity. In our method, first, we ask the trained recommendation model for n recommendation for each interacted product in the session and we create a recommendation list R_{i_p}. Later, we get n recommendation for the last interacted item in the session and create a recommendation list R_{i_l}. Finally we measure

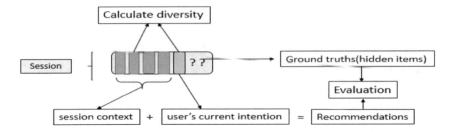

Fig. 2. Diversity calculation and its contribution to last recommendation list

the common recommended items between R_{i_p} and R_{i_l}. We keep common recommended items list in R_C (Eq. (1)).

$$R_C = R_{i_p} \cap R_{i_l} \tag{1}$$

So, similarity between these two items is calculated as in Eq. (2)

$$Sim_{(i_p, i_l)} = \frac{|R_C|}{n} \tag{2}$$

We have seen that when we include higher similar items with last interacted item in the session, the final recommendation list $FinalList$ in Eq. (4) will be almost the same with recommendation list R_{i_l} gathered from last interacted item. Therefore, overall, the contribution of the previously seen items will be less. However, when we give more contribution weight to recommendation list created from items which have more diversity level with the current intention, they can be ranked in the top-n recommendation list as their contribution scores are increased. So we calculate a diversity-aware contribution weight for the previously seen items in the session as in Eq. (3)

$$W_p = 1 - Sim_{(i_p, i_l)} \tag{3}$$

In Eq. (4), the final recommendation list is calculated. T is the number of interacted items in the session s. I is the whole items in the train dataset. $Score_p$ and $Score_l$ are confident scores of top-n neighbours of item p and item l, which are given from the recommendation model.

$$Final\ List = top_n(\sum_{p}^{T-1} W_p * Score_p(i_p, I)) + top_n(Score_l(i_l, I)) \tag{4}$$

In the Eq. (4), we choose n parameter as 100 to select the most 100 (n) similar *neighbours*. In this phase, n is used to calculate common recommended items between last interacted item and previously seen items. However as seen in Eq. (4), we do not give a contribution weight to the last interacted item.

3.4   Phase 4

In this phase, we evaluate the performance of our approach using *Recall@n* and *Precision@n* metrics [12,24]. n means in the evaluation context is selecting top-n high scored items in the ranked final recommendation list (Eq. (4)). In this work, we choose $n \in \{10, 20, 30\}$ since users are mostly interested in the recommendations in the top of the list.

4   Dataset Analysing

In this section, we give statistical details of the dataset we use in this paper. The dataset collected for a two weeks period from an e-commerce platform in the UK (Fresh relevance[1]). Before pre-processing the dataset, most of the sessions have a few item interactions as seen in Fig. 3. We eliminate the sessions which have less than 2 items interaction to deal with cold start problem. Also, having more interactions in the session can help to create better similarity correlations between items.

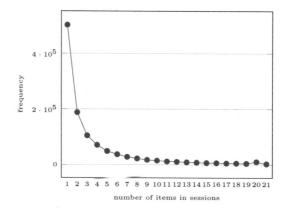

Fig. 3. Item frequency in the sessions

Before processing and after processing the dataset statistics is shown in Table 1.

Table 1. Dataset statistics

dataset	Sessions	Items	Interactions
fresh relevance before	961653	42363	2844362
fresh relevance after	466818	37537	2349527

---

[1] https://www.freshrelevance.com/.

5   Experimental Evaluation

We conduct four experiments to see the effect of integration of diversity to SBRS.

Experiment 1. In the first experiment, we create recommendations from last interacted item only. So, the given recommendation from this model does not take into consideration session context. We call this model as Item-Item KNN (Fig. 4).

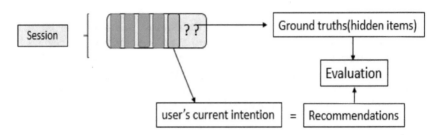

Fig. 4. Only user's current intention and its contribution to last recommendation list

Experiment 2. In this experiment, we consider the contribution of the session context. We call this model as C-Item KNN (Fig. 5).

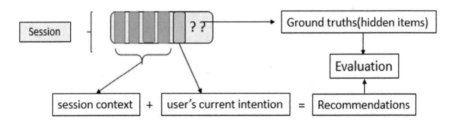

Fig. 5. Only Session context awareness and its contribution to last recommendation list

Experiment 3. The aim of this experiment is to see effect of diversity awareness in the context on SBRS performance. This model has interaction and diversity awareness. We call this model CD-Item KNN (Fig. 2).

Experiment 4. We aim in this experiment to see the effect of decreasing diversity on recommender system quality. In other words, more similar items to last interacted item will have more contribution weight. So, we updated Eq. (3) as in Eq. (5).

$$W_p = Sim(i_p, i_l) \tag{5}$$

In this final experiment, the last recommendation list will be created as a consequence of similar item interactions. We call this model SI-Item KNN.

Evaluation Method. To evaluate the experiments we hide a part of the inter-acted items from the session. The final recommendation list created in each experiment is ranked. From the ranked list, we choose the top-n ($n \in \{10, 20, 30\}$) best ranked recommendations since users mostly interested in recommendations which are in the top of the list. The model tries to predict hidden interacted items from the session which we call ground truths. We use the $Recall@n$ (Eq. (6)) and $Precision@n$ (Eq. (7)) metrics to measure how the models performing on pre-dicting missing interactions in the session.

$$Recall@n = \frac{|groudtruth \cap recommendations@n|}{|groudtruth|} \quad (6)$$

$$Precision@n = \frac{|groudtruth \cap recommendations@n|}{|recommendations@n|} \quad (7)$$

We tested the models with 1000 sessions. To choose test sessions, we iterated selection process ten times over the test dataset and in each iteration, 100 sessions are randomly selected from dataset. Overlapping of new selected 100 sessions with previously selected sessions are prevented to minimize the risk that the obtained results are specific to the single test split as shown in Fig. 6. Finally, the performance measurements of all splits are averaged.

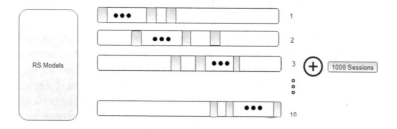

Fig. 6. Selecting test sessions to validate CDA framework

5.1 Results and Discussion

The experiment results (Table 2) show that adding diversity awareness to the contribution of the interacted items in the session context leads to good per-formance in terms of Recall and Precision in comparison to when diversity and context awareness are not included in the models. Moreover, the results confirm that context awareness in the RS domain, particularly in SBRS is an essential factor to have better-performing models. Increasing the diversity level of the items to build the final recommendation list helps by considering the different intentions of the user in the session. Moreover, remembering items which are diverse to users current intention and similar to session context help to have improve recommendations.

On the other hand, experiment result confirms that when the weight of con-tribution of similar interacted items with last interacted item is increased, the performance of SBRS becomes even worse than created recommendations from only last interacted item. Base on literature [5,19], CF-based RS show opposite performance results than this experiment. The reason for our result could be as a result of the specification of SBRS since in the sessions, users have a specific target. Therefore, browsed items will have a level of similarity. However, remembering the less similar products that are already seen in the session can match user behaviour in the session. Thus, more diversity level to users' last intention increases the RS performance.

Table 2. Performance comparision of models on Rec:Recall, Prec:Precision @$n \in$ $\{10, 20, 30\}$

Model	Rec@10	Prec@10	Rec@20	Prec@20	Rec@30	Prec@30
Item-Item KNN	0.2386	0.0239	0.3143	0.0157	0.3571	0.0119
C-Item KNN	0.3171	0.0317	0.4186	0.0209	0.4700	0.0157
DC-Item KNN	**0.3514**	**0.0351**	**0.4371**	**0.0219**	**0.4800**	**0.0160**
SI-Item KNN	0.2171	0.0217	0.2986	0.0149	0.3386	0.0113

6   Conclusion and Future Works

In this paper, we applied diversity adjustment to session context to create SBRS recommendation list. For diversity calculation, we developed a novel equation and proposed an algorithm to integrate it into the SBRS model. We applied the algorithm on session based dataset. In order to test the proposed algorithm, we created an efficient diversity and context-aware (CDA) SBRS framework. In this framework, interacted item contribution is increased when diversity increases between previously interacted item and last interacted item in a session to build up final recommendation list. Also, we give the highest contribution to last interacted item to define the final recommendation list since last interacted item will show a higher level of user intention than previously seen items [15,27]. Also, we used Item-Item KNN and modified Item-Item KNN models in the framework for our experiments. Our results showed that integrating dynamic item diversity in SBRS can improve recommendation performance in terms of precision and recall metrics.

For future works, the proposed diversity algorithm can be applied to other SBRS models, including deep learning, BPR, MC and Association Rule-based models. Also, the diversity level can be restricted to a certain level. For instance, items which have above a certain level of item diversity with users' current intention can be disregarded while building the final recommendation list. Lastly, we will test our algorithm on other session-based datasets to see the performance of our approach.

Another future direction is that calculated diversity can be used in machine learning classifications, including text classification [25] which can help to predict feature specifications of products from the user search query that the user might click or browse in the session. However, in these type of classification approaches, class imbalance is the main challenge [2,3] and to process data in real-time and in parallel, tools such as Apache Spark [7,8], need to be considered, in which recommendation process need to be done within a few seconds.

References

1. Adomavicius, G., Tuzhilin, A.: Toward the next generation of recommender systems: a survey of the state-of-the-art and possible extensions. IEEE Trans. Knowl. Data Eng. **6**, 734–749 (2005)
2. Bader-El-Den, M.: Self-adaptive heterogeneous random forest. In: 2014 IEEE/ACS 11th International Conference on Computer Systems and Applications (AICCSA), pp. 640–646. IEEE (2014)
3. Bader-El-Den, M., Teitei, E., Perry, T.: Biased random forest for dealing with the class imbalance problem. IEEE Trans. Neural Networks Learn. Syst. (2018)
4. Besbes, O., Gur, Y., Zeevi, A.: Optimization in online content recommendation services: beyond click-through rates. Manuf. Serv. Oper. Manag. **18**(1), 15–33 (2015)
5. Castagnos, S., Brun, A., Boyer, A.: When diversity is needed... but not expected! In: International Conference on Advances in Information Mining and Management, pp. 44–50. IARIA XPS Press (2013)
6. Di Noia, T., Ostuni, V.C., Rosati, J., Tomeo, P., Di Sciascio, E.: An analysis of users' propensity toward diversity in recommendations. In: Proceedings of the 8th ACM Conference on Recommender Systems, pp. 285–288. ACM (2014)
7. Diedhiou, C., Carpenter, B., Esmeli, R.: Comparison of platforms for recommender algorithm on large datasets. In: 7th Imperial College Computing Student Workshop, pp. 4–1. Schloss Dagstuhl–Leibniz Center for Informatics (2019)
8. Diedhiou, C., Carpenter, B., Shafi, A., Sarkar, S., Esmeli, R., Gadsdon, R.: Performance comparison of a parallel recommender algorithm across three hadoop-based frameworks. In: 2018 30th International Symposium on Computer Architecture and High Performance Computing (SBAC-PAD), pp. 380–387. IEEE (2018)
9. Fleder, D.M., Hosanagar, K.: Recommender systems and their impact on sales diversity. In: Proceedings of the 8th ACM Conference on Electronic Commerce, pp. 192–199. ACM (2007)
10. Hamedani, E.M., Kaedi, M.: Recommending the long tail items through personalized diversification. Knowl.-Based Syst. **164**, 348–357 (2019)
11. He, R., McAuley, J.: Fusing similarity models with markov chains for sparse sequential recommendation. In: 2016 IEEE 16th International Conference on Data Mining (ICDM), pp. 191–200. IEEE (2016)
12. Herlocker, J.L., Konstan, J.A., Terveen, L.G., Riedl, J.T.: Evaluating collaborative filtering recommender systems. ACM Trans. Inf. Syst. (TOIS) **22**(1), 5–53 (2004)
13. Hidasi, B., Karatzoglou, A., Baltrunas, L., Tikk, D.: Session-based recommendations with recurrent neural networks. arXiv preprint arXiv:1511.06939 (2015)
14. Hidasi, B., Quadrana, M., Karatzoglou, A., Tikk, D.: Parallel recurrent neural network architectures for feature-rich session-based recommendations. In: Proceedings of the 10th ACM Conference on Recommender Systems, pp. 241–248. ACM (2016)
15. Jannach, D., Lerche, L., Jugovac, M.: Adaptation and evaluation of recommendations for short-term shopping goals. In: Proceedings of the 9th ACM Conference on Recommender Systems, pp. 211–218. ACM (2015)

16. Jannach, D., Ludewig, M.: When recurrent neural networks meet the neighborhood for session-based recommendation. In: Proceedings of the Eleventh ACM Conference on Recommender Systems, pp. 306–310. ACM (2017)
17. Kabbur, S., Ning, X., Karypis, G.: Fism: factored item similarity models for top-n recommender systems. In: Proceedings of the 19th ACM SIGKDD International Conference on Knowledge Discovery and Data Mining, pp. 659–667. ACM (2013)
18. Karabadji, N.E.I., Beldjoudi, S., Seridi, H., Aridhi, S., Dhifli, W.: Improving memory-based user collaborative filtering with evolutionary multi-objective optimization. Expert Syst. Appl. **98**, 153–165 (2018)
19. Kunaver, M., Požrl, T.: Diversity in recommender systems-a survey. Knowl.-Based Syst. **123**, 154–162 (2017)
20. Lathia, N., Hailes, S., Capra, L., Amatriain, X.: Temporal diversity in recommender systems. In: Proceedings of the 33rd International ACM SIGIR Conference on Research and Development in Information Retrieval, pp. 210–217. ACM (2010)
21. Lika, B., Kolomvatsos, K., Hadjiefthymiades, S.: Facing the cold start problem in recommender systems. Expert Syst. Appl. **41**(4), 2065–2073 (2014)
22. Linden, G., Smith, B., York, J.: Amazon. com recommendations: item-to-item collaborative filtering. IEEE Internet Comput. (1), 76–80 (2003)
23. Liu, J., Dolan, P., Pedersen, E.R.: Personalized news recommendation based on click behavior. In: Proceedings of the 15th International Conference on Intelligent User Interfaces, pp. 31–40. ACM (2010)
24. Ludewig, M., Jannach, D.: Evaluation of session-based recommendation algorithms. User Model. User-Adap. Inter. **28**(4–5), 331–390 (2018)
25. Mohasseb, A., Bader-El-Den, M., Cocea, M.: Question categorization and classification using grammar based approach. Inf. Process. Manag. **54**(6), 1228–1243 (2018)
26. Moreira, G.D.S.P., Jannach, D., da Cunha, A.M.: Contextual hybrid session-based news recommendation with recurrent neural networks. arXiv preprint arXiv:1904.10367 (2019)
27. Quadrana, M., Cremonesi, P., Jannach, D.: Sequence-aware recommender systems. ACM Comput. Surv. (CSUR) **51**(4), 66 (2018)
28. Vargas, S.: New approaches to diversity and novelty in recommender systems. In: Fourth BCS-IRSG symposium on Future Directions in Information Access (FDIA 2011), Koblenz, vol. 31 (2011)
29. Wang, W., Yin, H., Sadiq, S., Chen, L., Xie, M., Zhou, X.: Spore: A sequential personalized spatial item recommender system. In: 2016 IEEE 32nd International Conference on Data Engineering (ICDE), pp. 954–965. IEEE (2016)

Fault Prognosis Method of Industrial Process Based on PSO-SVR

Yu Yao[1(✉)], Dongliang Cheng[2], Gang Peng[2], and Xuejuan Huang[3]

[1] Wuhan University, Wuhan 430072, China
yaoyu@whu.edu.cn
[2] Huazhong University of Science and Technology, Wuhan 430079, China
penggang@hust.edu.cn
[3] Wuhan Sports University, Wuhan 430079, China

Abstract. For latent faults or situations where the pre-failure characteristics are not obvious, fault prognosis techniques are needed. This work proposes a fault prognosis method based on support vector regression (SVR), in which particle swarm optimization (PSO) algorithm is utilized to optimize the parameters to improve the prediction accuracy. The SVR algorithm and grey prediction are tested on benchmark data taken from Tennessee-Eastman process and the "NASA prognosis data repository", and the experiments compare the prediction accuracy difference between the two algorithms.

Keywords: Particle Swarm Optimization · PSO-SVR · Fault prognosis

1 Introduction

With the development towards large-scale, continuous, and integrated modern industrial processes, it is essential to effectively monitor the plant-wide operations that cover decision, cooperative control, and base-level control. Fault prognosis, which determines whether a fault is impending and estimates how soon and how likely a fault will occur, will play an important role in condition-based maintenance and form a basis for further maintenance decision [1–3]. There are mainly two types of methods of fault prognosis: physics 'model'-based and data-driven i.e. statistical and artificial intelligence (AI) [4].

As a classic machine learning algorithm, SVR has obvious advantages in data processing of high-dimensional and nonlinear small sample data sets, and has good performance in data-driven fault prognosis [5]. The accuracy of the predictive model based on SVR is highly correlated with the model's parameters, so it is necessary to optimize to obtain the appropriate model parameters. In this paper, the SVR method is applied to predict the fault of industrial data, and the PSO algorithm is utilized to optimize the SVR parameters. And in experiments we employ the TE process data and the NASA prognosis data to simulate the algorithm. The first section briefly introduces SVR fault prognosis method used in this paper, and second section introduces the SVR parameter optimization by PSO algorithm. The simulation experiment and the experimental result discussion are proposed in the third section.

Z. Ju et al. (Eds.): UKCI 2019, AISC 1043, pp. 331–341, 2020.
https://doi.org/10.1007/978-3-030-29933-0_28

2   The Basic Theory of SVR Fault Prognosis Approach

2.1   Support Vector Regression Algorithm

SVM is based on statistical learning theory and structural risk minimization (SRM) guidelines, and SVM algorithm is applied for many machine learning tasks such as pattern recognition, object classification, and in the case of time series prediction, regression analysis [7]. And the algorithm in terms of efficiency and accuracy is over traditional learning algorithms or with comparable.

Given a training set of instance-label pairs $D = \{(x_1, y_1), \ldots, (x_m, y_m)\}, x_i \in \mathbb{R}^n, y_i \in \mathbb{R}$, the support vector regression (SVR) require the solution of the following optimization problem [6]:

$$\max_{\alpha, \alpha^*} \sum_{i=1}^{m} y_i \left(\alpha_i^* - \alpha_i\right) - \varepsilon \sum_{i=1}^{m} \left(\alpha_i^* + \alpha_i\right) - \frac{1}{2} \sum_{i,j=1}^{m} \left(\alpha_i^* - \alpha_i\right)\left(\alpha_j^* - \alpha_j\right)\kappa \tag{1}$$
$$\text{s.t.} \quad \sum_{i=1}^{m} \left(\alpha_i^* - \alpha_i\right) = 0, \quad 0 \le \alpha_i^*, \alpha_i \le C$$

Here $C > 0$ is the penalty parameter of the error term, ε is an insensitive parameter, and $\kappa(x_i, x_j) = \phi(x_i)^T \phi(x_j)$ is the kernel function. Training vectors x_i are mapped into a higher dimensional space by the function ϕ. SVR finds a linear separating hyper plane with the maximal margin in this higher dimensional space. The solution of SVR is as follows: $f(x) = \sum_{i=1}^{m} \left(\alpha_i^* - \alpha_i\right)\kappa(x, x_i) + b$. The application of SVMs to general regression analysis case is Support Vector Regression (SVR), and SVR is characterized by the application of kernel functions for spatial transformation without the need to obtain actual nonlinear transformation functions, and it will reduce the complexity of the calculation. According to the actual problem Gaussian kernel function (RBF) is often chosen, whose formula is shown as: $\kappa(x, x_i) = \exp\left(\frac{-\|x-x_i\|^2}{2\sigma^2}\right)$.

2.2   Fault Prognosis Model

SVR for Time Series Prediction. SVR approach for fault prognosis in the industrial production process is a data-driven method. The data of the production process changes over time. The continuous time interval data constitutes a time series, and SVR is vital for many of the time series prediction applications from financial market prediction to electric utility load forecasting to medical and other scientific fields [6].

Assuming that the time series $\{x_1, \ldots, x_n\}$ is known and the predicted target value is $\{x_n\}$, a mapping f between input $x = \{x_{n-m}, x_{n-m+1}, \ldots, x_{n-1}\}$ and output $y = \{x_n\}$ needs to be established, which should be defined as $f : \mathbb{R}^m \to \mathbb{R}$. To apply SVR method requires that the original time series be divided into multiple training samples sets, and the dimensionality of the training samples is m. The training samples are expressed as follows: $X = \begin{bmatrix} x_1 & \cdots & x_m \\ \vdots & \ddots & \vdots \\ x_{n-m} & \cdots & x_{n-1} \end{bmatrix}, Y = \begin{bmatrix} x_{m+1} \\ \vdots \\ x_n \end{bmatrix}$. Here, the number of samples is $n - m$, and the feature dimensionality of the samples is m. According to the basic theory of SVR, after using the sample data for training, the following regression function is obtained:

$$y_t = \sum_{i=1}^{n-m} \left(\alpha_i^* - \alpha_i\right)\kappa(X_i, X_{n-m+1}) + b, (t = m+1, \ldots, n) \qquad (2)$$

And the regression function can be used to predict subsequent target values in the time series. The predictive expression for single-step prediction is: $\hat{x} = \sum_{i=1}^{n-m} \left(\alpha_i^* - \alpha_i\right)\kappa(X_i, X_{n-m+1}) + b$. Suppose that the predicted value x_{n+1} at time point $t = n + 1$ has been calculated, the last item in the original time series $\{x_1, \ldots, x_n\}$ should be replaced with x_{n+1}, and the sequence length is kept constant, so the new series will be $\{x_2, \ldots, x_{n+1}\}$. According to the formula (2), the next predicted value x_{n+2} will be calculated, and so forth. This process is repeated to achieve continuous prediction.

Steps of SVR Fault Prognosis. Regression prediction using SVM, usually has the following steps:

A. *Data preprocessing and feature extraction.* Multiple feature extraction methods can be applied in terms of different types of data, such as dynamic mean deviation and variance processing (DMDVP), vibration signal feature extraction based on wavelet decomposition, and so on [8].

B. *Samples selection and partition.* Choose consecutive data feature construction time series according to time sequence. Select the first n consecutive feature data in the data set to construct the training samples. As the mapping dimensionality is m, therefore $n - m + 1$ training samples can be constructed. And the rest data are used as testing samples to evaluate the accuracy of the prediction.

C. *Parameter optimization and SVR model training.* The optimized parameters C, ε and σ are selected by using the method of PSO algorithm (proposed in the next section). And train SVR model using the optimized parameters.

D. *Regression and prediction.* Then the SVR model is applied to perform the prediction of the given data set, and the deviation between the predicted value and the actual value is analyzed to evaluate the prediction accuracy of the algorithm.

3   PSO-SVR Algorithm

Particle Swarm Optimization (PSO) is based on the flocking behavior and social co-operation of birds and fish [9]. The basic idea of PSO algorithm can be described as: all particles get the fitness value according to the adaptive function to judge the current position of the particle. Each particle has a memory function, which can remember the closest position to the "food" in the search process respectively. And particle can adjust its current position and velocity dynamically based on memory information of itself and other particles. PSO has been widely used in treating ill-structured continuous/discrete, constrained as well as unconstrained function optimization problems. The inner workings of the PSO make sufficient use of probabilistic transition rules to make parallel searches of the solution hyperspace without explicit assumption of derivative information [10].

3.1   SVR Parameters Optimization

Up to now, it is difficult to decide SVR kernel functions types for specific data patterns. The Gaussian RBF kernel executes easily and often been chosen for the feature spaces whose attributes are not well known. From formulation (1), there are at least three free parameters have been identified: penalty factor C, insensitive loss function ε, and Gaussian kernel parameter σ. These are user defined parameters that must be tuned to minimize time series prediction error and produce the best fitting result. C represents the degree of punishment for samples outside the margin, and the value is related to the degree of tolerance of error. A larger value indicates that the error tolerance is smaller. The Gaussian kernel bandwidth σ is related to the input spatial extent of the training samples. The larger σ value suitable for the larger training sample space. Parameter ε controls the width of the ϵ-insensitive zone, used to fit the training data. The bigger ε, the fewer support vectors are selected.

Tuning the optimal parameters guarantees the high accuracy of SVR prediction and reduces the error between the predicted value and the actual value. Especially the accuracy of multi-step prediction directly determines the long-term prediction ability of the fault prognosis system. Second, the fault prognosis system is to adapt to many types of sample data. According to the empirical given value, it is not guaranteed to be optimal.

In next subsection a PSO-SVR algorithm is proposed to obtain optimal regression model parameters and better accuracy for fault prognosis.

3.2   PSO-SVR Algorithm

Initialization parameters. Set the initial value of maximum iterations number G_{max}, population size N, and the range of particle position and velocity as follows: $\left[X_{min,C}, X_{max,C}\right]$, $\left[X_{min,\sigma}, X_{max,\sigma}\right]$, $\left[X_{min,\varepsilon}, X_{max,\varepsilon}\right]$, $\left[-V_{max,C}, V_{max,C}\right]$, $\left[-V_{max,\sigma}, V_{max,\sigma}\right]$, $\left[-V_{max,\varepsilon}, X_{max,\varepsilon}\right]$. Set the initial velocity $\left(v_{i,C}, v_{i,\sigma}, v_{i,\varepsilon}\right)$ and position $\left(x_{i,C}, x_{i,\sigma}, x_{i,\varepsilon}\right)$ of particle i randomly, and $x_{i,C}, x_{i,\sigma}, x_{i,\varepsilon}$ represent the value of penalty factor C, Gaussian kernel parameter σ and ε-intensive parameter respectively

The SVR model is trained with $\left(x_{i,C}, x_{i,\sigma}, x_{i,\varepsilon}\right)$ as parameters, and the cross-validation prediction accuracy is used as the fitness value. Assuming that the number of samples to be predicted is n, the fitness of the particle is measured by the mean square error (MSE) of prediction accuracy. The smaller the MSE value indicates the higher the prediction accuracy. Variable $pbest_i = \left(p_{i,C}, p_{i,\sigma}, p_{i,\varepsilon}\right)$ is the best position that particle i has ever found, and is to utilized to record the best parameter value of the particle i. And $gbest = \left(g_C, g_\sigma, g_\varepsilon\right)$ record the best position of the whole particle swarm ever passed, i.e. the best solution at the present (Fig. 1).

Update the value of position and velocity of particle i according to the following formula (3). If the particle's position and velocity values exceed the set value range,

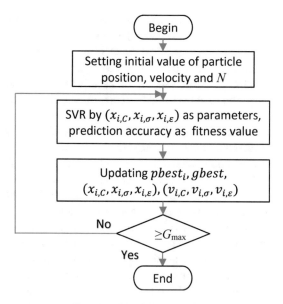

Fig. 1. PSO-SVR algorithm.

take the boundary value to limit the particle velocity and position. The algorithm terminates until the number of iterations is equal to the maximum G_{max}.

$$\begin{cases} v_{i,C}^{k} = wv_{i,C}^{k-1} + c_1r_1\left(p_{i,C} - x_{i,C}^{k-1}\right) + c_2r_2\left(g_C - x_{i,C}^{k-1}\right) \\ v_{i,\sigma}^{k} = wv_{i,\sigma}^{k-1} + c_1r_1\left(p_{i,\sigma} - x_{i,\sigma}^{k-1}\right) + c_2r_2\left(g_\sigma - x_{i,\sigma}^{k-1}\right) \\ v_{i,\varepsilon}^{k} = wv_{i,\varepsilon}^{k-1} + c_1r_1\left(p_{i,\varepsilon} - x_{i,\varepsilon}^{k-1}\right) + c_2r_2\left(g_\varepsilon - x_{i,\varepsilon}^{k-1}\right) \end{cases} \qquad \begin{cases} x_{i,C}^{k} = x_{i,C}^{k-1} + v_{i,C}^{k-1} \\ x_{i,\sigma}^{k} = x_{i,\sigma}^{k-1} + v_{i,\sigma}^{k-1} \\ x_{i,\varepsilon}^{k} = x_{i,\varepsilon}^{k-1} + v_{i,\varepsilon}^{k-1} \end{cases}$$

$$(3)$$

4 Experiment

In order to verify the prediction performance of the PSO-SVR approach, the comparative experiments applied on TE process dataset and NASA prognosis dataset were conducted.

4.1 Data Set

TE Process Dataset. Tennessee Eastman (TE) process model is a realistic simulation program of a chemical plant which is widely accepted as a benchmark for control and monitoring studies [11]. The process model involves a total of 52 variables, 41 of which are measurement variables, consisting of 22 continuous variables and 19 component measurement variables. These variables represent the material flow rate, pressure, liquid level, etc. of the chemical reaction unit. And the other 11 are control variables. The entire process model contains data corresponding to 21 kinds of process fault of 6 different types. The sampling interval is 3 min. The training set collected 500 samples of data in normal mode and 480 samples of 21 kinds of faults, and the fault occurred 1 h later. Data in the testing set are consecutive 960 samples of normal mode and the 21 fault modes. The fault occurred 8 h later, that is, starting from the 160th data sample.

It can be speculated that the degree of variety of each variable of the process model under different fault modes is different respectively. In order to measure this discrepancy, we make use of dynamic mean deviation and variance processing (DMDVP) for data preprocessing [8]. The basic idea is: Select consecutive n values of a variable from a certain time, and calculate the average deviation between these n values and the mean of the variable in normal mode. And in addition, calculate the variance of these n values. The dimensionality of the original 52-dimensional feature data becomes 104 through this preprocessing. The experimental results verify the above speculation that the variability of each variable is significant, and this preprocessing enhances this variability. Further, the final comparative experiment shows that the processed data has a direct impact on the accuracy of fault prognosis [8]. The sliding window size of DMDVP is 20, that is, the feature is acquired by DMDVP-20. The continuous samples are structured into a time series with a time interval of 3 min. Time series prediction is performed simultaneously on multiple variables. Figure 2 shows the change with time of the first variable in the 104-dimensional feature variable under fault mode 1.

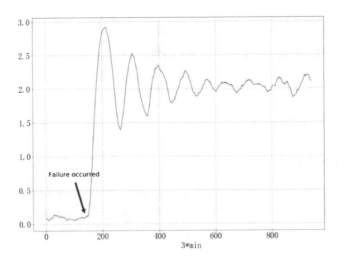

Fig. 2. The first variable in the 104-dimensional feature data (fault mode 1) vs time

NASA Prognosis Dataset. The second dataset we used in this work is extracted from NASA's prognostics data repository [12]. The data were obtained from the bearing experiments in which four bearings were tested under constant conditions. The angular velocity was kept constant at 2000 rpm, and a 6000 lb radial load was applied onto the shaft and bearings. On each bearing, two accelerometers were installed for a total of 8 accelerometers (one vertical, and one horizontal) to register the accelerations generated by the vibrations, with a sampling frequency equal to 20 kHz. Intercept the data of 1200 items in each data file and then carry out wavelet decomposition at 6-layer for the original data [8]. Calculate the energy of each segment according to wavelet spectrum energy formula, and finally obtain the curve of energy value with time, as shown in Fig. 3.

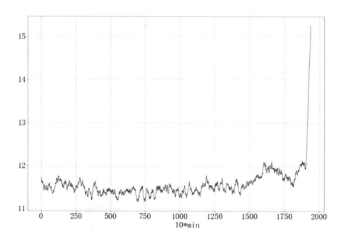

Fig. 3. Bearing vibration energy vs time

Figure 3's data is from the vibration signal collected by the *Ch*.3 channel in the first bearing experiment, and then the feature extraction is performed to obtain the time-varying energy curve. It can be seen from the figure that the bearing energy value increases significantly after 1900, which is caused by the inner ring fault of the bearing corresponding to the *Ch*.3 channel in the later stage of the test. In order to verify the prediction performance of the algorithm before/after the fault occurs, a total of 1900 (0–1899) continuous values are selected to construct a time series for training. As the mapping dimensionality is 5, the number of training samples is 1895. In addition, a total of 15 items (1900–1914) are used as testing sample.

4.2   Results Discussion

Comparison on TE Process Dataset. The first comparison experiment applied PSO-SVR algorithm and GM (1, 1) model (Grey Forecast) [13] on the TE process dataset. Set the value range of the SVR: $C \in [1, 300], \sigma \in [10^{-1}, 10^2], \varepsilon \in [10^{-3}, 1]$. The population size of particle swarm optimization is 20, and the number of iterations is 50. The optimal parameter combination obtained by the PSO algorithm is $(C, \sigma, \varepsilon) = (169, 0.325873, 0.001788)$. From the data in Table 1, the maximum relative error between the predicted and actual values of PSO-SVR is 12.9651%, and the relative error is 0.0468%. The MES of PSO-SVR prediction is 6.1106×10^{-5}; The maximum relative error between the predicted and actual values of GM (1, 1) is 26.4295%, and the minimum relative error is −0.5445%, the mean square error predicted by GM (1, 1) model is 9.2173×10^{-4}. From the comparison of the MSE of the predicted value of the PSO-SVR and the GM (1, 1) method, it is shown that the PSO-SVR method has a relatively high prediction accuracy on the TE process dataset.

Table 1. Comparison of predicted value of PSO-SVR and GM (1, 1) on TE process dataset

Item no.	Actual value	Predicted value by PSO-SVR	Relative error of PSO-SVR	Predicted value by GM (1, 1)	Relative error of GM (1, 1)
160	0.0990	0.0969	2.2017	0.0868	12.3098
161	0.1077	0.1019	5.3966	0.0902	16.1986
162	0.1156	0.1145	1.0043	0.0959	17.0637
163	0.1124	0.1202	−6.9925	0.1035	7.9345
164	0.1091	0.1086	0.4632	0.1097	−0.5445
165	0.1078	0.1080	0.2056	0.1146	−6.3659
166	0.1057	0.1057	0.0468	0.1170	−10.6632
167	0.1091	0.1042	4.4416	0.1166	−6.9191
168	0.1115	0.1120	−0.4495	0.1171	−5.0390
169	0.1275	0.1117	12.4466	0.1180	7.4598
170	0.1430	0.1381	3.3998	0.1233	13.7513
171	0.1740	0.1515	12.9651	0.1325	23.8361
172	0.2049	0.2037	0.5951	0.1508	26.4295
173	0.2399	0.2420	−0.8849	0.1775	25.9959
174	0.2747	0.2741	0.2171	0.2145	21.9058
		$MSE = 6.1106 \times 10^{-5}$		$MSE = 9.2173 \times 10^{-4}$	

Figure 4 are plots of predicted values plotted in accordance with data in Table 1. It can be intuitively found that the yellow plot closer to the green plot than the red one, indicating that the PSO-SVR predicts that the 15 values are closer to the actual value plot, especially after the number 170. In summary, in the experiment of predicting TE process variable, both PSO-SVR prediction and GM (1, 1) prediction could track the trend of real value, and PSO-SVR approach showed better prediction accuracy in the experiment.

Comparison on NASA Prognosis Dataset. The second experiment was performed on the NASA prognosis dataset, and the parameters settings were the same as the first one. The optimal parameter combination obtained by the PSO-SVR algorithm is $(C, \sigma, \varepsilon) = (244, 14.569448, 0.001464)$. Table 2 and Fig. 5 show the results of the experiment, and the characteristics are very similar to those of the first experiment.

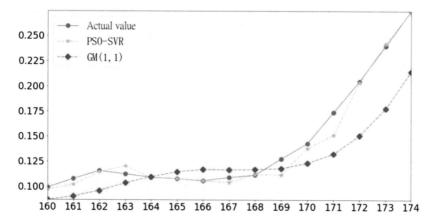

Fig. 4. Comparison of PSO-SVR and GM (1, 1) on TE process dataset

Table 2. Comparison of MSE of PSO-SVR and GM (1, 1) on TE process dataset

Item no.	Actual value	Predicted value by PSO-SVR	Relative error of PSO-SVR	Predicted value by GM (1, 1)	Relative error of GM (1, 1)
1900	11.9578	11.9257	0.2690	11.9427	0.1263
1901	11.9236	11.9585	−0.2926	11.9367	−0.1097
1902	11.9328	11.9227	0.0844	11.9180	0.1243
1903	11.9166	11.9305	−0.1165	11.9075	0.0765
1904	11.9965	11.9189	0.6469	11.8921	0.8700
1905	12.0679	11.9956	0.5991	11.8989	1.4002
1906	12.0471	12.0701	−0.1915	11.9383	0.9026
1907	12.1485	12.0427	0.8708	11.9749	1.4289
1908	12.2145	12.1495	0.5319	12.0385	1.4409
1909	12.3323	12.2226	0.8895	12.1089	1.8110
1910	12.4021	12.3359	0.5341	12.2064	1.5779
1911	12.4433	12.4160	0.2196	12.3095	1.0749
1912	12.5209	12.4643	0.4517	12.4044	0.9303
1913	12.6191	12.5538	0.5170	12.5008	0.9371
1914	12.7665	12.6689	0.7646	12.6053	1.2625
		$MSE = 4.2598 \times 10^{-3}$		$MSE = 1.8179 \times 10^{-2}$	

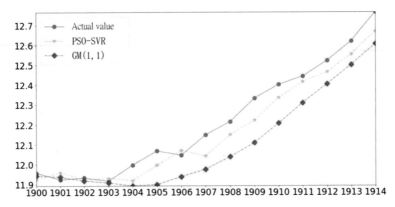

Fig. 5. Comparison of PSO-SVR and GM (1, 1) on NASA prognosis dataset

5 Conclusions

This paper proposes an SVR approach for industrial process fault prognosis using RBF as SVR kernel function. And for the core parameter optimization problem (penalty coefficient C, Gaussian kernel parameter σ and ε-insensitivity) of the SVR model, the PSO algorithm is utilized for optimization. The simulation experiment analyzes the prediction performance of PSO-SVR on benchmark TE process dataset and NASA prognosis dataset, and compares PSO-SVR with the prediction of grey GM (1, 1) model. The comparison results show that PSO-SVR method has an improved accuracy of the fault prognosis of industrial production process.

References

1. Liu, Q., Zhuo, J., Lang, Z., Qin, S.: Perspectives on data-driven operation monitoring and self-optimization of industrial processes. Acta Automatica Sinica **44**(11), 1944–1955 (2018)
2. Li, G., Qin, S., Ji, Y., Zhou, H.: Reconstruction based fault prognosis for continuous processes. Control Eng. Pract. **18**(10), 1211–1219 (2010)
3. Jardine, A.K.S., Lin, D., Banjevic, D.: A review on machinery diagnostics and prognostics implementing condition-based maintenance. Mech. Syst. Signal Process. **20**(7), 1483–1510 (2006)
4. El-Thalji, I., Jantunen, E.: A summary of fault modelling and predictive health monitoring of rolling element bearings. Mech. Syst. Signal Process. **60–61**, 252–272 (2015)
5. Smola, A.J., Scholfoph, B.: A tutorial on support vector regression. Stat. Comput. **14**(3), 199–222 (2004)
6. Zhou, Z.: Machine Learning. Tsinghua University Press, Beijing (2016)
7. Sapankevych, N.I., Sankar, R.: Time series prediction using support vector machines: a survey. IEEE Comput. Intell. Mag. **4**(2), 24–38 (2009)
8. Cheng, D.: Research on fault diagnosis and prediction based on data driven. Master's thesis. Huazhong University of Science and Technology, Wuhan, China (2018)
9. Kennedy, J., Eberhart, R.: Particle swarm optimization. In: Proceedings of IEEE International Conference on Neural Networks, IV, 1942–1948 (1995)

10. Sengupta, S., Basak, S.: Particle swarm optimization: a survey of historical and recent developments with hybridization perspectives. Mach. Learn. Knowl. Extr. 1(1), 157–191 (2018)
11. Downs, J., Fogel, E.: A plant-wide industrial process control problem. Comput. Chem. Eng. 17, 245–255 (1993)
12. Lee, J., Qiu, H., Yu, G., Lin, J.: Rexnord technical services. Bearing Data Set. IMS, University of Cincinnati. NASA Ames Prognostics Data Repository, NASA Ames Research Center, Moffett Field (2007). http://ti.arc.nasa.gov/project/prognostic-data-repository. Accessed 21 Jan 2018
13. Kayacan, E., Ulutas, B., Kaynak, O.: Grey system theory-based models in time series prediction. Expert Syst. Appl. 37, 1784–1789 (2010)

Preference Learning Based Decision Map Algebra: Specification and Implementation

Ahmed M. Abubahia[1]([⊠]), Salem Chakhar[2,3], and Mihaela Cocea[1]

[1] School of Computing, Faculty of Technology, University of Portsmouth, Portsmouth, UK
{ahmed.abubahia,mihaela.cocea}@port.ac.uk
[2] Portsmouth Business School, University of Portsmouth, Portsmouth, UK
salem.chakhar@port.ac.uk
[3] Centre for Operational Research and Logistics, University of Portsmouth, Portsmouth, UK

Abstract. Decision Map Algebra (DMA) is a generic and context independent algebra, especially devoted to spatial multicriteria modelling. The algebra defines a set of operations which formalises spatial multicriteria modelling and analysis. The main concept in DMA is decision map, which is a planar subdivision of the study area represented as a set of non-overlapping polygonal spatial units that are assigned, using a multicriteria classification model, into an ordered set of classes. Different methods can be used in the multicriteria classification step. In this paper, the multicriteria classification step relies on the Dominance-based Rough Set Approach (DRSA), which is a preference learning method that extends the classical rough set theory to multicriteria classification. The paper first introduces a preference learning based approach to decision map construction. Then it proposes a formal specification of DMA. Finally, it briefly presents an object oriented implementation of DMA.

Keywords: Decision Map Algebra · Preference learning · Dominance-based Rough Set Approach · Spatial modelling · Multicriteria modelling

1 Introduction

The GIS (Geographic Information System) is a powerful tool for collecting, storing, retrieving and analysing spatially-referenced data. Although GIS technology provides a large set of spatial analysis capabilities [10], it is still limited with respect to spatial multicriteria modeling where different decision alternatives and conflicting evaluation criteria and objectives need to be considered [7,23]. One possible solution to add spatial multicriteria modeling capabilities to the GIS is to develop a generic and context-independent language such as Tomlin's map algebra [36] or other similar tools such as image algebra developed by [30], the

Z. Ju et al. (Eds.): UKCI 2019, AISC 1043, pp. 342–353, 2020.
https://doi.org/10.1007/978-3-030-29933-0_29

work of Serra on mathematical morphology [33], and the work of van Deursen and the PCRaster group on dynamic modeling [20,37].

Building on these pioneer works on map algebra, several new and domain-specific algebra have been proposed in the literature [2,5,15,32]. However none of initial and new map algebra are suitable to spatial multicriteria modelling. To the best knowledge of the authors, the Decision Map Algebra (DMA) proposed in [6] is the first map algebra especially devoted to spatial multicriteria modeling within GIS technology. In this paper, we propose an object oriented implementation of the DMA within GIS technology, which constitutes the first step towards the development of a script-like spatial multicriteria modeling language. Furthermore, the development of abstract and generic framework for GIS-based multicriteria modeling is an important challenge as underlined in [3,21,26]

The rest of the paper is organized as follows. Section 2 introduces the decision map concept. Section 3 presents DMA. Section 4 reports on DMA modeling and implementation. Section 5 discusses some related work. Section 6 concludes the paper.

2   Preference Learning Based Decision Map

2.1   Definition and Construction of Decision Map

A decision map is a planar subdivision of the study area represented as a set of non-overlapping polygonal spatial units that are assigned, using a multicriteria classification model Γ, into an ordered set of classes or categories. More formally, a decision map $\mathbf{M}$ is defined as $\mathbf{M} = \{(u, \Gamma(u)) : u \in U\}$, where U is a set of homogenous spatial units and $\Gamma : U \to E$, where E is an ordinal measurement scale. The be useful, a decision map should be composed of non-overlapping spatial units.

The decision map construction procedure is composed of three main steps: (i) construction of criteria maps; (ii) overlay of these maps; and (iii) multicriteria Classification. The objective of the first step is to construct a set of criteria maps $c_1, c_2, \cdots, c_m$. The construction of criteria maps generally takes as input one or several basic maps. Then, a series of basic spatial operation (see [10]) are applied to combine these input maps into a new criterion map c defined such that each spatial unit in this map is characterized by a single evaluation $g(u)$ with respect to the criterion function g associated with the criterion map c.

The second step looks to overly the criterion maps $c_1, c_2, \cdots, c_m$, which leads to a multicriteria map composed of a new set of spatial units that result from the intersection of the boundaries of the features in the criteria maps. The multicriteria map may be described by the set $\{(u, g(u)) : u \in U\}$ with $g(u) = ((g_1(u), \cdots, g_m(u))$. This last vector represents the evaluations of spatial unit u with respect to evaluation criteria $g_1, g_2, \cdots, g_m$ associated with the criteria maps $c_1, c_2, \cdots, c_m$.

The aim of multicriteria classification is to apply a multicriteria classification model Γ on the multicriteria map obtained in terms of the previous step. The output is a decision map $\mathbf{M} = \{(u, \Gamma(u)) : u \in U\}$, where U is a set of homogenous spatial units.

We note that the criteria maps must represent the same territory and must be defined according to the same spatial scale and the same coordinate system. In addition, we mention that criteria maps must be polygonal ones. Non-polygonal input datasets can be easily transformed into polygonal ones using basic GIS analysis capabilities (see, e.g., [10,13]). It is also important to note that the overlay operation generally leads to a new set of spatial units resulting from the intersection of the boundaries of the spatial objects contained in the criteria maps. Finally, we note also that the overlay operation may generate silver polygons which should be eliminated.

Different methods can be used in the multicriteria classification step. In this paper, we advise to use a preference learning oriented method. The main argument beyond this proposition is to reduce the cognitive effort required from the experts and policymakers since they are not called to provide any preference parameter as with most of classical multicriteria classification methods. The principles of the preference learning method used in this paper are introduced in the rest of this section.

2.2   Principles of Preference Learning

Preference learning methods are primarily used to assess objects where decisions have been made and extract rules. Typically, this will be by taking an initial set of objects with known decisions (the learning set), applying an algorithm to extract rules and then applying these rules to predict the decision of new objects. The advantage of this approach are that the decisions can be predicted for the new objects without an extensive decision making process.

In this paper, we support the use of Dominance-based Rough Set Approach (DRSA) [16]. The DRSA is a preference learning method that extends classical rough set theory [28] to multicriteria classification. Rough set theory is a way of addressing analysis of imperfect data by taking lower (definitely belong) and upper approximations (possibly belong) of commonly held attributes between two objects. Figure 1 shows a rough set M, its lower approximation M_* and its upper approximation M^*. The set difference $Bn = M^* \setminus M_*$ between M^* and M_* is called the boundary.

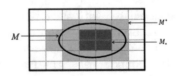

Fig. 1. Lower and upper approximations of rough set M [28]

In contrary to classical rough set theory and machine learning methods, DRSA assumes that attributes are preference-ordered. Preference-ordered attributes are commonly called criteria in multicriteria analysis. Furthermore, decision classes in DRSA are preference-ordered, while this is not a requirement in classical rough set theory and machine learning methods. In order to handle the monotonic dependency between criteria and decision classes, the DRSA uses two collections of union of classes defined as follows: (i) $Cl_t^{\geq} = \cup_{s \geq t} Cl_s$: upward union of classes; and (ii) $Cl_t^{\leq} = \cup_{s \leq t} Cl_s$: downward union of classes.

Then, DRSA determines the lower approximation for each union of classes (corresponding to objects, which according to their description certainly belong to the union of classes) and the upper approximation (corresponding to objects, which according to their description possibly belong to the union of classes). Boundary is the difference between lower and upper approximations. Therefore, rough set theory, especially DRSA, clearly separates certain and uncertain information [34]. Decision rules can then be extracted from the obtained approximations.

The input of DRSA is a learning dataset representing the description of a set of objects with respect to a set of criteria. The main output of DRSA is a collection of decision rules. A decision rule is a consequence relation $E \rightarrow H$ (read as If E, then H) where E is a condition (evidence or premise) and H is a conclusion (decision, hypothesis). Each elementary condition is built upon a single criterion while a consequence is defined based on a decision class. The obtained decision rules can then be applied to classify unseen objects. Figure 2 illustrates this working principle.

Fig. 2. Principles of DRSA

3   Decision Map Algebra

3.1   Specification of Decision Map Algebra

Symbols and Primitives. To specify DMA, we adopt the classical algebraic specification method of [18, 19]. This specification method consists of two parts: the syntactic specification and a set of axioms. For the purpose of an example, we define the types criterion_map and decision-map. A criterion_map is a mono-valued map layer where each spatial unit is characterized by one value representing the evaluation of this element with respect to a given criterion.

```
Type: criterion_map
set:   map_layer, spatial_unit, criterion_function, value
```

A decision_map is a planar subdivision of the study area represented as a set of non-overlapping polygonal spatial units that are assigned into a set of preference-ordered classes through a multicriteria classification.

> *Type*: **decision-map**
> *set*:   multicriteria_map, criterion_map, learning_map, validation_map, decision_rules,
>   spatial_unit, value

Syntax. The first part of the specification defines the syntax for the operators of the data type. In the example below, we have three such operators that are associated with decision_map data type. The MAKE operator creates a decision map as the intersection of a set of criterion maps. The MERGE operator groups two or more adjacent spatial units of a given decision_map. The GROUP operator takes a decision_map as input and generates a new decision_map by merging all adjacent spatial units that are assigned to the same class.

> MAKE criterion_map ×···× criterion_map → decision_map
> MERGE decision_map × spatial_unit × ··· × spatial_unit → spatial_unit
> GROUP decision_map → decision_map

Axioms. The third part of the specification is to define the behavior of the different operators. Following Guttag, there are two implicit axioms that must be present in any specification. The first set of axioms states that each operator returns "error" if any of its arguments does not belong to the domain of the operator. The second axiom states that an operator returns error if any of its arguments is "error".

The specifications of some operators relative to decision_map data type are given below. The MAKE operator uses the INTERSECT operator to combine a set of criteria_maps.

The MERGE operator groups two or more adjacent spatial units. It uses the MAKE operator, inherited from the basic polygon data type, to create a new spatial unit. The argument of MAKE is the boundary of the new spatial unit obtained by the union of the initial spatial units minus the common part (i.e. intersection of the boundaries of the initial spatial units). The evaluations of the new spatial unit with respect to all criteria are obtained by aggregating the evaluations associated with the initial spatial units. The operator ASSIGN of spatial_unit data type is used to assign the new evaluation to newly created spatial unit.

> d: decision_map; u, u_1, u_2:spatial_unit; $c_1, \cdots, c_m, g$: criterion_map;
> r: decision_rules
>
> MAKE($c_1, \cdots, c_m$)
> = INTERSECT($c_1, \cdots, c_m$)
>
> MERGE(d, u_1, u_2, f, op)
> =u.make(d, [BOUNDARIES(u_1) ∪ BOUNDARIES(u_2)]\
>   [BOUNDARIES(INTERSECTION(u_1, u_2)]
> $\forall(g)(g \in f)$[ASSIGN(u, g, op.combine(SCORE(u_1, g),SCORE(u_2, g)))]
>
> GROUP(d, op)
> = $\forall (u_1)(u_2)(u_1 \in d)(u_2 \in d) \wedge (u_1 <> u_2)$
>   [if ADJACENT(u_1, u_2) u_1.class = u_2.class then
>   MERGE(d, u_1, u_2, op, f)]

The specification of the MERGE operator is shown for two spatial units. The generalization to more than two spatial units is straightforward.

The GROUP operator takes as input a decision_map and generates a new decision_map by merging all adjacent spatial units that are assigned to the same class. The operator ADJACENT is used to test the adjacency of the spatial units in input. If the two spatial units are adjacent and have the same evaluation, then the operator MERGE is applied to merge them.

3.2 DMA Spatial Abstract Data Types

The spatial Abstract Data Types (ADT) supported by DMA are: criterion_map, weighted_criterion_map, multicriteria_map, learning_map, validation_map, decision_map andvalternatives_map. Each of these data types has a collection of proprieties and methods. Some of these methods permit to set or get the descriptive information of the corresponding data type while some others are devoted to set or get the spatial information of these data types. The criterion_map and decision_map data types have been introduced earlier.

The weighted_criterion_map is a specific version of criterion_map defined such that each spatial unit has a spatial weight. The multicriteria_map is obtained by overlying a set of $n > 1$ criteria maps or weighted criteria maps. Each spatial unit in the multicriteria_map is characterized by n scores corresponding to the n criteria maps in input.

The learning_map and validation_map are spatial representations of training and validation subsets often used in the application of preference learning methods. The training subset is labelled with known decisions and used to train the method. The validation subset is another subset of the input data with known decisions. We apply the preference learning method to this subset to see how accurately it identifies the known decisions.

The decision_map can used as it is to support suitability analysis as in [25]. In more complex decision problems, the use of a decision_map requires the definition of appropriate tools to generate solutions to the considered decision problem, which will lead to one or more alternatives_maps. Some formal solutions to generate alternatives_map have been proposed and used in [1,8,9]. These solutions rely largely on graph theory algorithms.

In the rest of this section, we present a summary of some properties and methods associated with criterion_map data type. The basic properties associated with a criterion_map are:

- NAME: a name that uniquely identifies the criterion_map,
- DESCRIPTION: a textual description of the criterion_map,
- DATA TYPE: the type of data represented in the criterion_map. It may be nominal, symbolic, ordinal, integer or continuous.
- POSSIBLE DATA VALUES: For nominal and symbolic data types, we need also to indicate the set of possible values.
- SCORE: the score of a spatial unit of the criterion_map.
- WEIGHT: the weight of the criterion_map.

- PREFERENCE: it indicates the direction of preference. Three cases are possible: (i) gain: an increase on the criterion value, will lead to a higher attractiveness; (ii) cost: an increase on the criterion value, will lead to a lower attractiveness; or (iii) none: this is for nominal or symbolic criteria with no preference structure.
- REFERENCE: it represents the geographic coordinate system used.
- MAP SCALE: it is the map scale, which is the ratio of a distance on the map to the corresponding distance on the ground.

The basic methods associated with a criterion_map data type are given in Table 1. A criterion_map supports two basic methods, namely SET and GET, permitting to set or get the value of any property. Each of these methods has two different syntaxes. The first one applies to all the above cited-properties, except SCORE for which a specific syntax for accessing the score of a spatial unit is used. The STANDARDIZE method permits to process the initial data represented by the criterion_map into form appropriate to multicriteria modelling. This method requires the specification of a standardization techniques.

Table 1. Some methods associated with criterion_map data type

Name	Syntax
SET	criterion_map $\times$ property $\times$ value $\rightarrow$ criterion_map
	criterion_map $\times$ spatial_unit $\times$ value $\rightarrow$ spatial_unit
GET	criterion_map $\times$ property $\rightarrow$ value
	criterion_map $\times$ spatial_unit $\rightarrow$ value
STANDARDIZE	criterion_map $\times$ std_procedure $\rightarrow$ criterion_map

4   Object Oriented Modeling and Implementation

In this section, we report an object oriented modeling of DMA. The adoption of an object oriented modeling formalism relies on the following reasons: (i) recent works on spatial data models are more and more oriented towards the object oriented modelling [12,22,31,32]; (ii) this approach seems to be in accordance with human perception of geographic space, often seen as "populated" with objects [11]; (iii) an object formalism is more adequate for developing and implementing abstract data types devoted to spatial multicriteria modelling.

Intuitively, each data type in DMA is defined as a class and the operators associated with this data type are defined as methods for this class. A simple version of the UML model associated with DMA is given in Fig. 3. In addition to classes implementing the spatial ADT introduced earlier, the UML model contains a new class called Classifier which represents a set of rules.

Objects of Classifier class are the result of calling the function Infer on learning_map object. The classify method of Classifier class takes as input a learning_map and a validation_map and generates a decision_map. Classifier class contains several other basic methods:

(1) *lhs* and *rhs* that are used to access the left-hand-sides or right-hand-sides of the rules.
(2) a set of methods permitting to access different performance measures of decision rules such as support, confidence and strength.
(3) *covers* which permits to check if a given rule covers a given spatial unit.
(4) *supports* which is used to check if a a given spatial unit supports a given rule.

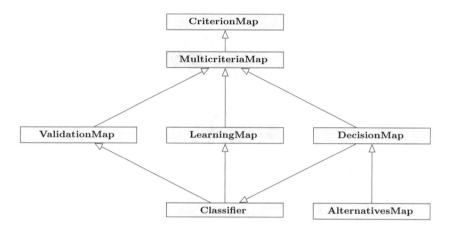

Fig. 3. UML model

The DMA is being implemented using Python and QGIS platform. For illustration, we provide in Fig. 4 three maps concerning seasonal influenza risk assessment in the northwest region of Algeria. The learning and validation maps are given in Figs. 4(a) and (b), respectively, while the final risk map is shown in Fig. 4(c). A four level risk scale ranging from 'Low' (light grey) to 'Very High' (dark red) has been used in the three maps. The assignment of the districts in learning and validation maps have been specified by the experts while the assignments of the districts in the final risk map have computed using method classify of Classifier class.

5   Related Work

The Map Analysis Package (MAP) [36] was the first comprehensive collection of analytical and spatial operations on the basis of regular tessellations. MAP has been extended in area ranging from cellular automata [35], to environmental modeling [29], to topographic analysis [4], to spatio-temporal analysis [24]. An important limitation of MAP is related to the fact that it describes map overlay operations textually, without applying the mathematical rigor necessary to

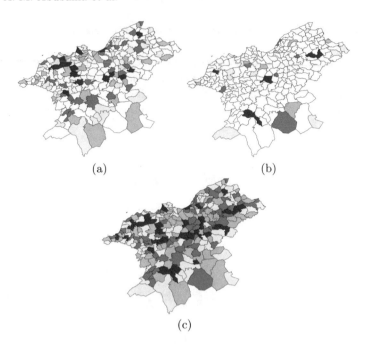

(c)

Fig. 4. Illustrative example

analyze the behavior of the operations. A second important problem of MAP is the strong link between geographical datatypes and data structures.

Another proposal that have inspired DMA is the one proposed [14]. A major finding of this paper is the use of decision table concept, which is adopted in DMA. More recent works are essentially devoted to develop script-like programming languages [29], to support spatial-temporal analysis [24], visual spatial modeling [27] and to the development of web-based map algebra-like frameworks [17].

There are also several new and domain-specific algebra that have been proposed in the literature [2,5,15,32]. For instance, authors in [32] propose an integrated modelling framework that provides descriptive means to specify (1) model components with conventional map algebra, and (2) interactions between model components with model algebra. A prototype implementation in a high-level scripting language supports the building of integrated spatio-temporal models is also proposed.

The authors in [5] design and implement a framework that uses compiler techniques to automatically speed up raster spatial analysis. In this way, users simply write sequential map algebra scripts in Python, which are translated into a graph where optimizations are applied.

In [2], the authors presents an algebra that extends the Systems Dynamics paradigm to the development of spatially explicit models of continuous change. The proposed algebra provides types and operators to represent flows of energy and matter between heterogeneous regions of geographic space. To this end, algebraic sets of operations similar to those in Map Algebras are introduced, allowing the representation of local, focal and zonal flows.

6   Conclusion

The paper provides an object oriented modelling and implementation of Decision Map Algebra (DMA). This is constitutes the first step towards the development of a script-like spatial multicriteria modeling language. From theoretical point of view, the paper mainly enhances DMA through the use of preference learning based approach to decision map construction. This will naturally reduce the cognitive effort required from the experts and policymakers since they are not called to provide any preference parameter.

Our current research concerns the full implementation of the proposed data types. We also intend to design and implement a script language for spatial multicriteria modeling. We are also concerned by the design and development of a graphical version of the script language.

References

1. Aissi, H., Chakhar, S., Mousseau, V.: Gis-based multicriteria evaluation approach for corridor siting. Environ. Plan. B Plan. Des. **39**(2), 287–307 (2012)
2. Amâncio, A., Carneiro, T.: An algebra for modeling and simulation of continuous spatial changes. J. Inf. Data Manag. **9**(3), 275–290 (2018)
3. Brauner, J., Foerster, T., Schaeffer, B., Baranski, B.: Towards a research agenda for geoprocessing services. In: Haunert, J., Kieler, B., Milde, J. (eds.) 12th AGILE International Conference on Geographic Information Science, pp. 1–12. IKG, Leibniz, University of Hanover, Hanover, Germany (2009)
4. Caldwell, D.: Extending map algebra with flag operators. In: Proceedings of GeoComputation 2000 (2000)
5. Carabaño, J., Westerholm, J., Sarjakoski, T.: A compiler approach to map algebra: automatic parallelization, locality optimization, and gpu acceleration of raster spatial analysis. GeoInformatica **22**(2), 211–235 (2018)
6. Chakhar, S., Mousseau, V.: An algebra for multicriteria spatial modelling. Comput. Environ. Urban Syst. **31**(5), 572–596 (2007)
7. Chakhar, S., Mousseau, V.: Spatial multicriteria decision making. In: Shehkar, S., Xiong, H. (eds.) Encyclopedia of Geographic Information Science, pp. 747–753. Springer, Boston (2007)
8. Chakhar, S., Mousseau, V.: GIS-based multicriteria spatial modeling generic framework. Int. J. Geogr. Inf. Sci. **22**(11), 1159–1196 (2008)
9. Chakhar, S., Mousseau, V.: Generation of spatial decision alternatives based on a planar subdivision of the study area. In: Proceedings The 2006 ACM/IEEE International Conference on Signal-Image Technology and Internet-Based Systems (SITIS 2006). Lecture Notes in Computer Science, vol. 4879, pp. 137–148. Springer (2009)
10. Chrisman, N.: Transformations. In: Exploring Geographic Information Systems, 2nd edn, pp. 217–242. Wiley, New York (2002)

11. Couclelis, H.: People manipulate objects (but cultivate fields): Beyond the raster vector debate in GIS. In Frank, A., Campari, I. (eds.) Theories and Methods of Spatio-Temporal Reasoning in Geographic Space. Lecture Notes in Computer Science, vol. 639, 65–77. Springer, Berlin (1992)

12. Cova, T., Goodchild, M.: Extending geographical representation to include fields of spatial objects. Int. J. Geogr. Inf. Sci. **16**(6), 509–532 (2002)

13. de Floriani, L., Magillo, P., Puppo, E.: Applications of computational geometry to geographic information systems. In: Sack, J.-R., Urrutia, J. (eds.) Handbook of Computational Geometry, vol. 7, pp. 333–388. Elsevier Science B.V, Amsterdam (1999)

14. Dorenbeck, C., Egenhofer, M.: Algebraic optimization of combined overlay operations. In: Mark, D., White, D. (eds.) Auto-Carto 10: Technical Papers of the 1991 ACSM-ASPRS Annual Convention, Baltimore, Maryland, USA, ACSM-ASPRS, pp. 296–312, March 1991

15. Graciano, A., Rueda, A., Feito, F.: A formal framework for the representation of stack-based terrains. Int. J. Geogr. Inf. Sci. **32**(10), 1999–2022 (2018)

16. Greco, S., Matarazzo, B., Słowiński, R.: Rough sets theory for multicriteria decision analysis. Eur. J. Oper. Res. **129**(1), 1–47 (2001)

17. Grunberg, W., Dale, J., Haseltine, M., Lerman, N., Olsson, A., Orr, B.: Web-based map-algebra challenges: A polygon solution. In: Proceedings of the 24th Annual ESRI International User Conference, San Diego, California, 9-10 August 2004

18. Guttag, J.: Abstract data types and the development of data structures. Commun. ACM **20**(6), 396–404 (1977)

19. Guttag, J., Horning, J.: The algebraic specification of abstract data types. Acta Inform. **10**(1), 27–52 (1978)

20. Karssenberg, D., Burrough, P., Sluiter, R., Jong, K.: The PCRaster software and course materials for teaching numerical modelling in the environmental sciences. Trans. GIS **5**, 99–110 (2001)

21. Le Roux, W., Du Preez, M., De Klerk, L., Hohls, D., McFerren, G., Van der Merwe, A.V.M., Van Zyl, T.: EO2HEAVEN: Mitigating environmental health risks. In: Fraunhofer-Gesellschaft, pp. 1–146 (2013)

22. Leung, Y., Leung, K., He, J.: A generic concept-based object-oriented geographical information system. Int. J. Geogr. Inf. Sci. **13**(5), 475–498 (2002)

23. Malczewski, J.: Multiple criteria decision analysis and geographic information systems. In: Trends in Multiple Criteria Decision Analysis, pp. 369–395. Springer, Boston (2010)

24. Mennis, J., Viger, R., Tomlin, C.: Cubic map algebra functions for spatio-temporal analysis. Cartogr. Geogr. Inf. Sci. **32**(1), 17–32 (2005)

25. Mercat-Rommens, C., Chakhar, S., Chojnacki, E., Mousseau, V.: Coupling GIS and multi-criteria modeling to support post-accident nuclear risk evaluation. In: Raymond, B., Dias, L., Meyer, P., Mousseau, V., Pirlot, M. (eds.) Evaluation and Decision Models with Multiple Criteria. International Handbooks on Information Systems, pp. 401–428. Springer, Heidelberg (2015)

26. Müller, M., Bernard, L., Vogel, R.: Multi-criteria evaluation for emergency management in spatial data infrastructures. In: Geographic Information and Cartography for Risk and Crisis Management: Towards Better Solutions, pp. 273–286. Springer, Heidelberg (2010)

27. Murray, S., Breslin, P., Ormsby, T., Miller, B.: A visual framework for spatial modeling. In: Proceedings of the 4th International Conference on Integrating GIS and Environmental Modeling (GIS/EM4): Problems, Prospects and Research Needs, Banff, Alberta, Canada (2000)

28. Pawlak, Z.: Rough Set: Theoretical Aspects of Reasoning About Data. Kluwer Academic Publishers, Dordrecht (1991)

29. Pullar, D.: MapScript: A map algebra programming language incorporating neighborhood analysis. GeoInformatica **5**(2), 145–163 (2001)

30. Ritter, G., Wilson, J., Davidson, J.: Image algebra: An overview. Comput. Vis. Graph. Image Process. **49**, 297–331 (1990)

31. Schmitz, O., de Kok, J., Karssenberg, D.: A software framework for process flow execution of stochastic multi-scale integrated models. Ecol. Inform. **32**, 124–133 (2016)

32. Schmitz, O., Karssenberg, D., de Jong, K., de Kok, J.L., de Jong, S.: Map algebra and model algebra for integrated model building. Environ. Model. Softw. **48**, 113–128 (2013)

33. Serra, J.: Image Analysis and Mathematical Morphology. Academic Press, London (1982)

34. Stefanowski, J., Vanderpooten, D.: Induction of decision rules in classification and discovery-oriented perspectives. Int. J. Intell. Syst. **16**(1), 13–27 (2001)

35. Takeyama, M., Couclelis, H.: Map dynamics: Integrating cellular automata and GIS through geo-algebra. Int. J. Geogr. Inf. Sci. **11**(1), 73–91 (1997)

36. Tomlin, C.: Geographic Information Systems and Cartographic Modeling. Prentice Hall, Englewood Cliffs (1990)

37. Wesseling, C., Karssenberg, D., Deursen, W., Burrough, P.: Integrating dynamic environmental models in GIS: The development of a dynamic modelling language. Trans. GIS **1**, 40–48 (1996)

Hybrid Methods

"Parallel-Tempering"-Assisted Hybrid Monte Carlo Algorithm for Bayesian Inference in Dynamical Systems

Shengjie Sun[1] and Yuan Shen[2($\boxtimes$)]

[1] Department of Mathematical Sciences, Xi'an Jiaotong-Liverpool University,
Suzhou, China
[2] School of Science and Technology, Nottingham-Trent University,
Nottingham, UK
`Yuan.Shen@ntu.ac.uk`

Abstract. The aim of this work is to tackle the problem of sampling from multi-modal distributions when Hybrid Monte Carlo (HMC) algorithm is employed for performing Bayesian inference in dynamical systems. Hybrid Monte Carlo is a powerful Markov Chain Monte Carlo (MCMC) algorithm but it still suffers from the "multiple peaks" problem. Due to non-trivial structure in the space of (a class of) dynamical systems, posterior distribution of its model parameters could exhibit complicated structures such as multiple ridges. We examined a MCMC algorithm combining HMC with so-called Parallel Tempering (PT) - a well-known strategy for tackling the problem highlighted above. The new algorithm is referred to as PT-HMC. Our numerical experiment demonstrated that when compared to the ground truth, the posterior distributions derived from PT-HMC samples is more accurate than those from HMC.

Keywords: Multi-modal distribution · Hybrid Monte Carlo ·
Parallel tempering · Dynamical systems

1 Introduction

Dynamical systems, in the form of ordinary differential equations, have been widely used for mechanistic modeling of temporal data in a variety of real-world applications (for example, biological pathway model in omics data analysis) [1, 7–9]. Not only is it a natural approach to incorporate domain experts' knowledge in time series analysis, but also such time-continuous model could serve as an appropriate representation of those time series data. It is of particular importance when the time series data are sparsely and/or irregularly sampled. Due to this sparsity as well as noise that contaminates the observations, model uncertainty needs to be accounted for adequately. Consequently, individual posterior distributions over (the class of) dynamical systems are used to represent the

© Springer Nature Switzerland AG 2020
Z. Ju et al. (Eds.): UKCI 2019, AISC 1043, pp. 357–368, 2020.
https://doi.org/10.1007/978-3-030-29933-0_30

corresponding data sets. Therefore, accurate estimation of those posterior distributions are very important.

Given a class of parameterized dynamical systems and a time series data set that is assumed to be generated by one instance of that model class, Markov Chain Monte Carlo is the computational tool for computing the corresponding posterior distribution. However, such posterior distributions over multiple model parameters could exhibit very complicated structure. This is because (1) for a number of (or even infinitely many) combinations of individual model parameters, their corresponding trajectories could be very similar; (2) even when there exist subtle differences among those trajectories, the observed time series don't differ from each other due to sparse observation and/or noise in the observations. All these result in multi-modal or multi-ridge posterior distributions. There are various strategies that have been proposed to deal with such problems. Examples are parallel tempering, simulated tempering, auxiliary variable, and delayed rejection. Among them, parallel tempering is the most straightforward approach. In theory, HMC is a variant of auxiliary-variable method. The research question we address in this paper is two-fold: (1) to which extent HMC can cope with the problem of sampling from multi-modal distributions? (2) How much can PT-HMC further improve HMC's performance.

2   The MCMC Algorithms

2.1   The HMC Algorithm

HMC is based on an analogy with (a class of) physical systems while it can also be interpreted as an instance of (a class of) MCMC algorithms that adopt so-called "auxiliary variable"-strategy for facilitating sampling in high dimensions [2].

The physical system involved in HMC is so-called Hamiltonian system. This class of systems is represented by a d-dimensional "position" vector, $\boldsymbol{\theta}$ and a d-dimensional "momentum" vector, $\mathbf{p}$. Further, time evolution of their state variables, $(\boldsymbol{\theta}, \mathbf{p})$, is governed by so-called Hamilton's equations as follows:

$$\frac{d\theta_i}{dt} = \frac{\partial \boldsymbol{H}}{\partial p_i} \tag{1}$$

$$\frac{dp_i}{dt} = -\frac{\partial \boldsymbol{H}}{\partial \theta_i} \tag{2}$$

for $i = 1, \cdots, d$, where function $\boldsymbol{H}$ is called Hamiltonian that defines the system. In HMC, we consider Hamiltonian functions that can be written as the sum of potential energy, $U(\boldsymbol{\theta})$ and kinetic energy, $K(\mathbf{p})$, that is,

$$\boldsymbol{H}(\boldsymbol{\theta}, \mathbf{p}) = U(\boldsymbol{\theta}) + K(\mathbf{p}) \tag{3}$$

According to the concept of a canonical distribution from statistical mechanics, there exists a theoretical relationship between Hamiltonian equation and the joint probability density of the position and momentum variables $p(\boldsymbol{\theta}, \mathbf{p})$ given by [10]:

$$p(\boldsymbol{\theta}, \mathbf{p}) = \frac{1}{Z_H} e^{-H(\boldsymbol{\theta}, \mathbf{p})} = \left(\frac{1}{Z_U} e^{-U(\boldsymbol{\theta})} \right) \left(\frac{1}{Z_K} e^{-K(\mathbf{p})} \right) \qquad (4)$$

where Z_H, Z_U, Z_K in (4) are the normalizing constants of their corresponding exponentials. That is, the numerical solution to an Hamiltonian equation is equivalent to sampling from the density given by Eq. 4.

The relationship highlighted above allows us to sample from a given probability distribution by solving its corresponding Hamiltonian equations numerically. To sample from a target density $\pi(\boldsymbol{\theta})$, we connect it with the marginal density $p(\boldsymbol{\theta})$ derived from $p(\boldsymbol{\theta}, \mathbf{p})$ by setting

$$\pi(\boldsymbol{\theta}) = \frac{1}{Z_U} e^{-U(\boldsymbol{\theta})} \qquad (5)$$

$$\Leftrightarrow \qquad U(\boldsymbol{\theta}) = -\log \pi(\boldsymbol{\theta}) - \log Z_U. \qquad (6)$$

As a result, the trajectory of $\boldsymbol{\theta}$ obtained from the Hamiltonian equation can be used as samples from $\pi(\boldsymbol{\theta})$.

As $\mathbf{p}$ is not the variable of interest in HMC, $K(\mathbf{p})$ can be defined arbitrarily. Usually, it defined as: $K(\mathbf{p}) = \dfrac{\mathbf{p}^T M \mathbf{p}}{2}$. Here M is a symmetric, positive-definite "mass matrix". This form for $K(\mathbf{p})$ corresponds to minus the log probability density (plus a constant) of the zero-mean Gaussian distribution with covariance matrix, M. The resulting joint probability density is given as $p(\boldsymbol{\theta}, \mathbf{p}) = \pi(\boldsymbol{\theta}) \mathcal{N}(\mathbf{p}; 0, M)$, and its corresponding Hamilton's equations (1) and (2) can be written in a vectorized form as follows:

$$\frac{d\boldsymbol{\theta}}{dt} = M^{-1} \mathbf{p} \qquad (7)$$

$$\frac{d\mathbf{p}}{dt} = -\nabla_{\boldsymbol{\theta}} U(\boldsymbol{\theta}) \qquad (8)$$

In HMC, so-called Leapfrog method [5] is employed to obtain the numerical solution to Eqs. 1 and 2. It works as follows:

$$p_i(t + \epsilon/2) = p_i(t) - (\epsilon/2) \frac{\partial U}{\partial \theta_i}(\theta(t)) \qquad (9)$$

$$\theta_i(t + \epsilon) = \theta_i(t) + \epsilon \frac{\partial K}{\partial p_i}(p(t + \epsilon/2)) \qquad (10)$$

$$p_i(t + \epsilon) = p_i(t + \epsilon/2) - (\epsilon/2) \frac{\partial U}{\partial \theta_i}(\theta(t + \epsilon)) \qquad (11)$$

where ϵ denotes the time increment used. In contrast to Euler method, Leapfrog method updates "position" and "momentum" variables at t differently. First, the "position" variables are updated at $t + \epsilon/2$, instead of $t + \epsilon$, by using Eq. 1. This should increase the accuracy of the numerical solution. Next, it updates the "momentum" variables at $t + \epsilon$ as Euler method but the derivatives of $\mathbf{p}$ w.r.t t is evaluated at $t + \epsilon/2$. This means that a better $\mathbf{p}$ is used for computing $\boldsymbol{\theta}(t + \epsilon)$, that is, $\mathbf{p}(t + \epsilon/2)$ instead of $\mathbf{p}(t)$. Similarly, a better $\boldsymbol{\theta}$ is used for computing $\mathbf{p}(t + \epsilon)$, that is, $\boldsymbol{\theta}(t + \epsilon)$ instead of $\boldsymbol{\theta}(t)$.

Recall that all state variables along the trajectory can be used as samples. For subsequent statistical analysis such as computing mean and covariance, however, these samples need to be independent. Therefore, an appropriate time interval between two consecutive samples need to be chosen, say s. Equivalently, we need to take $L = s/\epsilon$ leapfrog steps. However, the discretization errors resulted from leapfrog could be accumulated. Consequently, we need an accept-reject mechanism using Metropolis-Hasting ratio to ensure that the resulting samples are indeed drawn from the required distribution.

To sum up, each iteration of the HMC algorithm consists of two steps:

- **Step 1** First, draw a new sample $\mathbf{p}$ from $\mathcal{N}(\mathbf{p}; 0, M)$. Then, use this $\mathbf{p}$ and previous value of $\boldsymbol{\theta}$ to initiate next leapfrog cycle of L "frog leaps";
- **Step 2** Use the values of $\boldsymbol{\theta}$ and $\mathbf{p}$ at the last leap as the proposal candidates in the MH algorithm with target density $p(\boldsymbol{\theta}, \mathbf{p})$.

As a variant of MH MCMC algorithm. efficiency of HMC could be reasoned as follows: Given an initial pair of $\boldsymbol{\theta}$ and $\mathbf{p}$, the trajectory moves on a hypersurface of constant (joint) probability density. In other word, the sampler can move freely on this equi-density surface. Every time when we generate a new $\mathbf{p}$, the sampler jumps to another this equi-density surface. This could improve the mixing property of a MCMC sampler. This is related to the concept of auxiliary variables. However, it is an unanswered question whether HMC can cope with multi-modal problems.

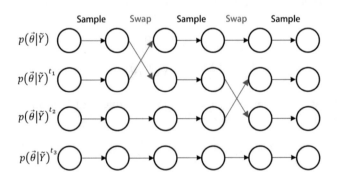

Fig. 1. Illustration of the idea of parallel tempering

2.2 Parallel Tempering

We adopt a strategy called parallel tempering (PT) to increase the chance of the jumping between local peaks while using HMC sampler to sample from a multi-modal probability distribution. Given a probability density $p(\mathbf{x})$, a tempered version of $p(\boldsymbol{\theta})$ is defined by unnormalized density $p(\boldsymbol{\theta})^t$ with $0 < t < 1$.

Compared to $p(\boldsymbol{\theta})$, the surface of $p(\mathbf{x})^t$ is increasingly flatter with decreasing t. A MCMC sampler can move in the state space much more freely with $p(\boldsymbol{\theta})^t$ than $p(\boldsymbol{\theta})$. The concept of PT is illustrated in Fig. 1.

In PT assisted HMC (PT-HMC), several HMC samplers work in parallel. One of them samples from the target probability density while each of the remaining samplers samples from one of several tempered versions of the target one. The resulting Markov chains are ordered by the values of their corresponding tempering parameter t. After several HMC steps, two neighboring chains try to swap their current states and this swap could be accepted or rejected probabilistically based on MH criteria. The swaps take place only between two neighboring Markov chains because their target densities are similar and the chance of being accepted is relatively larger. The swaps could help the chain with larger t values more freely in the state space than in a stand-alone setting. When a sampler gets stuck in one mode, it can get out this trap only by getting a proposal from other modes. But the proposal mechanism in HMC makes this event almost impossible. Finally, the accept-reject mechanism ensures that all chains will converge to their target densities. The swap schedule adopted in this paper is illustrated in Fig. 2.

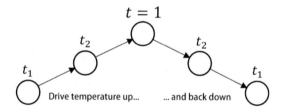

Fig. 2. The swapping order we adapted

3   Illustrative Experiment

Both HMC and HMC-PT algorithm were tested by numerical experiments with a nonlinear ODE system modelling circadian oscillation. This is because given a set of observations of this system, the posterior distribution of its model parameters exhibits multi-ridges structure.

3.1   Model of Circadian Oscillator

This simple model of Circadian Oscillator is originally described as follows [6] as follows:

$$\frac{dx_1}{dt} = \frac{k_1}{36 + k_2 x_2} - k3 \qquad (12)$$

$$\frac{dx_2}{dt} = k_4 x_1 - k_5 \qquad (13)$$

where $k_1 = 72$, $k_2 = 1$, $k_3 = 2$, $k_4 = 1$ and $k_5 = 1$, and the initial values are set to $x_1(0) = 7$ and $x_2(0) = -10$. For our numerical experiments, only k_3 and k_4 are considered as free parameters that need to be estimated from the data. That is, $\boldsymbol{\theta} = [k_3 \ k_4]$ and the data are generated from the system with $\boldsymbol{\theta} = [2 \ 1]$.

3.2   Derivation of HMC for the Toy System

First, we formulate the mathematical expression of the posterior density over $\boldsymbol{\theta}$ given a data set denoted by $\mathcal{D}$. Then, $U(\boldsymbol{\theta})$ can be derived from Eq. (6) straightforwardly. After that, we derive the formula of $\nabla_{\boldsymbol{\theta}} U(\boldsymbol{\theta})$ in Eq. 8. Also, we define the mathematical expression of both the state and the data.

Let $\mathbf{x}$ denote D-dimensional state vector of a dynamical system and $\boldsymbol{\theta}$ its d-dimensional parameter vector, that is, $\mathbf{x} \in \mathbb{R}^D$ and $\boldsymbol{\theta} \in \mathbb{R}^d$. We assume that this dynamical system is observed at the following time points: $t_1 = 0$, $t_1 = \Delta$, $t_3 = 2\Delta t$, ..., $t_T = (T-1)\Delta t$ where Δt denotes the time interval between two consecutive observations. Note that the time increments used for numerically solving the ODE systems is usually much smaller than Δt. Let $\mathbf{x}_i$ denote the state vector at the i-th observation time (that is, $\mathbf{x}_i = \mathbf{x}(t = t_i)$) and $\mathbf{y}_i$ the i-th observation. When all state variables are observed, we have $\mathbf{y}_i \in \mathbb{R}^D$.

The noise in the observations is modeled by $\mathbf{y}_i = \mathbf{x}_i + \eta_i$ where noise η_1, ..., η_T are identically and independently Gaussian distributed with zero mean and covariance matrix $\Sigma \in \mathbb{R}^{D \times D}$. For clarity in formulating $p(\boldsymbol{\theta}|\mathcal{D})$, we define: $\widetilde{X} = \mathbf{vec}([\mathbf{x}_1...\mathbf{x}_T]^\mathsf{T})$ and $\widetilde{Y} = \mathbf{vec}([\mathbf{y}_1...\mathbf{y}_T]^\mathsf{T})$ where $\widetilde{X} \in \mathbb{R}^{TD}$ and $\widetilde{Y} \in \mathbb{R}^{TD}$. Accordingly, the error covariance matrix of $\widetilde{Y}$ denoted by $\widetilde{\Sigma} \in \mathbb{R}^{TD \times TD}$ is given by $\widetilde{\Sigma} = \mathbb{1}_T \otimes \Sigma$ where $\mathbb{1}_T$ denotes an identity matrix of size $T \times T$ and $\otimes$ represents Kronecker's product of two matrices. Note that the k-th component of vector $\widetilde{X}$ represent the $(j+1)$-the component of $\mathbf{x}$ at the i-th time point where j is modulo k of T and i is the remainder. It is the same for $\widetilde{Y}$.

Combining Eqs. (6) and (8), noting that we adopt an uniform distribution as of model parameters as their prior. That is, $p(\boldsymbol{\theta}|\widetilde{Y}) = p(\widetilde{Y}|\boldsymbol{\theta})$ and we obtain:

$$\frac{d\mathbf{p}}{dt} = -\nabla_{\boldsymbol{\theta}} U(\boldsymbol{\theta}) = \frac{1}{p(\widetilde{Y}|\boldsymbol{\theta})} \nabla_{\boldsymbol{\theta}} \left[p(\widetilde{Y}|\boldsymbol{\theta}) \right] \tag{14}$$

Due to the i.i.d. property of observation noise, the likelihood function and thus the posterior density is written as

$$p(\widetilde{Y}|\boldsymbol{\theta}) = \frac{1}{(2\pi)^{\frac{TD}{2}} |\Sigma|^{\frac{1}{2}}} \left\{ exp(-\frac{1}{2}(\widetilde{Y} - \widetilde{X})^T \Sigma^{-1} (\widetilde{Y} - \widetilde{X})) \right\}. \tag{15}$$

To specify the HMC algorithm we devise, the derivative of momentum vector $\mathbf{p}$ with respect to algorithmic time $\tilde{t}$ needs to be derived. Here, we distinguish the algorithmic time variable used in HMC $\tilde{t}$ from the physical time t variable used in ODE. First, we derive the gradient of $p(\widetilde{Y}|\boldsymbol{\theta})$ w.r.t $\boldsymbol{\theta}$ as follows:

$$\nabla_{\boldsymbol{\theta}} \left[p(\widetilde{Y}|\boldsymbol{\theta}) \right] = p(\widetilde{Y}|\boldsymbol{\theta}) \nabla_{\boldsymbol{\theta}} \left[-\frac{1}{2}(\widetilde{Y} - \widetilde{X})^T \Sigma^{-1} (\widetilde{Y} - \widetilde{X}) \right] \tag{16}$$

Combining Eqs. (14), (15) and (16), we obtain

$$\frac{d\mathbf{p}}{d\tilde{t}} = \nabla_{\boldsymbol{\theta}} \left[-\frac{1}{2}(\tilde{Y} - \tilde{X})^T \Sigma^{-1} (\tilde{Y} - \tilde{X}) \right]. \tag{17}$$

Then, the i-component of the gradient with $i = 1, 2, \cdots d$ is given by

$$\begin{aligned}
\left[\frac{d\mathbf{p}}{dt} \right]_i &= \frac{d}{d\theta_i} \left[-\frac{1}{2}(\tilde{Y} - \tilde{X})^T \Sigma^{-1} (\tilde{Y} - \tilde{X}) \right] \\
&= -\frac{1}{2} \left[\nabla_{\tilde{X}} \left[(\tilde{Y} - \tilde{X})^T \Sigma^{-1} (\tilde{Y} - \tilde{X}) \right] \right]^T \frac{d\tilde{X}}{d\theta_i} \\
&= \left[\Sigma^{-1} (\tilde{Y} - \tilde{X}) \right]^T \frac{d\tilde{X}}{d\theta_i}
\end{aligned} \tag{18}$$

To derive $\dfrac{d\tilde{X}}{d\theta_i}$, we first need to derive $\dfrac{d\mathbf{x}_t}{d\theta_i}$. We denote it by $\mathbf{S}_t(\theta_i)$ as it actually is the sensitivity vector of $\mathbf{x}_t$ w.r.t. θ_i. Recall that an ODE system can be generally formulated as

$$\mathbf{x} \in \mathbb{R}^D \qquad \frac{d\mathbf{x}}{dt} = \mathbf{f}(\mathbf{x}; \boldsymbol{\theta}) \qquad \mathbf{f} : \mathbb{R}^D \to \mathbb{R}^D \tag{19}$$

Time evolution of $\mathbf{S}_t(\theta_i)$ is governed by

$$\begin{aligned}
\frac{d\mathbf{S}(\theta_i)}{dt} &= \frac{d}{dt} \left(\frac{d\mathbf{x}}{d\theta_i} \right) = \frac{d}{d\theta_i} \left(\frac{d\mathbf{x}}{dt} \right) = \frac{d\mathbf{f}(\mathbf{x})}{d\theta_i} \\
&= \mathbf{J} \frac{d\mathbf{x}}{d\theta_i} + \frac{\partial \mathbf{f}(\mathbf{x})}{\partial \theta_i} = \mathbf{J}\mathbf{S}(\theta_i) + \frac{\partial \mathbf{f}(\mathbf{x})}{\partial \theta_i}
\end{aligned} \tag{20}$$

where we have $\mathbf{S}_0(\theta_i) = 0$ and $\mathbf{J}$ is the Jacobian (that is, $\mathbf{J}_{ij} = \frac{\partial f_i(\mathbf{x})}{\partial x_j}$).

For the ODE system given by Eqs. (12) and (13), we obtain

$$\frac{d\mathbf{S}(\theta_1)}{dt} = \mathbf{J}\mathbf{S}(\theta_1) + \begin{bmatrix} -1 \\ 0 \end{bmatrix} \tag{21}$$

$$\frac{d\mathbf{S}(\theta_2)}{dt} = \mathbf{J}\mathbf{S}(\theta_2) + \begin{bmatrix} 0 \\ x_1 \end{bmatrix} \tag{22}$$

where $\mathbf{J} = \begin{bmatrix} 0 & -\frac{72}{(36+x_2)^2} \\ \theta_2 & 0 \end{bmatrix}$. To reduce the computational cost incurred by numerical computation of the sensitivity equations, we adopted a crude approximation as follows: $\mathbf{s}_{t(\theta_1)} \approx \begin{bmatrix} -1 \\ 0 \end{bmatrix} \Delta T$ and $\mathbf{s}_{t(\theta_2)} \approx \begin{bmatrix} 0 \\ x_1(t) \end{bmatrix} \Delta T$ where ΔT is the time increment used in solving the given ODE systems and is therefore very small. Unarguably, this approximation strategy can hardly be justified on theoretical ground. But the resulting algorithm worked very well as demonstrated in Sect. 4. Further investigation is required for a better understanding.

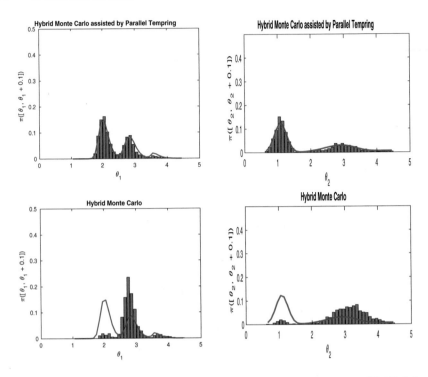

Fig. 3. Performance comparison between PT-HMC and HMC when HMC fails to explore the parameter space adequately. In each panel. a marginal posterior distribution either over θ_1 (Left column) or over θ_2 (Right column) is compared with it corresponding ground truth (Red curve). Upper row: PT-HMC; Lower row: HMC.

4   Comparison of HMC and PT-HMC

To validate the proposed PT-HMC algorithm, we generated synthetic time series data by sampling the first component of the simulated "circadian oscillator"-model's trajectory at every 10 time units in a 60 time units long observation window. Further, we add a random real number sampled from a Gaussian distribution with zero mean and variance $\sigma^2 = 121$ for each of those noise-free observations in an i.i.d. fashion.

We first evaluated the estimated posterior distributions over the model parameter θ_1 and θ_2 (that is, $\pi(\theta_1, \theta_2)$) by visual inspection of the corresponding three-dimensional plots of those distributions. This qualitative analysis has already showed that the $\pi(\theta_1, \theta_2)$s estimated by PT-HMC are much more accurate than those estimated by HMC in terms of (1) whether one or more peaks exhibited in the ground-truth $\pi(\theta_1, \theta_2)$ is missing; and (2) whether the proportions of heights of those observed peaks are similar to those for the ground-truth $\pi(\theta_1, \theta_2)$. However, such comparison is very subjective and fails to provide adequate details.

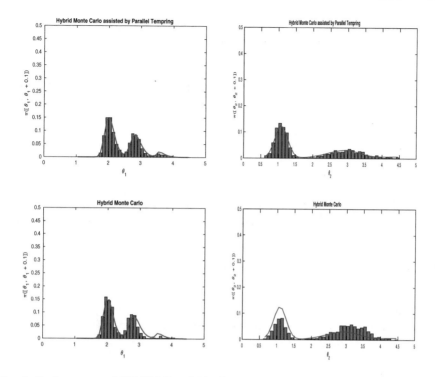

Fig. 4. Performance of PT-HMC and HMC when HMC returns a wrong proportion among between peaks

We now compare the marginal posterior distributions over θ_1 or θ_2 between the ground-truth, the MCMC estimates obtained by HMC and those estimates obtained by PT-HMC. In both Figs. 3 and 4, we compare three different estimates of $\pi(\theta_1)$ in the left columns and those of $\pi(\theta_2)$ in the right columns. In each of the four columns, the curves displayed in red represent their respective ground-truth marginal posterior probability distributions over θ_1 for the left columns and over θ_2 for the right columns. The values of $\pi(\theta)$ represent the probability of a parameter value being in the interval $[\theta, \theta + 0.01]$. The bars displayed in blue represent those probabilities estimated either by using PT-HMC samples (upper panels) or by using HMC samples (lower panels). The upper panels in Figs. 3 and 4 showed that the marginal posterior distributions estimated by using PT-HMC samples did capture all three modes exhibited in the ground-truth $\pi(\theta_1)$ as well as both modes exhibited in the ground-truth $\pi(\theta_2)$. However, such level of accuracy is not observed for the marginal posterior distributions estimated by using HMC samples (see in the lower panels in Figs. 3 and 4). Figure 3 illustrated a case where the HMC sampler got stuck to a local peak while completely missing the global peak around $(2, 1)$. Figure 4 illustrated another case where the HMC sampler did manage to explore the major modes but the estimated probability masses of these modes (that is, the areas under the corresponding peaks) clearly deviate from the ground-truth ones.

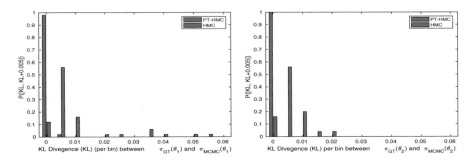

Fig. 5. Kullback Leiber divergence of distributions of θ_1 and θ_2 generated by HMC and PT-HMC between ground truth posterior

To quantitatively assess the accuracy of those estimated posterior distributions, we compute the Kullback-Leiber (KL) divergence between the ground-truth and the estimated probability distributions. The KL value between two probability distributions P and Q is defined as follows

$$D_{KL}(P\|Q) = \sum_{i \in I} P(i) \log \frac{P(i)}{Q(i)}$$

where I represents a collection of all possible random events. In the case considered here, such event means the occurrence of a parameter value in an interval given as $[\theta, \theta + \delta\theta]$ and set I represents all bins we used to approximately compute the marginal posterior distributions $\pi(\theta_1)$ and $\pi(\theta_2)$. Note that for both θ_1 and θ_2, we set $\delta\theta$ to 0.1. The number of bins used for computing $\pi(\theta_1)$ and $\pi(\theta_2)$ is 35 for θ_1 and 39 for θ_2. This can be straightforwardly extended to the joint posterior distribution $\pi(\theta_1, \theta_2)$, and the total number of (two-dimensional) bins is 35×39.

To make possible conclusion with statistical robustness, we randomly generated 50 independent initialization of (θ_1, θ_2). For each of these 50 initial parameter pairs, we run both HMC and PT-HMC to generate a set of HMC samples and another set of PT-HMC samples. For each of these two sets, we computed an estimated marginal distribution over θ_1, an estimated marginal distribution over θ_2, and an estimated joint distribution over (θ_1, θ_2). For each of these three distributions, we compute a KL divergence value (KL) that measures the discrepancy between that distribution and its corresponding ground truth. Totally, we computed 300 KL values. First, we quantitatively compare the performance of HMC and PT-HMC in terms of inferring the marginal distribution over θ_1. For this purpose, we have a set of 50 KL-values for HMC and a set of 50 KL-values for PT-HMC. For each set, we computed a probability distribution

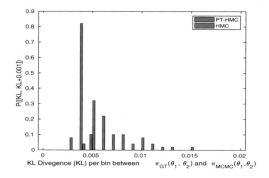

Fig. 6. Kullback Leiber divergence of distributions of θ_1 and θ_2 generated by HMC and PT-HMC between ground truth posterior

over KL values for a KL value being in the interval $[KL, KL + \delta KL]$. The resulting two probability distributions over KL are displayed in the left panel of Fig. 5. We found that the KL-values for HMC is significantly larger than those for PT-HMC, which indicates that PT-HMC outperforms HMC in terms of inferring accurate marginal posterior distributions over θ_1. Further, we found that the KL values for PT-HMC is narrowly peaked at a value close to zero whereas those for HMC are distributed over a considerable range of KL values. This indicates that the HMC performance strongly depends on the initialization when compared to PT-HMC. We obtained similar observations for the marginal posterior distribution over θ_2 (see the right panel of Fig. 5) as well as for the joint posterior distribution over θ_1 and θ_2 (see Fig. 6).

To sum up, the parallel tempered Hamiltonian Monte Carlo algorithms works well to draw samples from a distribution exhibiting multi-ridges structure.

5   Discussion

The aim of this work is to develop a MCMC algorithm that is capable of reliably sampling from posterior distribution over model parameter vector of an ODE system. In the following, we briefly discuss the state-of-the-arts in this research sub-field.

As highlighted in Sect. 1 multi-mode problem is a big obstacle for this inferential task. Parallel tempering is one of the most popular strategy for dealing with this problem. But it is worth of noting that other strategies are also available for this purpose, for example, Wormhole HMC [11] or Thermostat-assisted Continuous Tempering [12]. These strategies haven't yet been tested on ODE systems.

Another big challenge in Bayesian inference in ODE systems is so-called scalability. If the state and/or parameter dimensional is high, forward simulation of an ODE system will become computationally very expensive, left alone the simulation of its sensitivity equations for gradient computation. A Gaussian process approximation strategy has been adopted to meet this challenge [3,4]. It hasn't been tested whether such acceleration can be incorporated into HMC.

Efficient and robust MCMC should play an important role in machine learning. Traditionally, machine learning algorithms operate on numerical vectors. However, in the frontiers of machine learning research, those algorithms are being extended so that they can operate in the space of probability distributions over models [13]. It is because in some application domains, thi is the best way to represent sparse and/or noisy data. However, such probability distributions first have to be accurately inferred from the data using MCMC algorithms.

References

1. Anderson, B.: Optimal Filtering. Dover Publications Inc., United States (2005)
2. Andrieu, C.: An introduction to mcmc for machine learning. Mach. Learn. **50**, 5–43 (2003)
3. Barber, D.: Gaussian processes for Bayesian estimation in ordinary differential equations. In: International Conference on Machine Learning, pp. 1485–1493 (2014)
4. Calderhead, B.: Accelerating bayesian inference over nonlinear differential equations with gaussian processes. In: Advances in Neural Information Processing Systems, pp. 217–224 (2009)
5. Duane, S.: Hybrid monte carlo. Phys. Lett. B **195**, 216–222 (1987)
6. Goodwin, B.: Oscillatory behavior in enzymatic control processes. Adv. Enzyme Regul. **3**, 425–438 (1965)
7. Honerkamp, J.: Stochastic Dynamical Systems. VCH Publishers Inc., New York (1994)
8. Wilkinson, D.J.: Stochastic Modelling for System Biology. Chapman & Hall/CRC Press, Boca Raton (2006)
9. Kalnay, E.: Atmospheric Modeling, Data Assimilation and Predictability. Cambridge University Press, Cambridge University Press (2003)
10. Kuzmanovska, I.: Markov chain monte carlo methods in biological mechanistic models (2012). https://www.research-collection.ethz.ch/handle/20.500.11850/153590
11. Lan, S.: Wormhole hamiltonian monte carlo. In: Proceedings of the Twenty-Eighth AAAI Conference on Artificial Intelligence, pp. 217–224 (2015)
12. Luo, R.: Thermostat-assisted continuously-tempered hamiltonian monte carlo for bayesian learning. In: Proceedings of the 32nd International Conference on Neural Information Processing Systems, pp. 10696–10705 (2018)
13. Shen, Y.: A classification framework for partially observed dynamical systems. Phys. Rev. E **95**, 043303 (2017)

A Hybrid Regression Model for Mixed Numerical and Categorical Data

Nouf Alghanmi[✉] and Xiao-Jun Zeng[✉]

School of Computer Science, The University of Manchester,
Manchester M13 9PL, UK
nouf.alghanmi@postgrad.manchester.ac.uk, x.zeng@manchester.ac.uk

Abstract. It is noticeable in different heterogeneity types that complexity is inherent in heterogeneous data, and regression analysis methods are well defined and exhibit high-accuracy performance with numeric data. However, real-world problems contain non-numerical variables. There are two main approaches to handling mixed-type data sets in regression analyses. The first approach is unifying data types for all the variables (such as continuous numerical data) and then applying the regression analysis. However, this approach degrades the data quality, as some original data types are converted to other types in the learning stage. The second approach is to apply some similarity measurements, which can be highly complex in some situations. To overcome these limitations, we propose a tree-based regression model to effectively handle the mixed-type data sets without using a dummy code or a similarity measurement.

Keywords: Decision tree · Regression · Mixed data · Hybrid model

1 Introduction

One of the most important techniques in data mining and machine learning methods is a supervised learning regression analysis [5,9,11]. The objective of regression models is to explore the relationship between a continues outcome and one or more independent variables by defining a functional relationship $y = f(x) + e$ from the training samples. In many cases, the dependent variable in the regression analysis can only be one numerical interval variable while there are more than one independent variables of nominal/ordinal, i.e., categorical data or numerical interval data. It is easy for regression models to model the relation between independent and dependent variables when both their types are numerical. In contrast, when the training sample contains qualitative attributes, it is not possible to perform arithmetic operations on categorical variables.

To overcome this problem, one may apply the regression model to only the numerical data and may ignore the categorical data, but this may cause information loss and a lower-accuracy prediction model. One possible solution is to unify data types by converting the categorical variables to numerical data through the

© Springer Nature Switzerland AG 2020
Z. Ju et al. (Eds.): UKCI 2019, AISC 1043, pp. 369–376, 2020.
https://doi.org/10.1007/978-3-030-29933-0_31

use of various coding systems such as effect coding, dummy coding and contrast coding to handle the existence of categorical predictors [3, 10]. Another possible solution is to define similarity or dissimilarity measures between categorical and numerical variables using one of the distance-based regression algorithms. First, a distance configuration matrix is computed from the original data after defining a similarity or dissimilarity function on both categorical and numerical variables. Then, any distance-based regression algorithm can be applied to the converted matrix [2, 4, 7].

However, there are some limitations to these approaches. Coding systems may lead to an extreme number of predictors when the number of nominal variables or the number of categories among the nominal variables is significant. In order to reduce the number of predictors, the coding systems include further steps [6, 8]. Additionally, defining a suitable similarity or dissimilarity function between categorical and numerical variables constitutes a complex challenge [1].

However, recent work by Kim et al. [1] addressed this issue and proposed a hybrid decision-tree algorithm in order to process mixed categorical and numerical data sets in a regression analysis. Their algorithm measured the effectiveness of each data type in the final estimation value. They measured the effectiveness of the categorical feature, used these nominal attributes to construct a decision tree, and then predicted a value $\hat{y}$. This value $\hat{y}$ was then subtracted from the real value y to become y_{tree}. The numerical data were used to predict a value y_{reg} using one of the many regression models. The final predicted value was the sum of the two values y_{tree} and y_{reg}. To evaluate their model and the effectiveness of categorical variables on the final prediction value, they built two other models: $M1$ and $M2$. $M1$ used the typical method of dealing with mixed-type data sets, converting categorical variables into a continuous variable by applying dummy coding. For $M2$, they discarded the categorical variables and used numerical variables to build the regression models. Their model has been compared to $M5$ and evtree. The results showed that the model was much faster than these two (although in some cases, the model achieved a slightly worse performance).

The main limitation of the previous hybrid model was that it did not take into consideration the interaction between categorical and numerical variables. To overcome this limitation, we propose a tree-based regression model constructed using a decision-tree algorithm. In the proposed model, mean square error is the splitting criteria, the categorical variables are the tree node and their corresponding numerical features are used to fit the regression model and compute mean square error. The candidate internal tree node will be the categorical feature that has the minimum MSE value. The final result of the proposed model will be various regression models in the terminal nodes. The rest of this paper is organized as follows: Sect. 2 details the proposed algorithm, Sect. 3 reports the results and Sect. 4 discusses the conclusions and future work.

2   The Proposed Algorithm

The main idea behind the creation of this model is to build a hybrid regression model without using dummy coding and similarity measurements. A tree is constructed using the categorical features in as the nodes and different regression models as the terminal nodes and final predication models; So, rather than predicting a single value, it is predicting a regression model. To construct the tree, we followed these steps:

1. The training error of applying a linear regression model to the numerical features is computed.
2. For each categorical feature, a linear model is fitted to its sub-numerical data, and a training error is computed.
3. The feature that has the smallest training error, which significantly improves the initial error, is chosen for further splitting. A detailed description of the model as follows:

Let S be a data set consisting of K observations $S = (X_i, Y_i), i = 1, 2, \ldots, K$. Each observation i includes a scalar response y_i and a column vector x_i of p predictors x_{ij} for $j = 1, 2, \ldots, p$. As S is a mixed data set, the predictors have M categorical features and N numerical features such that $p = M + N$. We define the categorical features as follows:

$$X_i^{cm} = \{X_i^{c1}, X_i^{c2}, \ldots, X_i^{cm}\} \tag{1}$$

And this is how we define the numerical features:

$$X_i^{Nn} = \{X_i^{N1}, X_i^{N2}, \ldots, X_i^{Nn}\} \tag{2}$$

This definition will split the data set into $S^N = \{(X_i^{Nn}, Y_i^N)|i = 1, 2, \ldots, K\}$, and $S^C = \{(X_i^{Cm}, Y_i^C)|i = 1, 2, \ldots, K\}$.

The proposed algorithm works as follows:

– **Zero step**: An initial mean squared error MSE_0 is computed by applying a linear regression model on the data sets S by using the numerical features only as in Eq. 3.

$$MSE_0 = \frac{1}{k} \sum_{i=1}^{K} (Y_i^N - \hat{Y}_i)^2 \tag{3}$$

– **First step**: The purpose of this step is to split the tree by selecting the categorical feature that minimizes training error. In order to do that, for each nominal feature C_m with a predefined finite set of av labels such that $C_m = a1, a2, \ldots, av$ which represents different feature values. For each label av the corresponding sub data with numerical values is extracted as follow:

$$S_j^N = \{(X_j^N, Y_j^N)|X_j^{Cm} = X_i^{av}, j = 1, 2, \ldots, v\}. \tag{4}$$

This subset is then fitted to a linear regression model and a training $MSE(S_j^N)$ error is computed. After computing the $MSE(S_j^N)$ training error for all C_m labels, the categorical C_m training error is calculated by averaging the $MSE(S_j^N)$ value as in Eq. 5.

$$MSE(C_m) = \frac{1}{j} \sum_{j=1}^{v} MSE(S_j^N) \tag{5}$$

– **Second step**: After repeating the previous step for each categorical attribute in C_m set, the feature that has a minimum training error and is less than the initial MSE_0 is chosen as the root node for the tree. Its mean squared error values become the node training error as noted by MSE_p.

The first and the second steps are then repeated to construct the tree by replacing the initial MSE_0 with the parent node MSE_p as follows:

$$C_m = \{C_m | (\mathrm{argmin}(MSE(C_m) | m = 1, 2, \ldots, M) \wedge (MSE(C_m) < MSE_p)\} \tag{6}$$

The learning process continues until one of the following stopping criteria is satisfied: first, all the categorical features C_m have been examined in a single path. second, there is not a significant improvement in the MSE value; or third, the number of samples in a subset is not significant for the application of a regression model. Then, the node is tagged as a terminal node, and a predictive regression model is computed from its numerical subset as a final prediction model of the tree. The key point of our algorithm is to use categorical features as the tree nodes and the numerical attributes for fitting the linear regression. For each of the categorical features, the corresponding numerical variables are used to apply the regression model. The final prediction of the tree model is multiple regression models at the leaf nodes, as shown in Fig. 1.

3   Experiments

3.1   Data Description and Preparation

The main purpose of this algorithm to effectively process both categorical and numerical mixed-type data sets in real-world regression problems. The algorithm has been compared to one introduced by Kim et al. in [1]. Therefore, for a fair comparison, the same data sets were used. Four mixed data sets were obtained from a competitive online platform Kaggle [13] to evaluate the model.

3.2   Experiment Procedure

The proposed algorithm connected the tree with the ordinary linear regression algorithm. The numerical variables were standardized with a standard deviation of 1, a mean of 0. 5-fold cross-validation was used and ten iterations of the validation process were executed. For a fair comparison with work introduced by Kim et al. in [1], the same evaluation processes and metrics were followed.

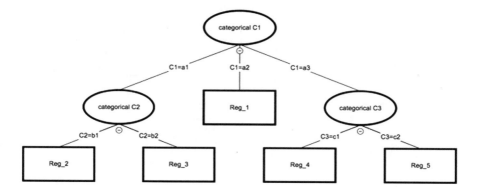

Fig. 1. Proposed tree model

To evaluate the model and the effectiveness of categorical variables on the final prediction value, two different aspects of the model were compared with other models. First, we compared standard methods of dealing with mixed-type data sets in regression problems. One model used for comparison was $M1$, which converts categorical variables into discrete variables by applying dummy coding. The other model used for comparison was $M2$, which discarded the categorical variables and used numerical variables to build $M2$. Figure 2 illustrated the experiment procedure for a single iteration. Second, we compared the proposed method to what was presented in [1] as well as their result of $M5$ and the evolutionary tree (evtree) [14,15].

The experiment of the proposed model was performed on a computer with Intel Core i5(2.7 GHz) and 8 GB of memory.

3.3 Experiment Results

The results are described below as the $MSE's$ average value across 10 iterations and the standard deviation value of the MSE values. The comparison is made in two ways. The first comparison involves measuring the effectiveness of the original categorical features in the regression analysis, so our model is compared to $M1$ and $M2$. Descriptions of their algorithms appear in the previous subsection. Second, the model is compared to other tree models: evtree, $M5$ and the hybrid model. The result of the hybrid model showed it was best to combine the model with five regression models: linear, lasso, ridge, support vector regression, and k-nearest neighbor. The experiment was performed on four mixed-type data sets. The overall result showed that, despite the simplicity of the proposed model, it outperformed complex models such as the hybrid model in some cases and had comparable performance in other cases. The detailed results of the experiment are shown in the following data sets:

- **House:** This data set had 1,460 samples with 35 numerical variables and 44 categorical features. The model was used to predict the selling price of a

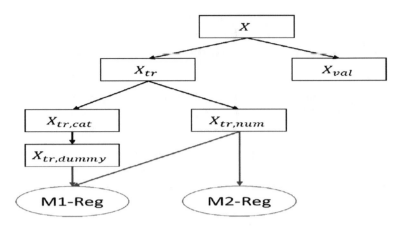

Fig. 2. Experiment procedure [1]

house. In analysing this set, the proposed model outperformed $M1$ and $M2$ with a validation error of $(14.9 \pm 7.73(\times 10^8))$ for the proposed model and $(4.55 \pm 8.70(\times 10^9))$ and $(14.97 \pm 7.73(\times 10^8))$ for $M1$ and $M2$ respectively. The performance was comparable to the others models in terms of the validation error, with $M5$ $(1.30 \pm 0.22(\times 10^9))$, evtree $(1.22 \pm 0.80(\times 10^9))$ and the hybrid model $1.08 \pm 0.00(\times 10^9)$.

– **Horse:** This data set consisted of 73,596 observations with 6 numerical features and 10 categorical variables. The model was used to predict a horse's finish time in a race. The proposed model significantly outperformed $M1$ and $M2$ with (1.46 ± 0.19) for the proposed model, $(1.72 \pm 11.95(\times 10^{18}))$)for $M1$ and (1.88 ± 0.19) for $M2$. In comparison with the other three models, the result was very close, and evtree outperformed all the other models. $M5$ scored (1.15 ± 0.00), while evtree and the hybrid model received scores of (1.12 ± 0.00) and (1.13 ± 0.00) respectively.

– **Nashville:** This data set had 24,162 samples with 10 numerical features and 8 categorical variables. The model was used to predict a house's sale price based on different features. The proposed model also showed outstanding performance compared to $M1$ and $M2$ with the following results: $(27.17 \pm 9.60(\times 10^9))$ for $M1$, $(29.40 \pm 9.53(\times 10^9))$ for $M2$ and $(22.4 \pm 0.13(\times 10^9))$ for the proposed model. The proposed model showed the best result when compared with $M5$, evtree and the hybrid model. The proposed model had a validation error of $(22.4 \pm 0.13(\times 10^9))$, $M5$ had $(2.75 \pm 0.01(\times 10^{10}))$, evtree had $(3.50 \pm 0.05(\times 10^{10}))$ and the hybrid model had $(2.77 \pm 0.00(\times 10^{10}))$.

– **Autos:** This data set consisted of 65,689 samples with 4 numerical features and 6 categorical features. In this data set, the objective was to predict the selling price of an automobile. The result showed there was no difference between applying a regression model by using numerical features with dummy coding for categorical ones $(M1)$ or by just applying a regression model with numerical features $(M2)$. They both had the same MSE values for this data set $(1.18 \pm 1.43(\times 10^{11}))$. On the other hand, when applying the proposed model to the data set, the result significantly improved to (0.75 ± 0.99) and was best when compared with $M5$, evtree and the hybrid model, which had the following results: $(3.51 \pm 0.01(\times 10^6))$ for M5, $(3.63 \pm 0.11(\times 10^6))$ for evtree, $(4.84 \pm 0.01(\times 10^6))$ for the hybrid model, and (0.75 ± 0.99) for the proposed model.

4   Conclusion

This study proposed a hybrid regression algorithm that can effectively deal with both numerical and categorical variables. The primary concept of this algorithm is to utilize a decision tree from the categorical features, fit the regression model to their numerical features and predict different regression models at terminal nodes. The splitting criteria depended on the MSE value computed from numerical data and related to each specific categorical feature.

The proposed model constructed a decision tree with the inclusion of a linear regression model. This model was then compared to other ordinary linear regression models in two different ways based on the inclusion of categorical features. The experiments were performed on seven mixed numerical and categorical data sets. In most cases, using the proposed model achieved a significant improvement in MSE values over $M1$ and $M2$. Compared to other decision-tree models such as $M5$, evtree and the hybrid model, the proposed decision tree achieved better or at least comparable performance.

In sum, the proposed model has several advantages over ordinary linear regression models: first, the proposed model analyses mixed-type data sets without using a coding system or similarity measurement. Second, the model is easy to implement and interpret. Finally, it predicts a regression model rather than a single value. Though the initial experiment with the proposed model is based on a limited number of data sets, the result is promising. Future work should consider the inclusion of more data sets and should also focus on building a deeper regression model that can effectively handle data heterogeneity.

References

1. Kim, K., Hong, J.S.: A hybrid decision tree algorithm for mixed numeric and categorical data in regression analysis. Pattern Recogn. Lett. **98**, 39–45 (2017). https://doi.org/10.1016/j.patrec.2017.08.011
2. Cuadras, C.M., Arenas, C.: A distance-based regression model for prediction with mixed data. Commun. Stat. - Theor. Meth. **19**, 2261–2279 (1990). https://doi.org/10.1080/03610929008830319
3. Hardy, M.A.: Regression with Dummy Variables, vol. 93. Sage Publications, Newbury Park (1993)
4. Cuadras, C.M., Areans, C., Fortiana, J.: Some computational aspects of a distance-based model for prediction. Commun. Stat. - Simul. Comput. **25**, 593–609 (1996). https://doi.org/10.1080/03610919608813332

5. Marcoulides, G.A.: Discovering knowledge in data: an introduction to data mining. J. Am. Stat. Assoc. **100**, 1465–1465 (2005)
6. Yuan, M., Lin, Y.: Model selection and estimation in regression with grouped variables. J. Royal Stat. Soc.: Ser. B (Statistical Methodology) **68**, 49–67 (2006). https://doi.org/10.1111/j.1467-9868.2005.00532.x
7. Boj Del Val, E., Claramunt Bielsa, M.M., Fortiana, J.: Selection of predictors in distance-based regression. Commun. Stat. - Simul. Comput. **36**, 87–98 (2007). https://doi.org/10.1080/03610910601096312
8. Meier, L., Van De Geer, S., Bühlmann, P.: The group lasso for logistic regression. J. Royal Stat. Soc.: Ser. B (Statistical Methodology) **70**, 53–71 (2008). https://doi.org/10.1111/j.1467-9868.2007.00627.x
9. Han, J., Kamber, M., Pei, J.: Data Mining: Moncepts and Techniques. Morgan Kaufmann, San Francisco (2012)
10. Cohen, J., Cohen, P., West, S.G., Aiken, L.S.: Applied Multiple Regression/Correlation Analysis for the Behavioral Sciences. Routledge, New York (2003)
11. Witten, I.H., Frank, E., Hall, M.A.: Data Mining: Practical Machine Learning Tools and Techniques. Morgan Kaufmann, San Francisco (2011). https://doi.org/10.1016/C2009-0-19715-5
12. UCI Machine Learning repository. https://archive.ics.uci.edu/ml/index.php
13. Kaggle. https://www.kaggle.com
14. Quinlan, J.R.: Learning with continuous classes. In: 5th Australian Joint Conference on Artificial Intelligence, vol. 92, pp. 343-348. Singapore (1992)
15. Grubinger, T., Zeileis, A., Pfeiffer, K.-P.: Evtree : evolutionary learning of globally optimal classification and regression trees in R. J. Stat. Softw. **61**, 1–29 (2014). https://doi.org/10.18637/jss.v061.i01

Integration of Interpolation and Inference with Multi-antecedent Rules

Nitin Naik[1]([⊠]) and Qiang Shen[2]

[1] Defence School of Communications and Information Systems, Ministry of Defence,
Blandford Forum, UK
nitin.naik100@mod.gov.uk
[2] Department of Computer Science, Aberystwyth University, Aberystwyth, UK
qqs@aber.ac.uk

Abstract. The efficacious fuzzy rule based systems perform their tasks with either a dense rule base or a sparse rule base. The nature of the rule base decides on whether compositional rule of inference (CRI) or fuzzy rule interpolation (FRI) should be applied. Given a dense rule base where at least one rule exists for every observation, CRI can be effectively and sufficiently employed. For a sparse rule base where rules do not cover all possible observations, FRI is required. Nonetheless, certain observations may be matched partly or completely with any of the existing rules in the sparse rule-base. Such observations can be directly dealt with using CRI and the conclusion can be inferred via firing the matched rule, thereby avoiding extra overheads of interpolation. If no such matching can be found then correct rules should be selected to ensure the accuracy while performing FRI. This paper proposes a generalised approach for the integration of FRI and CRI. It utilises the notion of alpha-cut overlapping to determine the matching degree between rule antecedents and a given observation in order to determine if CRI is to be applied. In the event of no matching rules, the nearest rules will be chosen to derive conclusion using FRI based on the best suitable distance metric among possible alternatives such as the Centre of Gravity, Hausdorff Distance and Earth Mover's Distance. Comparative results are presented to demonstrate the effectiveness of this integrated approach.

Keywords: Computational rule of inference · Rule interpolation · Rule extrapolation · Integration of interpolation and inference · Multi-antecedent rules

1 Introduction

Fuzzy systems infer the results based on the use of either dense or sparse rule bases. For a sparse rule base where rules do not cover possible observations, fuzzy rule interpolation (FRI) is the most popular way of approximating a conclusion [2,4]. Nonetheless, interpolation generally incurs more overheads for generating approximate conclusions [8,10]. Despite the sparsity of a rule base, on many

© Crown 2020
Z. Ju et al. (Eds.): UKCI 2019, AISC 1043, pp. 377–391, 2020.
https://doi.org/10.1007/978-3-030-29933-0_32

occasions, observations may still match with a certain existing rule partly or completely, which may avoid the need of conducting rule interpolation [9,11]. This can be ensured using a suitable pre-interpolation inference technique such as compositional rule of inference (CRI) [22]. If this pre-interpolation inference technique could not find any matching rule then interpolation would be the next operation to obtain the result [12]. This leads the way for finding a simple and fast mechanism to determine when to apply CRI or FRI given a sparse rule base [18,19].

Fuzzy rule interpolation and extrapolation requires only few closest rules to infer results [5]. Therefore, the selection of appropriate closest rules determines the correctness of results generated by interpolation or extrapolation. Consequently, if the selected rules are not realistically the closest ones then the results may be inaccurate irrespective of the FRI approach employed, even though it may be a generally powerful technique. This reveals the fact that the selection of what distance metric to use is critical in finding the correct closest rules for the given observation [6]. Many rule interpolation or extrapolation methods use the most popular distance metric based on the Centre of Gravity (COG) of the membership functions concerned [7]. Unfortunately, the COG values may be the same for two completely different fuzzy sets, which can be seen in Fig. 2. This leads to present further investigation for effective distance metrics in an effort to improve the accuracy of FRI approach, especially those transformation based techniques [3].

The potential efficiency gains of running CRI prior to interpolation and the effective use of distance metrics in FRI have both motivated the development of an integrated solution for inference with fuzzy rule-based systems [13,15–17]. This paper presents an integration of interpolation and inference to obtain the best of both for systems with a sparse fuzzy rule base. This integrated approach determines the possibility of applying CRI based on exact or partial matching rule(s) in the sparse rule base, whereby minimising the interpolation overheads [20]. If no matching rule is found in the rule base then certain efficacious distance metric is applied to obtain the best closest rules. Here, alongside with COG, two alternatives, namely the Hausdorff Distance (HD) and the Earth Mover's Distance (EMD) are introduced, which are tested and compared against the use of COG to determine the more correct nearest rules for interpolation/extrapolation. The HD calculates the proximity of two fuzzy sets and provides the scalar score of the similarity between them, while the EMD calculates the similarity between the two multi-dimensional distributions over a region. As a result, HD and EMD distances can find the closest rules for a given observation more precisely than the COG based distance metric. This helps increase the overall accuracy of the integrated approach.

This paper is organised as follows. Section 2 presents an overview of the proposed method for integrating inference and interpolation. It works by exploiting the α-cut threshold in an effort to decide on the suitability of whether to run CRI or FRI, and on the applicability of which preferred distance metric to employ for selecting the closest rules to perform interpolation or extrapolation. Section 3

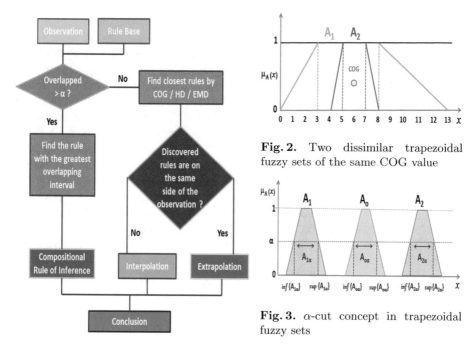

Fig. 2. Two dissimilar trapezoidal fuzzy sets of the same COG value

Fig. 3. α-cut concept in trapezoidal fuzzy sets

Fig. 1. Integrated system with both interpolation and inference

elucidates the implementation procedure of the integrated system and its components. Section 4 reports on the experimental results with detailed comparison among the use of COG, HD and EMD. Section 5 summarises the proposed approach and discusses relevant further research.

2 Integrating Interpolation and Inference

Rule inference (CRI) and rule interpolation (FRI) have been a vital part of developing fuzzy rule-based systems. The use of any particular method is dependent on the type of the rule base employed in the system. CRI is useful for dense rule base, where an observation matches with any existing rule partly or completely [21]. However, in real-life applications, the design of the dense rule base is expensive or unachievable [14]. In such situations, sparse rule base is the default options and thus FRI. FRI only generates approximate conclusions, which may not be necessarily equally accurate as the CRI result which may be achieved via partial or complete rule matching. The integration of the two methods can provide the best of both whilst compensating the limitations of each other. A particular implementation of this idea is proposed here.

Figure 1 outlines the operational procedure of the proposed integrated system. For any given observation, first it checks the applicability of the CRI by finding a matching rule in the existing rule base based on the use of a pre-determined α-cut threshold. If one or more matching rules are found to be above the pre-determined α-cut threshold, it selects the most overlapped rule amongst all the matched ones and infer the conclusion using CRI. However, if it fails to find any matching rule above the pre-determined α-cut threshold, the interpolation (or extrapolation) becomes the natural choice to infer the conclusion [7]. For improving the results of such interpolation, it utilises one of the possible alternative distance metrics, taken from the set of COG, HD and EMD, to identify the closest rules to the observation. The choice of which distance metric to use depends on its suitability for a given problem domain. From the above, depending on the locations of the closest rules, the system performs the relevant operation of interpolation or extrapolation. In this work, owing to the popularity and availability, the scale and move transformation interpolation (T-FRI) method [3] is applied to compute the interpolated conclusions. Nevertheless, if desired, any other FRI approach may be used in place of the T-FRI method.

3   Implementation of Proposed Approach

The integrated inference and interpolation system as proposed above has two distinct sub-systems: an α-cut matching component for CRI and a distance metric (COG, HD or EMD) based component for selecting closest rules for interpolation/extrapolation. For this implementation, by following the conventional T-FRI implementations, trapezoidal fuzzy sets are considered. Also, each rule is entitled to involve multiple antecedent variables.

3.1 α-Cut Matching for Inference

An α-cut level is used to acquire a crisp set from a fuzzy set based on the pre-determined threshold of α-cut. Suppose that in the universe of discourse X, A is a fuzzy set with the membership function $\mu_A(x)$. Then, for $\alpha \in [0, 1]$, the α-cut of A is [4]:

$$A_\alpha = \{x \in X | \mu_A(x) \geq \alpha\} \tag{1}$$

This notion of α-cuts for two rule antecedents that take on trapezoidal fuzzy sets A_1 and A_2 and that for an observed trapezoidal fuzzy set A_o are together shown in Fig. 3, where inf and sup stand for $infimum$ and $supremum$ operators respectively, and α is the given α-cut threshold. In computing α-cut matching, an observation $A_{o,j} = (a_{o,0}, a_{o,1}, a_{o,2}, a_{o,3})$ over a certain variable X_j is compared with the possible antecedent values $A_{i,j} = (a_{i,j,0}, a_{i,j,1}, a_{i,j,2}, a_{i,j,3})$, based on the given α level, where $i = 1, 2, ..., n$ (indicating that there are n rules which involve x_j) and $j = 1, 2, ..., N$, (indicating that there are N antecedent variables in the problem). Suppose that $[inf\{A_{o,j}\}, sup\{A_{o,j}\}]$ and $[inf\{A_{i,j}\}, sup\{A_{i,j}\}]$ denote the α-cut of $A_{o,j}$ and that of $A_{i,j}$ respectively. Then the check for α-cut matching is simply implemented by assessing whether either of the following holds:

$$inf\{A_{o,j}\} \leq inf\{A_{i,j}\} \leq sup\{A_{o,j}\} \text{ or } inf\{A_{o,j}\} \leq sup\{A_{i,j}\} \leq sup\{A_{o,j}\}.$$

Finding the matching rules requires computation only above the pre-determined threshold of α-cut. As such, it saves a significant amount of computation for otherwise firing all those rules below the α-cut threshold and improves the efficiency of the system. In particular, if there is only one existing rule matching with the observation with all antecedents above the threshold of α-cut, thereupon, the conclusion is derived from this matched rule using CRI. In case of more than one matched rule, the rule with the greatest degree of matching is selected to derive the conclusion using CRI.

In implementation, the total degree of matching between one rule and an observation is calculated as follows: it sums up all areas constructed by the overlap between the two α-cut fuzzy sets involved in each pair of the fuzzy sets that describe a certain antecedent variable and its counterpart in the observation. The selected rule should have the largest sum of the overlapping areas amongst all the matched rules. Consequently, interpolation is avoided for sparse rule base in such α-cut matching situations. However, if no rule is matched with at least the α-cut level, then fuzzy rule interpolation becomes necessary.

3.2   Closest Rule Selection for Interpolation

When no rule from the existing rule base is matched with the given observation above the α-cut threshold, FRI is the most preferred choice to infer the conclusion. However, its accuracy is dependent on the selected closest rules to derive the conclusion. This highlights the importance of a suitable distance metric to determine the correct closest rules for FRI. Most FRI techniques utilise the centre of gravity (COG) distance metric, which performs well for certain situations but not in many others. When it measures the distances incorrectly, inaccurate or even incorrect interpolation/extrapolation results. To address this issue, in this paper, the Hausdorff Distance (HD) metric and Earth Mover's Distance (EMD) are implemented with FRI as potential alternatives to the COG-based metric. For completeness, more details of these metrics are given below.

Centre of Gravity (COG). COG is computed regarding an imaginary point in a physical object of matter where, for convenience in certain calculations, the total weight of the object is deemed to be concentrated an average of the masses factored by their distances from a unique reference point. The closeness of two fuzzy sets can be determined based on their unique COG points. In particular, given an observation O: $A_{o,j} = (A_{o,j,1}, A_{o,j,2}, A_{o,j,3}, A_{o,j,4})$ and the antecedent value of the corresponding variable x_j within the ith rule R_i: $A_{i,j} = (A_{i,j,1}, A_{i,j,2}, A_{i,j,3}, A_{i,j,4})$, the COG distance between these two fuzzy sets is defined by:

$$COG(R_i, O) = \sum_{j=1}^{N} \frac{d(COG(A_{o,j}), COG(A_{i,j}))}{range_{x_j}} \qquad (2)$$

where $COG(A_{o,j})$ and $COG(A_{i,j})$ are the COGs of the sets $A_{o,j}$ and $A_{i,j}$ respectively, and $range_{x_j} = \max x_j - \min x_j$ is defined over the domain of the variable x_j.

Hausdorff Distance (HD). HD computes the proximity of two arbitrary subsets (sets of points) of a metric space, returning a scalar score of similarity between the two sets of points [1]. It measures the maximum distance of one set $A_{o,j}$ from the closest point of the other set $A_{i,j}$. For the proposed work herein, given an observation O: $A_{o,j} = (A_{o,j,1}, A_{o,j,2}, A_{o,j,3}, A_{o,j,4})$ and the antecedent value of the corresponding variable x_j within the ith rule R_i: $A_{i,j} = (A_{i,j,1}, A_{i,j,2}, A_{i,j,3}, A_{i,j,4})$, the HD metric is defined by:

$$HD(R_i, O) = \sum_{j=1}^{N} \frac{\max\limits_{l \in \{1,2,3,4\}} \left\{ \min\limits_{k \in \{1,2,3,4\}} \{d_{j,kl}(A_{o,j,l}, A_{i,j,k})\} \right\}}{range_{x_j}} \tag{3}$$

where $range_{x_j} = \max x_j - \min x_j$, and $d_{j,kl}$ is any conventional distance metric between the two points involved.

Earth Mover's Distance (EMD). EMD calculates the similarity between two multi-dimensional distributions over a region. The EMD is the minimum amount of work required to transform one distribution into another distribution. The distance is measured by the minimum amount of computation needed to transform a set $A_{i,j}$ to another $A_{o,j}$. Here, a unit of computation corresponds to the cost required to calculate the ground distance (i.e., the base distance metric employed as follows) [23]. For this work, given an observation O: $A_{o,j} = (A_{o,j,1}, A_{o,j,2}, A_{o,j,3}, A_{o,j,4})$ and the antecedent value of the corresponding variable x_j within the ith rule R_i: $A_{i,j} = (A_{i,j,1}, A_{i,j,2}, A_{i,j,3}, A_{i,j,4})$, the EMD metric is defined by:

$$EMD(R_i, O) = \sum_{j=1}^{N} \frac{\min_{F_j} \frac{\sum_{k=1}^{m} \sum_{l=1}^{n} f_{j,kl} d_{j,kl}}{\sum_{k=1}^{m} \sum_{l=1}^{n} f_{j,kl}}}{range_{x_j}} \tag{4}$$

where $f_{j,kl}$ is the amount of mass transported from $A_{i,j,k}$ to $A_{o,j,l}$ for morphing $A_{i,j}$ into $A_{o,j}$, and $d_{j,kl}$ is the base distance metric that may be implemented by any standard distance measure.

Depending on the problem domain and computational requirements as discussed next, the best suitable metric out of COG, HD and EMD can be selected and the closest rules discovered for performing interpolation or extrapolation. When all the selected neighbouring rules are on one side of the given observation, extrapolation is carried out to infer the conclusion, if not then interpolation is performed. Here, for simplicity, only two closest rules are considered to perform interpolation or extrapolation using the T-FRI mechanism as per the approach reported in [3]. However, the number of the closest rules needed can be increased if it is desirable for a certain problem domain. The algorithms required to implement the integrated system are shown in Figs. 4, 5 and 6.

IntegratedSystem(R_i, O, α)
R_i: $A_{i,j} = (A_{i,j,1}, A_{i,j,2}, A_{i,j,3}, A_{i,j,4})$,
O: $A_{o,j} = (A_{o,j,1}, A_{o,j,2}, A_{o,j,3}, A_{o,j,4})$,
α, α-cut threshold.

1: $R_{overlap} = \alpha$-CutOverlapping(R_i, O, α);
2: **if** $R_{overlap} \neq NULL$ **then**
3:     $B_o = \text{CRI}(R_{overlap}, O)$;
4: **else**
5:     $R_{close1} = \text{COG/HD/EMD\_Closest}(R, O)$;
6:     $R_{close2} = \text{COG/HD/EMD\_Closest}(R - R_{close1}, O)$;
7:     **if** $R_{close1} < O < R_{close2}$ **or** $R_{close2} < O < R_{close1}$ **then**
8:         $B_o = \text{Interpolation}(R_{close1}, O, R_{close2})$;
9:     **else**
10:         $B_o = \text{Extrapolation}(R_{close1}, R_{close2}, O)$;
11:     **end if**
12: **end if**

Fig. 4. Integrated approach to interpolation and inference

α−CutOverlapping(R_i, O, α)
$inf\{A_{o,j}\}$, infimum value of crisp set $A_{o,j}$
$sup\{A_{o\,j}\}$, supremum value of crisp set $A_{o,j}$

1: $maxArea = 0, maxIndex = -1$;
2: **for each** R_i in R **do**
3:     **if** $inf\{A_{o,j}\} \leq inf\{A_{i,j}\} \leq sup\{A_{o,j}\}$ **or** $inf\{A_{o,j}\} \leq sup\{A_{i,j}\} \leq sup\{A_{o,j}\}$ **then**
4:         $overlap$ = overlapping area of $A_{o,j}$ and $A_{i,j}$ above α;
5:         **if** $overlap > maxArea$ **then**
6:             $maxArea = overlap$;
7:             $maxIndex = i$;
8:         **end if**
9:     **end if**
10: **end for**
11: **if** $maxIndex == -1$ **then**
12:     **return** NULL;
13: **else**
14:     **return** $R_{maxIndex}$;
15: **end if**

Fig. 5. α-cut matching approach

3.3 General Node on Choice of Appropriate Distance Metric

The likelihood of successfully selecting the most appropriate distance metric for use in a given application depends on multiple factors. Experimental evaluation

COG/HD/EMD_Closest(R, O)

```
1: closeDist = Max_Value, closeIndex = −1;
2: for each Rᵢ in R do
3:     dist = COG/HD/EMD(Rᵢ, O);
4:     if dist < closeDist then
5:         closeDist = dist;
6:         closenIndex = i;
7:     end if
8: end for
9: return R_closeIndex;
```

Fig. 6. COG/HD/EMD distance metric for finding the closet rules

is expected in general to reach such a decision. However, COG can be employed if computational complexity is the main consideration for the problem domain. The complexity of COG is $O(nm)$ for the two polygons having n and m vertices respectively, whereas the complexity of HD is $O(nm^2)$. The complexity of using EMD is $O(N^3 logN)$ for an N-bin histogram. Both HD and EMD may be particularly suitable for multidimensional environments to achieve higher accuracy as empirically proven (see later). In addition, EMD is often better for matching perceptual similarity where the ground distance is perceptually meaningful, whilst HD can be more effective and utilised for those applications where the presence of noise or occlusion is significant.

4   Illustrative and Experimental Case Studies

This section shows the experimental results for the integrated approach. This is based on the use of a sparse rule base, as shown in Table 1 consisting of eight rules each involving four antecedents. The confidence (i.e. α-cut) level $\alpha = 0.5$ is utilised for α-cut matching to check the suitability of CRI and also, for comparing the performances of COG, HD and EMD in their respective action for identifying the best neighbouring rules to perform either interpolation or extrapolation.

4.1 α-Cut Overlapping Operation for Inference

This case demonstrates how the α-cut overlapping procedure is used to determine the applicability of CRI for a sparse rule base, avoiding the use of FRI and saving computational overheads. Table 4 shows the considered five observations, which overlap with the existing rules of the sparse rule base. The first observation O_1: $A_{o,1} = (12.6, 14.3, 15.6, 16.7)$, $A_{o,2} = (14.6, 16.3, 17.6, 18.7)$, $A_{o,3} = (16.6, 18.3, 19.6, 20.7)$, and $A_{o,4} = (19.6, 11.3, 22.6, 23.7)$, overlaps with the existing rule-5 to rule-8 above the α-cut threshold. Where, rule-5 has the least and rule-7 has the most overlapping area with this observation. Therefore, rule-7 is chosen to infer the conclusion using CRI.

Similarly, the second observation O_2: $A_{o,1} = (3.8, 4.9, 5.9, 7.0)$, $A_{o,2} = (5.8, 6.9, 7.9, 9.0)$, $A_{o,3} = (8.8, 9.9, 10.9, 12.0)$, and $A_{o,4} = (10.8, 11.9, 12.9, 14.0)$, overlaps with the existing rule-1 to rule-4 above the α-cut threshold. Where, rule-1 has the least and rule-4 has the most overlapping area with this observation. Therefore, rule-4 is chosen for firing to infer the conclusion using CRI. The third observation O_3: $A_{o,1} = (11.2, 12.3, 13.2, 13.7)$, $A_{o,2} = (13.2, 14.3, 15.2, 15.7)$, $A_{o,3} = (16.2, 17.3, 18.2, 18.7)$, and $A_{o,4} = (18.2, 19.3, 20.2, 20.7)$, overlaps with the existing rule-5 to rule-8 above the α-cut threshold. Where, rule-7 has the least and rule-5 has the most overlapping area with this observation. Therefore, rule-5 is chosen to infer the conclusion using CRI. For the fourth observation O_4: $A_{o,1} = (2.5, 3.8, 4.7, 7.3)$, $A_{o,2} = (4.5, 5.8, 6.7, 9.3)$, $A_{o,3} = (7.5, 8.8, 9.7, 12.3)$, and $A_{o,4} = (9.5, 10.8, 11.7, 14.3)$, overlaps with the existing rule-1 to rule-4 above the α-cut threshold. Where, rule-3 has the least and rule-1 has the most overlapping area with this observation. Therefore, rule-1 is chosen for firing to derive the conclusion using CRI. The final and fifth observation O_5: $A_{o,1} = (11.6, 13.1, 14.5, 15.4)$, $A_{o,2} = (13.6, 15.1, 16.5, 17.4)$, $A_{o,3} = (16.6, 18.1, 19.5, 20.4)$, and $A_{o,4} = (17.1, 18.6, 20.0, 20.9)$, overlaps with the existing rule-5 to rule-8 above the α-cut threshold. Where, rule-5 has the least and rule-8 has the most overlapping area with this observation. Therefore, rule-8 is chosen to infer the conclusion using CRI.

4.2   Closest Rules Selection Using COG, HD or EMD

This experiment exhibits and compares the performances of using COG, HD or EMD for closest rules selection. Table 2 shows all given observations and their selected closest rules for interpolation or extrapolation, based on these three distance metrics. Here, the results for most of the observations are completely different for the three distance metrics. Surprisingly, for the first observation O_1: $A_{o,1} = (7.1, 8.4, 9.8, 11.6)$, $A_{o,2} = (9.1, 10.4, 11.8, 13.6)$, $A_{o,3} = (12.1, 13.4, 14.8, 17.6)$, and $A_{o,4} = (14.1, 15.4, 16.8, 19.6)$, the closest rule (i.e., rule-7) selected by HD and EMD is the furthest rule found by COG as shown in Table 3. The HD and EMD metrics have determined most of the closest rules, where the shapes of the membership functions of their antecedent variables are quite similar to their counterparts in the given observation. This is helpful in maintaining the interpretability of the integrated system.

In this experiment, following conventional T-FRI approaches and empirical results obtained elsewhere [5], only two nearest rules are chosen by either of COG, HD and EMD distance metrics to perform interpolation/extrapolation. However, more than two nearest rules can also be chosen if preferred. For all the observations, these three distance metrics could not find one similar result. For every observation, these distance metrics selected different nearest rules, which lead to different inference results. Interestingly, the experiment shows a distinctive pattern of selecting the nearest rules by every distance metric. The use of COG has regularly chosen rule-3, rule-4 and rule-5; that of HD has regularly chosen rule-1, rule-2 and rule-7; and that of EMD has regularly chosen rule-7, rule-4 and rule-2. Also, examining the outcomes of using EMD, it can be seen

Table 1. Sparse fuzzy rule base

No	Antecedents $R\{A_{i,1}, A_{i,2}, A_{i,3}, A_{i,4}\}$	Consequent B_i
R1	$(2.4, 3.7, 4.6, 7.1),$ $(4.4, 5.7, 6.6, 9.1),$ $(7.4, 8.7, 9.6, 12.1),$ $(9.4, 10.7, 11.6, 14.1)$	$(5.9, 7.2, 8.1, 10.6)$
R2	$(3.1, 3.7, 4.8, 7.0),$ $(5.1, 5.7, 6.8, 9.0),$ $(8.1, 8.7, 9.8, 12.0),$ $(10.1, 10.7, 11.8, 14.0)$	$(6.6, 7.2, 8.3, 10.5)$
R3	$(4.3, 4.6, 5.8, 6.8),$ $(6.3, 6.6, 7.8, 8.8),$ $(9.3, 9.6, 10.8, 11.8),$ $(11.3, 11.6, 12.8, 13.8)$	$(7.8, 8.1, 9.3, 10.3)$
R4	$(3.5, 4.8, 6.1, 6.9),$ $(5.6, 6.8, 8.1, 8.9),$ $(8.6, 9.8, 11.1, 11.9),$ $(10.6, 11.8, 13.1, 13.9)$	$(7.1, 8.3, 9.6, 10.4)$
R5	$(11.9, 12.5, 13.4, 14.0),$ $(13.9, 14.5, 15.4, 16.0),$ $(16.9, 17.5, 18.4, 19.0),$ $(18.9, 19.5, 20.4, 21.0)$	$(15.4, 16.0, 16.9, 17.5)$
R6	$(11.8, 13.2, 14.1, 14.8),$ $(13.8, 15.2, 16.1, 16.8),$ $(16.8, 18.2, 19.1, 19.8),$ $(18.8, 20.2, 21.1, 21.8)$	$(15.3, 16.7, 17.6, 18.3)$
R7	$(11.5, 14.4, 15.2, 16.0),$ $(13.5, 16.4, 17.2, 18.0),$ $(16.5, 19.4, 20.2, 21.0),$ $(18.5, 21.4, 22.2, 23.0)$	$(15.3, 17.7, 18.9, 19.5)$
R8	$(11.7, 13.1, 14.4, 15.3),$ $(13.7, 15.1, 16.4, 17.3),$ $(16.7, 18.1, 19.4, 20.3),$ $(18.7, 20.1, 21.4, 22.3)$	$(15.2, 16.6, 17.9, 18.8)$

that employing EMD yields a greater sensitivity than utilising the other two, as it changes the nearest rules with any minor change in the value of the observation.

4.3   Selection of Interpolation/Extrapolation Mechanism

There are a range of variations in selecting the type of inference method in response to the use of different distance metrics. Indeed, while one selects interpolation and the others may select extrapolation as illustrated in Table 6. Interestingly, however, despite different nearest rules may be chosen, the selected

Table 2. Two closest rules determined by COG, HD and EMD

No.	Observation $O\{A_{o,1}, A_{o,2}, A_{o,3}, A_{o,4}\}$	Closest rules by COG	Closest rules by HD	Closest rules by EMD
O_1	$(7.1, 8.4, 9.8, 11.6)$, $(9.1, 10.4, 11.8, 13.6)$,	$R5$	$R7$	$R7$
	$(12.1, 13.4, 14.8, 17.6)$, $(14.1, 15.4, 16.8, 19.6)$	$R3$	$R1$	$R4$
O_2	$(7.1, 7.8, 8.6, 9.5)$, $(9.1, 9.8, 10.6, 11.5)$,	$R3$	$R1$	$R4$
	$(12.1, 12.8, 13.6, 14.5)$, $(14.1, 14.8, 15.6, 16.5)$	$R4$	$R2$	$R2$
O_3	$(7.7, 8.6, 9.6, 11.3)$, $(9.7, 10.6, 11.6, 13.3)$,	$R5$	$R7$	$R7$
	$(12.7, 13.6, 14.6, 16.3)$, $(14.7, 15.6, 16.6, 18.3)$	$R3$	$R8$	$R4$
O_4	$(9.7, 10.6, 11.3, 12.0)$, $(11.7, 12.6, 13.3, 14.0)$,	$R6$	$R7$	$R5$
	$(14.7, 15.6, 16.3, 17.0)$, $(16.7, 17.6, 18.3, 19.0)$	$R8$	$R8$	$R6$
O_5	$(8.4, 8.9, 9.8, 10.3)$, $(10.4, 10.9, 11.8, 12.3)$,	$R5$	$R7$	$R2$
	$(13.4, 13.9, 14.8, 15.3)$, $(15.4, 15.9, 16.8, 17.3)$	$R3$	$R1$	$R7$
O_6	$(8.0, 8.5, 9.0, 9.7)$, $(10.0, 10.5, 11.0, 11.7)$,	$R3$	$R1$	$R4$
	$(13.0, 13.5, 14.0, 14.7)$, $(15.0, 15.5, 16.0, 16.7)$	$R4$	$R2$	$R7$
O_7	$(7.5, 8.5, 9.0, 10.3)$, $(9.5, 10.5, 11.0, 12.3)$,	$R3$	$R1$	$R4$
	$(12.5, 13.5, 14.0, 15.3)$, $(14.5, 15.5, 16.0, 17.3)$	$R4$	$R2$	$R3$
O_8	$(8.1, 8.8, 9.6, 10.3)$, $(10.1, 10.8, 11.6, 12.3)$,	$R5$	$R1$	$R2$
	$(13.1, 13.8, 14.6, 15.3)$, $(15.1, 15.8, 16.6, 17.3)$	$R3$	$R2$	$R1$
O_9	$(8.8, 9.3, 10.1, 10.5)$, $(10.8, 11.3, 12.1, 12.5)$,	$R5$	$R7$	$R7$
	$(13.8, 14.3, 15.1, 15.5)$, $(15.8, 16.3, 17.1, 17.5)$	$R6$	$R8$	$R2$
O_{10}	$(8.6, 9.0, 9.5, 10.6)$, $(10.6, 11.0, 11.5, 12.6)$,	$R5$	$R7$	$R7$
	$(13.6, 14.0, 14.5, 15.6)$, $(15.6, 16.0, 16.5, 17.6)$	$R3$	$R8$	$R6$

Table 3. Distance measures between rule and observation using COG, HD and EMD

Observation	Rules	Order of rules by COG	Order of rules by HD	Order of rules by EMD
O	R	$d_{COG}(R_i, O)$	$d_{HD}(R_i, O)$	$d_{EMD}(R_i, O)$
	R1	4.78	4.50	3.51
	R2	4.57	4.60	3.34
$(7.1, 8.4, 9.8, 11.6)$,	R3	3.85	4.80	3.58
$(9.1, 10.4, 11.8, 13.6)$,	R4	3.88	4.69	3.21
$(12.1, 13.4, 14.8, 17.6)$,	R5	**3.72**	4.80	3.40
$(14.1, 15.4, 16.8, 19.6)$	R6	4.25	4.70	3.33
	R7	5.05	**4.40**	**2.89**
	R8	4.40	4.60	3.82

Table 4. Results for α-cut matching for inference

Observation $O\{A_{o,1}, A_{o,2}, A_{o,3}, A_{o,4}\}$	Best rule
$(12.6, 14.3, 15.6, 16.7)$, $(14.6, 16.3, 17.6, 18.7)$, $(16.6, 18.3, 19.6, 20.7)$, $(19.6, 11.3, 22.6, 23.7)$	R7
$(3.8, 4.9, 5.9, 7.0)$, $(5.8, 6.9, 7.9, 9.0)$, $(8.8, 9.9, 10.9, 12.0)$, $(10.8, 11.9, 12.9, 14.0)$	R4
$(11.2, 12.3, 13.2, 13.7)$, $(13.2, 14.3, 15.2, 15.7)$, $(16.2, 17.3, 18.2, 18.7)$, $(18.2, 19.3, 20.2, 20.7)$	R5
$(2.5, 3.8, 4.7, 7.3)$, $(4.5, 5.8, 6.7, 9.3)$, $(7.5, 8.8, 9.7, 12.3)$, $(9.5, 10.8, 11.7, 14.3)$	R1
$(11.6, 13.1, 14.5, 15.4)$, $(13.6, 15.1, 16.5, 17.4)$, $(16.6, 18.1, 19.5, 20.4)$, $(17.1, 18.6, 20.0, 20.9)$	R8

Table 5. Interpolation/extrapolation results

Obs O_i	Result based on COG $Consequent_{COG}(B_{o,i})$	Result based on HD $Consequent_{HD}(B_{o,i})$	Result based on EMD $Consequent_{EMD}(B_{o,i})$
O_1	$(10.66, 12.81, 12.16, 14.16)$	$(10.73, 11.97, 13.31, 15.09)$	$(10.73, 11.99, 13.29, 15.07)$
O_2	$(10.62, 11.78, 11.54, 13.02)$	$(10.60, 11.3, 12.1, 13.0)$	$(10.62, 11.67, 11.66, 13.02)$
O_3	$(11.27, 12.75, 12.15, 14.77)$	$(11.36, 12.57, 12.53, 14.74)$	$(11.36, 12.58, 12.52, 14.74)$
O_4	$(13.18, 14.51, 14.45, 15.48)$	$(13.29, 14.17, 14.81, 15.52)$	$(13.18, 14.55, 14.42, 14.48)$
O_5	$(11.90, 12.40, 13.30, 13.8)$	$(11.99, 12.46, 13.33, 13.83)$	$(11.99, 13.46, 13.33, 13.83)$
O_6	$(11.52, 12.33, 12.09, 13.22)$	$(11.5, 12.0, 12.5, 13.2)$	$(11.59, 12.28, 12.24, 13.23)$
O_7	$(11.03, 12.55, 11.84, 13.83)$	$(11.01, 11.58, 12.14, 13.81)$	$(11.03, 12.55, 11.84, 13.83)$
O_8	$(11.51, 12.53, 12.51, 13.80)$	$(11.60, 12.30, 13.10, 13.80)$	$(10.1, 10.8, 11.6, 12.3)$
O_9	$(12.30, 12.80, 13.60, 14.0)$	$(12.38, 12.86, 13.63, 14.02)$	$(12.40, 12.86, 13.63, 14.03)$
O_{10}	$(12.07, 12.81, 12.40, 14.17)$	$(12.23, 12.75, 12.68, 14.16)$	$(12.23, 12.74, 12.71, 14.15)$

inference operation can be the same for almost half of the given observations. Specially, the use of COG or EMD appears somewhat similar in signifying the type of inference method. Nonetheless, HD is rather different from COG and EMD as its use leads to two interpolation and eight extrapolation operations.

4.4 Accuracy of Interpolation/Extrapolation Results

Table 5 shows the interpolation/extrapolation results with respect to all the given observations using T-FRI based on COG, HD and EMD. These results are further compared against the underlying ground truth values, as reflected in Table 7.

Table 6. Suggested inference operation using COG, HD or EMD

Obs O_i	Inference method Based on COG	Inference method Based on HD	Inference method Based on EMD
O_1	Interpolation	Interpolation	Interpolation
O_2	Extrapolation	Extrapolation	Extrapolation
O_3	Interpolation	Extrapolation	Interpolation
O_4	Extrapolation	Extrapolation	Extrapolation
O_5	Interpolation	Interpolation	Interpolation
O_6	Extrapolation	Extrapolation	Interpolation
O_7	Extrapolation	Extrapolation	Extrapolation
O_8	Interpolation	Extrapolation	Extrapolation
O_9	Extrapolation	Extrapolation	Interpolation
O_{10}	Interpolation	Extrapolation	Extrapolation

Table 7. Accuracy of interpolation/extrapolation in relation to the use of COG, HD and EMD

Metrics	Ground truth vs. Result based on COG $\epsilon_{\%COG}$	Ground truth vs. Result based on HD $\epsilon_{\%HD}$	Ground truth vs. Result based on EMD $\epsilon_{\%EMD}$
AVG	5.38	1.87	3.62
SD	3.46	2.26	2.34

Thus, those closest rules selected by HD have produced the best results (over this set of observations). These outcomes confirm the intuition that the use of a carefully chosen distance metric can help improve (or otherwise, affect adversely) the interpolation/extrapolation result. However, it requires further evaluation of the HD and EMD distance metrics on different types of rule and other observations, in order to better verify the accuracy and consistency of their use.

5   Conclusions

This paper has presented an integrated system to perform interpolation and inference effectively and efficiently. Initially, it performs a pre-interpolation inference assessment to determine whether to use compositional rule of inference (CRI). For this, it utilises the α-cut operation to determine the applicability of CRI despite the sparsity of the given rule base. This helps avoid any unnecessary interpolation/extrapolation while attaining better accuracy. Consequently, interpolation/extrapolation is only applied when no match is found for the given observation. For enhancing the accuracy of interpolated/extrapolated results, this paper has introduced a method which utilises the HD or EMD metric to decide on the nearest rules for interpolation or extrapolation. Both HD and EMD metrics are very effective in multidimensional environment, facilitating efficient computation of the required distance measures between values represented by fuzzy sets. Experimental results have shown that the use of either HD or EMD leads to moderately better results for this implementation than the utilisation of the conventional COG-based metric. In future, it is worthwhile to investigate the effectiveness of this approach for a large rule base and multiple rule selection for interpolation/extrapolation.

References

1. Chaudhuri, B.B., Rosenfeld, A.: A modified Hausdorff distance between fuzzy sets. Inf. Sci. **118**, 159–171 (1999)
2. Dubois, D., Prade, H.: On fuzzy interpolation. In: 3rd International Conference on Fuzzy Logic & Neural Networks, pp. 353–354 (1995)
3. Huang, Z., Shen, Q.: Fuzzy interpolative reasoning via scale and move transformations. IEEE Trans. Fuzzy Syst. **14**(2), 340–359 (2006)
4. Koczy, L.T., Hirota, K.: Rule interpolation by α-level sets in fuzzy approximate reasoning. J. BUSEFAL, Automne, URA-CNRS **46**, 115–123 (1991)
5. Li, F., Li, Y., Shang, C., Shen, Q.: Fuzzy knowledge-based prediction through weighted rule interpolation. IEEE Trans. Cybern. (2019)
6. McCulloch, J.C., Hinde, C.J., Wagner, C., Aickelin, U.: A fuzzy directional distance measure. In: 2014 IEEE International Conference on Fuzzy Systems (FUZZ-IEEE) (2014)
7. Naik, N.: Dynamic fuzzy rule interpolation. Ph.D. Thesis, Department of Computer Science, Institute of Mathematics, Physics and Computer Science, Aberystwyth University, UK (2015)
8. Naik, N., Diao, R., Quek, C., Shen, Q.: Towards dynamic fuzzy rule interpolation. In: IEEE International Conference on Fuzzy Systems (FUZZ-IEEE), pp. 1–7 (2013)
9. Naik, N., Diao, R., Shang, C., Shen, Q., Jenkins, P.: D-FRI-WinFirewall: dynamic fuzzy rule interpolation for windows firewall. In: 2017 IEEE International Conference on Fuzzy Systems (FUZZ-IEEE) (2017)
10. Naik, N., Diao, R., Shen, Q.: Genetic algorithm-aided dynamic fuzzy rule interpolation. In: IEEE International Conference on Fuzzy Systems (FUZZ-IEEE), pp. 2198–2205 (2014)
11. Naik, N., Diao, R., Shen, Q.: Application of dynamic fuzzy rule interpolation for intrusion detection: D-FRI-Snort. In: IEEE International Conference on Fuzzy Systems (FUZZ-IEEE), pp. 78–85 (2016)
12. Naik, N., Diao, R., Shen, Q.: Dynamic fuzzy rule interpolation and its application to intrusion detection. IEEE Trans. Fuzzy Syst. **26**(4), 1878–1892 (2018)
13. Naik, N., Jenkins, P.: Enhancing windows firewall security using fuzzy reasoning. In: IEEE International Conference on Dependable, Autonomic and Secure Computing (2016)
14. Naik, N., Jenkins, P.: Fuzzy reasoning based windows firewall for preventing denial of service attack. In: IEEE International Conference on Fuzzy Systems, pp. 759–766 (2016)
15. Naik, N., Jenkins, P.: A fuzzy approach for detecting and defending against spoofing attacks on low interaction honeypots. In: 21st International Conference on Information Fusion, pp. 904–910. IEEE (2018)
16. Naik, N., Jenkins, P., Cooke, R., Ball, D., Foster, A., Jin, Y.: Augmented windows fuzzy firewall for preventing denial of service attack. In: 2017 IEEE International Conference on Fuzzy Systems (FUZZ-IEEE) (2017)
17. Naik, N., Jenkins, P., Kerby, B., Sloane, J., Yang, L.: Fuzzy logic aided intelligent threat detection in Cisco adaptive security appliance 5500 series firewalls. In: 2018 IEEE International Conference on Fuzzy Systems (FUZZ-IEEE) (2018)
18. Naik, N., Shang, C., Shen, Q., Jenkins, P.: Intelligent dynamic honeypot enabled by dynamic fuzzy rule interpolation. In: The 4th IEEE International Conference on Data Science and Systems (DSS-2018), pp. 1520–1527. IEEE (2018)

19. Naik, N., Shang, C., Shen, Q., Jenkins, P.: Vigilant dynamic honeypot assisted by dynamic fuzzy rule interpolation. In: IEEE Symposium Series on Computational Intelligence (2018)
20. Naik, N., Shang, C., Shen, Q., Jenkins, P.: D-FRI-CiscoFirewall: dynamic fuzzy rule interpolation for Cisco ASA Firewall. In: IEEE International Conference on Fuzzy Systems (FUZZ-IEEE). IEEE (2019)
21. Naik, N.: Fuzzy inference based intrusion detection system: FI-Snort. In: IEEE International Conference on Dependable, Autonomic and Secure Computing, pp. 2062–2067 (2015)
22. Naik, N., Su, P., Shen, Q.: Integration of interpolation and inference. In: UK Workshop on Computational Intelligence, pp. 1–7 (2012)
23. Rubner, Y., Tomasi, C., Guibas, L.J.: A metric for distributions with applications to image databases. In: IEEE International Conference on Computer Vision, pp. 59–66 (1998)

Emerging Intelligence

iBuilding: Artificial Intelligence in Intelligent Buildings

Will Serrano[(✉)]

Intelligent Systems and Networks Group, Electrical and Electronic Engineering,
Alumni Imperial College London, London, UK
g.serrano11@alumni.imperial.ac.uk

Abstract. This paper presents iBuilding: Artificial Intelligence embedded into Intelligent Buildings that adapts to the external environment and the different building users. Buildings are becoming more intelligent in the way they monitor the usage of its assets, functionality and space; the more efficient a building can be monitored or predicted, the more return of investment can deliver as unused space or energy can be redeveloped or commercialized, reducing energy consumption while increasing functionality. This paper proposes Artificial Intelligence embedded into a Building based on a simple Deep Learning structure and Reinforcement Learning algorithm. Sensorial neurons are dispersed through the Intelligent Building to gather and filter environment information whereas Management Sensors based on Reinforcement Learning algorithm make predictions about values and trends in order for building managers or developers to make commercial or operational informed decisions. The proposed iBuilding is validated with a research dataset. The results show that Artificial Intelligence embedded into the Intelligent Building enables real time monitoring and successful predictions about its variables; although there is further research to improve the algorithm's performance as the results are not optimum.

Keywords: Intelligent Building · Smart city · Reinforcement Learning · Smart energy · Artificial Intelligence

1 Introduction

1.1 Definition of Intelligent Building

Intelligent Buildings are becoming increasingly on demand due to their potential for applying innovative and sustainable design initiatives in addition to emerging technologies that optimizes its users' comfort, well-being, space, energy and operational management [1]. However, various definitions, interpretations and key performance indicators of Intelligent Buildings have been proposed in different contexts [2] where a systems view of a building is the starting point for considering a building business, space and management: an Intelligent Building enables its owner and tenant to fulfil their objectives by supporting the management of these resources thus increasing the effectiveness and efficiency of the building business eco-system while adapting to social and technological change driven by human and business needs. Multi-criteria frameworks composed by primary factors such as energy, environment, space

© Springer Nature Switzerland AG 2020
Z. Ju et al. (Eds.): UKCI 2019, AISC 1043, pp. 395–408, 2020.
https://doi.org/10.1007/978-3-030-29933-0_33

flexibility, cost-effectiveness, client comfort, working efficiency, safety, culture, and technology have been proposed as a comprehensive tool for the selective categorization of Intelligent Buildings [3].

1.2 Coworking or Wework

Recent years have witnessed the creation and fast development of "coworking" spaces that are disrupting traditional models of work spaces and businesses and the way people work and collaborate. This additional autonomy and empowerment have raised human, social, managerial and organizational issues [4]; coworking implies a new form of work organization that enables collaboration opportunities and encourages a sense of community inside a shared space, joining together workers from different companies or even freelancers and contractors with different profiles and objectives. Sharing activities requires no only spatial proximity, it also requires enablement, human behavior and education [5]; the socio-spatial dimensions of sharing space are modeled through three vectors on different spatial scales: urban sharing, sharing a living space, and shared social spaces. Innovation is more likely to materialize when there are shared practices and spaces as it builds on openness and collaboration [6]; however, the process, relation and outcomes between collaboration and innovation, or the "sparking idea generator" is still not very well understood.

1.3 Research Motivation

This research proposes Artificial Intelligence embedded into Intelligent Buildings that enables their adaptation to its environment by learning from its users and monitoring its usage and functionality in terms of assets, space and energy therefore assisting building managers or developers to make informed investment or commercial decisions. Sensorial neurons are dispersed through the Building to gather and filter building environment information (space, energy, environment, assets) whereas Management Sensors based on Reinforcement Learning algorithm make predictions about values and trends (upwards, downwards, equal) that enables the Intelligent Building to adapt to future demand of its space or energy. Future work will include the addition of Deep Reinforcement Learning to increase the accuracy of its predictions, data structures, additional sensors such as occupancy and Artificial Intelligence to codify and transmit building information to interconnect iBuildings to generate clusters.

1.4 Research Proposal

Related work covering energy management, usage, building management, space occupancy, building models, Machine Learning and Artificial Intelligence are described in Sect. 2. iBuilding definition is detailed on Sect. 3 whereas iBuilding model is validated on Sect. 4 where experimental results are also presented, finally, Sect. 5 shares the conclusions of this research.

2 Related Work

2.1 User Behaviour

Occupant behaviour is one of the key factors that influences building energy consumption and contributes to uncertainty in building energy use prediction and simulation [7]; advancements in data collection techniques, analytical and modelling methods and simulation applications provide insights in modelling occupant behavior to quantify its impact on building energy use and energy savings. While most studies focus on energy savings during occupied hours, energy is also wasted during non-occupied hours in commercial buildings [8]; actually, more energy is used during non-working hours (56%) than during working hours (44%) mostly from occupants' behaviour and partly due to inadequate zoning and controls. Occupant behaviour is not well understood and it is often oversimplified in the building life cycle due to its stochastic, diverse, complex, and interdisciplinary nature [9]; the use of simplified methods to quantify the impact of occupant behaviour in building performance simulations significantly contributes to performance gaps between simulated models and the actual building energy consumption.

2.2 Building and Energy Management

Buildings consume a significant amount of energy, approximately one-third of the total primary energy resources [10]; building energy efficiency is a complex problem based on the limitation from the occupants' comfort level and user behaviour where intelligent control systems for energy and comfort management in smart energy buildings include intelligent methods, simulation tools, occupants' behaviour and preferences for various building types. Energy unaware behaviour increases one-third to a building's designed energy performance [11] where user activity and behaviour is considered as a key element on energy saving potential and it has long been used for control of various assets such as artificial light, heating, ventilation, air conditioning and power sockets.

2.3 Space Occupancy

There is potential of using occupancy information to develop a more energy efficient building climate control based on office buildings equipped with integrated room automation that integrates the control of heating, ventilation, air conditioning as well as lighting and blind positioning of a building zone or room [12]; the evaluation of the energy saving potential cover different types of occupancy information used in a model predictive control framework. Commercial office buildings represent the largest floor area in most developed countries; this makes office buildings a target for occupant-driven demand control measures [13], however the application of occupant-driven demand control measures in buildings, most especially in the control of thermal, visual and indoor air quality providing systems, is hindered due to the lack of comprehensive detailed occupancy information.

2.4   Building Models

Many retrofit projects are being carried out in existing buildings to reduce energy consumption; although the energy consumption after retrofit can be determined through measurement, the energy consumption before retrofit is more difficult to assess [14]; dynamic simulations or regression models could be used to estimate the energy consumption of buildings before retrofit, however, existing regression models are not able to calibrate the model if it is inaccurate. One of the largest users of electricity in the average household are appliances, which when aggregated, also accounts for approximately 30% of electricity used in the residential building sector [15]; modelling the usage energy of appliances and what causes variation in their use is becoming more relevant to control the demand on the electricity grid infrastructure.

2.5   Machine Learning and Artificial Intelligence

Several machine learning regression methods that develop a predictive model are examined and applied [16] to predict the hourly full load electrical power output of a combined cycle power plant where the base load operation is influenced by four main parameters: ambient temperature, atmospheric pressure, relative humidity and exhaust steam pressure; these parameters are used as input variables in the dataset that affect the electrical power output, which is considered as the target variable. The prediction of Building energy usage has an important role in building energy management and conservation as it assists in the evaluation of the building energy efficiency, conduct building commissioning, detect and diagnose building system faults [17]; Artificial Intelligence based approach for building energy use prediction applies historical data and methods such as multiple linear regression, artificial neural networks and support vector regression. A statistical machine learning framework studies the effects of eight input variables (relative compactness, surface area, wall area, roof area, overall height, orientation, glazing area, glazing area distribution) on two output variables: heating load and cooling load of residential buildings based on a classical linear regression approach [18]; the model is compared against a state of the art nonlinear nonparametric method based on random forests.

Forecasting the energy consumption in homes is an important aspect for the smart grid where the prediction of energy consumption in housing is very dependent on inhabitants' behavior; [19] a stochastic prediction method segments data based on patterns in energy consumption and aggregates it using the k-means clustering algorithm. Data-driven predictive models for the energy use of appliances include measurements of temperature and humidity sensors from a wireless network, weather from a nearby airport station and energy use [20]; data is filtered to remove non-predictive parameters and feature ranking where four statistical models are trained and evaluated with repeated cross validation: multiple linear regression, support vector machine with radial kernel, random forest and gradient boosting machines. Machine Learning and Artificial Intelligence have been used in emergency navigation in a cloud environment to reduce device energy consumption [21] and without static ad hoc networks such as wireless sensor network infrastructure [22].

3   iBuilding Definition

3.1   iBuilding Mathematical Model

iBuilding abstracts the underlying Digital Infrastructure of the Intelligent Building Physical Infrastructure providing a higher layer that enables Artificial Intelligence to measure, manage and virtualize any Intelligent Building. iBuilding is accessed via a common platform tailored and adapted to the Building functionality and role enabling a flexible, expandable and modular interoperable solution where independent Physical Buildings (PB) will be integrated. iBuilding model (Fig. 1) is defined as:

Sensor (i_{Sensor})	Management $(i_{Management})$	Transmission $(i_{Transmission})$
$ai = \{i_{Sensor}, i_{Management}, i_{Transmission}\}$		
Artificial Intelligence $\{ai_1, ai_2, ai_n\}$		
Physical Building $\{P_{Building-1}, P_{Building-2}, P_{Building-m}\}$		
iBuilding Artificial Intelligence in Intelligent Building		

Fig. 1.   iBuilding mathematical definition

- PB = $\{P_{Building-1}, P_{Building-2}, \dots P_{Building-m}\}$ as a set of m Physical Buildings
- AI = $\{ai_1, ai_2, \dots ai_n\}$ as a set of n Artificial Intelligence vectors associated to one or several Physical Buildings $P_{Building}$ that corresponds to the different building variables such as temperature, humidity, CO_2, lighting, heating, cooling, air conditioning, occupation, energy usage, user location or asset status
- ai = $\{i_{Sensor}, i_{Management}, i_{Transmission}\}$ as a set that consists of the three layers of Artificial Intelligence that will monitor and manage $P_{Building}$
- iSensor = $(i_{Sensor-1}, i_{Sensor-2}, i_{Sensor-p})$ as a P dimensional vector that represents the sensorial neurons that collect physical information related to a $P_{Building}$ variable such as temperature or humidity
- iManagement = $(i_{Management-1}, i_{Management-2}, i_{Management-s})$ as a S Dimensional vector that represents the different Artificial Intelligence management algorithms and autonomous decision making methods for each $i_{Sensor-p}$
- iTransmission = $(i_{Transmission-1}, i_{Transmission-2}, i_{Transmission-q})$ as a Q Dimensional vector that represents the different Artificial Intelligence data transmission methods that filter, compress, codify and finally transmit $P_{Building}$ information to a central entity. iTransmission will effectively multiplex different m $P_{Building}$ therefore creating iBuilding: a cluster of Intelligent Buildings that will be remotely managed

3.2   iBuilding Neural Model

iBuilding model for this research proposal (Fig. 2) consists on a layer of sensor neurons that collect Building information and feed that data to their respective management sensor (iSensor) and a management layer that makes predictions about the iSensor values and trends (iManagement). iTransmission, based on genetic algorithms, will be defined in future additional research work.

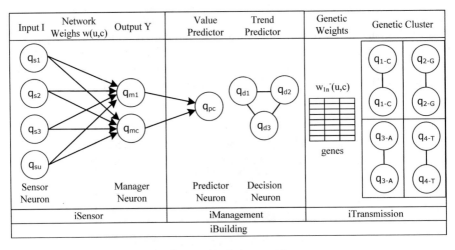

Fig. 2. iBuilding model

iSensor

iSensor consists of u input neurons that take iBuilding measurements of a specific variable such as temperature or humidity related to a precise area or floor connected to their management sensor by network weights. iSensor defines:

- $I = (q_{s1}, q_{s2}, \ldots, q_{su})$, a U-dimensional vector $I \in [0, 1]^U$ that represents the input state q_s for the sensor neuron u;
- $w(u, c)$ is the $U \times C$ matrix of network weights from the U sensor neurons q_{su} to their C manager neurons q_{mc};
- $Y = (q_{m1}, q_{m2}, \ldots, q_{mc})$, a C-dimensional vector $Y \in [0, 1]^C$ that represents the output state q_m for the manager neuron c.

The manager neuron q_{mc} stores the sensors q_{su} averaged or weighted values based on $w(u, c)$ that can be fixed or variable to include feedback on performance.

iManagement

iManagement takes predictions on iSensor values and trends based on a Random Neural Network with a modified Reinforcement Learning algorithm [23–29]. Given some Goal G that the iBuilding has to achieve as a function to be optimized, reward R is a consequence of the interaction with the environment; Reward R_l $l = 1, 2 \ldots$ is computed based successive measured values provided by the Y manager neurons q_{mc} where these are used to compute the neural network weights updates threshold T_l:

$$R_1 = \beta(Y_1 - Y_{1-1})$$
$$T_1 = \alpha T_{1-1} + (1 - \alpha)R_1 \tag{1}$$

where α represents the Threshold memory, $0 < \alpha < 1$, and β are variables that can be statically assigned or dynamically updated based on the external observations. A Random Neural Network (RNN) [30–32] with at least as many nodes as the number of decisions to be taken is generated where neurons are numbered 1, ..., j, ..., n; therefore for any decision i, there is some neuron i. Decisions in the RL algorithm with the RNN are taken by selecting the decision j for which the corresponding neuron is the most excited, the one which has the largest value of q_j. The state q_j is the probability that it is excited, these quantities satisfy the following system of non-linear equations:

$$q_j = \frac{\lambda^+(j)}{r(j) + \lambda^-(j)}$$
$$\lambda^+(j) = \sum_{i=1}^{n}[q_i r(i)p^+(i,j)] + \Lambda(j) \tag{2}$$
$$\lambda^-(j) = \sum_{i=1}^{n}[q_i r(i)p^-(i,j)] + \lambda(j)$$

This research uses Reinforcement Learning to make three trend decisions with three associated neurons and an c independent neurons that makes predictions of the values of the Y manager neurons q_{mc}:

- q_{d0} predicts that the trend of iSensor is to go upwards or Up;
- q_{d1} predicts that the trend of iSensor is to go downwards or Down;
- q_{d2} predicts that the trend of iSensor is to keep its value or Equal;
- q_{pc} predicts the value of the Y manager neurons q_{mc}.

iBuilding takes the l_{th} trend decision which corresponds to the highest potential neuron j and then the l_{th} reward R_1 is measured is calculated where the network weighs are updated with the Threshold T_1 as follows for all neurons $i \neq j$.

Reward if trend is correct: ($R_1 > 0$ and $j = 0$ or $R_1 < 0$ and $j = 1$ or $R_1 = 0$ and $j = 2$)

- $w^+(i, j) = w^+(i, j) + T_1$
- $w^-(i, k) = w^-(i, k) + $ if $k \neq j$

else Penalise:

- $w^+(i, k) = w^+(i, k) + $ if $k \neq j$
- $w^-(i, j) = w^-(i, j) + T_1$

On the above equations, W_{+ij} is the rate at which neuron i transmits excitation spikes to neuron j and w_{-ij} is the rate at which neuron i transmits inhibitory spikes to neuron j in both situations when neuron i is excited. Λ_i and λ_i are the rates of external excitatory and inhibitory signals respectively (Fig. 3). In addition to the Reinforcement Leaning algorithm for trends, q_{dc} makes predictions on the future values of the Y manager neurons q_{mc} based on the previous prediction and current measurement:

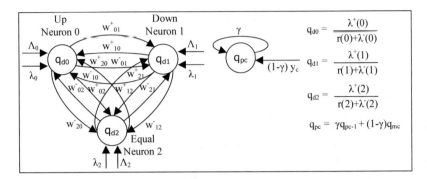

Fig. 3. iManagement Reinforcement Learning algorithm

$$q_{pc} = \gamma q_{pc-1} + (1 - \gamma)q_{mc} \qquad (4)$$

where γ is a variable $0 < \gamma < 1$, that can be statically or dynamically assigned that represents the prediction memory and q_{pc-1} is the previous value of the predictor neuron q_{pc}.

4 iBuilding Validation

iBuilding is validated with a research dataset [20] based on a house with electric metering with M-BUS energy counters that measures the energy consumption of appliances, electric baseboard heaters and lighting (Fig. 4). The house temperature and humidity were monitored with a ZigBee wireless sensor network located in 9 different zones, in addition, the temperature and humidity of an external weather station is also included. Information was collected every ten minutes for 137 days (4.5 months), from 11/01/2016 17:00:00 to 27/05/2016 18:00:00 with 19736 measurements in total.

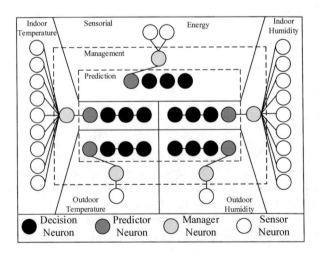

Fig. 4. iBuilding Validation

iSensor Validation

iBuilding consists a configuration of 5 iSensor networks with a neuron per external sensor, 22 sensor neurons in total, connected to with 5 manager neurons associated to the Energy consumption, Indoor and Outdoor Temperature and Humidity respectively. On this validation, w(u, c) is fixed in order to either add the energy or make the average of the temperature and humidity sensors, as shown on Table 1.

Table 1. iSensor Validation

iSensor	Measurement	Sensor I u	w(u c)	Sensor Y c
1	Energy (W)	2	(1.0, 1.0)	1
2	Indoor Temperature (C)	9	(1/9, 1/9, 1/9, 1/9, 1/9, 1/9, /1/9, 1/9, 1/9)	1
3	Indoor Humidity (H)	9	(1/9, 1/9, 1/9, 1/9, 1/9, 1/9, /1/9, 1/9, 1/9)	1
4	Outdoor Temperature (C)	1	(1.0)	1
5	Outdoor Humidity (H)	1	(1.0)	1

Figure 5 shows the value of the five iSensor neuron managers q_{mc} for the complete dataset (top) and zoomed into only one day (bottom).

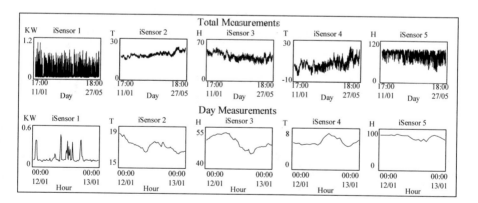

Fig. 5. iSensor Validation

As expected, the value of the q_{mc} for Temperature and Humidity follows a continuous trend however, the q_{mc} for energy consumption is alternative.

iManagement Validation

Equally, there are 5 iManagement networks that make trend and value predictions using Reinforcement Learning algorithm. Table 2 shows the number of Rewards (R) or success, Penalizations (P) or misses, and Accuracy (A) for different values of the Threshold memory α across the 19736 data measurements.

Table 2. iManagement Validation: trend predictor - α value

α	iManagement 1	iManagement 2	iManagement 3	iManagement 4	iManagement 5	Total values
0.1	R: 6419	R: 14459	R: 14268	R: 17548	R: 17333	R: 70027
	P: 13317	P: 5277	P: 5468	P: 2188	P: 2403	P: 28653
	A: 32.52%	A: 73.26%	A: 72.29%	A: 88.91%	A: 87.82%	A: 70.96%
0.5	R: 6553	R: 14103	R: 14190	R: 17609	R: 17231	R: 69686
	P: 13183	P: 5633	P: 5546	P: 2127	P: 2505	P: 28994
	A: 33.20%	A: 71.46%	A: 71.90%	A: 89.22%	A: 87.31%	A: 70.62%
0.9	R: 6588	R: 14032	R: 14158	R: 17694	R: 17294	R: 69766
	P: 13148	P: 5704	P: 5578	P: 2042	P: 2442	P: 28914
	A: 33.38%	A: 71.10%	A: 71.74%	A: 89.65%	A: 87.63%	A: 70.70%

Table 2 reflects that the minor the Threshold memory α, the higher the accuracy although its variation does not have a great impact. iManagement 1 provides worse results due its alternative values. Table 3 shows the root-mean-square error of the predicted values against the real measurements for different values of the prediction memory γ across the 19736 data measurements.

Table 3. iManagement Validation: value predictor - γ value

γ	iManagement 1	iManagement 2	iManagement 3	iManagement 4	iManagement 5	Total values
0.1	5.14E − 01	3.34E − 04	2.36E − 03	1.23E − 03	6.85E − 03	5.25E − 01
0.5	5.29E − 01	5.30E − 04	3.27E − 03	2.09E − 03	1.14E − 02	5.46E − 01
0.9	6.09E − 01	1.90E − 03	8.69E − 03	7.88E − 03	3.89E − 02	6.67E − 01

The smallest prediction memory generates the lowest error, as shown in Table 3. Figure 6 represents the iManagement 1 energy prediction q_{pc} for only one day (144 data measurements) across different values of γ where $\gamma = 0.1$ provides the closest prediction to the real measurement.

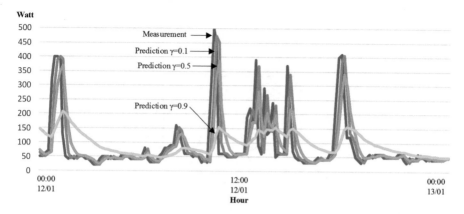

Fig. 6. iManagement Validation: value predictor - γ value

5   Conclusions

This research has proposed iBuilding: Artificial Intelligence embedded into Intelligent Buildings that enable their adaptation to the external environment, learning from its users and monitoring its functionality in terms of assets, space and energy therefore assisting building managers or developers to make commercial or operational decisions. Sensorial neurons are dispersed through the Building to gather and filter building environment information whereas Management Sensors based on Reinforcement Learning algorithm make predictions about values and trends (upwards, downwards and equal) that enable the Intelligent Building to adapt to future demand of its space, environmental conditions or energy.

Future work will include the addition of Deep Reinforcement Learning to increase the accuracy of its trend and value predictions, mostly when the measurements are alternating rather than continuous, and the addition of a data structure that analyses and correlates information from different sensors. Additionally, iTransmission, or Artificial Intelligence that codifies and transmits Building information, will be included to interconnect iBuildings with each other in order to generate a cluster of Intelligent Buildings.

Appendix: iBuilding Neural Schematic

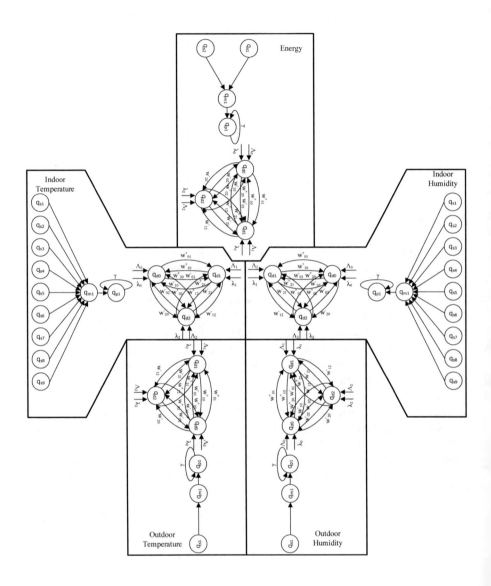

References

1. Ghaffarianhoseini, A., Berardi, U., AlWaer, H., Chang, S., Halawa, E., Ghaffarianhoseini, A., Clements, D.: What is an intelligent building? Analysis of recent interpretations from an international perspective. Arch. Sci. Rev. **59**(5), 338–357 (2016)
2. Clements, D.: What do we mean by intelligent buildings? Autom. Construction. **6**(5–6), 395–400 (1997)
3. Omar, O.: Intelligent building, definitions, factors and evaluation criteria of selection. Alex. Eng. J. **57**, 2903–2910 (2018)
4. Leclercq, A., Isaac, H.: The new office: how coworking changes the work concept. J. Bus. Strat. **37**(6), 3–9 (2016)
5. Hui, J.K., Zhang, Y.: Sharing space: urban sharing, sharing a living space, and shared social spaces. Space Cult., 1–13 (2018)
6. Yacoub, G.: How do collaborative practices emerge in coworking spaces? Evidence from fintech start-ups. Acad. Manag. Proc. **1**, 1–10 (2018)
7. Honga, T., Taylor-Langea, S., D'Ocab, S., Yanc, D., Corgnati, S.: Advances in research and applications of energy-related occupant behaviour in buildings. Energy Build. **116**, 694–702 (2016)
8. Masoso, O.T., Grobler, L.J.: The dark side of occupants' behaviour on building energy use. Energy Build. **42**, 173–177 (2010)
9. Hong, T., Yan, D., D'Oca, S., Chen, C.: Ten questions concerning occupant behaviour in buildings: the big picture. Build. Environ. **114**, 518–530 (2017)
10. Nguyen, T., Aiello, M.: Energy intelligent buildings based on user activity: a survey. Energy Build. **56**, 244–257 (2013)
11. Shaikh, P., Nor, N., Nallagownden, P., Elamvazuthi, I., Ibrahim, T.: A review on optimized control systems for building energy and comfort management of smart sustainable buildings. Renew. Sustain. Energy Rev. **34**, 409–429 (2014)
12. Oldewurtel, F., Sturzenegger, D., Morari, M.: Importance of occupancy information for building climate control. Appl. Energy **101**, 521–532 (2013)
13. Labeodan, T., Zeiler, W., Boxem, G., Zhao, Y.: Occupancy measurement in commercial office buildings fordemand-driven control applications—A survey and detection system evaluation. Energy Build. **93**, 303–314 (2015)
14. Ko, J.H., Kong, D.S., Huh, J.H.: Baseline building energy modeling of cluster inverse model by using daily energy consumption in office buildings. Energy Build. **140**, 317–323 (2017)
15. Cetina, K.S., Tabares-Velascob, P.C., Novoselac, A.: Appliance daily energy use in new residential buildings: use profiles and variation in time-of-use. Energy Build. **84**, 716–726 (2014)
16. Tüfekci, P.: Prediction of full load electrical power output of a base load operated combined cycle power plant using machine learning methods. Electr. Power Energy Syst. **60**, 126–140 (2014)
17. Wang, Z., Srinivasan, R.S.: A review of artificial intelligence based building energy use prediction: Contrasting the capabilities of single and ensemble prediction models. Renew. Sustain. Energy Rev. **75**, 796–808 (2017)
18. Tsanas, A., Xifara, A.: Accurate quantitative estimation of energy performance of residential buildings using statistical machine learning tools. Energy Build. **49**, 560–567 (2012)
19. Arghira, N., Hawarah, L., Ploix, S., Jacomino, M.: Prediction of appliances energy use in smart homes. Energy **48**, 128–134 (2012)
20. Candanedo, L.M., Feldheim, V., Deramaix, D.: Data driven prediction models of energy use of appliances in a low-energy house. Energy Build. **140**, 81–97 (2017)

21. Bi, H., Gelenbe, E.: A cooperative emergency navigation framework using mobile cloud computing. In: International Symposium Computer and Information Sciences, pp. 41–48 (2014)
22. Gelenbe, E., Bi, H.: Emergency navigation without an infrastructure. Sensors **14**(8), 15142–15162 (2014)
23. Gelenbe, E.: Cognitive Packet Network. Patent US 6804201 B1 (2004)
24. Gelenbe, E., Xu, Z., Seref, E.: Cognitive packet networks. In: International Conference on Tools with Artificial Intelligence, pp. 47–54 (1999)
25. Gelenbe, E., Lent, R., Xu, Z.: Networks with cognitive packets. In: IEEE International Symposium on the Modeling, Analysis, and Simulation of Computer and Telecommunication Systems, pp. 3–10 (2000)
26. Gelenbe, E., Lent, R., Xu, Z.: Measurement and performance of a cognitive packet network. Comput. Netw. **37**(6), 691–701 (2001)
27. Gelenbe, E., Lent, R., Montuori, A., Xu, Z.: Cognitive packet networks: QoS and performance. In: IEEE International Symposium on the Modeling, Analysis, and Simulation of Computer and Telecommunication Systems, pp. 3–9 (2002)
28. Bi, H., Desmet, A., Gelenbe, E.: Routing emergency evacuees with cognitive packet networks. In: International Symposium on Computer and Information Sciences, pp. 295–303. (2013)
29. Bi, H., Gelenbe, E.: Routing diverse evacuees with cognitive packets. In: IEEE Computer Society PerCom Workshops, pp. 291–296 (2014)
30. Gelenbe, E.: Random neural networks with negative and positive signals and product form solution. Neural Comput. **1**, 502–510 (1989)
31. Gelenbe, E.: Learning in the recurrent random neural network. Neural Comput. **5**, 154–164 (1993)
32. Gelenbe, E.: G-networks with triggered customer movement. J. Appl. Probab. **30**, 742–748 (1993)

5G Cybersecurity Based on the Blockchain Random Neural Network in Intelligent Buildings

Will Serrano[✉]

Intelligent Systems and Networks Group, Electrical and Electronic Engineering,
Alumni Imperial College London, London, UK
g.serrano11@alumni.imperial.ac.uk

Abstract. 5G promises much faster Internet transmission rates at minimum latencies with indoor and outdoor coverage; 5G potentially could replace traditional Wi-Fi for network connectivity and Bluetooth technology for geolocation with a seamless radio coverage and network backbone that will accelerate new services such as the Internet of Things (IoT). New infrastructure applications will depend on 5G as a mobile Internet service provider therefore eliminating the need to deploy additional private network infrastructure or mobile networks to connect devices; however, this will increase cybersecurity risks as radio networks and mobile access channels will be shared between independent services. To address this issue, this paper presents a digital channel authentication method based on the Blockchain Random Neural Network to increase Cybersecurity against rogue 5G nodes; in addition, the proposed solution is applied to Physical Infrastructure: an Intelligent Building. The validation results demonstrate that the addition of the Blockchain Neural Network provides a cybersecure channel access control algorithm that identifies 5G rogue nodes where 5G node identities are kept cryptographic and decentralized.

Keywords: Neural network · Intelligent buildings · Blockchain ·
Cybersecurity · 5G · Access credentials · Smart infrastructure

1 Introduction

5G will be an innovative mobile solution that consists of very high carrier frequencies with great bandwidth (mmWave), extreme node and device densities (ultra-densification), and unprecedented numbers of MIMO antennas (massive multiple-input multiple-output) [1]; potential new 5G application are divided into Enhanced Mobile Broadband, Ultra Reliable Low Latency Communications, and Massive Machine Type Communications. 5G will be also highly integrative: the connection to a 5G air interface and spectrum together with LTE and WiFi will enable global, reliable, scalable, available and cost-efficient connectivity solution that provides high rate coverage and a seamless user experience; both are considered as a potentially key driver for the yet to stablish global IoT [2]. To achieve these additional requirements, the 5G core network will also have to reach extraordinary levels of flexibility and intelligence to increase node capacity that supports high data rates, extremely low

© Springer Nature Switzerland AG 2020
Z. Ju et al. (Eds.): UKCI 2019, AISC 1043, pp. 409–422, 2020.
https://doi.org/10.1007/978-3-030-29933-0_34

latency, and a significant improvement in users' perceived Quality of Service (QoS) where energy and cost efficiencies will become even more critical considerations [3].

This high levels of network integration are provided at a cybersecurity cost, 5G rogue nodes could gather user IoT data leaving users or devices incapable to use a different transmission network such as Wi-Fi. Recent Blockchain solutions enable the digitalization of contracts as it provides authentication between parties and information encryption of data that gradually increments while it is processed in a decentralized network such as the IoT [4]. Due to these features, Blockchain has been already applied in Cryptocurrency [5], Smart Contracts [6], Intelligent Transport Systems [7] and Smart Cities [8].

1.1 Research Motivation

To address the increased cybersecurity risk of 5G, this paper proposes a digital channel authentication method based on the Blockchain Random Neural Network [9, 10]. The Blockchain Neural Network connects neurons in a chain configuration that provides an additional layer of resilience against Cybersecurity rogue 5G nodes in the IoT. The 5G Cybersecurity application presented on this paper can be generalized to emulate an Authentication, Authorization and Accounting (AAA) server where 5G node identity is encrypted in the neural weights and stored decentralized servers.

The Blockchain Neural Network solution is equivalent to the Blockchain with the same properties: 5G node identity authentication, data encryption and decentralization where 5G node identities are gradually incremented and learned while are becoming operational within the Intelligent Building. The Neural Network configuration have analogue biological properties as the Blockchain where neurons are gradually incremented and chained through synapses as variable 5G node identities increase; in addition, information is stored and codified in decentralized neural networks weights. The main advantage of this research proposal is the biological simplicity of the solution however it suffers high computational cost when neurons increase.

1.2 Research Proposal

5G and the Blockchain related work is described in Sect. 2. This paper defines a decentralized solution that emulates the Blockchain validation process: mining the input neurons until the neural network solution is found as presented in Sect. 3. The proposed 5G node identity authentication mathematical model is described in Sect. 4; 5G nodes are authenticated before they are enabled into the Intelligent Building where the International Mobile Subscriber Identity (IMSI) associated to the Intelligent Building provides the Private Key and there is no need for a Public Key. Experimental results in Sect. 5 show that the additional Blockchain neural network increases cybersecurity resilience and decentralized confidentiality to 5G node identity authentication. The main conclusion presented in Sect. 6 proves that the 5G node identify is kept secret codified in the neural weights while detecting rogue 5G nodes.

2   Related Work

2.1   Cybersecurity in 5G

5G will support a wide range of industry diverse business models and use cases addressing the motivations from different stakeholders such as mobile network operators, infrastructure and cloud service providers and tenants as presented by Adam et al. [11]; these diverse devices and applications will lead to different cybersecurity requirements that shall be considered, in addition to performance requirements from future applications as well as security and regulatory compliance with Service Level Agreements in multi-tenant cloud environments. SHIELD is a novel design and development cybersecurity framework which offers Security-as-a-Service in 5G as presented by Katsianis et al. [12]; SHIELD framework leverages network functions virtualization and Software Defined Networks for virtualization and dynamic placement of virtual network security functions, Big Data analytics for real-time incident detection and mitigation as well as trusted computing attestation techniques for securing both the infrastructure and its services.

Cyber physical systems create new online networking services and applications in the IoT through their defined interactions and automated decisions as exposed by Atat et al. [13]; future 5G cellular networks will facilitate the physical systems communications through different technologies such as device-to-device communications that need to be protected from eavesdropping, especially with the large amount of traffic that will constantly flow through the network. The connectivity of many stand-alone IoT systems through the Internet or 5G Networks introduces numerous cybersecurity challenges as sensitive information is prone to be exposed to malicious users as studied by Mozzaquatro et al. [14]; appropriate security services adapted to the threats improves IoT cybersecurity from an ontological analysis. The next wireless strategy is centered on creative 5G and IoT, as confirmed by Chang et al. [15]; while the confidence of the future mobile technology will help innovate governments, workforce, industry and social media, the key threat to future mobile wireless development is the growing concern of its security, privacy and anomaly detection. Fog and mobile edge computing will play a key role in the upcoming 5G mobile networks to support decentralized applications, data analytics and self network management by using a highly distributed computing model, as stated by Fernández et al. [16]; user-centric cybersecurity solutions particularly require the collection, process and analyses of significantly large amount of data traffic and huge number of network connections in 5G networks in real-time and in an autonomous way using Deep Learning techniques to detect network anomalies as presented by Fernández et al. [17].

The protection of the IoT is a challenging task due to system security is the foundation for the development of IoT, as researched by Lu et al. [18]; the key factors of the security model are the protection and integration of heterogeneous smart devices and Information Communication Technologies (ICT). Mobile cloud computing is applied in multiple industries to obtain cloud-based services by leveraging mobile technologies. With the development of 5G, defensive mitigations against threats from wireless communications have been playing a remarkable role in the Web security domain and Intrusion Detection Systems (IDS) as shown by Gai et al. [19]; a high level framework for implementing secure mobile cloud computing that adopts IDS techniques in 5G networks based on mobile cloud-based solutions.

2.2   Neural Networks and Cryptography

Neural Networks have been already applied to Cryptography; Pointcheval [20] presents a linear scheme based on the Perceptron problem, or N-P problem, suited for smart cards applications. Kinzel et al. [21] train two multilayer neural networks on their mutual output bits with discrete weights to achieve a synchronization that can be applied to secret key exchange over a public channel; Klimov et al. [22] propose three cryptanalytic attacks (genetic, geometric and probabilistic) to the above neural network. Volna et al. [23] apply feed forward neural networks as an encryption and decryption algorithm with a permanently changing key. Yayık et al. [24] present a two-stage cryptography multilayered neural network where the first stage generates neural network-based pseudo random numbers and the second stage encrypts information based on the non-linearity of the model. Schmidt et al. [25] present a review of the use of artificial neural networks in cryptography.

2.3   Blockchain and Security

Currently; there is a great research effort in blockchain algorithms applied to security applications. Xu et al. [26] propose a punishment scheme based on the action record on the blockchain to suppress the attack motivation of the edge servers and the mobile devices in the edge network. Cha et al. [27] utilize a blockchain network as the underlying communication architecture to construct an ISO/IEC 15408-2 compliant security auditing system. Gai et al. [28] propose a conceptual model for fusing blockchains and cloud computing over three deployment modes: Cloud over Blockchain, Blockchain over Cloud and Mixed Blockchain-Cloud. Gupta et al. [29] propose a Blockchain consensus model for implementing IoT security. Agrawal et al. [30] present a Blockchain mechanism that continuously evaluates legitimate presence of user in valid IoT-Zones without user intervention.

3   Blockchain Neural Network in 5G Cybersecurity

Blockchain [5] is based on cryptographic concepts which can be applied similarly by the use of Neural Networks. Information in the Blockchain is contained in blocks that also include a timestamp, the number of attempts to mine the block and the previous block hash. Decentralized miners then calculate the hash of the current block to validate it. Information contained in the Blockchain consists of transactions which are authenticated by a signature that uses the Intelligent Building private key, transaction origin, destination and value (Fig. 1).

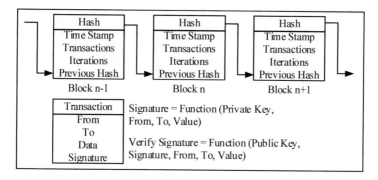

Fig. 1. Blockchain model

3.1   The Random Neural Network

The proposed Blockchain configuration is based on the Random Neural Network (RNN) [31–33] which is a spiking neuronal model that represents the signals transmitted in biological neural networks, where they travel as spikes or impulses, rather than as analogue signal levels. The RNN is a spiking recurrent stochastic model for neural networks where its main analytical properties are the "product form" and the existence of the unique network steady state solution.

3.2   The Random Neural Network with Blockchain Configuration

The Random Neural Network with Blockchain configuration consists of L Input Neurons, M hidden neurons and N output neurons Network (Fig. 2). Information in this model is contained networks weights $w^+(j, i)$ and $w^-(j, i)$ rather than neurons x_L, z_M, y_N.

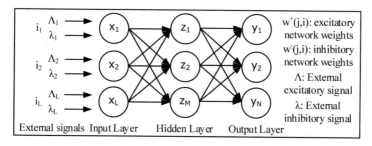

Fig. 2. The Random Neural Network structure

- $I = (\Lambda_L, \lambda_L)$, a variable L-dimensional input vector $I \in [-1, 1]^L$ represents the pair of excitatory and inhibitory signals entering each input neuron respectively; where scalar L values range $1 < L < \infty$;
- $X = (x_1, x_2, \dots, x_L)$, a variable L-dimensional vector $X \in [0, 1]^L$ represents the input state q_L for the neuron L; where scalar L values range $1 < L < \infty$;

- $Z = (z_1, z_2, \ldots, z_M)$, a M-dimensional vector $Z \in [0, 1]^M$ that represents the hidden neuron state q_M for the neuron M; where scalar M values range $1 < M < \infty$;
- $Y = (y_1, y_2, \ldots, y_N)$, a N-dimensional vector $Y \in [0, 1]^N$ that represents the neuron output state q_N for the neuron N; where scalar N values range $1 < N < \infty$;
- $w^+(j, i)$ is the $(L + M + N) \times (L + M + N)$ matrix of weights that represents from the excitatory spike emission from neuron i to neuron j; where $i \in [x_L, z_M, y_N]$ and $j \in [x_L, z_M, y_N]$;
- $w^-(j, i)$ is the $(L + M + N) \times (L + M + N)$ matrix of weights that represents from the inhibitory spike emission from neuron i to neuron j; where $i \in [x_L, z_M, y_N]$ and $j \in [x_L, z_M, y_N]$.

The main concept of the Random Neural Network Blockchain configuration is that the neuron vector sizes, L, M and N are variable instead of fixed. Neurons or blocks are iteratively added where the value of the additional neurons consists on both the value of the additional information and the value of previous neurons therefore forming a neural chain. Information in this model is transmitted in the matrixes of network weighs, $w^+(j, i)$ and $w^-(j, i)$ rather than in the neurons. The input layer X represents the Intelligent Building's incremental verification data; the hidden layer Z represents the values of the chain and the output layer Y represents the Intelligent Building Private Key.

4   5G Cybersecurity Neural Network Blockchain Model

The 5G Cybersecurity Neural Network Blockchain model described in this section is based on the main concepts shown on Fig. 3:

- Private key, y_N;
- Authentication, A(t) and Verification, V;
- Neural Chain network and Mining;
- Decentralized information, $w^+(j, i)$ and $w^-(j, i)$.

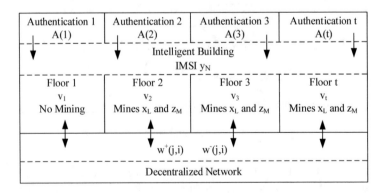

Fig. 3. 5G Cybersecurity Neural Network Blockchain model

4.1   Private Key

The private key $Y = (y_1, y_2, \ldots, y_N)$ consists on the Intelligent Building International Mobile Subscriber Identity (IMSI). The private key is presented by the Intelligent Building every time a 5G node becomes operational and its identity credential requires authorization and verification from the Intelligent Building in order to enable its digital channel (Table 1).

Table 1. IMSI private key y_N

Identifier	Code	Digits	Bits	Name	Key
IMSI	MCC	3	12	Mobile Country Code	y_3
	MNC	3	12	Mobile Network Code	y_2
	MSIN	9	36	Mobile Subscription Identification Number	y_1

4.2   Authentication and Verification

Let's define Authentication and Verification as:

- Authentication, $A(t) = \{A(1), A(2), \ldots A(t)\}$ as a variable vector where t is the 5G node authentication stage number;
- Verification, $V = \{v_1, v_2, \ldots v_t\}$ as a set of t I-vectors where $v_o = (e_{o1}, e_{o2}, \ldots e_{oI})$ and e_o are the I different dimensions for $o = 1, 2, \ldots t$.

The first Authentication $A(1)$ has associated an input state $X = x_I$ which corresponds v_1 representing the 5G node identity data. The output state $Y = y_N$ corresponds to the Intelligent Building Private Key and the hidden layer $Z = z_M$ corresponds to the value of the neural chain that will be inserted in the input layer for the next authentication.

The second Authentication $A(2)$ has associated an input state $X = x_I$ which corresponds to the 5G node identity data v_1 for the first Authentication $A(1)$, the chain, or the value of the hidden layer z_M and the additional 5G node identity data v_2. The output state $Y = y_N$ still corresponds the Intelligent Building Private Key and the hidden layer $Z = z_M$ corresponds to the value of the neural chain for the next transaction. This process iterates as more 5G node identity data is inserted in a staged enabling process. The neural chain can be formed of the values of the entire hidden layer neurons, a selection of neurons, or any other combination to avoid the reverse engineering of the Intelligent Building identity from the stored neural weights.

4.3   Neural Chain Network and Mining

The first Authentication $A(1)$ calculates the Random Neural Network neural weights with an $E_k < Y$ for the input data $I = (\Lambda_L, \lambda_L)$, and the Intelligent Building private key $Y = y_N$. The calculated network weights $w^+(j, i)$ and $w^-(j, i)$ are stored in the decentralized network and retrieved in the mining process. After the first Authentication; the 5G node requires to be validated at each additional Authentication with the Intelligent

Building private key where the 5G node verification data is validated and verification data v_t from Authentication A(t) are added to the Intelligent Building.

Verification data is validated or mined by calculating the outputs of the Random Neural Network using the transmitted network weighs, $w^+(j, i)$ and $w^-(j, i)$ at variable random inputs i, or following any other method. The solution is found or mined when quadratic error function E_k is lesser than determined minimum error or threshold T.

$$E_k = \frac{1}{2}\sum_{n=1}^{N}(y'_n - y_n)^2 < T \qquad (1)$$

where E_k is the minimum error or threshold, y' is the output of the Random Neural Network with mining or random input I and y_n is the Intelligent Building Private Key. The mining complexity can be tuned by adjusting E_k. The Random Neural Network with Blockchain configuration is mined when an Input I is found that delivers an output Y with an error E_k lesser than a threshold T for the retrieved Intelligent Building network weights $w^+(j, i)$ and $w^-(j, i)$.

When the solution is found, the Intelligent Building data can be processed, the potential value of the neural hidden layer $Z = z_M$ is added to form the Neural Chain where more Intelligent Building data is added. Once the solution is found or mined, the values of the hidden layer are used in the input of the next transaction, along with the new data; where the Random Neural Network with gradient descent learning algorithm is calculated again to generate the new network matrixes $w^+(j, i)$ and $w^-(j, i)$. The more authentication and verification data; the validation or mining process increases on complexity.

Finally, the system calculates the Random Neural Network with Gradient descent algorithm for the new pair (I, Y) where the new calculated network weights $w^+(j, i)$ and $w^-(j, i)$ are stored in the decentralized network.

4.4   Decentralized Information

The Intelligent Building network weights $w^+(j, i)$ and $w^-(j, i)$ are stored in the decentralized network rather than its data I directly from which are calculated. The network weights expand as more verification data is inserted, therefore creating an adaptable method. In addition; only the Intelligent Building Data can be extracted when the IMSI private key is presented therefore making secure to store information in a decentralized system.

5   Neural Blockchain in 5G Cybersecurity Validation

This section proposes a practical validation of the Neural Blockchain model in the 5G Cybersecurity using the network simulator Omnet ++ with Java for a network of five 5G nodes with associated passive Distributed Antenna Systems. The experiment will emulate a gradual implementation of a 5G network in an Intelligent Building were 5G nodes are increasingly enabled into the Smart Building (Table 2).

Table 2. Neural Blockchain in cybersecurity validation – 5G node values

Simulation	Application	Cell	Coverage	5G node identity Vt	Intelligent building IMSI Y_N
Five 5G Nodes	Intelligent Building	Pico	100 m 0.314 km²	60 bits 234-151-234512340-4	60 bits 234-151-234512351

A 5G Network generic Performance Specification is shown on Table 3:

Table 3. 5G network performance specification

Data rate	Frequency	Latency	Modulation	Cell
20 Gbit/s	3.5 GHz–26 GHz	<1 ms	Non-Orthogonal Multiple Access	Pico Small

The Intelligent Building is assigned to a IMSI private key y_N whereas the gradual addition of 5G nodes require validation before they are enabled to transmit information into the Intelligent Building. When an 5G node is ready to become operational, its identity is validated by the Intelligent Building; the decentralized system retrieves the neural weights associated to the private key; mines the block, adds the node code and stores back the network weights in the decentralized system. Mining for this validation is considered as the selection of random neuron values until $E_k < T$. When more 5G nodes are ready to become operational, the Intelligent Building uses its IMSI private key y_N and the 5G node identity v_t is added to the neural chain once it is mined. Each bit is codified as a neuron however instead of the binary 0–1, neuron potential is codified as 0.25–0.75 (Fig. 4); this approach removes overfitting in the learning algorithm as neurons only represent binary values.

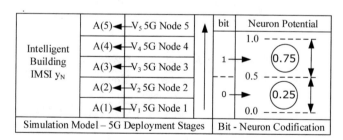

Fig. 4. Neural Blockchain in 5G Cybersecurity validation

The simulations are run 100 times for a five 5G node Network (Table 4). The information shown is the number of iterations the Random Neuron Network with Blockchain configuration requires to achieve an $E_k < 1.0E{-}10$; the error E_k, the number of iterations to mine the Blockchain and the number of neurons for each layer; input x_L, hidden z_M and output y_N.

Table 4. 5G network simulation – learning and mining

5G node	Learning iteration	Learning error	Mining iteration	Mining threshold	Mining error E_k	Number of neurons (x_L, z_M, y_N)
1	267.00	9.50E−11	106.87	1.00E−02	4.54E−03	60-4-60
2	219.73	9.48E−11	10230.04	1.00E−02	6.63E−03	124-4-60
3	194.76	9.38E−11	104456.1	1.00E−02	8.03E−03	188-4-60
4	178.91	9.29E−11	632011.8	1.00E−02	8.47E−03	252-4-60
5	167.22	9.45E−11	570045.3	1.00E−02	8.87E−03	316-4-60

With four neurons in the hidden layer, the number of learning iterations gradually decreases while the number of input neurons increases due to the additional information added activating 5G nodes. The results for the mining iteration are not as linear as expected because mining is performed using random values (Fig. 5). Surprisingly; mining is easier in the final authentication stage when it would have been expected harder as the number of neurons increases.

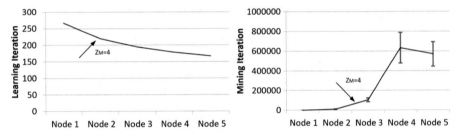

Fig. 5. 5G network simulation – learning and mining iterations

The Blockchain Random Neural Network algorithm must detect 5G rogue nodes to be effective (Table 5); Δ represents the number of bit changes in the Node identity v_t for different values hidden neurons Z_M.

Table 5. 5G network simulation – rogue node tampering error

5G node	Δ	Neurons (x_L, z_M, y_N)	Error	Neurons (x_L, z_M, y_N)	Error	Neurons (x_L, z_M, y_N)	Error
1	0.0	60-4-60	9.50E−11	60-60-60	9.97E−11	60-60-60	9.97E−11
	1.0		1.83E−03		1.53E−03		1.53E−03
2	0.0	124-4-60	9.84E−11	124-60-60	9.93E−11	124-124-60	9.98E−11
	1.0		2.43E−04		1.40E−04		1.40E−04
3	0.0	188-4-60	9.01E−11	188-60-60	9.93E−11	188-188-60	9.98E−11
	1.0		6.76E−05		3.57E−05		3.78E−05
4	0.0	252-4-60	8.91E−11	252-60-60	9.95E−11	252-252-60	9.99E−11
	1.0		2.58E−05		1.30E−05		1.48E−05
5	0.0	316-4-60	8.88E−11	316-60-60	9.95E−11	316-316-60	9.99E−11
	1.0		1.57E−05		6.08E−06		4.54E−06

The addition of a 5G rogue node into the Intelligent Building is detected by the learning algorithm Neural Block Chain (Fig. 6) even when the identity value only differ in a bit, $\Delta = 1.0$.

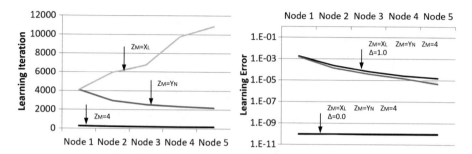

Fig. 6. 5G network simulation – rogue node tampering error

The simulation results show that the increment of neurons in the hidden layer requires additional learning iterations, however this increment is not reflected in higher accuracy to detect input errors or 5G rogue nodes, therefore $Z_M = 4$ is the most optimum configuration for this model.

6 Conclusions

This paper has presented the application of the Blockchain Random Neural Network in 5G Cybersecurity where neurons are gradually incremented as 5G node identity authentication data increases through their gradual addition into an Intelligent Building. This configuration provides the proposed algorithm the same properties as the Blockchain: security and decentralization with the same validation process: mining the input neurons until the neural network solution is found.

The Random Neural Network in Blockchain configuration has been applied to an 5G node Authentication server; experimental results show that Blockchain applications can be successfully implemented using neural networks where mining effort can be gradually increased, user authentication and data encryption in a decentralized network therefore removing centralized validation mechanisms.

This paper has proposed a Cybersecurity application in 5G where the addition of a 5G node to the Intelligent Building requires prior authentication and verification between decentralized parties. 5G node identity data is encrypted, information is decentralized where a 5G rogue node attackers can be identified if it is present in the Intelligent Building.

Future work will include the authentication and verification of additional 5G nodes and the impact of the number of chain neurons in relation to the learning and mining process. In addition, the mining threshold will be assessed against the required mining iterations that find the network solution.

Appendix

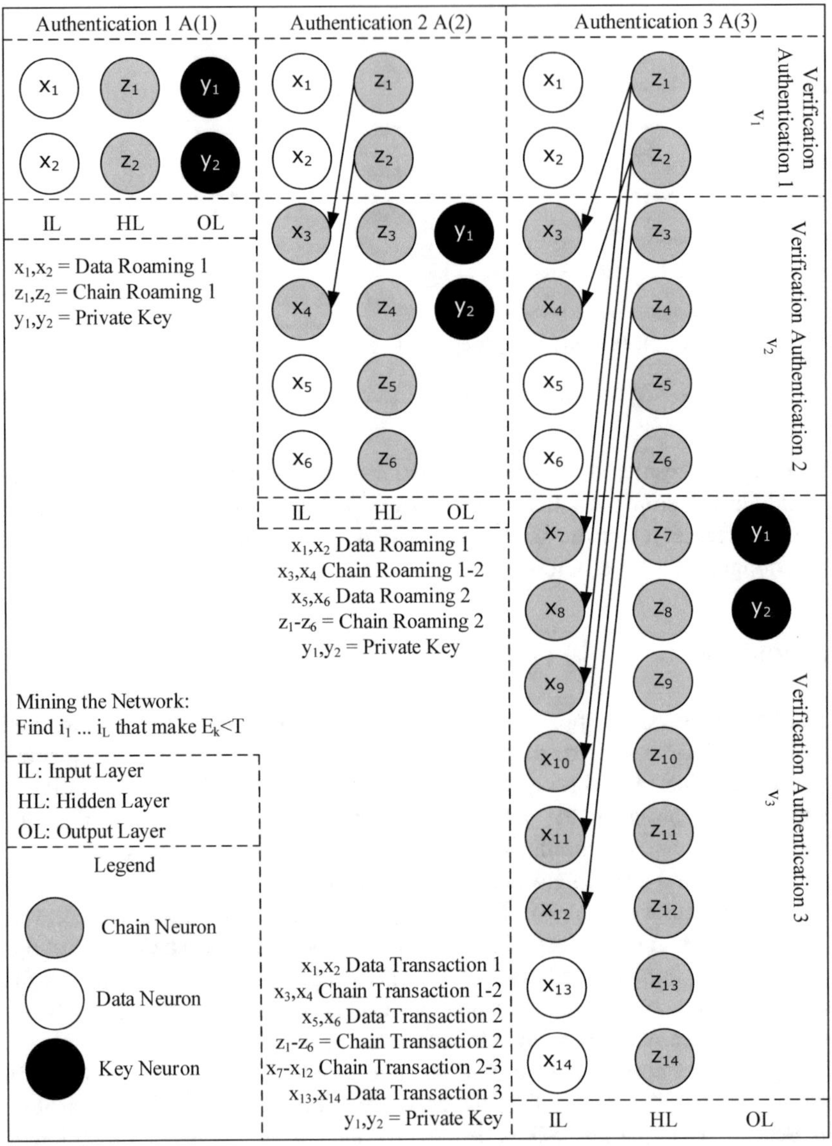

Authentication 1 A(1)
Authentication 2 A(2)
Authentication 3 A(3)
Verification Authentication 1
v1
Verification Authentication 2
v2
Verification Authentication 3
v3
IL HL OL
x1,x2 = Data Roaming 1
z1,z2 = Chain Roaming 1
y1,y2 = Private Key
x1,x2 Data Roaming 1
x3,x4 Chain Roaming 1-2
x5,x6 Data Roaming 2
z1-z6 = Chain Roaming 2
y1,y2 = Private Key
Mining the Network:
Find i1 ... iL that make Ek<T
IL: Input Layer
HL: Hidden Layer
OL: Output Layer
Legend
Chain Neuron
Data Neuron
Key Neuron
x1,x2 Data Transaction 1
x3,x4 Chain Transaction 1-2
x5,x6 Data Transaction 2
z1-z6 = Chain Transaction 2
x7-x12 Chain Transaction 2-3
x13,x14 Data Transaction 3
y1,y2 = Private Key

References

1. Palattella, M., Dohler, M., Grieco, A., Rizzo, G., Torsner, J., Engel, T., Ladid, L.: Internet of Things in the 5G era: enablers, architecture, and business models. IEEE J. Sel. Areas Commun. **34**(3), 510–527 (2016)
2. Andrews, J., Buzzi, S., Choi, W., Hanly, S., Lozano, A., Soong, A., Zhang, J.: What Will 5G Be? IEEE J. Sel. Areas Commun. **32**(6), 1065–1082 (2014)
3. Agiwal, M., Roy, A., Saxena, N.: Next generation 5G wireless networks: a comprehensive survey. IEEE Commun. Surv. Tutor. **18**(3), 1617–1655 (2016)
4. Huh, S., Cho, S., Kim, S.: Managing IoT devices using blockchain platform. In: International Conference on Advanced Communication Technology, pp. 464–467 (2017)
5. Nakamoto, S.: Bitcoin: a peer-to-peer electronic cash system - Bitcoin.org, pp. 1–9 (2008)
6. Watanabe, H., Fujimura, S., Nakadaira, A., Miyazaki, Y., Akutsu, A., Kishigami, J.: Blockchain contract: securing a blockchain applied to smart contracts. In: International Conference on Consumer Electronics, pp. 467–468 (2016)
7. Yuan, Y., Wang, F.-Y.: Towards blockchain-based intelligent transportation systems. In: International Conference on Intelligent Transportation Systems, pp. 2663–2668 (2016)
8. Biswas, K., Muthukkumarasamy, V.: Securing smart cities using blockchain technology. In: International Conference High Performance Computing and Communications/Smart City/Data Science and Systems, pp. 1392–1393 (2016)
9. Serrano, W.: The random neural network with a blockchain configuration in digital documentation. In: International Symposium on Computer and Information Sciences, pp. 196–210 (2018)
10. Serrano, W.: The blockchain random neural network in cybersecurity and the Internet of Things. In: Artificial Intelligence Applications and Innovations, pp. 50–63 (2019)
11. Adam, I., Ping, J.: Framework for security event management in 5G. In: International Conference on Availability, Reliability and Security, vol. 51, pp. 1–10 (2018)
12. Katsianis, D., Neokosmidis, I., Pastor, A., Jacquin, L., Gardikis, G.: Factors influencing market adoption and evolution of NFV/SDN cybersecurity solutions. Evidence from SHIELD Project. In: European Conference on Networks and Communications, pp. 261–265 (2018)
13. Atat, R., Liu, L., Chen, H., Wu, J., Li, H., Yi, Y.: Enabling cyber-physical communication in 5G cellular networks: challenges, spatial spectrum sensing, and cyber-security. IET Cyber-Phys. Syst. Theory Appl. **2**(1), 49–54 (2017)
14. Mozzaquatro, B., Agostinho, C., Goncalves, D., Martins, J., Jardim, R.: An ontology-based cybersecurity framework for the Internet of Things. Sensors **18**(9), 3053 (2018). 1-20
15. Chang, E., Gottwalt, F., Zhang, Y.: Cyber situational awareness for CPS, 5G and IoT. Front. Electron. Technol. **433**, 147–161 (2017)
16. Fernández, L., Huertas, A., Gil Pérez, M., Garcia, F., Martinez, G.: Dynamic management of a deep learning-based anomaly detection system for 5G networks. J. Ambient. Intell. Hum. Comput., 1–15 (2018)
17. Fernandez, L., Perales, A., Garcia, F., Gil, M., Martinez, G.: A self-adaptive deep learning-based system for anomaly detection in 5G networks. IEEE Access **6**, 7700–7712 (2018)
18. Lu, Y., Da-Xu, L.: Internet of Things (IoT) cybersecurity research: a review of current research topics. IEEE Internet Things J., 1–13 (2018)
19. Gai, K., Qiu, M., Tao, L., Zhu, Y.: Intrusion detection techniques for mobile cloud computing in heterogeneous 5G. Secur. Commun. Netw. **9**, 3049–3058 (2016)
20. Pointcheval, D.: Neural networks and their cryptographic applications. Livre des resumes Eurocode Institute for Research in Computer Science and Automation, pp. 1–7 (1994)

21. Kinzel, W., Kanter, I.: Interacting neural networks and cryptography secure exchange of information by synchronization of neural networks. In: Advances in Solid State Physic, vol. 42, pp. 383–391 (2002)

22. Klimov, A., Mityagin, A., Shamir, A.: Analysis of neural cryptography. In: International Conference on the Theory and Application of Cryptology and Information Security, vol. 2501, pp. 288–298 (2002)

23. Volna, E., Kotyrba, M., Kocian, V., Janosek, M.: Cryptography based on the neural network. In: European Conference on Modelling and Simulation, pp. 1–6 (2012)

24. Yayık, A., Kutlu, Y.: Neural Network based cryptography. Int. J. Neural Mass-Parallel Comput. Inf. Syst. **24**(2), 177–192 (2014)

25. Schmidt, T., Rahnama, H., Sadeghian, A.: A review of applications of artificial neural networks in cryptosystems. In: World Automation Congress, pp. 1–6 (2008)

26. Xu, D., Xiao, L., Sun, L., Lei, M.: Game theoretic study on blockchain based secure edge networks. In: IEEE International Conference on Communications in China, pp. 1–5 (2017)

27. Cha, S.-C., Yeh, K.-H.: An ISO/IEC 15408-2 compliant security auditing system with blockchain technology. In: IEEE Conference on Communications and Network Security, pp. 1–2 (2018)

28. Gai, K., Raymond, K.-K., Zhu, L.: Blockchain-enabled reengineering of cloud datacenters. IEEE Cloud Comput. **5**(6), 21–25 (2018)

29. Gupta, Y., Shorey, R., Kulkarni, D., Tew, J.: The applicability of blockchain in the Internet of Things. In: IEEE International Conference on Communication Systems & Networks, pp. 561–564 (2018)

30. Agrawal, R., Verma, P., Sonanis, R., Goel, U., De, A., Anirudh, S., Shekhar, S.: Continuous security in IoT using blockchain. In: IEEE International Conference on Acoustics, Speech and Signal Processing, pp. 6423–6427 (2018)

31. Gelenbe, E.: Random neural networks with negative and positive signals and product form solution. Neural Comput. **1**, 502–510 (1989)

32. Gelenbe, E.: Learning in the recurrent random neural network. Neural Comput. **5**, 154–164 (1993)

33. Gelenbe, E.: G-networks with triggered customer movement. J. Appl. Probab. **30**, 742–748 (1993)

Beautified QR Code with Security Based on Data Hiding

Huili Cai[1,2], Xiaofeng Liu[1(✉)], and Bin Yan[2]

[1] College of Internet of Things Engineering, Hohai University,
Nanjing 213022, People's Republic of China
2602676056@qq.com, xfliu@hhu.edu.cn
[2] College of Electronic and Information Engineering,
Shandong University of Science and Technology,
Qingdao 266590, People's Republic of China
yanbinhit@hotmail.com

Abstract. In this paper, we combine beautified QR (Quick Response) code with data hiding algorithm. The QR code beautification algorithm based on error correction mechanism is used to generate the beautified QR code. In the process of generating the beautified QR code, the QR code modules are removed with the error correction capability allowed. The verified QR code is added to the beautified QR code to prevent the tampering of the QR code. Thus, we can enhance the security of the QR code when user scans it. The original image covered by the QR code and the verified QR code are embedded in the beautified QR code by using the LSB (Least Significant Bit) data hiding method. The experimental results show that the beautified QR code generated by the proposed algorithm is superior to the reference method in visual quality and security.

Keywords: QR code · Data hiding · Error correction mechanism · Visual quality · Security

1 Introduction

The Quick Response Code, also known as the QR code, is an open two-dimensional code that was developed in the 1990s by DENSO WAVE [8]. As shown in Fig. 1, in the QR code symbol, there are two parts: the function patterns and an encoding region. The function patterns include the finder patterns, the separators, the alignment patterns and the timing patterns. The function patterns are designed for the decoder to accurately read the QR code. The shape and size of the function patterns are independent of the information to be encoded by the user. Conversely, the encoding region is used to store information encoded according to the coding standard, and consists of data and error correction codewords, format information and version information. According to the QR code encoding standard [7], the QR code has a certain error correction

© Springer Nature Switzerland AG 2020
Z. Ju et al. (Eds.): UKCI 2019, AISC 1043, pp. 423–432, 2020.
https://doi.org/10.1007/978-3-030-29933-0_35

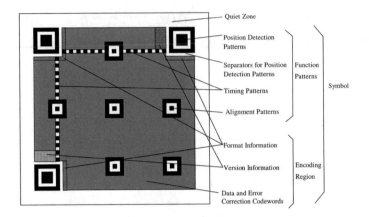

Fig. 1. Structure of standard QR Code symbol

capability. The QR code has a total of four error correction levels, corresponding to four error correction capacities. In the encoding theory of QR code, the Reed-Solomon code (RS code) error correction mechanism is widely used [11]. After the data codeword are being formed, the encoding mechanism will form corresponding error correcting codewords according to the RS error correction mechanism.

In recent years, for the visual optimization of QR code, scholars have proposed a number of algorithms. Some studies take advantage of the error correction ability of QR codes, at the expense of the readability of some modules, and replace them with commercial logos or other images [1–3,10,12]. Lin *et al.* modified the RS code to make the QR code present a specific appearance [9]. Chu *et al.* studied the method of constructing a similarity matrix to determine the best replacement rule of modules [5]. So that the QR code can present a specific image appearance. However, the quality of the generated QR code image is not good. Meanwhile, this algorithm is limited to binary images. More research has optimized the visual quality of QR codes through an effective combination of various methods.

With the increasing demand for QR code visual effects in the field of media advertising, the QR code and data hiding are combined to decode the hidden image, which not only can access the relevant links, but also restore the original image [6]. However, in this way, the standard QR code completely covers a part of the background image, which greatly destroys the integrity of the background image. Moreover, the image quality with the secret information is poor. In response to the above shortcomings, Chen proposed a data hiding algorithm based on visual QR code [4]. But Chen's algorithm has the following two problems. First, Chen directly combines the standard QR code with the original image, and the visual effect needs to be improved. Second, when user scans the beautified QR code, there is no security to be guaranteed.

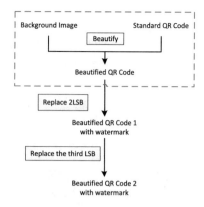

Fig. 2. The overall framework of the watermark embedding.

In this paper, we have improved Chen's algorithm. We consider the beautification and security at the same time. The QR code beautification algorithm based on the error correction mechanism is used to generate the beautified QR code. The verified QR code is added to the beautified QR code to prevent the tampering of the QR code and increase the security of the QR code that scanned by users.

2   The Proposed Algorithm

In this paper, we combine beautified QR codes with data hiding algorithm. The partial image information covered by the QR code and the verified QR code are embedded into the beautified QR code. The overall framework of the watermark embedding is shown in Fig. 2. First, the beautified QR code is generated according to the beautification algorithm proposed in Sect. 2.1. Then, the image information covered by the QR code are embedded in the 2LSB of the beautified QR code to generate the watermarked beautified QR code 1. Finally, the verified QR code is embedded in the third LSB of the beautified QR code to generate a watermarked beautified QR code 2. The QR code in the beautified QR code can be directly scanned and the extracted verified QR code is used for information authentication. The overall framework of watermark extraction is shown in Fig. 3. First, the third LSB of the received beautified QR code is extracted, and the verified QR code is obtained for decoding. Then we can authenticate with the content of the QR code directly by scanning the received beautified QR code. The 2LSB of the received beautified QR code are then extracted to recover the original image.

2.1   Generation of Beautified QR Code

The beautified QR code proposed in this paper has the following two optimizations: 1. Optimization of basic function patterns; 2. Elimination of some modules.

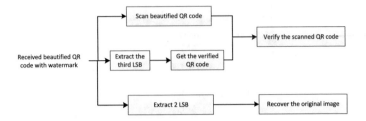

Fig. 3. The overall framework of the watermark extraction.

Fig. 4. Basic function patterns of QR code.

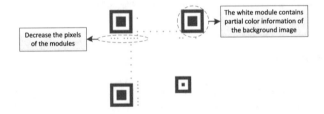

Fig. 5. Modified basic function patterns of QR code

These two optimizations herein are within the error correctable range of the QR code. The basic function patterns of the standard QR code are shown in Fig. 4. In the beautification algorithm, the basic function patterns of the QR code are modified, and the results are shown in Fig. 5. In the modified QR code basic function patterns, the white modules of the three finder patterns and alignment pattern contain partial color information of the background image, and the function pattern other than the three finder patterns are reduced the number of pixels of the module.

The steps to generate the beautified QR code are as follows:

- Generate a standard QR code (version: 3; error correction level: L; encoded content: http://www.lenna.org).
- Reduce the pixels of the basic function patterns and data modules. Embed the modules of QR code $Q(i, j)$ into the background image $I(i, j)$ according to Eq. 1 and generate the beautified QR code $I_Q(i, j)$,

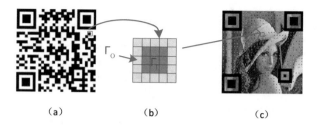

Fig. 6. Embedding the module of the QR code into a image. (a) Standard QR code, (b) One module, (c) Beautified QR code

Fig. 7. Beautified QR code that eliminated some modules

$$I_Q(i,j) = \begin{cases} I(i,j), & \forall(i,j) \in M_O \\ Q(i,j), & \forall(i,j) \in M_I \end{cases} \tag{1}$$

$$M = M_I \cup M_O \tag{2}$$

where M is a module in the QR code, M_I is the inner region of the module, M_O is the outter region of the module, as shown in Fig. 6.
- According to the error correction mechanism of the QR code, some modules of the QR code can be eliminated, and the generated result image is shown in Fig. 7.

2.2 Watermark Embedding

Traditionally, the watermarked image generated by using LSB replacement algorithm can be described according to Eq. 3,

$$\hat{H} = R(H; W) \tag{3}$$

where $H = (h_1, h_2, \cdots, h_n)$, H is the host image, $W = (w_1, w_2, \cdots, w_n)$, W is the watermark information, and R is the LSB replacement algorithm, $\hat{H}$ is the generated watermark image. First, we extract the kth LSB of h_i by using Eq. 4,

$$h_{i,k} \triangleq \frac{h_i}{2^{k-1}} \bmod 2 \tag{4}$$

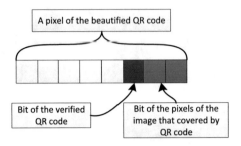

Fig. 8. Embedding of the QR code and the image information covered by the QR code.

where $h_{i,k}$ is the kth LSB of h_i, h_i is the ith element of H, $h_{i,k}$ is in the binary form. Then, we replace the LSB of h_i with w_i according to Eq. 5.

$$\hat{h}_{i,1} = w_i \tag{5}$$

Finally, we get h_i with watermark information according to Eq. 6,

$$\hat{h}_i = \sum_{k=1}^{8} \hat{h}_{i,k} \times 2^{k-1} \tag{6}$$

where $\hat{h}_i$ is in the decimal form.

According to the above method, we use Eq. 7 to generate the beautified QR code image with watermark information,

$$\hat{I}(i,j) = \mathrm{R}\left(\underbrace{I(i,j)}_{H}; \underbrace{I(i,j), (i,j) \in U}_{W}\right) \tag{7}$$

where U is the union of all M_I for all modules in the QR code. In this paper, the image information covered by the QR code is embedded in the 2LSB, and the verified QR code image is embedded in the third LSB, as shown in Fig. 8.

The embedding steps of the watermark are as follows:

- The image information covered by the QR code (black and white in Fig. 9) is embedded into the 2LSB of the beautified QR code to generate a watermarked beautified QR code 1, and the result is shown in Fig. 10(a).
- Embed the verified QR code (version: 3; error correction level: L; code content: http://www.lenna.org) into the third LSB of the watermarked beautified QR code 1 (used to verify the QR code in the beautified QR code). A watermarked beautified QR code 2 is generated, and the result is shown in Fig. 10(b).

Fig. 9. Image information covered by QR code (black and white)

(a)   (b)

Fig. 10. Watermark embedding. (a) Watermarked beautified QR code 1, (b) Watermarked beautified QR code 2.

2.3 Watermark Extraction

The steps for watermark extraction are as follows:

- Extract the third LSB of the received image to obtain the embedded verified QR code, as shown in Fig. 11(a). Then we can authenticate the content of the QR code by directly scanning the received image.
- Extract the 2LSB of the received image to restore the original image, as shown in Fig. 11(b).

3 Experimental Results

In the experiment, we select 30 images for testing. The experimental parameters are the same as [4]. The average value of PSNR of the beautified QR code generated in this paper is 32.83 dB without embedding the verified QR code. The average value of PSNR of the QR code generated by [4] is 32.54 dB. It can be seen that the image quality of the beautified QR code generated by proposed algorithm is better than that of [4]. The average value of PSNR of the recovered

(a)   (b)

Fig. 11. Results of watermark extraction. (a) Extracted verified QR code, (b) Recovered original image (PSNR $= 37.28$ dB).

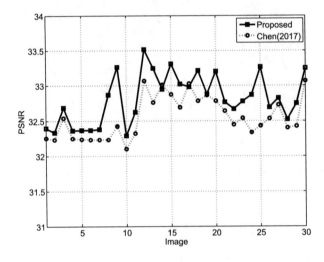

Fig. 12. Comparison of the PSNR between Chen (2017) [4] and proposed algorithm.

image in this paper is 40.88 dB and the highest is 43.68 dB. As shown in Fig. 12, the worst image quality is the test image 10, that is, 'Baboon'. Because the texture structure of the image is complicated, changing the LSB of the image will have a large impact on the image. The highest image quality is the test image 12, that is, 'Geometry'. Because the texture structure of the image is relatively simple, changing the LSB of the image does not have a large impact on the image. Because the data hiding method selected in this paper is LSB algorithm, which is a fragile watermark, the method in this paper has certain security (Figs. 13 and 14).

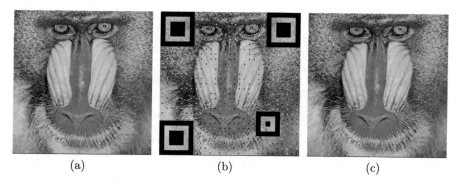

(a)   (b)   (c)

Fig. 13. Test image 10: 'Baboon'. (a) Original image, (b) Beautified QR code image, (c) Recovered original image.

(a)   (b)   (c)

Fig. 14. Test image 12: 'Geometry'. (a) Original image, (b) Beautified QR code image, (c) Recovered original image.

4   Conclusion

In this paper, we have combined visual QR codes with data hiding techniques. The QR code beautification algorithm based on the error correction mechanism is used to generate the beautified QR code. The verified QR code is added to the beautified QR code to prevent the tampering of the QR code and increase the security of the QR code when user scans it. The experimental results show that the visual quality and security of the beautified QR code generated by the proposed algorithm is better than that of [4].

Acknowledgment. This work is supported in part by the Key Research and Development Program of Jiangsu under grants BE2017071, BE2017647 and BE2018004-04, the Projects of International Cooperation and Exchanges of Changzhou under grant CZ20170018, the Fundamental Research Funds for the Central Universities under grant 2018B47114, the Open Research Fund of State Key Laboratory of Bioelectronics, Southeast University under grant 2019005, and the State Key Laboratory of Integrated Management of Pest Insects and Rodents under grant IPM1914.

References

1. Logoq.net. http://logoq.net
2. Visualead.com. http://www.visualead.com
3. Bhardwaj, N., Kumar, R., Verma, R., Jindal, A., Bhondekar, A.P.: Decoding algorithm for color QR code: a mobile scanner application. In: 2016 International Conference on Recent Trends in Information Technology (ICRTIT), pp. 1–6, April 2016
4. Chen, S.K.: Auto-recovery from photo QR code. In: 13th International Conference on Intelligent Information Hiding and Multimedia Signal Processing, pp. 290–295, August 2017
5. Chu, H.K., Chang, C.S., Lee, R.R., Mitra, N.J.: Halftone QR codes. ACM Trans. Graph. **32**(6), 1–8 (2013)
6. Huang, H.C., Chang, F.C., Fang, W.C.: Reversible data hiding with histogram-based difference expansion for QR code applications. IEEE Trans. Consum. Electron. **57**(2), 779–787 (2011)
7. ISO/IEC Standard 18004: Information Technology – Automatic Identification and Data Capture Techniques – Bar Code Symbology – QR Code (2000)
8. Kan, T.W., Teng, C.H., Chou, W.S.: Applying QR code in augmented reality applications. In: Proceedings of the 8th International Conference on Virtual Reality Continuum and its Applications in Industry, pp. 253–257, December 2009
9. Lin, S.S., Hu, M.C., Lee, C.H., Lee, T.Y.: Efficient QR code beautification with high quality visual content. IEEE Trans. Multimed. **17**(9), 1515–1524 (2015)
10. Liu, S.J., Zhang, J., Pan, J.S., Weng, C.J.: SVQR: a novel secure visual quick response code and its anti-counterfeiting solution. J. Inf. Hiding Multimed. Signal Process. **8**(5), 1132–1140 (2017)
11. Liu, S.J., Zhang, J., Pan, J.S., Weng, C.J.: A novel information embedding and recovering method for QR code based on module subdivision. J. Inf. Hiding Multimed. Signal Process. **9**(2), 515–522 (2018)
12. Tkachenko, I., Puech, W., Destruel, C., Strauss, O., Gaudin, J.M., Guichard, C.: Two-level QR code for private message sharing and document authentication. IEEE Trans. Inf. Forensics Secur. **11**(3), 571–583 (2016)

A Macro Human Resource Management Platform Enabled by Big Data Technology

Hongwei Wang[1(✉)], Yichun Yang[2], and Yufei Zhang[3]

[1] ZJU-UIUC Institute, Zhejiang University, Haining, China
hongweiwang@zju.edu.cn
[2] School of Journalism and Communication, Xiamen University, Xiamen, China
[3] School of Mechanical Engineering, Zhejiang University of Technology, Hangzhou, China

Abstract. In the era of big data, digital technologies are increasingly used to solve a range of problems, with the support of the Internet, Cloud computing and Internet of things. Human resource management is a traditional discipline that faces the challenge of ever-increasing diversity and complexity of requirements – something that can be solved by applying new technologies. This paper first introduces the development of big data technology and the importance of human resource management in enterprises and governments. Then, it proposes a novel solution for developing a macro human resource management service platform which not only helps collect data about human resources requirements from both enterprises and local governments but also provides support of data classification, analysis and prediction. This paper describes the solutions mentioned above in detail and discusses their evaluation in the development of a prototype system and a case study. The results show that new technologies have great potential in improving effectiveness and efficiency in traditional management disciplines like human resource management.

Keywords: Human resource management · Digital administration · Big data · Data classification and prediction · Data visualization

1 Introduction

As the wide application of modern information technologies and intelligent devices, such as Internet, cloud computing and Internet of things, data and information resources grow at an exponential rate, both in life and at work. The traditional data processing methods cannot make full use of the effective information carried by these data, so new data processing modes are needed to process and analyze a large amount of data [1]. Big data technology has become more and more mature with the advent of the big data era. The concept of big data was first proposed in the McKinsey&Company's report 'Big data: The next Frontier for Innovation, Competition, and Productivity'. Big data refers to the use of the latest data processing methods to mine, process and manage the rapidly growing and massive data sets. Big data technology adopts the latest data processing technology, which can explore potential logical associations from massive data resources and mine the changing pattern of data so as to accurately predict

© Springer Nature Switzerland AG 2020
Z. Ju et al. (Eds.): UKCI 2019, AISC 1043, pp. 433–445, 2020.
https://doi.org/10.1007/978-3-030-29933-0_36

the future trend of data [2]. With the development of digitalization in a range of industries, big data provides a new concept of promoting the development of modern society as well as scientific support for governments and social organizations to realize intelligent decision-making.

The rapid development of the IT industry in China has accumulated abundant data resources. In order to seize the development opportunity of the big data market, the State Council issued the Platform for Action for Great Data Development in August 2015, and raised the development and application of large data to the national strategic level. As a consequence, implementation of the national big data strategy is proposed (1) to expand the network economic space; (2) to promote the open sharing of the data resources; (3) to implement the national big data strategy; (4) and to advance the next generation of the Internet and to improve the timeliness and accuracy of information about economic operations. As the main force of science and technology change in the new era, big data technology can analyze massive data resources and play a guiding role in decision-making, both in life and at work. It has been applied to all aspects of enterprise management, business activities and daily life, such as medical treatment, retail, education, management and other fields [3].

This paper first analyzes the influence of big data on human resource management, and establishes the service platform of the macro human resource management with big data technology as the core. Next, the data collected from a survey of human resource status of large companies in Diankou Town, are analyzed and processed. Last, the paper analyzers the regional human resource situation and proves the feasibility of the service platform.

2   Big Data in the Field of Human Resources Management

With the trend of economic globalization, it has become increasingly important for enterprises to improve effectiveness and efficiency to survive in global competition, and human resource management has become a very important part of this competition. In the past few decades, human resource management has become the focus of enterprise development, and innovation of human resources management is a necessary measure to improve the competitiveness of enterprises [4]. At present, the main challenges of traditional human resources management are as follows: (1) in the era of knowledge, enterprises and the regional government are under the pressure of talent competition, and there is a lack of effective talent incentive mechanism; (2) the purpose of traditional human resource management is to exert the maximum potential of enterprise personnel and create more benefits. The existing management system is not good enough for the long-term development of enterprises; (3) human resources management departments recruit more employees only to fill job gaps, not to make the best use of their talents; and (4) it is impossible to predict the development trend of human resources in the future. Big data technology has promoted the innovation of human resource management. By using big data technology, more rational decisions can be made to allocate human resources to real needs and improve utilization of these resources. The advantages of big data in human resources management are as follows:

1. Big data technology can be used for unified analysis and processing of global data and the dimension of data is extensible [5]. The problem can be set up and analyzed in different contexts, and the island problem between different sources of data can be solved. In the era of big data, what is important is correlation rather than causation. There may be no direct causality between different sources of data, but the correlation between data can be concluded by effective analysis of the data.
2. Big data technology can use the collected human resources data for data prediction so as to identify issues early and formulate related countermeasures for these issues. For example, the forecast of employee demand can predict the total number of talents needed by enterprises in a region, and can on this basis formulate the corresponding talent policy and strengthen recruitment efforts. In this way, effective recruitment plans can be made for the region and the enterprises alike.
3. Big data technology can realize the visual analysis of data, so as to improve the understandability of data and improve the efficiency of information sharing.

Big data provides forward-looking analysis of human resources management, which can predict the future trends of human resources development based on a large amount of historical data. Additionally, it provides sufficient basic data support for the decision-making and measurement management of human resources [6].

2.1 Micro Human Resource Management

As the name suggests, micro human resource management refers to the management of human resource within an enterprise or organization, which mainly involves personnel performance appraisal, staff training, personnel transfer, talent selection, etc.

Digital management enabled by big data analysis is central to micro human resource management, which can achieve accurate analysis of individual personnel in terms of the knowledge structure, ability and expertise of individual. In the process of human resource management in enterprises, hospitals and universities, big data management thinking and relative methods are used in the recruitment and training of professionals, as well as the evaluation of performance and the handling of labor relations, and so on.

2.2 Macro Human Resource Management

Macro human resource management refers to the overall human resources management of a country or region according to the specific development of its economy. It mainly includes establishment and introduction of relevant policies by national governments, the optimal allocation of human resources within a country, and so on.

The change of macro human resource management, such as macro human resources planning, the overall design of the welfare system and retirement system arrangement, should adopt big data as comprehensively as possible, so as to promote more appropriate reforms of national and government personnel system in an effective way.

The government makes a comparative analysis according to the job demand and the situation of job-seekers through statistical analysis of regional human resources data. The analysis of the employment situation can be made from numerous factors such as composition of the employment group, average salaries of the industries concerned, stability of staff, etc. The analysis of unemployment situation can be based on the causes of unemployment, the average time of unemployment, the amount of social benefits and the number of unemployed, etc.

3   A Macro Human Resource Service System

When making decisions on regional human resources, governments need tremendous data such as those about current situations and those about future trends. The arrival of the big data era has opened up a new opportunity for the macroscopic human resource management of governments. Big data enabled human resource service platform can provide data support for local and regional governments to implement appropriate talent policy and recruitment strategy.

3.1   Overall Architecture of the Platform

Architecture of the proposed macro human resource management service platform mainly comprises the service platform layer, the data acquisition layer, the data storage layer, the data analysis layer and the data visualization layer.

Specifically, the service platform layer includes network equipment, storage equipment and the operating system; the data acquisition layer include human resource data provided by enterprises; the data processing layer involves analysis and processing (e.g. prediction and classification) of the acquired data; the data storage layer involves storage of the processed data in databases and the data visualization layer involves presentation of data using visualization techniques (Fig. 1).

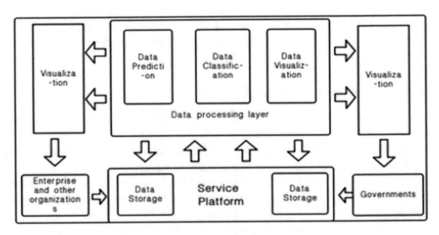

Fig. 1. Architecture of the service platform.

3.2   Data Analysis

Data Prediction. Exponential Smoothing (ES) can predict the required result with less data, and the result of the prediction is more accurate with respect to other prediction methods, because the previous data will not be abandoned, but the influence of the data can be controlled by the size of the weight value. The most commonly used ES methods include single ES, quadratic ES and cubic ES and higher order ES. Among these methods, the quadratic ES takes into account not only all the historical data but also the changing trend of time series. So this method is used in this research to carry out data prediction, with the mathematical model below:

Given smoothing coefficient α, the single ES is represented as follows:

$$S_t^{(1)} = \alpha x_{t-1} + (1 - \alpha)S_{t-1}^{(1)} \tag{1}$$

Then the formula for calculating the quadratic ES is:

$$S_t^{(2)} = \alpha S_t^{(1)} + (1 - \alpha)S_{t-1}^{(2)} \tag{2}$$

The calculation formula for predicting the value x_{t+T} of the future T is as follows:

$$x_{t+T} = M_T + N_T T \tag{3}$$

Where

$$M_T = 2S_t^{(1)} - S_t^{(2)} \tag{4}$$

$$N_T = \left(\frac{\alpha}{1 - \alpha}\right)\left(S_t^{(1)} - S_t^{(2)}\right) \tag{5}$$

Data Classification. In traditional statistical methods, parameters of the hypothetical distribution probability model are deduced back while the naive Bayesian classification algorithm takes a different approach of establishing mathematical models according to actual reasoning. Naive Bayes obtain the prior probability and conditional probability through the training set in the method of direct generation, and the classification results can be obtained more accurately. Naive Bayes method is a classification method based on Bays theorem and independent hypothesis of characteristic conditions. The basic principles of naive Bayesian classification algorithm are as follows:

The probability of occurrence of events X and Y is given, which are independent of each other. And their conditional independence formulas are obtained:

$$P(X, Y) = P(X)P(Y) \tag{6}$$

The probability of event X occurring under the condition that event Y occurs is known as the conditional probability of event X under event Y, recorded as P(X|Y). Two representations of conditional probability formula are given below:

$$P(X|Y) = P(X, Y)P(Y) \tag{7}$$

$$P(Y|X) = P(X, Y)P(X) \tag{8}$$

That is:

$$P(X|Y) = P(Y|X)P(X)P(Y) \tag{9}$$

Total probability formula is:

$$P(Y) = \sum_{k}^{n} P(Y|X = X_k)P(X_k) \tag{10}$$

Among,

$$\sum_{k}^{n} P(X_k) = 1 \tag{11}$$

Bayes formula is:

$$P(X_k|Y) = \frac{P(Y|X_k)P(X_k)}{\sum_{k}^{n} P(Y|X = X_k)P(X_k)} \tag{12}$$

Visualized Analysis. In this research, managers can upload the processed results to the platform for display. The results input interface is shown in Fig. 2.

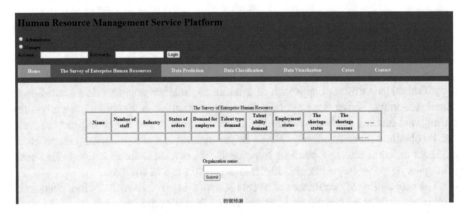

Fig. 2. Data collection interface

4 A Case Study

The proposed method and human resource management platform are described in detail in this section through a case study of data obtained from a survey done with 184 companies in Diankou Town, Zhejiang Province, China.

4.1 Forecast of Talent Demand

Using the talent demand data in Table 1, the calculation steps below can be used:

Table 1. The number of employees in Diankou Town from 2008 to 2017.

Year	Serving officers
2008	216.83
2009	222.99
2010	245.17
2011	281.22
2012	299.46
2013	296.42
2014	294.48
2015	319.51
2016	349.35
2017	355.03

1. Select smoothing index $\alpha = 0.5$ and calculate the single ES value.

$$S_{2008}^{(1)} = \frac{x_{2008} + x_{2009} + x_{2010}}{3} \tag{13}$$

$$S_{2009}^{(1)} = \alpha x_{2008} + (1 - \alpha)S_{2008}^{(1)} \tag{14}$$

$$S_t^{(1)} = \alpha x_{t-1} + (1 - \alpha)S_{t-1}^{(1)} \tag{15}$$

Calculate it successively to t = 2017.

2. Calculate the quadratic ES value.

$$S_{2008}^{(2)} = \frac{S_{2008}^{(1)} + S_{2009}^{(1)} + S_{2010}^{(1)}}{3} \tag{16}$$

$$S_{2009}^{(2)} = \alpha S_{2008}^{(1)} + (1 - \alpha)S_{2008}^{(2)} \tag{17}$$

$$S_t^{(2)} = \alpha S_t^{(1)} + (1 - \alpha)S_{t-1}^{(2)} \tag{18}$$

Calculate it successively to $t = 2017$.

3. Forecast of talent demand in the next four years.

$$x_{t+T} = M_T + N_T T \tag{19}$$

Where, $T = 1, 2, 3, 4$.

$$M_T = 2S_t^{(1)} - S_t^{(2)} \tag{20}$$

$$N_T = \left(\frac{\alpha}{1 - \alpha}\right)\left(S_t^{(1)} - S_t^{(2)}\right) \tag{21}$$

4.2 The Relationship Between Talent Type and Enterprise Industry

Table 2. Data of different talent types and the industries they are concerned with.

Professional talents	General-purpose talent	Innovative talent	Skilled people	Ordinary workers	Couple talents	Industry
1	1	1	1	1	1	Auto parts
0	0	0	0	1	0	Auto parts
1	0	0	1	0	1	Hardware machinery
1	0	1	1	1	0	Hardware machinery
0	1	0	1	0	0	Hardware machinery
1	0	0	0	0	1	New pipe
1	1	1	1	1	1	Copper fabrication
1	0	1	1	0	0	Auto parts
1	0	0	0	0	0	New pipe
1	1	1	1	1	0	copper fabrication
1	1	0	1	0	0	The auto parts

Using the data in Table 2, the main flow of naive Bayes calculation is:

1. Import raw data;
2. Division of training set and test set;
3. Train the training set data to generate Bayesian classifiers;
4. Put the test set data into the classifier and compare it with the expected results.

Suppose that the given variable Y is the enterprise type and x_i is an independent feature variable, the equation below can be derived from the Bayes theorem:

$$P(Y|x_1, x_2, \ldots \ldots, x_n) = \frac{P(Y)P(x_1, x_2, \ldots \ldots, x_n|Y)}{P(x_1, x_2, \ldots \ldots, x_n)} \qquad (22)$$

Where $i = 1, 2, \cdots, n$; $n = 184$.

According to the assumptions of the Bayesian formula for the prior conditions, each feature is independent of each other, and the formula above can be converted to:

$$P(Y|x_1, x_2, \ldots \ldots, x_n) = \frac{P(Y) \prod_{i=1}^{n} P(x_i, Y)}{P(x_1, x_2, \ldots \ldots, x_n)} \qquad (23)$$

Where $i = 1, 2, \cdots, n$; $n = 184$.

4.3   Data Visualization

In this experiment, some of the data obtained are displayed using visualization techniques, including data of job opening, the demand of employees, the reasons for lack of work and enterprise recruitment mode.

5   Experiment Results and Analysis

5.1   Demand Forecast Results and Analysis

The prediction of personnel demand using the platform is shown in Fig. 3 where graphic results are used to indicate the trend. The data used are listed in Table 3.

Table 3. Comparison of raw data with predicted results

Year	Original value	Predicted value	Relative error
2008	216.83	0	
2009	222.99	228.33	−0.024
2010	245.17	225.66	0.080
2011	281.22	235.42	0.163
2012	299.46	258.32	0.137
2013	296.42	278.89	0.059
2014	294.48	287.65	0.023
2015	319.51	291.07	0.089
2016	249.35	305.29	−0.224
2017	255.03	327.32	−0.283
2018		341.17	
2019		325.94	
2020		310.71	
2021		295.48	

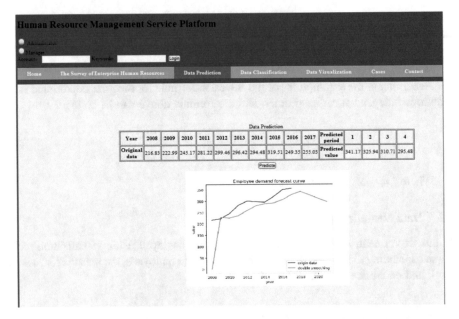

Fig. 3. Forecast results of personnel demand.

The employee demand forecast curve is shown in Fig. 4.

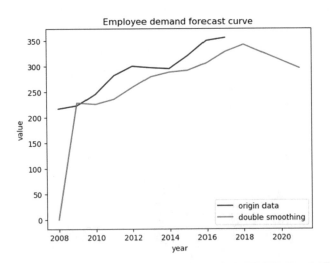

Fig. 4. Forecast of employee demand.

Both the relative error between the predicted value and the actual value is less than 0.3, and the trend of the prediction curve prove that the accuracy of the result of using the quadratic exponential smoothing is high. As such, the method is to a great extent feasible. From the personnel demand forecast curves, it can be seen that the overall number of employees needs does not significantly fluctuate in the coming years.

5.2 Data Classification Result Analysis

Through the classification of the industries the enterprises belong to and the varied types of employees they need, the visualized display of results are given in Fig. 5.

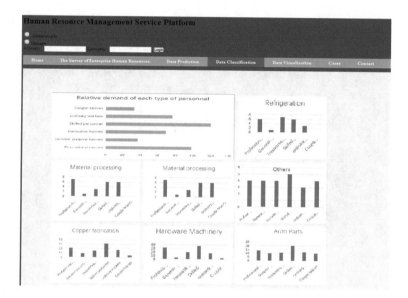

Fig. 5. The classification results

On the whole, there is more demand for skilled personnel in Diankou Town. Among them, industries of hardware machinery, copper processing and auto parts are relatively more in need of skilled personnel, focusing on the demand of senior technicians; refrigeration industry needs innovative talents; and the pipe industry needs ordinary workers. In Diankou Town, most of the enterprises belong to the hardware machinery or copper processing enterprises. Hence, formulating the talent policy aim to introducing talents for precision and realizing the accurate matching of people and posts, the government can increase the introduction of skilled senior technicians and attract more talents at home and abroad.

5.3 Data Visualization Analysis

In terms of personnel shortage, more enterprises, in Diankou Town, are lack of workers in the production line, followed by senior technicians. If this is combined with the demand obtained through the survey, it can be concluded that nearly half of the enterprise's demand for workers in the production line is still increasing. There are some difficulties such as shortage of worker and employee turnover, as discovered both in this research and the survey. The data visualization part shows clearly that the main reasons for the loss of the employees in the area are the higher wages offered by surrounding enterprises and the seasonal fluctuation of orders in a short term. In this context, more on-site recruitment and internal staff recommendation are adopted, resulting in the lack of information of job seekers. Enterprises need to improve the online recruitment and expand the source of employment personnel information. And the government needs to attract more foreign personnel at the macro level (Fig. 6).

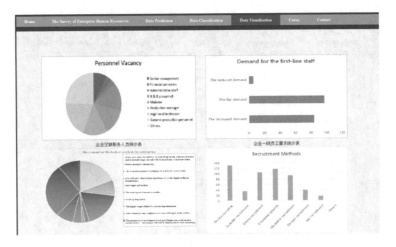

Fig. 6. Visualization of results

6 Conclusion

In the era of big data, the development of the regional economy depends more on the macro human resource management to gradually change the traditional human resource model and constantly improve the human resource management system. Using big data to analyze and process human resource data can improve the management efficiency and decision-making effectiveness of enterprises and governments alike. Prediction of data of talent demand can identify the future development trend of talent demand in advance, and provide data support for local and regional to make informed decisions. Classification of talent type and the industry enterprise further makes the talent policy more suitable and data visualization enables more straightforward display of analysis results and better readability of data. Through big data technology, governments can make relevant policies suitable for the economic development of local enterprises on the premise of understanding the local human resource situation and future trend and combining the characteristics of local enterprises. The proposed platform proves to be useful in the case study and more work will be done in the future to develop further functions for the prototype system.

References

1. Ren, J.: Research on management transforms of enterprise human resources in Big Data era. In: 4th International Conference on Business, Economics and Management, Shandong, China (2017)
2. Mauro, A.D., Greco, M., Grimaldi, M., Ritala, P.: Human resources for Big Data professions: a systematic classification of job roles and required skill sets. Inf. Process. Manage. **54**(5), 807–817 (2018)
3. Zang, S., Ye, M.: Human resource management in the era of Big Data. J. Hum. Resour. Sustain. Stud. **03**(01), 41 (2015)
4. Calvard, S.T., Jeske, D.: Developing human resource data risk management in the age of Big Data. Int. J. Inf. Manage. **43**, 159–164 (2018)
5. Zhang, K., Xu, P.: Research on transformation strategy of enterprise human resource management in Big Data era. In: 2018 International Conference on Management, Economics, Education and Social Sciences (MEESS 2018), Shanghai, China (2018)
6. Fang, W.: Research on the innovation of human resource management in the era of artificial intelligence and Big Data. In: 5th International Conference on Economics, Management and Humanities Science (ECOMHS 2019), Bangkok (2019)

Intelligent Healthcare

Classification of EEG Signals
Based on Image Representation
of Statistical Features

Jodie Ashford, Jordan J. Bird$^{(\boxtimes)}$, Felipe Campelo, and Diego R. Faria

School of Engineering and Applied Science, Aston University, Birmingham, UK
{ashfojsm,birdj1,f.campelo,d.faria}@aston.ac.uk

Abstract. This work presents an image classification approach to EEG brainwave classification. The proposed method is based on the representation of temporal and statistical features as a 2D image, which is then classified using a deep Convolutional Neural Network. A three-class mental state problem is investigated, in which subjects experience either relaxation, concentration, or neutral states. Using publicly available EEG data from a Muse Electroencephalography headband, a large number of features describing the wave are extracted, and subsequently reduced to 256 based on the Information Gain measure. These 256 features are then normalised and reshaped into a 16×16 grid, which can be expressed as a grayscale image. A deep Convolutional Neural Network is then trained on this data in order to classify the mental state of subjects. The proposed method obtained an out-of-sample classification accuracy of 89.38%, which is competitive with the 87.16% of the current best method from a previous work.

Keywords: Machine learning · Convolutional neural networks ·
Image recognition · Mental state classification · Electroencephalography

1 Introduction

Human-machine interaction is often considered a mirror of the human experience; sound and visuals constitute voice recognition, human activity classification, facial recognition, sentiment analysis and so on. Though, with the availability of sensors to gather data that the human body cannot, interaction with machines can often exceed the abilities of the natural human experience. An example of this is the consideration of electroencephalographic brainwaves. The brain, based on what a person is thinking, feeling, or doing, has a unique pattern of electrical activity that emerges as a consequence of the aggregate firing patterns of billions of individual neurones [1,2]. These electrical signals can, in principle, be detected and processed to infer the state of the brain and, by extension, the mental state

---

* J. Ashford and J. J. Bird—co-first authors.

Z. Ju et al. (Eds.): UKCI 2019, AISC 1043, pp. 449–460, 2020.
https://doi.org/10.1007/978-3-030-29933-0_37

of a given subject. Besides clinical applications, this possibility is also useful, e.g., for brain-machine interfacing.

More effective methods of feature extraction and classification are of utmost importance in brain-machine interaction, since better performing models can interpret human brain activity with higher accuracy. Previous works [3,4] suggest that static statistical descriptions of brainwaves present the information in the signals in a more machine learning-friendly shape than the raw waves themselves, even when temporally-aware machine learning methods are employed.

This work focuses on the process of feature extraction, selection and formatting in order to achieve improved classification accuracy of EEG signals. More specifically, the main contribution is a framework to perform classification of these signals, based on (i) the extraction of a large number of static statistical features of the data, followed by (ii) automated feature selection and (iii) representation of the selected attributes as a 2D matrix. The resulting matrices are (iv) interpreted as grayscale images, which allows the leveraging of the state-of-the-art performance of convolutional neural networks [5,6] as image classifiers.

The remainder of this paper is organised as follows. A brief presentation of the background concepts related to the present work is provided in Sect. 2, followed by the description of the proposed approach in Sect. 3. The results obtained by the proposed method are discussed in Sect. 4. Finally, conclusions and suggestions of further investigations are provided in Sect. 5.

2   Background

Electroencephalography (EEG) is a technique used to measure the electrical activity of a brain. The human brain contains billions of neurones, which each exhibit electrical activity in the form of nervous impulses [7, p. 31]. The electrical signal produced by a single neurone is difficult to detect, but the combined signal from the action of many neurones together can be measured using EEG [8, p. 4].

Typically, EEG involves placing electrodes onto the scalp of the subject. These electrodes measure the voltage fluctuations generated by thousands of active neurones in the brain. These signals are then digitised and amplified [9,10]. Possibly the main advantages of this method of measuring brain activity are that it is a non-invasive and inexpensive technique. Even less invasive techniques, such as the Muse headband, have extended the utility of EEG beyond medical examination alone, at the expense of sensitivity. Unlike imaging techniques such as MRI, EEG can measure fluctuations in electrical activity on the scale of milliseconds, which makes it an incredibly powerful tool for measuring real-time brain activity in response to stimulus [10].

The ability to infer human mental states is as important in human-machine interaction as a form of Affective Computing [11] as it is in natural human interaction. In the past, such techniques have used attributes available to humans: speech, gestures, facial expressions, etc. [12,13]. With the increasing development of non-invasive EEG technology, researchers can take advantage of sensors available only to machines to attempt to classify human emotions directly from the brain. Such analysis is less dependent on environmental factors or differences in somatic expression between individuals. It also offers a more seamless avenue of human-machine communication.

2.1   Related Work

Non-invasive EEG headsets have been used in previous works to analyse mental states. In a related example, data from the Muse headset was was found to be useful in the evaluation of the enjoyment levels of subjects playing two different video games [14]. The findings aligned with the current understanding of how waves detected by EEG (in this case, frontal theta frequencies) map to enjoyment. This is an example of how non-invasive techniques can provide uses of EEG outside of the medical setting, and provide data for emotional classification.

Previous works have also shown the excellent performance of convolutional neural networks (CNN) in EEG-based mental state classification. In 2017, using the DEAP dataset [5], EEG signals were classified using both deep (DNN) and convolutional (CNN) neural networks [6]. Two different classifications were performed: one for valence and one for arousal, classifying each as either high or low. The DNN achieved 75.78% accuracy for valence and 73.28% for arousal, while the CNN achieved 81.41% and 73.35% respectively [6].

Projection of EEG data onto a "visual" space is a fairly recent approach, with relatively little work as of yet performed into its exploration. Most of the relevant literature in this area [15, 16] relates to mapping the signal readings of electrodes to a spatial representation of the brain itself, interpolating intermediary points based on values from the nearest electrodes. Alternatively, some limited but successful work has explored the CNN classification of visual spectrograms produced by the signals [17]. Spectrograms produced by Limited Field Potentials have also found varying levels of success in classifying biological signals from rat brains [18,19]. In these works, a limited set of five features were extracted and machine learning approaches (decision trees, discriminant analysis, support vector machines, and nearest neighbour classifiers) were used to recognise patterns, producing results with accuracies ranging from 95.8% to 98.8%. These solutions, though effective, rarely consider statistical processing of the waves as a way to extract relevant data from the complex waveforms generated by EEG. Visual pixel-wise approaches and subsequent CNN applications have been successfully implemented in other biological domains such as image segmentation of electron microscopy images [20], with promising results for a variety of applications.

The solution suggested in this study, on the other hand, is based on extracting statistical features from EEG signal waves and maps them onto static 2D matrices, which are then represented as images and used for the classification of mental states using a convolutional neural network. This proposed methodology is detailed in the following section.

3   Proposed Approach

Firstly, an available training set of EEG signals is preprocessed. The data is assumed to contain the time series related to one or more electrodes, within a given experimental time frame, labelled in terms of three distinct mental states (relaxed, concentrating, and neutral) that the subjects were keeping during data collection [4]. From these signals a number of statistical features are extracted [3, 4], resulting in a high dimensional attribute space - in the case of this work, 1274 features are generated for each time window, as detailed in Sect. 3.1. To focus only on the most relevant ones for the classification process, feature selection is applied to the resulting features. Here, the $16^2 = 256$ most descriptive ones, based on the estimated information gain [21], are selected.

Finally, the selected features are converted into a 16×16 grid of numerical values normalised to the $[0, 1]$ range, which can be represented as a grayscale image. Figure 1 shows a number of samples of relaxed, neutral and concentrating brainwave data, using this particular image representation.

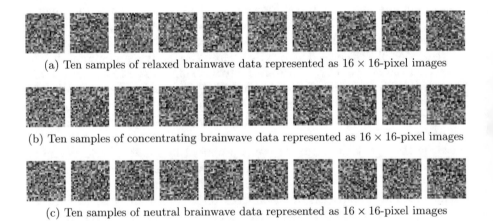

(a) Ten samples of relaxed brainwave data represented as 16×16-pixel images

(b) Ten samples of concentrating brainwave data represented as 16×16-pixel images

(c) Ten samples of neutral brainwave data represented as 16×16-pixel images

Fig. 1. Examples of image representations for each of the three mental states considered in this work.

The resulting set of images is then used to train a convolutional neural network (CNN) [22] as a classifier of the three mental states investigated in this particular work. The details of the CNN are provided in Sect. 3.2.

3.1   Feature Extraction

Due to the temporal, auto-correlated nature of the EEG waves, single-point features cannot generally provide enough information for good rules to be generated by machine learning models. In this work we follow the approach of extracting statistical features based on sliding time windows [3,4]. More specifically, the

EEG signal is divided into a sequence of windows of length one second, with consecutive windows overlapping by 0.5 s, e.g., $[(0s-1s), [0.5s-1.5s), [1s-2s),$ ...]).

Assume that each 1-second time window contains a sequence $\mathbf{x} = [x_1, \ldots, x_N]$ composed of N samples. Also let $\mathbf{x}_{h1}$ and $\mathbf{x}_{h2}$ denote the first and second halves of the window, and $\mathbf{x}_{q1}, \mathbf{x}_{q2}, \mathbf{x}_{q3}, \mathbf{x}_{q4}$ denote the four quarter-windows obtained by dividing the window into four (roughly) equal-sized parts, each composed of approximately $N/4$ samples.[1]

In this work the following statistical features were generated for each time window:

- Considering the full time window:
    - The sample mean and sample standard deviation of each signal (8 features).
    - The sample skewness and sample kurtosis of each signal [23] (8 features).
    - The maximum and minimum value of each signal (8 features).
    - The sample variances of each signal, plus the sample covariances of all signal pairs [24] (10 features).
    - The eigenvalues of the covariance matrix [25] (4 features).
    - The upper triangular elements of the matrix logarithm of the covariance matrix [26]. (10 features)
    - The magnitude of the frequency components of each signal, obtained using a Fast Fourier Transform (FFT) [27] (300 features).
    - The frequency values of the ten most energetic components of the FFT, for each signal (40 features).
- Considering the two half-windows:
    - The change in the sample means and in the sample standard deviations between the first and second half windows, for all signals (8 features).
    - The change in the maximum and minimum values between the first and second half-windows, for all signals (8 features).
- Considering the quarter-windows:
    - The sample mean of each each quarter-window, plus all paired differences of sample means between the quarter-windows, for all signals (56 features).
    - The maximum (minimum) values of each quarter-window, plus all paired differences of maximum (minimum) values between the quarter-windows, for all signals (112 features).

Regarding the representation of the signals in the frequency domain using FFT [27], two specific aspects were taken into account: first, the DC-component of the signals was filtered out prior to the application of the FFT, so the zero-frequency component was always set as zero. This was done to prevent the offset to completely dominate the power spectrum, even though it carries no relevant information for the classification task. The second aspect is that frequencies in

---

[1] In this work we standardised the number of samples within each window to $N = 150$. This means that quarter-windows have either $n = 37$ or $n = 38$ observations.

the range of (50 ± 1) Hz were also filtered out, to remove any contamination from the AC electrical distribution frequency, which could also skew the power spectrum of our signals.

Each window receives as features the vector of quantities computed above for both itself and the window that immediately precedes it (*1-lag window*). Features from the 1-lag window that were clearly redundant due to the half-window overlaps were removed prior to the composition of the feature vector, namely the sample means, maximum and minimum values of $\mathbf{x}_{q3}$ and $\mathbf{x}_{q4}$, as well as their respective differences. In the end a total of 989 features were generated for each time window (except the first, which was only used as the 1-lag for the second one).

After the statistical features were extracted the resulting dataset was composed of 2479 data objects, each represented by its corresponding 989 feature values plus a single class label. Feature selection was then performed based on the Information Gain of each feature, and the total number of features was reduced to 256 (plus class label). Due to privacy considerations the raw EEG data cannot be released, but the processed dataset is publicly available at https://www.kaggle.com/birdy654/eeg-mental-state-v2 as a UTF-8 encoded CSV with approximately 6 MB.

3.2   Convolutional Neural Network

Convolutional neural networks (CNN) [28] are a specialised kind of neural network for processing data that has a known grid-like topology, which makes them particularly suitable for dealing with data represented as time series or images [22]. The main distinguishing feature of these networks is the use of a convolution [29] instead of simple matrix multiplication in at least one of their layers [22]. Convolutional neural networks are generally very effective at image classification tasks [30–32], which motivates their use here. For more details on these networks, please refer to Ian Goodfellow *et al.*'s book on the subject [22].

In this particular work we have opted for using the CNN implementation available in the Keras Deep Learning Python library [33]. The network was trained on an Nvidia GTX1060 (1280 CUDA Cores, 6 GB 8 Gbps GDDR5 VRAM). The topology and hyperparameters of the convolutional neural network were defined based on preliminary, trial-and-error experimentation. Table 1 shows the resulting model for the classification of brainwave images.

Other design choices that were arbitrarily set in this experiment are the use of the ADAM optimiser [34] to train the network; and the use of a batch size of 100, trained for 400 epochs, with the loss calculated via categorical cross entropy at a 70/30 validation split:

$$CE = -\sum_{c=1}^{M} y_{o,c} \log(p_{o,c}), \qquad (1)$$

where M is the number of class labels (in this case, 3), y is a binary indication of a correct prediction (1 or 0), and p is the predicted probability of observation

Table 1. Network topology and parameters used. Please refer to the Keras documentation [33] for specific definitions.

Layer	Output	Params
Conv2d (ReLu)	(0, 14, 14, 32)	320
Conv2d (ReLu)	(0, 12, 12, 64)	18496
Max Pooling	(0, 6, 6, 64)	0
Dropout (0.25)	(0, 6, 6, 64)	0
Flatten	(0, 2304)	0
Dense (ReLu)	(0, 512)	1180160
Dropout (0.5)	(0, 512)	0
Dense (Softmax)	(0, 3)	1539

o of class c. The entropy of each class within the testing split is calculated and added for a final, overall result. In this case, this is the entropy of the three classes of mental state - relaxed, neutral, and concentrating.

4   Results

In this section the results for the experiments are presented. The experiments were performed three times, the difference between the three runs being random seeds set at the start of the experiment. The overall final score always resulted within 400 epochs. Accuracy and loss per-epoch are illustrated for the first run.

4.1   Results Obtained

Figure 2 illustrates the accuracy and loss of the network, for both training and testing data from the validation split. The overall out-of-sample accuracy of the CNN in classifying the dataset was 89.38% (665/744 correct classifications, $CI_{0.95} = [86.94, 91.50]\%$). As can be observed, the accuracy curve saturates after about 50 epochs, after which the loss starts increasing. This suggests that

Table 2. Comparison with related studies using the same dataset as this experiment. Column *Accuracy* also provides 95% confidence intervals for the accuracy.

Study	Method	Validation	Focus	Accuracy
This study	Inf. Gain Selection, CNN	70/30 Split	Accuracy	89.38% [86.94, 91.50]
[3]	OneR Selection, Random Forest	10-fold	Accuracy	87.2% [85.7, 88.6]
[35]	Evol. Selection, DEvoMLP	5-fold	Accuracy, Resource Usage	79.8% [78.1, 81.5]

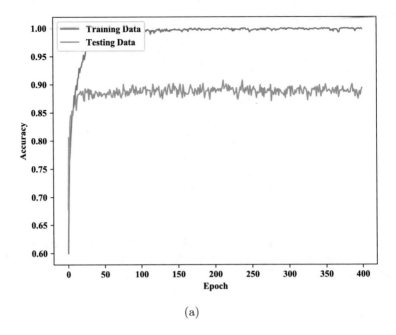

(a)

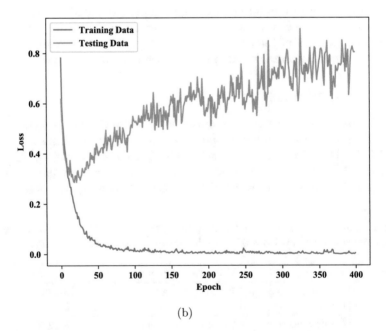

(b)

Fig. 2. (a) Accuracy and (b) Loss of the CNN for Training and Testing Data across 400 epochs.

computational resources are essentially wasted after this point, and more parsimonious training can be employed in the future.

Table 2 contrasts the results obtained in this paper with previous works dealing with the same mental state dataset. It is worth mentioning that only one of the compared experiments had the single goal of maximising accuracy, while the other was also focused on minimising computational effort. Another noteworthy point is that both previous works used cross-validation instead of a split set in order to estimate accuracy. With these factors in mind, the approach used in the present work has provided results that seem to be very competitive, with a point estimate of the accuracy that is approximately 2.18% greater (in absolute terms) than the one reported in the 2018 study. This difference is not, however, statistically significant at the 95% confidence level ($p = 0.129$ using the chi-squared test for equality of two binomial proportions [24]). Despite not clearly outperforming the current state of the art, this result suggests that the proposed approach of coupling an image-based representation of the data with CNN-based classification may represent an effective new strategy for performing EEG classification, with potential extensions to classification in the context of general time series data.

5   Future Work and Conclusions

In this work, a new approach for classification of EEG signals has been presented, based on the sequential application of statistical feature extraction and selection, normalisation and subsequent projection of the selected features as small images, and classification based on a convolutional neural network. The results obtained for this method have been shown to be very competitive with the best known results to date for the available dataset.

Possibly the most clear limitation of the present work is related to the question of generality. Since a single dataset is used, it remains to be seen how well the proposed methodology generalises not only to larger, possibly richer EEG data, but also - and more interestingly - to other similar time series. In this regard, further testing and statistical assessment of the proposed methodology are fundamental next steps as this line of research progresses.

Due to the limited available resources, the experiment reported in this work used a simple 70/30 data split instead of the more usual (but more computationally demanding) cross-validation, which should be used in future experiments whenever possible so as to obtain better estimates of out-of-sample accuracy [36]. Two other aspects related to the issue of limited resources were present. The first was the lack of a principled parameter tuning approach for both the structure and other parameters of the network, which can be optimised using, e.g., iterated racing [37], hyperheuristics [38], or topology-specific tuning methods [39–41]. Even under more constrained computational budgets, traditional design and analysis of experiments approaches [42] can be useful in defining the best network for this particular problem. The second issue is related with the selection of only 256 features to compose the image to be used in the training of the CNN. Future work in this direction should concern the testing of varying image sizes in order to better fine-tune the attribute selection process. In addition, further methods of feature extraction should be investigated and compared, rather than focusing solely on Information Gain as this study has done. The investigation of other CNN architectures, which have shown much promise in other contexts [43], is also an interesting point for further development.

Regardless of the possible improvements discussed above, we argue that the proposed framework of projecting selected features onto a 2D matrix and subsequent image recognition through a Convolutional Neural Network already constitutes a competitive approach for brainwave data classification. The results obtained are promising, as compared to current scientific standards, and further exploration is strongly suggested to advance the results beyond the preliminary outcome presented in this paper.

References

1. Caton, R.: The electric currents of the brain. Am. J. EEG Technol. **10**(1), 12–14 (1970)
2. Llinás, R.R.: Intrinsic electrical properties of mammalian neurons and cns function: a historical perspective. Front. Cell. Neurosci. **8**, 320 (2014)
3. Bird, J.J., Manso, L.J., Ribiero, E.P., Ekart, A., Faria, D.R.: A study on mental state classification using EEG-based brain-machine interface. In: 9th International Conference on Intelligent Systems, IEEE (2018)
4. Bird, J.J., Ekart, A., Buckingham, C.D., Faria, D.R.: Mental emotional sentiment classification with an EEG-based brain-machine interface. In: The International Conference on Digital Image and Signal Processing (DISP 2019). Springer, (2019)
5. Koelstra, S., Muhl, C., Soleymani, M., Lee, J.-S., Yazdani, A., Ebrahimi, T., Pun, T., Nijholt, A., Patras, I.: Deap: a database for emotion analysis; using physiological signals. IEEE Trans. Affect. Comput. **3**(1), 18–31 (2012)
6. Tripathi, S., Acharya, S., Sharma, R.D., Mittal, S., Bhattacharya, S.: Using deep and convolutional neural networks for accurate emotion classification on deap dataset. In: Twenty-Ninth IAAI Conference (2017)
7. Purves, D., Augustine, G., Fitzpatrick, D., Hall, W., LaMantia, A., McNamara, J., Williams, S.: Neuroscience. Sinauer Associates, Sunderland (2004)
8. Britton, J.W., Frey, L.C., Hopp, J., Korb, P., Koubeissi, M., Lievens, W., Pestana-Knight, E., St, E.L.: Electroencephalography (EEG): An introductory text and atlas of normal and abnormal findings in adults, children, and infants. American Epilepsy Society, Chicago (2016)
9. Buzsáki, G., Anastassiou, C.A., Koch, C.: The origin of extracellular fields and currents–EEG, ECOG, LFP and spikes. Nat. Rev. Neurosci. **13**(6), 407 (2012)
10. Cohen, M.X.: Analyzing Neural Time Series Data: Theory and Practice. MIT press, Cambridge (2014)
11. Picard, R.W.: Affective Computing. MIT press, Cambridge (2000)
12. Pantic, M., Rothkrantz, L.J.: Toward an affect-sensitive multimodal human-computer interaction. Proc. IEEE **91**(9), 1370–1390 (2003)
13. Rouast, P.V., Adam, M., Chiong, R.: Deep learning for human affect recognition: insights and new developments. In: IEEE Transactions on Affective Computing (2019)

14. Abujelala, M., Abellanoza, C., Sharma, A., Makedon, F.: Brain-EE: Brain enjoyment evaluation using commercial EEG headband. In: Proceedings of the 9th ACM International Conference on Pervasive Technologies Related to Assistive Environments, p. 33. ACM (2016)
15. Abhang, P.A., Gawali, B.W.: Correlation of EEG images and speech signals for emotion analysis. Br. J. Appl. Sci. Technol. **10**(5), 1–13 (2015)
16. Gevins, A., Smith, M.E., McEvoy, L., Yu, D.: High-resolution EEG mapping of cortical activation related to working memory: effects of task difficulty, type of processing, and practice. Cerebral cortex (New York, NY: 1991), vol. 7, no. 4, pp. 374–385 (1997)
17. Zhang, X., Wu, D.: On the vulnerability of cnn classifiers in EEG-based BCIS. IEEE Trans. Neural Syst. Rehabil. Eng. **27**(5), 814–825 (2019)
18. Wang, X., Magno, M., Cavigelli, L., Mahmud, M., Cecchetto, C., Vassanelli, S., Benini, L.: Embedded classification of local field potentials recorded from rat barrel cortex with implanted multi-electrode array. In: 2018 IEEE Biomedical Circuits and Systems Conference (BioCAS), pp. 1–4. IEEE (2018)
19. Wang, X., Magno, M., Cavigelli, L., Mahmud, M., Cecchetto, C., Vassanelli, S., Benini, L.: Rat cortical layers classification extracting evoked local field potential images with implanted multi-electrode sensor. In: 2018 IEEE 20th International Conference on e-Health Networking, Applications and Services (Healthcom), pp. 1–6, IEEE (2018)
20. Mahmud, M., Kaiser, M.S., Hussain, A., Vassanelli, S.: Applications of deep learning and reinforcement learning to biological data. IEEE Trans. Neural Netw. Learn. Syst. **29**(6), 2063–2079 (2018)
21. Tan, P.-N.: Introduction to Data Mining. Pearson Education India, Chennai (2018)
22. Goodfellow, I., Bengio, Y., Courville, A.: Deep Learning. MIT Press, Cambridge (2016). http://www.deeplearningbook.org
23. Zwillinger, D., Kokoska, S.: CRC Standard Probability and Statistics Tables and Formulae. Chapman & Hall, London (2000)
24. Montgomery, D.C., Runger, G.C.: Applied Statistics and Probability for Engineers. John Wiley & Sons, New Jersey (2010)
25. Strang, G.: Linear Algebra and its Applications. Brooks Cole, California (2006)
26. Chiu, T.Y., Leonard, T., Tsui, K.-W.: The matrix-logarithmic covariance model. J. Am. Stat. Assoc. **91**(433), 198–210 (1996)
27. Van Loan, C.: Computational frameworks for the fast Fourier transform, vol. 10, Siam (1992)
28. LeCun, Y., Boser, B.E., Denker, J.S., Henderson, D., Howard, R.E., Hubbard, W.E., Jackel, L.D.: Handwritten digit recognition with a back-propagation network. In: Advances in Neural Information Processing Systems, pp. 396–404 (1990)
29. Oppenheim, A.V., Willsky, A.S., Nawab, S.: Signals and Systems. Prentice Hall, New Jersey (1996)
30. Ciresan, D., Meier, U., Schmidhuber, J.: Multi-column deep neural networks for image classification. In: 2012 IEEE Conference on Computer Vision and Pattern Recognition, pp. 3642–3649 (2012)
31. Russakovsky, O., Deng, J., Su, H., Krause, J., Satheesh, S., Ma, S., Huang, Z., Karpathy, A., Khosla, A., Bernstein, M., Berg, A.C., Fei-Fei, L.: ImageNet large scale visual recognition challenge. Int. J. Comput. Vis. **115**(3), 211–252 (2015)
32. Szegedy, C., Liu, W., Jia, Y., Sermanet, P., Reed, S., Anguelov, D., Erhan, D., Vanhoucke, V., Rabinovich, A.: Going deeper with convolutions. In: Proceedings of the IEEE Conference on Computer Vision and Pattern Recognition, pp. 1–9 (2015)

33. Chollet, F., et al.: Keras. https://keras.io (2015)
34. Kingma, D.P., Ba, J.: Adam: a method for stochastic optimization. arXiv e-prints, p. arXiv:1412.6980, Dec 2014
35. Bird, J.J., Faria, D.R., Manso, L.J., Ekart, A., Buckingham, C.D.: A deep evolutionary approach to bioinspired classifier optimisation for brain-machine interaction. Complexity **2019**, 14 (2019)
36. Kohavi, R.: A study of cross-validation and bootstrap for accuracy estimation and model selection. In: Proceedings of the 14th International Joint Conference on Artificial Intelligence, vol. 2, IJCAI 1995, pp. 1137–1143 (1995)
37. López-Ibáñez, M., Dubois-Lacoste, J., Pérez Cáceres, L., Stützle, T., Birattari, M.: The irace package: iterated racing for automatic algorithm configuration. Oper. Res. Perspect. **3**, 43–58 (2016)
38. Burke, E.K., Gendreau, M., Hyde, M., Kendall, G., Ochoa, G., Özcan, E., Qu, R.: Hyper-heuristics: a survey of the state of the art. J. Oper. Res. Soc. **64**(12), 1695–1724 (2013)
39. Martín, A., Lara-Cabrera, R., Fuentes-Hurtado, F., Naranjo, V., Camacho, D.: Evodeep: a new evolutionary approach for automatic deep neural networks parametrisation. J. Parallel Distrib. Comput. **117**, 180–191 (2018)
40. Assunçao, F., Lourenço, N., Machado, P., Ribeiro, B.: Denser: deep evolutionary network structured representation. arXiv preprint arXiv:1801.01563 (2018)
41. Bird, J.J., Ekart, A., Faria, D.R.: Evolutionary optimisation of fully connected artificial neural network topology. In: SAI Computing Conference 2019, SAI (2019)
42. Montgomery, D.C.: Design and Analysis of Experiments, 8th edn. John Wiley & Sons, New Jersey (2012)
43. Ji, S., Xu, W., Yang, M., Yu, K.: 3D convolutional neural networks for human action recognition. IEEE Trans. Pattern Anal. Mach. Intell. **35**(1), 221–231 (2013)

A Non-invasive Subtle Pulse Rate Extraction Method Based on Eulerian Video Magnification

Yang Wei[1]([⊠])(iD), Nadezhda Gracheva[2], and John Tudor[2]

[1] Nottingham Trent University, Nottingham NG11 8NS, UK
yang.wei@ntu.ac.uk
[2] University of Southampton, Southampton SO17 1BJ, UK

Abstract. Measuring pulse rate by means of video recording of the wrist area is a non-invasive approach. Eulerian Video Linear Magnification (EVLM) is used in this paper to magnify, and make visible, subtle pulse-induced wrist motions. A series of experiments are conducted to investigate the performance of ELVM under various conditions, such as light intensity, background colour and a set of video recording parameters. The results show that a light intensity of around 224 to 229 lx is optimal; excess or inadequate light significantly impairs the success of amplifying the skin movement resulting from the pulse. It is demonstrated that a white background colour enables both the radial and ulnar areas to be clearly visible in the recorded video, thus improving pulse measurement. In addition, it is shown that a female's pulse strength is approximately 40% weaker than that of a male averaged over the participants.

Keywords: Video magnification · Non-invasive · Pulse detection

1 Introduction

Video magnification (VM) is a technique that reveals subtle imperceptible motions by selectively amplifying them to be visible to the human eye. Video magnification has developed rapidly in the last few years with applications in visual vibrometry [1], medicine [2] and industrial engineering [3]. In [1], the material properties of visible objects are estimated by analysing subtle and otherwise imperceptible vibrations in a video and the use of VM offers a promising alternative to more specialised tools such as laser vibrometry. He *et al.* [2] extracted the subtle pulse information from the wrist area based on a video recorded using a digital camera to predict important cardio-vascular events. Wadhwa *et al.* [3] applied VM to estimate the subtle vibrations of large objects such as a crane swaying in the wind. It was found that, in a controlled environment, the estimated vibrations (i.e. acceleration) correlate well with the results from a commercially available accelerometer. A number of papers also report human pulse magnification based on colour and motion magnification. Magnification of subtle colour changes might in addition show evenness of blood flow, which could be useful for early diagnosis of arterial problems [4]. These studies investigate the reliability of the method compared to conventional procedures, such as manual measurement and pulse oximetry. In a video-based pulse extraction scenario, the wrist [2], the face [5] and the neck [6] are the areas typically exploited.

© Springer Nature Switzerland AG 2020
Z. Ju et al. (Eds.): UKCI 2019, AISC 1043, pp. 461–471, 2020.
https://doi.org/10.1007/978-3-030-29933-0_38

Eulerian Linear Video Magnification (EVLM) was first presented by Wu *et al.* [4], where both colour variations in the facial area, and motion magnification in the wrist area, were analysed. It was demonstrated that the algorithm amplifies skin colour changes for both light and dark skin complexions. This paper also proved that the pulse rate extracted using video magnification matches the results obtained using a photoplethysmogram. He *et al.* [5] measured the pulse transit time from colour changes at both the wrist and neck; an Arduino board and PulseSensors [7] are used with ELVM measurements. 10 subjects were tested but no results or statistics on the accuracy were reported. Miljkovic *et al.* [4] presented pulse rate measurement based on ELVM by amplifying colour changes around the facial area. Two volunteers were involved in the study and the results verified by electrocardiogram. It was demonstrated that ELVM was a reliable method for measuring pulse rate from the face but no data on accuracy was reported.

No study investigated the performance of ELVM under various ambient conditions, such as under different lighting intensities, skin complexions, camera selections and background colours. The contribution of this paper is to study the effect of these parameters on ELVM. This paper also presents an investigation of the environmental, equipment and subject variations in order to optimise ELVM based pulse amplification. Section 2 of this paper describes the background and compares different video magnification techniques. Section 3 details the experimental setup for using video magnification to acquire the pulse rate from the inner wrist area. Section 4 provides measurement results whilst varying the measurement conditions in terms of ambient light, background colour, camera type and participant gender, skin tone and texture. Finally, conclusions are given in Sect. 5 followed by acknowledgements and references.

2   Background

Lagrangian motion magnification was the first reported use of VM which was achieved by tracking the trajectories of pixels over time [8]. However, the computational time is long since it requires complex processing. For example, the magnification processing of a video with a resolution of 866×574 pixels, and a duration less than 30 s, requires 10 h. In contrast, the Eulerian approach does not track motion explicitly as the input video is spatially decomposed, temporally filtered and then the motion of interest is multiplied by an amplification factor α [6]. In comparison to the Lagrangian approach, Eulerian Video Magnification (EVM) is less computationally demanding and does not use complex feature tracking procedures. In addition, EVM produces a higher quality output video and also additionally magnifies colour variations.

In Table 1, four EVM techniques are presented [9–12]. All the EVM techniques employ spatial decomposition followed by temporal filtering. However, the various EVM techniques differ in the spatial decomposition approach and whether amplitude or phase based amplification is used. Computational time grows towards the bottom of the table. ELVM has the shortest computational time, but the output video is the lowest quality with higher noise. For the aim of this research, where the pulse magnification in

Table 1. Eulerian Video Magnification techniques.

Technique	Spatial decomposition	Colour space	Computational time (sec)
Eulerian Linear VM	Laplacian pyramids	YIQ	35.6
Phase - Based VM	Riesz pyramids	YIQ	75
Phase - Based VM	Complex steerable pyramids	YIQ	325.9
Dynamic VM	Decomposition into a foreground and background	Not provided	Not provided

a wrist area is of interest, ELVM is chosen as it offers the shortest computational time and so provides a fast pulse measurement.

To implement ELVM, the raw input video is first subjected to spatial decomposition. Generally, each video frame is represented by a Laplacian pyramid in which each spatial band is temporally processed to extract the motion of interest. The extracted motion is amplified by a given amplification factor α. Finally, the amplified motion is added back to the original video and the pyramid is collapsed to reconstruct the video. The temporal filtration employs a first-order Taylor series expansion analysis [13]. The relationship between temporal processing and motion magnification is represented by a 2D matrix reported in [6], shown in Eq. 1.

$$f(x) + (1+\alpha)\delta(t)\frac{\partial f(x)}{\partial x} \approx f(x + (1+\alpha)\delta(x)) \tag{1}$$

Where x and t are position and time, $\delta(t)$ is the displacement function and α is the amplification factor. The bounds for amplification factor given motion $\delta(t)$ is therefore:

$$(1+\alpha)\delta(t) < \frac{\lambda}{8} \tag{2}$$

For pulse magnification, the motion of interest is always a motion at relatively low spatial frequencies as the subtle motion itself is never large; this implies implicitly unconstrained parameters so that the amplification factor α is the only explicitly constrained parameter. More detailed analysis is included in [6].

3 Video Magnification Setup

Pulse can be measured by palpating arteries in the body. All arteries can be used for pulse measurements but it is easier to palpate the artery at particular places [14]:

- Where an artery is closer to the skin surface.
- When the artery is located just above firm tissue.

Two arteries, radial and ulnar, on the wrist can be used to measure the pulse but the radial artery is normally used for manual measurement as this location satisfies both the bulleted conditions above. However, for video magnification, the second condition is not required and therefore the ulnar artery may also be considered if it is close enough to the skin surface so that the subtle intensities changes can be captured. The original MATLAB code of Eulerian Linear Video Magnification is a free source that is available from MIT [15]. It executes C++ functions using MEX files.

3.1   Video Recording Parameters

Since microscopic intensity changes are to be detected and amplified in this research, the quality of the video is of great importance. The video clips of the wrist area in MOV format are used as the input due to its uncompressed quality compared to the MP4 format [16, 17].

The video clips are recorded in the laboratory using: (i) a Nikon D5500 camera, and (ii) an iPhone 6 camera. When using a professional camera, there are often two video standards available: (i) NTSC (60 Hz) [9], (ii) PAL (50 Hz) [18]. This difference is due to the differences in the alternating current (AC) frequency at different geographical areas in the world. This is important to note because light flickering will be clearly visible in the video if a camera is not synchronised with the AC frequency of the area where the video is recorded. Automatic flicker reduction is an option when a camera standard cannot meet the flicker requirements of the area. Since the video is recorded in the UK which is a PAL area, the flicker reduction option was used while using the NTSC video standard. The reasons for working with the NTSC standard are explained in the next paragraph.

The choice of the frame rate is based on the motion to be magnified. If more frequent motions occur, more frames per second are required in order to capture the desired signal with sufficient accuracy to produce an accurate magnification. The videos reported in [6] are recorded at 30 frames per second (fps) which is the option offered with the NTSC standard. A variety of frame rates were investigated in this research: 25 fps (PLA), 30 fps (NTSC), 50 fps (PAL) and 60 fps (NTSC); more frames require greater computational effort. This heavily affects the computational time without giving advantages in quality, so 25/30 fps are preferred over 50/60 fps. Nevertheless, 30 fps produces a result of higher quality compared to 25 fps. As a result, 30 fps NTSC was selected for this research.

A higher resolution provides more pixels and hence more detail in the area of interest. However, the computational time is much longer than with a lower resolution. For a resolution of 640×424, a full Laplacian pyramid contains 361, 852 elements for each colour layer, whereas a full Laplacian pyramid of a frame with a resolution of 1920×1080 has 2, 774, 885 elements. It is clear that computational effort grows dramatically for a greater resolution as processing of a video with a resolution of 1920×1080 takes up to 14 times longer compared to 640×424 (from 70 to 80 s to 20 min). Hence, a resolution of 640×424 at 30 fps was chosen as an efficient video specification offering satisfactory quality.

3.2   Algorithm Parameters

It is important to note that all the processing is achieved in the YIQ colour space and not RGB. The Y component represents the luminance information, and I and Q are the chrominance component on the orange-blue and purple-green axes, respectively. This colour space is a rotated version of RGB [19]. YIQ colour space is chosen to reduce the colour artefacts by attenuating the chrominance components.

The Laplacian pyramid is used for spatial decomposition as it provides access to different spatial frequencies without performing direct transformations in the frequency domain [20]. To construct a Laplacian pyramid, a Gaussian pyramid is required. A recursive process of building a Gaussian pyramid generally comprises two steps: (i) low-pass filtering an image and (ii) down sampling the result by a factor of 2. To obtain a Laplacian pyramid, a blurred version of the original image needs to be subtracted at each level of a Gaussian pyramid. In ELVM, a binomial filter is used to construct the Laplacian pyramid [21].

Temporal filtering is required to eliminate amplification of the motion at undesired frequencies. Such motion might be noise due to non-ideal camera sensors or involuntary subtle movements of a hand. For the pulse magnification, the most suitable filter is a relatively wide band pass at low frequencies to ensure that all the pulse frequencies fall within the pass band. A second-order infinite impulse response (IIR) band pass filter is then used to extract the pulse. In general, subtraction of two low pass filters produces a second-order band pass filter [8] and this is employed in the implementation of the ELVM. In practice, the ELVM would become extremely computationally demanding if filtering full Laplacian pyramids for every intensity of a pixel at all frames in a video clip. Therefore, an approximate IIR filter is used to minimise computational effort.

$$lowpass1 = (1 - r1) * pyramid + r1 * lowpass1; \qquad (3)$$

$$lowpass2 = (1 - r2) * pyramid + r2 * lowpass2; \qquad (4)$$

$$filtered = lowpass1 - lowpass2; \qquad (5)$$

In the above approximation, r1 is the parameter corresponding to an upper cut-off frequency $\omega1$ and r2 is the lower cut-off frequency $\omega2$ of the passband. Such an approximation imposes constraints on the frequency parameters, which should take values greater than 0 but less than 1. Under such conditions, the frequency band of 0.4 Hz to 5 Hz, used for pulse extraction, is approximated with a pair of parameters r2 = 0.04 and r1 = 0.5. This type of approximation restricts the range of frequencies that might be extracted but is sufficient for pulse magnification, while speeding up calculations.

For different sub-bands, the amplification factor might vary due to its dependency on spatial wavelength (λ) which controls the cut-off spatial frequency. The spatial wavelength is calculated for each frame and depends only on the resolution. It is clear that, with each down sampling step, λ becomes smaller by a factor of two. Hence, for each iteration, the current amplification factor αcur is recalculated considering the spatial wavelength λ and calculated displacement δ. The amplification factor decreases as the size of a sub-band becomes smaller.

4   Pulse Extraction Results

Lighting is crucial for recording video for the purposes of pulse magnification. Inadequate or excess light will impair the ability of ELVM to capture subtle intensity changes. The experiments were all conducted with the same lighting conditions arising from the ceiling lamps in the laboratory. However, to investigate how lighting affects the performance of the algorithm, two additional devices were used: a bench fluorescent lamp LC8076 (LightCraft Magnifier Lamp) providing a consistent shadow free light source and Light meter LX - 8809A.

The additional fluorescent lamp was fixed to the bench and positioned with its tube towards the plane on which a wrist is located. The experiments investigated the following light conditions:

- No additional light source.
- An additional light source at three different distances from a wrist.
- With artificial shadow which is introduced by a solid cover above the wrist area.

The detection performance is quantified using a Successful Pulse Magnification Rate (SPMR) which is defined as follows:

$$SPMR = \frac{N_s}{N_T} \tag{6}$$

Where Ns is the number of successful detection and Nt is the total number of participants. In every experiment, the illuminance was measured using the light meter, of which the sensor was positioned in the same plane as the wrist. 10 participants took part in the experiments and the illuminance measurements with the corresponding successful pulse magnification rate (SPMR) taken as an average over all participants, are shown in the Table 2.

Table 2. Light intensity measurement at various conditions.

Lighting condition	Illuminance (lx)	$SPMR_{average}$ (%)
With extra shadow	74–76	20
No extra light	224–229	86
Lamp at ≈1 m away	387–391	40
Lamp at ≈50 cm away	496–498	20
Lamp at ≈20 cm away	1179–1180	10

The results show that lighting condition is a crucial parameter and it has a great effect on the performance of ELVM. It is seen that ELVM produces 86% correct results within a range of 224 to 229 lx, while for lighting conditions around 74 to 76 lx, the SPMR drops below 20%. With the additional light source at less than 20 cm away, the SPMR was 0 and there is no pulse seen in the output video.

4.1 Recording Background Variation

Colours reflect light of different wavelengths so that the background may create distinct lighting conditions in a frame. In [6], a higher contrast background produces a higher SPMR. The hypothesis is therefore that the colour reflecting most light would produce a better environment for video magnification. It is anticipated that a white-coloured background is the most suitable one. To test this seven colours were investigated as the background as shown in the frames from each video clip in Fig. 1.

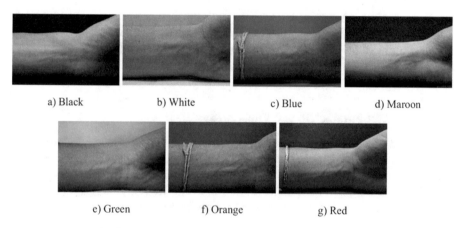

a) Black   b) White   c) Blue   d) Maroon

e) Green   f) Orange   g) Red

Fig. 1. Comparisons between backgrounds with different colours.

It is clear that each frame has a slightly different tone. Apart from the white background, the bottom side of the arm is covered in shadow and so is less visible and clear because of the different reflective properties of the different background colours. White is the only colour to provide a clear full wrist frame image. In Fig. 1, the radial artery is located at the upper part of the wrist and clearly visible in all frames. However, the ulnar artery is located at the bottom and is in shadow for all background colours except white. For some participants, the pulse occurs only in the area of the ulnar artery and therefore using the non-white backgrounds increases the risk of missing important pulse information.

4.2 Recording Camera Selection

Two subjects are tested with a Nikon digital single-lens reflex (DSLR) D5500 and an iPhone 6 camera to compare processing results. The first participant, P1, has a strong pulse, which is clearly magnified under different conditions in various experiments. The second participant, P2, has a moderate pulse, which might not be magnified if the conditions are not chosen carefully (Fig. 2).

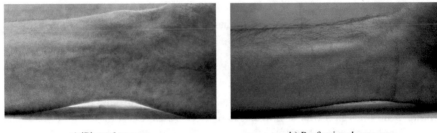

a) iPhone6 camera   b) Professional camera

Fig. 2. Output video frames recorded using (a) an iPhone 6 built-in camera and (b) a Nikon D5500.

Processing the iPhone 6 video with the same parameters as were successful for the Nikon does not reveal the magnified pulse. This indicates that the sensors of the phone camera do not sufficiently capture the tiny intensity changes. In addition, there is a greater noise level in the iPhone output video when compared with that of the professional Nikon camera.

4.3 Final Test Parameters and Experiments

10 male and 5 female participants with different physical skin textures and tones took part the experiments to optimise the ELVM parameters. All the participants were tested under the conditions of: a white background, an illuminance of around 225 lx and a distance to the camera of approximately 27 cm. The ELVM processing parameters, however, are tuned in order to investigate the performance in detail (Fig. 3).

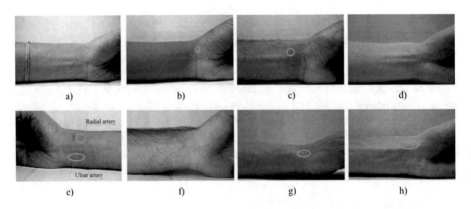

Fig. 3. Examples of areas where pulses appear to be magnified on different participants.

The pulses have different strengths and generally female participants have a weaker pulse than males. Two magnified pulses are demonstrated in Fig. 4.

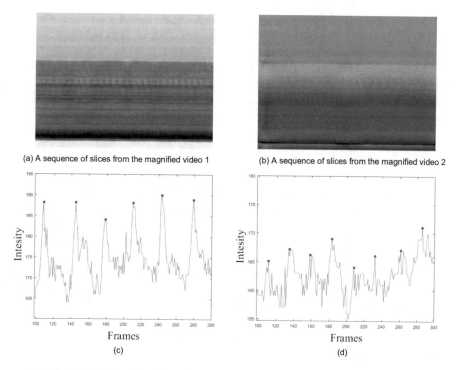

(a) A sequence of slices from the magnified video 1   (b) A sequence of slices from the magnified video 2

(c)   (d)

Fig. 4. Magnified pulse of a male participant (a, c) and a female participant (b, d).

A magnified pulse of a male is shown in Fig. 4(a) and the pulse as a pixel intensity series of a single pixel is given in Fig. 4(c). Peaks that correspond to the pulse are clearly visible in Fig. 4(c). Such a pulse is defined as strong as it is clearly magnified under various conditions with different processing parameters. The maximum-to-minimum difference for this intensity series is 25. A pulse of a female participant is demonstrated in (b) and (d). It is seen, in a sequence of slices and in the intensity series, that this pulse is weaker. The maximum-to-minimum difference for this series is 15, which is 40% less than that for the male pulse in (c). These graphs demonstrate the generally observed difference in the strength of male and female participants. To validate this information, the whole group of participants is summarised in Table 3.

Table 3. Summary of pulse strength.

Category	Participants		Percentage (%)	
	Male	Female	Male	Female
Weak	0	2	0	40
Moderate	3	2	30	40
Strong	7	1	70	20

The performance of the algorithm did not depend on complexion but only on the strength of the pulse. The use of ELVM is proven to work for both light and dark skin tones. The magnification pulse is visible at one of the arteries or at both.

The results also showed that Body Weight Index (BMI) also has an influence on the success of the ELVM algorithm as the pulse rates of those participants with a BMI higher than 25 were difficult to detect which is believed to be due to both the radial and ulnar arteries being covered by excess body fat.

5 Conclusions

The Eulerian Linear Video Magnification (ELVM) algorithm was evaluated by implementing it in MATLAB with various operation parameters and conditions for measuring the pulse rate from the inner wrist area of a human subject. A video clip from a camera was used as the source and the pulse at the radial artery was successfully amplified to reveal the subtle pulse variation. For operation parameters, environmental light intensity and the background colour were evaluated as well as the type of camera. The results show that a light intensity between 224 and 229 lx, with a white colour background, provide the highest magnification rate when the video is taken using a professional DSLR camera. It is also shown that the successful magnification rate on male subjects is much higher than on female subjects due to the weaker pulse strength of females.

Acknowledgement. We acknowledge support for this project as part of the partnership resource of the EPSRC Interdisciplinary Research Collaboration 'SPHERE' - a Sensor Platform for Healthcare in a Residential Environment. EPSRC grant number: EP/K031910/1.

References

1. Davis, A., Bouman, K., Chen, J., Rubinstein, M., Buyukozturk, O., Durand, F., Freeman, W.: Visual vibrometry: estimating material properties from small motions in video. IEEE Trans. Pattern Anal. Mach. Intell. **39**, 732–745 (2017)
2. He, X., Goubran, R., Liu, X.: Wrist pulse measurement and analysis using Eulerian video magnification. In: IEEE-EMBS International Conference on Biomedical and Health Informatics (BHI), Las Vegas, USA (2016)
3. Wadhwa, N., Rubinstein, M., Durand, F., Freeman, W.: Phase based video motion processing. ACM Trans. Graph. (TOG) **32**(80) (2013)
4. Miljkovic, N., Trifunovic, D.: Pulse rate assessment: Eulerian video magnification vs. electrocardiography recordings. In: 12th Symposium on Neural Network Applications in Electrical Engineering (NEUREL) (2014)
5. He, X., Goubran, R., Liu, X.: Using Eulerian video magnification framework to measure pulse transit time. In: IEEE International Symposium on Medical Measurements and Applications (MeMeA) (2014)
6. Wu, H., Rubinstein, M., Shih, E., Guttag, J., Durand, F., Freeman, W.: Eulerian video magnification for revealing subtle changes in the world. ACM Trans. Graph. (TOG) **31**(4) (2012)

7. Pulsesensor: https://pulsesensor.com. Accessed 14 Dec 2018
8. Liu, C., Torralba, A., Freeman, W., Durand, F., Adelson, E.: Motion magnification. ACM Trans. Graph. (TOG) **24**, 519–526 (2005)
9. Elgharib, M., Hefeeda, M., Durand, F., Freeman, W.: Video magnification in presence of large motions. In: Proceedings of the IEEE Conference on Computer Vision and Pattern Recognition (2015)
10. Liu, L., Lu, L., Luo, J., Zhang, J., Chen, X.: Enhanced Eulerian video magnification. In: Image and Signal Processing (CISP) (2014)
11. Wadhwa, N., Rubinstein, M., Durand, F., Freeman, W.: Riesz pyramids for fast phase-based video magnification. In: IEEE International Conference on Computational Photography (ICCP) (2014)
12. Wadhwa, N., Wu, H., David, A., Rubinstein, M., Shih, E., Mysore, G., Chen, J., Buyukozturk, O., Guttag, J., Freeman, W.: Eulerian video magnification and analysis. Commun. ACM **60**, 87–95 (2016)
13. Horn, B., Schunck, B.: Determining optical flow: a retrospective. Artif. Intell. **59**, 81–87 (1993)
14. Nichols, W., Rourke, M., Vlachopoulos, C.: McDonald's Blood Flow in Arteries: Theoretical, Experimental and Clinical Principles. CRC Press, Boca Raton (2011)
15. Wu, H., Rubinstein, M., Shih, E., Guttag, J., Durand, F., Freeman, W.: Video magnification. http://people.csail.mit.edu/mrub/vidmag. Accessed 03 May 2018
16. Rao, K., Kim, D., Hwang, J.: Video Coding Standards. Springer, Dordrecht (2014)
17. Sikora, T.: MPEG digital video-coding standards. IEEE Signal Process. Mag. **14**, 82–100 (1997)
18. Evans, B.: Satellite communication systems. IET (1999)
19. Ford, A., Roberts, A.: Colour space conversions. Westminster University (1998). http://www.photo-lovers.org/pdf/coloureq.pdf. Accessed 03 May 2018
20. Burt, P., Adelson, E.: The Laplacian pyramid as a compact image code. IEEE Trans. Commun. **31**, 532–540 (1983)
21. Kupce, E., Freeman, R.: Binomial filters. J. Magn. Reson. **99**, 644–651 (1992)

Classification of Fibrillation Subtypes with Single-Channel Surface Electrocardiogram

Xinyang Li, Balvinder S. Handa, Nicholas S. Peters, and Fu Siong Ng[(✉)]

National Heart & Lung Institute, Imperial College London, London, UK
{xinyang.li,balvinder.handa05,n.peters,f.ng}@imperial.ac.uk

Abstract. Atrial fibrillation (AF) and ventricular fibrillation (VF) are complex heart rhythm disorders with increasing prevalence. Mechanisms sustaining these arrhythmias are different, and subsequently, the required treatments options differ. Although many algorithms have been developed for differentiating fibrillation from normal sinus rhythm, very few methods exist to differentiate between different forms of AF and VF from surface electrocardiogram (ECG). To address the issue, we propose a novel ECG classification method to differentiate fibrillation that is completely chaotic from forms where it is organized with key driving sites. Differentiating fibrillation organisation from ECGs may aid patient selection, and identify those who may benefit from targeted ablation treatment. Evaluation using real-world data sets based on rat VF model shows that the proposed method could recognise the correct Fibrillation subtype from the single-channel electrocardiogram with an accuracy of 88.89%.

Keywords: Electrocardiogram · ECG · EGM · Fibrillation · Rotational activity

1 Introduction

Atrial fibrillation (AF) and ventricular fibrillation (VF) are complex heart arrhythmia with increasing prevalence. Modern signal processing and machine learning approaches have been widely applied for modelling and analyzing AF and VF recordings. One group of methods aims at automatically detecting fibrillation. For example, fibrillation could be differentiated from normal sinus rhythm by extracting irregular QRS patterns from body surface electrocardiogram (ECG) [1]. In [2], features including root mean square of successive RR differences (RMSSD), Turning Points Ratio (TPR) and Shannon entropy have been proposed for AF detection.

Besides fibrillation detection, another group of methods focus on uncovering the underlying arrhythmia mechanisms itself. Regions of rotational activity or rotors, defined as regions where propagating wavefronts perpetuate around a

© Springer Nature Switzerland AG 2020
Z. Ju et al. (Eds.): UKCI 2019, AISC 1043, pp. 472–479, 2020.
https://doi.org/10.1007/978-3-030-29933-0_39

phase singularity point which theoretically remains excitable have been demonstrated in optical mapping ex vivo studies. They are hypothesised to be responsible for initiating and maintaining fibrillation. This mechanism is of particular interest because the arrhythmia could potentially be terminated by using invasive internal catheters inside the heart to deliver controlled burns at the driving sites of the rotational activities, i.e., ablating [3,4].

Although, at present driver guided ablation remains controversial. There has been considerable interest in mapping sites in the heart localising rotational activity/drivers in both AF and VF. For instance, it has been found that pivot point of the rotational activities shows high-frequency activity, and as a result, could be identified with dominant frequency (DF) mapping [5,6]. Similar signal processing techniques used for the mapping include Shannon, Kurtosis and multiscale entropy [7].

There is a continuing debate in regards to the precise underlying mechanism of AF and VF. Whilst, stable rotational driver have been reported as driving both AF and VF, there is also evidence for existence of chaotic fibrillation states where wavefronts are randomly propagating and not driven by stable rotational drivers [8–10]. The therapeutic approach to a chaotic form of fibrillation would in theory be different from that driven by stable rotational drivers. With regards to chaotic forms of fibrillation it may be unnecessary and inefficient to conduct mapping.

Moreover, the ablation carries 2–3% risk of serious complications, which could at worst result in a requirement for urgent major cardiac surgery. Therefore, patient selections should ideally be conducted to identify those with targets that are potentially amenable to the therapy, to avoid the unnecessary ablation. At present, to our knowledge, all clinical studies that focused on ablating 'substrate' do not have a non-invasive means of differentiating chaotic fibrillation without drivers from more organized fibrillation sustained by discrete rotational driver(s). Certainly, ECG categorisation of AF or VF has not been used for this purpose. There is a lack of non-invasive analysis tools for differentiating fibrillation with different underlying mechanisms and identifying the fibrillation forms where the arrhythmia is sustained by rotational activity, where potentially targeted ablation therapy might be beneficial.

To address the issue, in this work, we propose to classify chaotic fibrillation with no driver regions from organised fibrillation with stable rotational drivers utilising ECG alone in a rat model of VF. The algorithm is developed for ECG, a form of routine non-invasive recordings of the heart electrical activity, so that the method is feasible for practical patient selection. One of the challenges in applying machine learning for fibrillation study lies in the lack of ground truth for model training. In this work, for each heart both optical signals and single-channel ECG are recorded. Given the high resolution of the optical signal, phase mapping is performed for localization of rotational drivers. The results of phase mapping are used as the ground truth to label the corresponding ECG data as chaotic or organized.

With the labeled ECG, two classification frameworks are proposed: one is based on temporal and spectrum feature extraction and support vector machine (SVM), and the other convolutional neural network (CNN). 9 sets of data, recorded from 9 hearts, are used to evaluate the proposed classification method. In the evaluation, the classification models are trained in a leave-one-out manner so that the data from the test sets have never been used for training. Results show that the proposed method could detect the right class from the single-channel ECG with an accuracy up to 88.89%.

2   Classification of Fibrillation Subtypes

2.1   Data Collection and Pre-processing

Optical mapping data and ECG data were recorded from 9 isolated rat hearts with a sampling rate of 1000 Hz. Phase mapping studies were applied for the optical mapping data to characterize the signal propagation and localize the rotational drivers, the methodology for which has been previously described by us in [11]. In this work, stable rotational were defined as those showing more than 2 rotations. Based on phase mapping, 4 sets of data are identified as organized, where VF is sustained by stable rotational drivers, whilst the remaining 5 sets of VF data are labelled chaotic, where VF is driven by chaotic wavefronts with no identifiable stable rotational drivers.

For each ECG data set, the data segments corresponding to fibrillation were identified visually by a clinical expert. The fibrillation segments were further windowed into multiple trials by a 2-second window with 50% overlapping. Those ECG trials were labeled as organized or chaotic depending on whether they were from an organized data set or chaotic data set. Examples of the ECG trials are illustrated in Fig. 1, where subfigures (a) and (b) correspond to the organized and chaotic classes, respectively. As shown by the figure, it is very difficult to discriminate the two types of fibrillation by visual evaluation.

2.2   Classifying Fibrillation with SVM

Autoregressive (AR) model has been widely used for modelling time-series, and it was adopted in this work to extract temporal features of ECG. Given the ECG trial $x(t)$, $t = 1, ..., n_t$, an AR model was applied as the following

$$\hat{a} = \arg\min \sum_{t=p+1}^{n_t} ||x(t) - \sum_{\tau=1}^{p} a_\tau x(t - \tau)||_2 \tag{1}$$

where a_τ is the τ-th element of the AR coefficient vector a and the order of the AR model p was set as 20 in this work.

8 temporal filters with a bandwidth of 4 Hz ranging from 2–34 Hz were constructed (2–6 Hz, 6–10 Hz, ..., 30–34 Hz), and the frequency range 2–34 Hz was used because it covered the dominant frequency range of the fibrillation signals

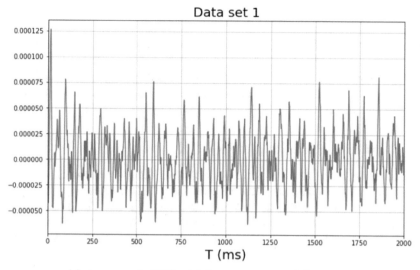

(a) An example ECG trial from an organized data set.

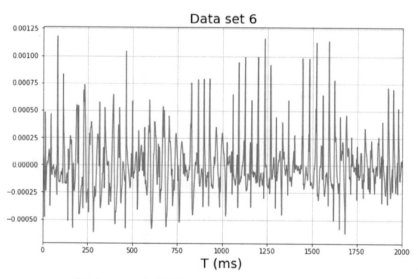

(b) An example ECG trial from a chaotic data set.

Fig. 1. Examples of ECG trials.

[11]. Then the powers of the ECG of the 8 bands were calculated as spectrum features.

The AR coefficient feature and band-power feature were concatenated as $\mathbf{f}$, for which feature selection was applied for. In the feature selection, the mutual information between each feature dimension $\mathbf{f}_j$ and class label $c \in \{+, -\}$ is calculated as

$$I(\mathbf{f}_j, c) = H(c) - H(c|\mathbf{f}_j) \qquad (2)$$

$H(c)$ is the entropy of c and $H(c|\mathbf{f}_j)$ is the conditional entropy of the c given $\mathbf{f}_j$, and details of the calculation of $I(\mathbf{f}_j, c)$ could be found in [12].

In this work, 14 features with the highest mutual information $I(\mathbf{f}_j, c)$ in 2 were selected, which was 50% of the total number of features. The selected features were then classified by SVM with radial basis function (RBF) kernel.

2.3 Classifying Fibrillation with CNN

In this method, an 1-second moving windows with 0.05-second window-shift were applied to each ECG trial, and the AR coefficient features was extracted for each window, which yields a 20×20 feature matrix for each ECG trial. Compared to the concatenated feature vector used for SVM, the AR feature matrix is more suitable for the CNN, since CNN has been proved to be a powerful tool in learning 2D information.

The CNN proposed in this work consisted of 3 layers, including 2 convolutional layers (Conv) and 1 fully-connected layer (FC), the architecture of which is illustrated in Fig. 2. Both the convolutional and max-pooling layers were done in 2D with the kernel size of 3 and 2, respectively. The strides for the convolutional and max-pooling layers are 1 and 2, respectively. The rectified linear units (ReLU) were adopted in the convolutional layers. For the network training, the batch size was 500, the number of epochs was 200000, and a drop-out rate of 0.5 is used to avoid over-fitting.

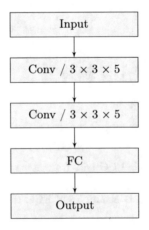

Fig. 2. The network architecture of the CNN

2.4 Model Evaluation

The proposed fibrillation subtype classification frameworks were evaluated in a leave-one-out manner so that data from the same heart for testing would not be used for training. The classification models were trained and tested at the trial level. In other words, for each 2-second trial, one est of features were extracted and classified, yielding one prediction result. In the practical clinical setting, only one diagnosis is needed per subject. Therefore, in this work, the final classification result for each data set was obtained in a voting manner, i.e., if more than half of the trials were classified as organized/chaotic, the data set would be classified as organized/chaotic.

3 Results

The ground truth and the classification results of the data type for each data set are summarized in Table 1, where "O" and "C" refer to data type organized and chaotic, respectively. SVM and CNN refer to the method based on SVM described in Sect. 2.2 and that based on CNN described in Sect. 2.3, respectively. Note that the features used in the two classification methods are different. Although for clinical applications, classification accuracies at the trial level (**acc.**) are not of much interest, a confidence level (**conf.**) could be derived from the **acc.** as the following

$$\textbf{conf.} = \max(\textbf{acc.}, 1 - \textbf{acc.}) \qquad (3)$$

which is the percentage of the trials being classified as the same as the final decision, and the higher the number the more confident for the classification result. The confidence level **conf.** is also included in Table 1.

As shown in Table 1, the group classification accuracies, which is the percentage of the data sets being classified correctly, are 89.89% and 77.78% for the SVM and CNN, respectively. In our work, the training set size in the leave-one-out setting ranged from 1994 to 2416 trials, which may not be enough for training a CNN processing the raw ECG trials with size of 2000. Thus, the CNN is proposed to process the 2D AR coefficient matrices rather than the raw data, and the CNN structure is not very deep with only 3 layers. These could be the reason that CNN has not outperformed the more conventional approach that is based on feature extraction, feature selection and SVM classifier.

For data set 8, all trials were classified as organized by framework 1, resulting in a confidence level of 100% but a wrong prediction. Framework 2 also fails to classify data set 8 correctly with a relatively high confidence level of 88.24%. This dataset was relatively noisier than other data sets, and only a small segment of data was adopted and labeled as VF, which could be the reason for the wrong classification results. For the data sets with more than 200 trials, the proposed classification frameworks could detect the data type correctly. Given that the data were segmented into trial with 2-second windows with 50% overlapping,

Table 1. Classification results

Data set	Data type	No. of trials	SVM		CNN	
			Prediction	**Conf.**	Prediction	**Conf.**
1	O	439	O	61.50%	C	72.44%
2	O	196	O	94.39%	O	72.96%
3	O	577	O	94.80%	O	89.94%
4	O	168	O	98.81%	O	98.21%
5	C	434	C	72.35%	C	74.65%
6	C	162	C	74.69%	C	79.62%
7	C	261	C	95.02%	C	99.61%
8	C	17	O	100.0%	O	88.24%
9	C	179	C	100.0%	C	56.98%
Group accuracy			**89.89%**		**77.78%**	

the required time length of the ECG recording would be around 4–5 min, which is feasible for practical application.

In our future work, more experimental data would be collected so that deeper CNN structures could be adopted and applied for the raw ECG data to capture more complex data patterns. In this work, we set typical parameters for both feature extraction and feature classification. With data set of a larger size, a certain amount of training data would be set aside for hyper-parameter optimization, and subsequently, CNN with different parameterizations could be explored. More importantly, the proposed methods will be applied for real-world human ECG. Proper domain adaptation and unsupervised learning may need to be developed if it is difficult to obtain ground truth for real-world human ECG.

4   Conclusion

In myocardial fibrillation studies, rotational activity mapping in VF and AF aims at locating the driving sites that are considered as potential targets for ablative therapies. However, for fibrillation characterized as chaotic by our methodology, the invasive mapping to localise rotational driver sites may be unnecessary, expose the patient to potential complications and be time-consuming.

Therefore, before the invasive mapping, there is a role for a non-invasive screening process with ECG utilisation to differentiate between chaotic and organised form of fibrillation to aid selection of patients with high probability of having rotational driver sites. This is usually not taken into consideration in conventional fibrillation detection methods.

To address this issue, in this work, we proposed two frameworks to differentiate fibrillation that are completely chaotic or organized with key driving sites using surface ECG. In the first framework, temporal and spectrum ECG features were extracted, selected using mutual information and classified by SVM.

In the second framework, a 2D AR feature matrix was classified by CNN. The proposed SVM and CNN frameworks were validated by 9 data sets based on the rat VF model in a leave-one-out manner, yielding the group classification accuracies of 88.89% and 77.78%, respectively. In our future work, we would continue improving the models with data set of larger size and human ECG.

References

1. Lee, J., Nam, Y., McManus, D.D., Chon, K.H.: IEEE Trans. Biomed. Eng. **60**(10), 2783 (2013). https://doi.org/10.1109/TBME.2013.2264721
2. Dash, S., Chon, K.H., Lu, S., Raeder, E.A.: Ann. Biomed. Eng. **37**(9), 1701 (2009)
3. Moe, G.K., Abildskov, J.A.: Am. Heart J. **58**(1), 59 (1959). https://doi.org/10.1016/0002-8703(59)90274-1. http://www.sciencedirect.com/science/article/pii/0002870359902741
4. Pandit, S.V., Jalife, J.: Circ. Res. **112**(5), 849 (2013)
5. Sanders, P., Berenfeld, O., Hocini, M., Jaïs, P.R.V., Hsu, L.F., Garrigue, S., Takahashi, Y., Rotter, F.S.M., Scavée, P.P.S.C., Jalife, J., Haissaguerre, M.: Circulation **112**(6), 798 (2005)
6. Salinet, J., Schlindwein, F.S., Stafford, P., Almeida, T.P., Li, X., Vanheusden, F.J., Guillem, M.S., Ng, G.A.: Heart Rhythm **14**(9), 1269 (2017). https://doi.org/10.1016/j.hrthm.2017.04.031. http://www.sciencedirect.com/science/article/pii/S1547527117304976. Focus Issue: Atrial Fibrillation
7. Annoni, E.M., Arunachalam, S.P., Kapa, S., Mulpuru, S.K., Friedman, P.A., Tolkacheva, E.G.: IEEE Trans. Biomed. Eng. **65**, 273 (2018)
8. Narayan, S.M., Krummen, D.E., Shivkumar, K., Clopton, P., Rappel, W.J., Miller, J.M.: J. Am. Coll. Cardiol. **60**(7), 628 (2012)
9. Haissaguerre, M., Hocini, M., Cheniti, G., Duchateau, J., Sacher, F., Puyo, S., Cochet, H., Takigawa, M., Denis, A., Martin, R., et al.: Circ. Arrhythm. Electrophysiol. **11**(7), e006120 (2018)
10. Krummen, D.E., Ho, G., Villongco, C.T., Hayase, J., Schricker, A.A.: Future Cardiol. **12**(3), 373 (2016)
11. Handa, B.S., Roney, C.H., Houston, C., Qureshi, N.A., Li, X., Pitcher, D.S., Chowdhury, R.A., Lim, P.B., Dupont, E., Niederer, S.A., Cantwell, C.D., Peters, N.S., Ng, F.S.: Computers in Biology and Medicine (2018). https://doi.org/10.1016/j.compbiomed.2018.07.008. http://www.sciencedirect.com/science/article/pii/S0010482518301999
12. Li, X., Guan, C., Zhang, H., Ang, K.K.: IEEE Trans. Neural Netw. Learn. Syst. **28**(11), 2727 (2016). https://doi.org/10.1109/TNNLS.2016.2601084

Engineering Data- and Model-Driven Applications

Vehicle Warranty Claim Prediction from Diagnostic Data Using Classification

Denis Torgunov[(✉)], Paul Trundle, Felician Campean, Daniel Neagu, and Andrew Sherratt

University of Bradford, Richmond Road, Bradford BD7 1DP, UK
D.Torgunov@bradford.ac.uk

Abstract. This paper presents an approach to predict warranty repair claims on automotive units based on joint on-board diagnostic and historic warranty repair data. The problem is framed as binary classification, facilitating the applicability of a variety of machine learning techniques. The approach allows automotive manufacturers to make better use of the operational and failure data collected from the field, allowing for better spend forecast and more targeted vehicle health management interventions and campaigns. The research evaluates the performance of Support Vector Machines, Random Forests and Decision Trees on the data set thus obtained is evaluated and the results are presented, highlighting the importance of hyper-parameter tuning for the problem considered. It is shown that the modelling methods employed demonstrate comparable performance, however the Decision Tree approach seems to perform the most consistently across the various target failure codes considered at this time.

Keywords: Machine learning · Support Vector Machines ·
Decision trees · Random forests · Warranty · On-board diagnostics

1 Introduction

Warranty repair and maintenance comprises a large part of many businesses in a variety of sectors. Often, warranty spend is used to make important, business critical decisions, such as unit recall, costly repair campaigns and evaluation of various models and units. Those decisions are frequently made based on historical warranty data, by fitting a variety of statistical models to predict future outcomes. Such models could also incorporate other, related data sources, such as maintenance logs, product history or diagnostic readings.

In the automotive industry, historical warranty records and diagnostic data from Electronic Control Units (ECUs) are available, providing opportunities for better modelling and prediction of future warranty claims, opening the opportunity of intervention to optimise costs. In this paper, we examine how diagnostic data can be used in conjunction with historical warranty record to enhance the

© Springer Nature Switzerland AG 2020
Z. Ju et al. (Eds.): UKCI 2019, AISC 1043, pp. 483–492, 2020.
https://doi.org/10.1007/978-3-030-29933-0_40

prediction of future warranty claims, by utilising Machine Learning (ML) in order to deal with the large volume of data being considered.

This paper frames this prediction as a classification problem, utilising a variety of ML methods for training binary classifiers aimed at predicting specific types of failures. The impact of training set size on various classifiers is examined, evaluating the overall fitness of several ML techniques for this task by comparing how a chosen accuracy measure scales with training set size.

The individual binary classifiers can serve as "detectors" for specific types of failures, as specified by the user. Those binary classifiers can then be utilised on a recent history of diagnostic data in order to make a decision on whether a particular kind of fault is likely to occur within a time period of interest or by comparing relative probabilities of different faults of interest.

This paper also demonstrates the importance of hyper-parameter tuning when choosing a model, by considering the impact of hyper-parameters on the chosen accuracy metric. It is shown that the considered algorithms benefit from hyper-parameter tuning as part of the training process when their accuracy is evaluated on a separate validation data set, which is not used during either training or tuning process.

The paper is structured as follows: Sect. 2 introduces relevant related work by other authors. Section 3 describes the data set used as a case study for this paper. Section 4 describes the methods employed by this paper in order to assign classes to diagnostic data in order to make classification algorithms applicable. Section 5 provides the summary of the results obtained by the different algorithms. Section 6 summarises the work done and outlines the direction for future research work.

2   Background

It is not uncommon to treat fault prediction as a classification problem when it comes to maintenance [1,10–13], as the problem of interest is usually related to whether intervention or repair is required before the next diagnostic cycle, which is a binary choice, or only several faulty states need to be discriminated between, leading to a finite number of classes to assign to a unit. As such, the classification problem could be stated in terms of a fault diagnosed and a *prediction horizon*. Predictive diagnostics in such a case is re-framed as a binary classification problem: whether a given fault is likely to occur before the specified prediction horizon. Such an approach can be seen in [12,13]. Often, classification is made difficult by the presence of a *class imbalance*, and it is generally the case that the under-represented classes are of high importance and need to be reliably classified [11].

Other sources that utilise diagnostic trouble codes for fault prediction have control over the sensor data collected and its frequency [15], and are therefore able to select channels of interest and apply various aggregation methods. The data available to us, however, is drawn from a historic database wherein the monitored channels are fixed and only the data at the moment of "fault" is

available, with no real signal of a consistent frequency, which poses a challenge
as a lot of the techniques cannot be readily applies.

Some related publications also make use of regulated fleets of vehicles where
the diagnostic data is obtained with consistent frequency and with an ability to
aggregate data from multiple vehicles [8,14]. This is not always possible with pri-
vately owned vehicles due to privacy concerns from customers and the fact that
often data can only be obtained when a vehicle is brought in for maintenance.

3   Case Study Data Sets

For the purposes of this work two main data sets from an automotive warranty
repair database have been utilised, namely: the warranty claims database and
the diagnostic trouble code data set.

The warranty database provided the information on repair dates for vehicles
as well as the broad type of repair performed, referred to as "failure code"
from here on. This information is collected from dealerships based on the fix
implemented, and does not necessarily reflect the actual fault.

The warranty database forms an n-by-3 matrix $\mathbf{W}$, where n is the total
number of warranty records:

$$
\mathbf{W} = \begin{pmatrix}
\text{VIN}_1 & \text{repairDate}_1 & \text{failureCode}_1 \\
\text{VIN}_2 & \text{repairDate}_2 & \text{failureCode}_2 \\
\vdots & \vdots & \vdots \\
\text{VIN}_i & \text{repairDate}_i & \text{failureCode}_i \\
\vdots & \vdots & \vdots \\
\text{VIN}_n & \text{repairDate}_n & \text{failureCode}_n
\end{pmatrix}
\tag{1}
$$

In addition to the date and type of repair, a Vehicle Identification Number
(VIN) is also recorded, allowing for cross-referencing with other databases as
necessary.

The diagnostic database consists of readouts from the on-board diagnostic
system and is logged whenever a system enters a pre-set state which is considered
"abnormal". Each such state is characterised by a designated "diagnostic trouble
code" (DTC) and is tied to a set of sensor measurements that triggers it. Along
with the DTC itself a collection of sensor values are also stored in this database.
Each DTC record is tagged with a "session time" which represents the time it
has been downloaded off the vehicle.

The diagnostic database forms an m-by-12 matrix $\mathbf{D}$, where m is the total
number of diagnostic records:

$$
\mathbf{D} = \begin{pmatrix}
\text{VIN}_1 & \text{sessionDate}_1 & \text{DTC}_1 & v1_1 & \dots & v9_1 \\
\text{VIN}_2 & \text{sessionDate}_2 & \text{DTC}_2 & v1_2 & \dots & v9_2 \\
\vdots & \vdots & \vdots & \vdots & \vdots & \vdots \\
\text{VIN}_i & \text{sessionDate}_i & \text{DTC}_i & v1_i & \dots & v9_i \\
\vdots & \vdots & \vdots & \vdots & \vdots & \vdots \\
\text{VIN}_m & \text{sessionDate}_m & \text{DTC}_m & v1_m & \dots & v9_m
\end{pmatrix}
\tag{2}
$$

The values $v1_i$ to $v9_i$ denote specific sensor information recorded together with the DTC, summarised in Table 1. The VIN in **D** corresponds to the related value in **W**, allowing the two databases to be joined.

Table 1. Sensor columns in diagnostic data

Variable	Recorded value
$v1$	Distance travelled
$v2$	CPU clock
$v3$	Battery voltage
$v4$	Engine status
$v5$	Inside temperature
$v6$	Outside temperature
$v7$	Engine power mode
$v8$	Engine speed
$v9$	Vehicle speed

For the experiments described here the size of the data set (extracted from warranty and recorded diagnostics database) consisted of roughly 18,000 data points.

4   Methodology

4.1   Prediction as Classification

In order to apply ML to the problem of warranty claim prediction based on diagnostic data, the problem has been framed as that of classification. In this paper we consider the simplest case of binary classification where classes are assigned based on two parameters: the intended prediction horizon and the failure code.

Before any classifier can be trained a collection of binary classification data sets is constructed. The inputs to training are formed from rows of the diagnostic database matrix **D** shown in Sect. 3.

Each binary classifier will determine whether, given a certain row $\mathbf{D}_{j,*}$, it is likely that a claim with a given `failureCode` will be recorded within the time frame of h months from the `sessionDate`. This time frame, the prediction horizon (PH), is determined at the time of training.

In order to express multiple such assignments of `failureCode` and h as necessary, we introduce a vector function $\mathbf{B}(\mathbf{D}, \mathbf{W}|h, f)$ for generating a list of binary outputs (i.e. 0 or 1) for classifying each row in **D** as containing a claims with failure code f within h months after the session time. Each such output is defined as:

$$\mathbf{B}_j(D, W | h, f) = \begin{cases} 1 & \text{if } \exists \mathbf{W}_{i,*} \in \mathbf{W} : \mathbf{W}_{i,1} = \mathbf{D}_{j,1} \\ & \wedge \, 0 \le \text{monthdiff}(\mathbf{W}_{i,2}, \mathbf{D}_{j,2}) \le h \wedge \mathbf{W}_{i,3} = f \\ 0 & \text{otherwise} \end{cases} \quad (3)$$

Where monthdiff(x, y) is defined as a function returning the difference in full calendar months between two dates x and y (with the result being positive if x has occurred after y and negative otherwise).

The resulting vector can then be used to determine intended classes for each row. For example, $\mathbf{B}(\mathbf{D}, \mathbf{W} | 3, \text{water pump})$ will label as True (1) those rows of $\mathbf{D}$ which have a water pump failure recorded within 3 months of the session date, according to $\mathbf{W}$.

4.2   Machine Learning Algorithms

Three different ML approaches are used for the experiments that follow, namely Support Vector Machines (SVMs) [6], Decision Trees [4] and Random Forests [3]. All three models can accommodate binary classification, although significant performance differences have been found, in particular with SVMs taking considerably longer to train that the other methods.

When evaluating the performance of different algorithms under different sets of hyper-parameters an accuracy measure is needed in order to evaluate how well a classifier performs on a given data set. Due to the unbalanced nature of the data set, wherein the number of items to be classified as positive is much smaller than the number of negative examples, a typical accuracy measure of the ratio of correctly classified items is a poor indicator of performance, as it is biased towards the majority class, which is the absence of a fault.

A better accuracy measure is provided by a value of F_β, defined as [7]:

$$F_\beta = (1 + \beta^2) \frac{\text{precision} \times \text{recall}}{\beta^2 \times \text{precision} + \text{recall}} \quad (4)$$

or, equivalently:

$$F_\beta = \frac{(1 + \beta^2) \cdot \text{true positive}}{(1 + \beta^2) \cdot \text{true positive} + \beta^2 \cdot \text{false negative} + \text{false positive}} \quad (5)$$

F_β is biased towards discovering true positives and is therefore of greater interest for the given data set. The results reported in this paper are obtained with $\beta = 1$.

4.3   Training Procedure and Selection

Before training begins, 20% of the data set withheld in order to be used for validation of the results. The remaining 20% is used for training hyper-parameter tuned models of each type. The hyper-parameter optimisation is performed using Randomised Grid Search [2] and 5-fold cross-validation is used when training the models and optimising the parameters.

It should be noted that hyper-parameter optimisation incurs a runtime cost [9], although the exact impact varies with different training methods, it is mostly dependent on the time it takes to train and evaluate an instance of a model. Indeed, it has been found that tuning significantly increases the runtime during training, with SVMs notibly taking a very long time to train with tuning. Moreover, overly aggressive tuning might result in over-fitting to the testing data used in the tuning process [5]. Much like how data can be over-fit to training data during normal training, by tuning the model parameters to essentially replicate the results present in the training set with little to no generalisation, the hyper-parameters could be similarly "over-tuned" to provide best results on the testing data set. This would manifest itself as the models generilising well to the testing data sets, but performing poorly on external validation data (not used during tuning) as well as on new data it is used to make predictions about.

Categorical parameters within the data set are one-hot encoded before the training begins, and missing values are filled by used the mean values of that attribute across the data set.

The resulting models are then filtered, selecting the best-performing model of each type for each considered value of h and f in $\mathbf{B}(\mathbf{D}, \mathbf{W}|h, f)$. The process is summarised in Fig. 1. Each point in the figure represents a particular assignment of hyper-parameters. The colour of the point represents the target it is trained for and its position on the y axis represents the mean F1 on the testing data set across 5-fold cross-validation applied during training with that particular set of hyper-parameters.

Each block in the figure represents either a database, a derived data set or a user-supplied set of parameters, with notation summarised in Table 2. The bottom portion of each block represents the dimensionality of the data, with each number to the right of the colon representing the size of the corresponding dimension of the data set. The number of dimensions corresponds to the lettered indices. Greek letters represent dimensions with no special labels attached to them, merely indexed by numbers ranging from 0 to $l-1$ where l is the size of the corresponding dimension. For example, the index ρ of $\mathbf{R}$ has the range $[0, 375)$. Latin-letter indices represent columns which have specific names assigned to them, with the mapping show in the bottom of the figure.

Table 2. Description of blocks in the work-flow diagram

Symbol	Meaning
☐	An external data set, for example a database query or export
▭	A derived data set, computed based on other data sets via a referentially transparent transformation
⌐ ⌐ ⌐	A data set supplied by the user before the process begins

The arrows represent referentially transparent transformations. This means that given the same input data sets these transformations will always produce the same results, ensuring replicability. This entails that any pseudo-random algorithms utilised by the transformations must necessarily rely on a random seed, which is supplied as part of the user-defined parameter vector $\mathbf{U}_p$.

After a collection of models $\mathbf{R}_{pst}$ is trained using hyper-parameter optimisation, shown in Fig. 2, the top model for each (h, f) pair in each class ("SVM", "Decision Tree", "Random Forest") is selected, yielding a reduced set of models $\mathbf{T}_{mst}$, which can be utilised for classifying new data.

5   Results

For each of the prediction targets, defined using the binary vector function $\mathbf{B}(\mathbf{D}, \mathbf{W}|h, f)$ introduced in Sect. 4.1, a selection of binary classifiers is constructed. The entirety of classifiers considered at this stage is presented in Fig. 2, out of which top classifier of each type is selected for each prediction target. Those classifiers can be used to construct multi-classifier systems which can discriminate between multiple classes, in addition to performing the simple binary classification.

The overall impact of tuning can be seen in Fig. 2. The tuning process itself produces a variety of classifiers with the mean value of F_1 for those classifiers distributed differently for various modelling methods.

The difference in accuracy between the methods considered appears to be less significant than the impact of hyper-parameter tuning, as evidenced by Fig. 2. Overall decision trees appear to perform the best regardless of failure mode considered.

SVMs appear to be a lot more heavily stratified based on hyper-parameter selection, with a significant portion of results clustered around the [0.60, 0.65] region, and only a few models reaching 0.75 accuracy. This, combined with the significantly longer training times for SVM models, makes them appear to be a poor fit for the data being used.

Random forests, much like other models, perform poorly when the hyper-parameter selection is sub-optimal, but increase by 0.10 with minimal tuning, reaching as high as 0.75 for some target assignments with gradually improving results, as opposed to the more sudden "jump" observed with SVMs.

Decision trees seem to perform the best, with the highest classifiers in the 0.78 range for certain failure codes. Much like with random forests, the results seem to suggest that poor tuning severely impacts accuracy, but even minimal hyper-parameter adjustment gives results over 0.60. It should be noted that, with decision trees, all targets achieve accuracy above the 0.70 line, whereas (9, "2G01") seems to be harder to predict using both SVM and random forests, with accuracies peaking at just below 0.70 for both.

However, given the unbalanced nature of the data set in question as well as the amount of data evaluated at this time, the initial results show promise for enhanced accuracy if the pool of training data is further expanded.

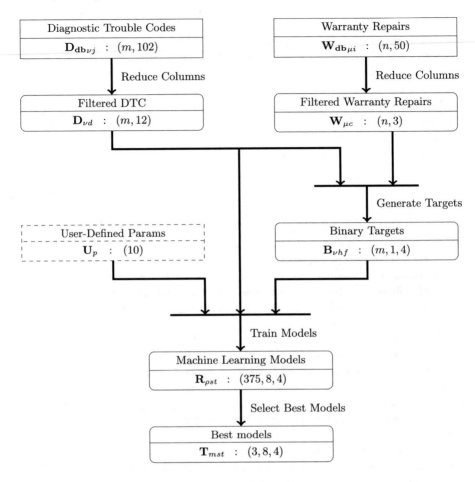

$c \mapsto \{\texttt{VIN}, \texttt{repairDate}, \texttt{failureCode}\}$
$d \mapsto \{\texttt{VIN}, \texttt{sessionDate}, \texttt{DTC}, v1, \ldots, v9\}$
$h \mapsto \{9\}$
$f \mapsto \{"7A01", "2G02", "2G01", "1F03"\}$
$s \mapsto \{"\text{Input}", "\text{Method}", "\text{ParamsInstance}", "\text{Seed}", "\text{Fold}", "\text{Run}", "\text{Model}", F_1\}$
$t \mapsto \{(9, "7A01"), (9, "2G02"), (9, "2G01"), (9, "1F03")\}$
$p \mapsto \{"\text{folds}", "\text{runs}", "\text{seed}", "\text{kernel}", "c", "\text{n\_estimators}",$
     $"\text{criteria}", "\text{splitter}", "\text{min\_samples}", "\text{min\_features}"\}$
$m \mapsto \{"\text{SVM}", "\text{Decision Tree}", "\text{Random Forest}"\}$

Fig. 1. The model training process

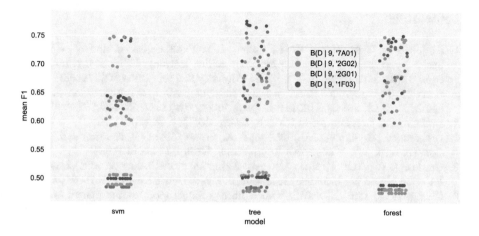

Fig. 2. Binary classifier mean F_1 metric

6   Conclusions and Future Work

This paper demonstrates how future warranty claims can be predicted from diagnostic trouble codes using binary classification based on failure types of interest and a finite prediction horizon to be considered. The initial results obtained demonstrate that the method of assigning binary targets to chosen failure codes of interest could be applied in order to predict future claims, although the accuracy of the results appears to be limited by the size of the data set. The exact impact of data set size on the accuracy of the model will be investigated in detail in a follow-up study.

However the results seem to suggest that with the data currently available decision trees perform the best across the considered range of failure codes, and the importance of the choice of hyper-parameters when modelling has been made clear. Furthermore it appears that SVM require the most attention when tuning hyper-parameters.

A follow-up study will consider the impact of data set size on the results and accuracy of the method across different ML modelling techniques, as well as the way in which the binary classifiers trained could be applied in practice to discriminate between multiple failure modes without the need for re-training the obtained classifiers. This could include construction of a hybrid classifier based on the best performing models $\mathbf{T}_{mst}$ or by combining the predictions across a range of target failure codes in some other way.

Acknowledgments. The research presented in this paper is funded by the Intelligent Personalised Powertrain Health Care research project, in collaboration with Jaguar Land Rover. The authors would like to thank the anonymous reviewers for their valuable feedback.

References

1. Abdelgayed, T.S., Morsi, W.G., Sidhu, T.S.: Fault detection and classification based on co-training of semisupervised machine learning. IEEE Trans. Ind. Electron. **65**(2), 1595–1605 (2018). https://doi.org/10.1109/TIE.2017.2726961
2. Bergstra, J., Bengio, Y.: Random search for hyperparameter optimization. J. Mach. Learn. Res. **13**, 281–305 (2012). https://doi.org/10.1162/153244303322533223
3. Breiman, L.: Random forests. Mach. Learn. **45**(1), 5–32 (2001)
4. Breiman, L., Friedman, J.H., Olshen, R.A., Stone, C.J.: Classification and Regression Trees, p. 432. Wadsworth International Group, Belmont (1984)
5. Cawley, G.C., Talbot, N.L.: On over-fitting in model selection and subsequent selection bias in performance evaluation. J. Mach. Learn. Res. **11**, 2079–2107 (2010)
6. Chang, C.C., Lin, C.J.: LIBSVM: a library for support vector machines. ACM Trans. Intell. Syst. Technol. **2**, 27:1–27:27 (2011). Software http://www.csie.ntu.edu.tw/~cjlin/libsvm
7. Chinchor, N.: MUC-4 evaluation metrics. In: Proceedings of the 4th Conference on Message Understanding, pp. 22–29. Association for Computational Linguistics (1992)
8. Fan, Y., Nowaczyk, S., Rögnvaldsson, T.S.: Incorporating expert knowledge into a self-organized approach for predicting compressor faults in a city bus fleet. In: SCAI, pp. 58–67 (2015)
9. Horváth, T., Mantovani, R.G., de Carvalho, A.C.: Effects of random sampling on SVM hyper-parameter tuning. In: International Conference on Intelligent Systems Design and Applications, pp. 268–278. Springer (2016)
10. Luo, B., Wang, H., Liu, H., Li, B., Peng, F.: Early fault detection of machine tools based on deep learning and dynamic identification. IEEE Trans. Ind. Electron. **66**(1), 509–518 (2018). https://doi.org/10.1109/TIE.2018.2807414
11. Mathew, J., Pang, C.K., Luo, M., Leong, W.H.: Classification of imbalanced data by oversampling in kernel space of support vector machines. IEEE Trans. Neural Netw. Learn. Syst. **29**(9), 4065–4076 (2018). https://doi.org/10.1109/TNNLS.2017.2751612
12. Nowaczyk, S., Prytz, R., Rögnvaldsson, T., Byttner, S.: Towards a machine learning algorithm for predicting truck compressor failures using logged vehicle data. Front. Artif. Intell. Appl. **257**, 205–214 (2013). https://doi.org/10.3233/978-1-61499-330-8-205
13. Prytz, R., Nowaczyk, S., Rögnvaldsson, T., Byttner, S.: Predicting the need for vehicle compressor repairs using maintenance records and logged vehicle data. Eng. Appl. Artif. Intell. **41**, 139–150 (2015). https://doi.org/10.1016/j.engappai.2015.02.009
14. Rögnvaldsson, T., Nowaczyk, S., Byttner, S., Prytz, R., Svensson, M.: Self-monitoring for maintenance of vehicle fleets (2018). https://doi.org/10.1007/s10618-017-0538-6
15. Shafi, U., Safi, A., Shahid, A.R., Ziauddin, S., Saleem, M.Q.: Vehicle remote health monitoring and prognostic maintenance system. J. Adv. Transp. **2018** (2018). https://doi.org/10.1155/2018/8061514

Feature Analysis for Emotional Content Comparison in Speech

Zied Mnasri[1,2(✉)], Stefano Rovetta[1], and Francesco Masulli[1,3]

[1] DIBRIS, University of Genova, Genoa, Italy
{stefano.rovetta,francesco.masulli}@unige.it
[2] Electrical Engineering Department, ENIT, University Tunis El Manar,
Tunis, Tunisia
zied.mnasri@enit.utm.tn
[3] Sbarro Institute for Cancer Research and Molecular Medicine,
Temple University, Philadelphia, USA

Abstract. Emotional content analysis is getting more and more present in speech-based human machine interaction, such as emotion recognition and expressive speech synthesis. In this framework, this paper aims to compare the emotional content of a pair of speech signals, uttered by different speakers and not necessarily having the same text. This exploratory work employs machine learning methods to analyze emotional content in speech from different angles: (a) Evaluate the relevance of the used features in the analysis of emotions, (b) Calculate the similarity of the emotional content independently from speakers and text. The final goal is to provide a metric to compare emotional content in speech. Such a metric would form the basis for higher-level tasks, such as clustering utterances by emotional content, or applying kernel methods for expressive speech analysis.

Keywords: Human-machine interaction ·
Emotional content comparison · Feature analysis · Similarity measure ·
Machine learning

1 Introduction

In spite of the considerable advances in speech processing, expressive speech is still struggling to catch up other speech technology applications. For instance, it is still problematic to accurately detect emotion change or similarity in a dialogue-type speech signal [13].

Typically emotion recognition from speech is considered as a pattern recognition problem. Hence a variety of models, feature sets and databases have been developed and tested since many years to enhance recognition rates. The used modeling tools include supervised learning techniques like HMM-GMM [18,22], ANN [12,17], SVM [23] and more recently DNN, LSTM and CNN [15]. In [7], it is reported that these models have been tested with approximately the same

© Springer Nature Switzerland AG 2020
Z. Ju et al. (Eds.): UKCI 2019, AISC 1043, pp. 493–503, 2020.
https://doi.org/10.1007/978-3-030-29933-0_41

accuracy. Another approach consists in using these classifiers in a combined scheme, either hierarchically, in series or in parallel. All these combined models were tested, giving nearly similar accuracy and outperforming single models, as mentioned in [7].

Though clustering was less employed than classification for emotion recognition, some works have proved that it could be successfully used to detect emotions. For instance, in [27] SOM (Self-organizing maps) were applied to detect emotions from audiobooks, based on articulatory features; whereas in [8], hierarchical k-means were used to detect emotions in a corpus for expressive speech synthesis, relying on a standard set of prosodic and acoustic features. Note that, in order to detect inherent clusters, the choice of features is more important in the unsupervised case than in the supervised one.

A variety of emotional speech databases were designed or recorded, covering more or less the classical psychological emotion categorizations proposed by [6, 20] and [19]. In [7], an inventory of emotional speech databases shows that the main differences between them lie in (a) the size, varying from a few tens of sentences to a few thousands [11], (b) the number of speakers, (c) the type of speech, whether acted, or spontaneous, and (d) the number of emotions, which depends on the psychological emotion model.

Feature analysis for emotion recognition has been the topic of several works, such as [14, 21, 29]. However most of these works (and other ones) were intended to evaluate the contribution of the selected features to recognize emotions or some emotional aspects like valence, activation and dominance [14]. In fact, the novelty of the proposed work consists in feature relevance analysis for the comparison of emotional content and some other related characteristics of expressive speech, such as neutrality, arousal and valence. The prospective goal is to find a metric for emotion comparison. Such a metric could be useful for a variety of tasks: for emotion clustering, but also for kernel methods, spectral embeddings and others.

More precisely, this work aims to analyse the effect of each subset of the standard Interspeech 2009 EC feature set for emotional content comparison of speech, independently from speaker and speech. Therefore, a large set of pairs of EMO-DB signals, having or not the same emotion, uttered or not by the same speaker and not necessarily having the same text, will be analyzed to check the contribution of every single LLD (Low-Level Descriptor) in emotion similarity for both utterances. In this work, similarity does not only refer to sharing the same emotion label, but also to other aspects of emotion perception, like arousal, valence and neutrality. Then the paper is organized as follows: Sect. 2 presents the state-of-the-art feature sets used in emotion recognition, with a focus on the Interspeech 2009 EC feature set, and the related work about feature analysis. Section 3 describes the speech material, whereas Sect. 4 details the methods and the experiments led in this work and the obtained results, with some comments on the findings. At last, the conclusion summarises the main ideas and results.

2   Feature Analysis for Emotion Recognition from Speech

Feature analysis has always been an important step in machine learning models, in particular pattern recognition. In the special case of emotion recognition, this could be more problematic since there is no evident relationship between emotion perception and speech parameters, in contrary to prosody perception for example, where rhythm, intonation and accentuation could be controlled by duration, pitch and energy. Therefore, several feature sets relying on prosodic and acoustic features have been proposed. Since a few years, the expressive speech community have been trying to unify their efforts to create standard feature sets for paralinguistic speech processing problems in general, and emotion recognition in particular.

2.1   Standard Feature Sets

The first attempt to build a standard feature set for emotion recognition started in 2006 with CEICES initiative [1], which aimed to gather the feature sets which had been used by different research teams in order to form a standard one. However, as reported in [9], the CEICES feature set was rather engineering-oriented, since all features which had been successfully used in prior works were simply collected, without any further deep analysis. A few years later, since 2009, standard feature sets have started to be more formalized, with the Interspeech emotion and paralinguistic challenges [24, 26]. More recently, Eyben et al. proposed the GeMAPS (Geneva Minimalistic Acoustic Parameter Set) which consists of a synthesis of years of research in emotional speech recognition by different teams, and where the proposed features were selected according to their theoretic significance and experimental relevance in detecting emotion change in voice [9]. The GeMAPS contains 18 LLD's related to frequency (pitch, jitter, formants 1–3 frequencies and formant 1 bandwidth), energy (shimmer, loudness, HNR (Harmonic-to-noise ratio)) and spectrum (alpha ratio, Hammerberg index, spectral slope at 0–500 Hz and 500–1500 Hz bands and formants 1–3 relative energy). Besides, an extended version, named eGeMAPS, contains more spectral LLD's (MFCC 1–4 and spectral flux) and more frequency LLD's (formant 2–3 bandwidth) [9].

2.2   Feature Analysis

Feature analysis, and in particular feature selection, aims to reduce the dimensionality of the feature space, so that only the most relevant features are kept. In [28], the most relevant features were selected by SFFS (Sequential format floating search) for an emotion recognition system based on a Bayesian classifier. The SFFS criterion is applied assuming that the features follow a multivariate Gaussian distribution. Then the variance of the correct classification rate of the Bayesian classifier during cross-validation is estimated.

PCA (Principal component analysis) has also been a popular feature analysis tool. PCA aims to reduce the dimensionality of the data into a subspace whose basis vectors maximise the variance in the original data [3]. It is reported in [7] that for emotion recognition, the classification accuracy increases when raising the number of principal components to a certain order, after which the accuracy starts to decrease.

LDA (Linear discriminant analysis) is another largely used technique for feature space dimension reduction. However it has the constraint that the reduced dimensionality cannot exceed the number of classes [5]. LDA was used in emotion recognition in a few works, but the results about the relevance of each group of features, i.e. pitch-related, energy-related and spectral features are not so coherent as reported in [7]. This may be due to the use of different databases and feature sets in each work.

For a comparison between PCA and LDA for emotion recognition, both techniques were used in [3] with ANN and SVM classifiers applied on BHUDES, a chinese emotional speech corpus [16]. The results show that for both classifiers, the use of LDA for feature selection gives better recognition rates than the use of PCA, either generally, for all classes, or particularly, for every single emotion.

Also, to deal with feature sparseness autoencoding neural networks were used. An autoencoder is a neural network, which outputs are the same than its inputs. It is generally used to discover latent data structures in the inputs. In [4], an autoencoder was used to resolve the problem of feature transfer learning in emotion recognition. Actually, an emotion classifier trained on some kind of data, e.g. adult speech, wouldn't be efficient when tested on another kind of data, e.g. children voices. The technique consisted in applying a single autoencoder for each class of targets. The reconstructed data was then used to build the emotion recognition system.

Finally, to conclude this state-of-the-art summary, the feature relevance was also assessed in real life conditions as studied in [8]. To achieve that, clean speech was corrupted by different noise level, before extracting the ComParE feature set [26]. Then the Pearson correlation coefficients (CC) of each feature with continuous target label was calculated. The first 400 features (amongst 6353 ones) having the best CC coefficients for arousal, valence and level of interest (LOI) tests were selected. However, the tests have shown that change in feature group relevance depend more on the individual tests than on the level of noise.

3   Speech Material

To perform this work, an emotional speech database was selected from the available speech corpora. In particular, EMO-DB [2] has been widely used and cited as a reference emotion recognition database. Besides, choosing a feature set was addressed with a special attention, since several feature sets have been proposed in the literature.

3.1   Speech Database

EMO-DB is a publicly available database of acted emotional speech [2]. In fact acted speech corpora differ from spontaneous speech, since they are carefully

Table 1. Interspeech 2009 EC LLD's and functionals [24]

Groups	LLD's	Functionals
Prosodic	(Δ)RMS energy	min, max, range
	$(\Delta)\ln F_0$	min rel. position
Signal-related	(Δ)ZCR	max rel. position
	(Δ)HNR	kurtosis, skewness
Spectral	(Δ)MFCC 1–12	arithmetic mean standard deviation Linear regression (offet, slope, MSE)

elaborated to represent all the related language phenomena in a balanced and normalized way. EMO-DB contains 10 German sentences (5 short and 5 long) uttered by 10 actors (5 male and 5 female). Every sentence was uttered by every actor in 7 emotions (neutral, anger, boredom, fear, disgust, joy and sadness) once (or twice in a few cases). The sentences were recorded in an anechoic chamber, at 16 KHz sampling rate. The database was labeled including the emotion of each sentence, the syllabic segmentation and the stress level of each syllable. It should be noted that EMO-DB has provided the highest emotion recognition rates using classical classifiers, such as HMM-GMM and SVM, as reported in [24].

3.2 Feature Set

Since emotion recognition is a pattern recognition problem, different sets of features were proposed to solve such a problem. Particularly, three feature sets have been widely used for emotion recognition, i.e. Interspeech 2009 EC feature set [24], ComParE [25] and GeMAPS [9]. Though the three feature sets share some common descriptors, like f_0, loudness, MFCC, the Interspeech 2009 EC was preferred to conduct this work, first because it has already the most reduced number of features (384 in total) amongst all Interspeech challenges feature sets which have been proposed since 2009, and secondly because it is focused only on emotion recognition, without any other paralinguistic side of speech. However an investigation of the aforementioned feature sets should be considered in the future. Then features were extracted using Opensmile toolkit [10]. Table 1 shows the complete set of features and the calculated statistics, so that 384 features ((16 LLD's + their 16 Δ-values) × 12 functionals) were extracted from each signal (Table 2).

3.3 Classes Related to Emotion Aspects

In this work, four binary classification problems are treated. For each problem, all pairs of EMO-DB signals were assigned a label. The labels were defined

Table 2. Emotional classes and content aspects

Emotion	Neutrality		Valence		Arousal	
	Yes	No	Low	High	Low	High
Neutral	X			X	X	
Anger		X	X			X
Boredom		X	X		X	
Disgust		X	X		X	
Fear		X	X			X
Joy		X		X		X
Sadness		X	X		X	

Table 3. Emotional content comparison criteria

Aspect	Label	Value	Proportion	# pairs
Classes of emotions	Different	0	50%	89386
	Identical	1	50%	
Neutrality	Different	0	50%	144096
	Identical	1	50%	
Arousal	Different	0	48.9%	143112
	Identical	1	51.1%	
Valence	Different	0	41.6%	143112
	Identical	1	58.4%	

Table 4. Emotional content classification accuracy

LLD group/Single LLD	Emotion identity	Neutrality	Arousal	Valence
Prosodic & signal-related	0.95	0.97	0.99	0.9
MFCC 1–12	1	1	1	1
All LLD's	1	1	1	1
RMS-Energy	0.92	0.92	0.90	0.78
ZCR	0.93	0.95	0.96	0.81
HNR	0.92	0.94	0.90	0.75
ln F_0	0.89	0.90	0.82	0.62
MFCC	0.92	0.95	0.82	0.77

following whether both signals of each pair of signals share (a) the same emotion class, (b) the same neutrality, (c) the same arousal, and (d) the same valence (cf. Table 3).

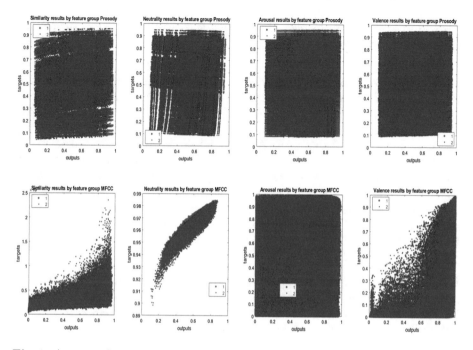

Fig. 1. Autoencoder results for LLD groups Prosody and MFCC 1–12 for different emotional content aspects comparison

4   Experiments

Along this section, the experimental protocol will be presented with details on the different experimental approaches and the obtained results.

4.1   Experimental Protocol

Since this study is interested in feature analysis from different angles of emotional content comparison, different experimental approaches were explored, as follows.

Feature Analysis for Emotion Recognition. This experiment aims to check whether the individual/grouped input features used in emotion recognition represent any visible or hidden structure related to the emotional content aspects, independently of speakers and the uttered text. For this purpose, the feature vectors of each pair of EMO-DB signals were concatenated into one feature vector. The matching label is the corresponding value for each aspect as mentioned in Table 3. To achieve this analysis, an autoencoder was trained on the set of input features corresponding to pairs of signals, using at each experiment, either all LLD's, or a group of LLD's or an individual LLD as input. Then the output of the hidden layers are plotted. Actually the goal of applying the autoencoder consists in discovering any structure linking each feature group to the studied emotional content aspect.

LLD Group Analysis for Emotional Content Comparison. The goal of this experiment is to assess the relevance of prosodic, signal-related and spectral features (cf. Table 1) in classifying each of the emotional content aspects of a pair of different speech signal, uttered by different speakers and not necessarily having the same text. Table 3 shows the different labels for each binary classification problem. In this experiment, a bottleneck DNN is trained using only one LLD group, as input, and the label vector as target. In all experiments, less than 40 nodes in the hidden layer were used, whatever the number of input nodes. The DNN was trained with a sigmoid transfer function, stochastic gradient descent algorithm, with 1000 epochs. The input data was randomly divided into 70% for training, 15% for validation and 15% for test sets.

Individual LLD Analysis for Emotional Content Comparison. The purpose of this experiment is to assess the individual relevance of features. Hence, the DNN training was performed using only one LLD (and it's Δ) as input. The output is the same as the previous experiment.

4.2   Results

The results are represented according to the order of experiments, as presented in the experimental protocol (cf. Sect. 4.1). First, regarding the features analysis by the autoencoder, Fig. 1 shows a typical example of the obtained results. Though these plots correspond to only prosodic and spectral LLD groups (cf. Table 1), the same phenomenon was noticed for all other individual LLD's or groups of LLD's. Actually, we could not obtain a clearly discriminated structure for any of them. As the hidden layer of the autoencoder contains two nodes, the ideal scheme would contain two separate distributions, each corresponding to a value of the target emotional aspect. However, it appears from this example, and for all other features as well, that either there is no specific structure which characterizes each of the studied aspects of emotional content, or there is a multi-dimensional structure which cannot be visualized using the autoencoder.

Regarding the second experiment, i.e. LLD-group analysis, the top part of Table 4 shows the accuracy rates for each aspect of the emotional content. It appears that the effects of all features of each LLD-group are accumulated to result in a high accuracy, reaching 100% in some cases.

For the third experiment, the analysis of the effect of each individual LLD (and it's Δ) is mentioned in the bottom part of Table 4. For MFCC LLD's, we opted for reporting the mean accuracy because accuracy values obtained by individual MFCC LLD's were very close for each aspect. With these results, the relevance of individual features looks more evident. Analyzing by aspect, emotion class identity and neutrality are modeled better by individual features than arousal and valence. Looking at individual features, all LLD's except $\ln(F_0)$ seem to model each aspect with nearly the same accuracy.

4.3   Discussion

This series of experiments led us to draw the following comments:

– Though prosodic and acoustic features give a very high accuracy for the classification of some aspects of emotional content in speech signal, it is not possible to determine exactly whether they are adapted for emotional content comparison. Actually, if they were so, a latent structure would emerge by autoencoding, for each binary aspect of emotional content.
– Though groups of LLD's provide a very high accuracy, this could be explained by the compensation effect that the good features, for each specific aspect, apply to cover the shortcoming of the other ones.
– Some individual features seem to be quite good in discriminating the emotional content aspects, especially for emotion class identity and neutrality. A speech signal having more than one emotion, or neutral and non-neutral parts, would certainly present different feature values than a signal with the same emotion or neutrality. This phenomenon could be investigated further, to propose a specific feature set for emotional content comparison in multi-speaker emotional speech signals.

5   Conclusion

Feature analysis for emotional content comparison was described throughout this paper. A combination of EMO-DB expressive signals into pairs, with different speakers, texts and emotion classes, was analyzed using Interspeech 2009 EC feature set, to detect some traits of emotional content, i.e. similarity of specific emotion aspects, namely: emotion class identity, neutrality, arousal and valence, between parts of each pair of signals. Different experiments were realized, according to an experimental protocol set to evaluate the relevance of each individual feature and each group of features. Though the used features have not shown a particularly discriminative structure, their individual and mutual contributions were varying from quite high to extremely high. As a perspective, this work will be continued to propose a metric for emotional content classification/clustering using kernel methods, spectral embeddings and other methods using relational representation.

Acknowlegments. This work was supported by the research grant "Fondi di ricerca di ateneo 2016" of the university of Genova.

References

1. Batliner, A., Steidl, S., Schuller, B., Seppi, D., Laskowski, K., Vogt, T., Devillers, L., Vidrascu, L., Amir, N., Kessous, L., et al.: Combining efforts for improving automatic classification of emotional user states. In: Proceedings of 5th Slovenian and 1st International Language Technologies Conference (IS LTC 2006), Ljubljana, Slovenia (2006)
2. Burkhardt, F., Paeschke, A., Rolfes, M., Sendlmeier, W.F., Weiss, B.: A database of German emotional speech. In: Ninth European Conference on Speech Communication and Technology (2005)
3. Chen, L., Mao, X., Xue, Y., Cheng, L.L.: Speech emotion recognition: features and classification models. Dig. Signal Process. **22**(6), 1154–1160 (2012)
4. Deng, J., Zhang, Z., Marchi, E., Schuller, B.: Sparse autoencoder-based feature transfer learning for speech emotion recognition. In: 2013 Humaine Association Conference on Affective Computing and Intelligent Interaction, pp. 511–516. IEEE (2013)
5. Duda, R., Hart, P., Stork, D.: Pattern Classification. Wiley, New York (2001)
6. Ekman, P.: An argument for basic emotions. Cogn. Emot. **6**(3–4), 169–200 (1992)
7. El Ayadi, M., Kamel, M.S., Karray, F.: Survey on speech emotion recognition: features, classification schemes, and databases. Pattern Recogn. **44**(3), 572–587 (2011)
8. Eyben, F., Buchholz, S., Braunschweiler, N., Latorre, J., Wan, V., Gales, M.J., Knill, K.: Unsupervised clustering of emotion and voice styles for expressive TTS. In: 2012 IEEE International Conference on Acoustics, Speech and Signal Processing (ICASSP 2012), pp. 4009–4012. IEEE (2012)
9. Eyben, F., Scherer, K.R., Schuller, B.W., Sundberg, J., André, E., Busso, C., Devillers, L.Y., Epps, J., Laukka, P., Narayanan, S.S., et al.: The geneva minimalistic acoustic parameter set (GeMAPS) for voice research and affective computing. IEEE Trans. Affect. Comput. **7**(2), 190–202 (2016)
10. Eyben, F., Wöllmer, M., Schuller, B.: openSMILE: the Munich versatile and fast open-source audio feature extractor. In: Proceedings of the 18th ACM International Conference on Multimedia, pp. 1459–1462. ACM (2010)
11. Hansen, J.H., Bou-Ghazale, S.E.: Getting started with SUSAS: a speech under simulated and actual stress database. In: Fifth European Conference on Speech Communication and Technology (1997)
12. Hozjan, V., Kačič, Z.: Context-independent multilingual emotion recognition from speech signals. Int. J. Speech Technol. **6**(3), 311–320 (2003)
13. Huang, Z.: An investigation of emotion changes from speech. In: 2015 International Conference on Affective Computing and Intelligent Interaction (ACII), pp. 733–736. IEEE (2015)
14. Kadiri, S.R., Gangamohan, P., Gangashetty, S.V., Yegnanarayana, B.: Analysis of excitation source features of speech for emotion recognition. In: Sixteenth Annual Conference of the International Speech Communication Association (2015)
15. Kim, J., Saurous, R.: Emotion recognition from human speech using temporal information and deep learning. In: Annual Conference of the International Speech Communication Association, Interspeech 2018 (2018)
16. Mao, X., Chen, L., Fu, L.: Multi-level speech emotion recognition based on HMM and ANN. In: 2009 WRI World Congress on Computer Science and Information Engineering, vol. 7, pp. 225–229. IEEE (2009)

17. Nicholson, J., Takahashi, K., Nakatsu, R.: Emotion recognition in speech using neural networks. Neural Comput. Appl. **9**(4), 290–296 (2000)
18. Nwe, T.L., Foo, S.W., De Silva, L.C.: Speech emotion recognition using hidden Markov models. Speech Commun. **41**(4), 603–623 (2003)
19. Plutchik, R.: The nature of emotions: human emotions have deep evolutionary roots, a fact that may explain their complexity and provide tools for clinical practice. Am. Sci. **89**(4), 344–350 (2001)
20. Russell, J.A.: A circumplex model of affect. J. Pers. Soc. Psychol. **39**(6), 1161 (1980)
21. Schuller, B., Rigoll, G.: Recognising interest in conversational speech-comparing bag of frames and supra-segmental features. In: Proceedings of Interspeech 2009, Brighton, UK, pp. 1999–2002 (2009)
22. Schuller, B., Rigoll, G., Lang, M.: Hidden Markov model-based speech emotion recognition. In: 2003 Proceedings of Acoustics, Speech, and Signal Processing (ICASSP 2003), vol. 2, pp. II–1. IEEE (2003)
23. Schuller, B., Rigoll, G., Lang, M.: Speech emotion recognition combining acoustic features and linguistic information in a hybrid support vector machine-belief network architecture. In: 2004 Proceedings of Acoustics, Speech, and Signal Processing, (ICASSP 2004), vol. 1, pp. I–577. IEEE (2004)
24. Schuller, B., Steidl, S., Batliner, A.: The INTERSPEECH 2009 emotion challenge. In: Tenth Annual Conference of the International Speech Communication Association (2009)
25. Schuller, B., Steidl, S., Batliner, A., Burkhardt, F., Devillers, L., Müller, C., Narayanan, S.: The INTERSPEECH 2010 paralinguistic challenge. In: 2010 Proceedings of INTERSPEECH, Makuhari, Japan, pp. 2794–2797 (2010)
26. Schuller, B., Steidl, S., Batliner, A., Vinciarelli, A., Scherer, K., Ringeval, F., Chetouani, M., Weninger, F., Eyben, F., Marchi, E., et al.: The INTERSPEECH 2013 computational paralinguistics challenge: social signals, conflict, emotion, autism. In: Proceedings of 14th Annual Conference of the International Speech Communication Association, INTERSPEECH 2013, Lyon, France (2013)
27. Székely, E., Cabral, J.P., Cahill, P., Carson-Berndsen, J.: Clustering expressive speech styles in audiobooks using glottal source parameters. In: 12th Annual Conference of the International Speech Communication Association (2011)
28. Ververidis, D., Kotropoulos, C.: Emotional speech recognition: resources, features, and methods. Speech Commun. **48**(9), 1162–1181 (2006)
29. Wu, D., Parsons, T.D., Narayanan, S.S.: Acoustic feature analysis in speech emotion primitives estimation. In: Eleventh Annual Conference of the International Speech Communication Association (2010)

Co-modelling Strategy for Development of Airpath Metamodel on Multi-physics Simulation Platform

Gaurav Pant[1($\boxtimes$)], Felician Campean[1], Aleksandrs Korsunovs[1],
Daniel Neagu[1], and Oscar Garcia-Afonso[2]

[1] University of Bradford, Bradford, UK
{g.pant, f.campean, a.korsunovs, d.neagu}@bradford.ac.uk
[2] University of La Laguna, Tenerife, Spain

Abstract. Dynamic modelling techniques are intensively studied to generate alternative solutions for engine mapping and calibration problem, aiming to address the need to increase productivity (reduce development time) and to develop better models for the actual behaviour of the engine under real-world conditions. There are many dynamic experiment and modelling techniques available in the literature and the trend is to select either a dynamic experiment or modelling technique in advance. The preselection of either a dynamic experiment or modelling technique does not allow for the analysis of the effect of such a choice on the modelling task and there exists a possibility that a different combination of experiment or modelling technique might perform better.

This paper presents an investigation of a co-modelling strategy which allows to select a signal and modelling technique combination suitable for the system modelling task. The proposed strategy was implemented on 2.0-L diesel engine using modelling techniques (neural network and neuro-fuzzy models) based on Multi-Physics simulation platform. The model selected via this strategy models the system behaviour accurately and enhances the real-time performance.

Keywords: Engine modelling · Local linear neuro-fuzzy models · Lolimot · Neural networks · Dynamic modelling

1 Introduction

Multiple engine operating modes (steady-state and transient) and challenges imposed by legislation, such as transient emission regulations, fuel economy reduction, optimising driveability for load changes, has led to increased interest in techniques for modelling dynamic behaviour. This has led to a rise in efforts placed on the investigation of dynamic calibration methodologies [1–4] and the application of dynamic experiments and modelling techniques for system modelling task [5–11]. The reason for these developments was underpinned by the possible advantages of these techniques: a faster data capture as no settling time is required; improved model fidelity by capturing dynamic behaviour, inherent interpolation, and also the fact that point-based

© Springer Nature Switzerland AG 2020
Z. Ju et al. (Eds.): UKCI 2019, AISC 1043, pp. 504–516, 2020.
https://doi.org/10.1007/978-3-030-29933-0_42

calibration process would be expensive to represent the transient behaviour, as data needs to be captured at an increased number of reference point.

Despite the availability of various modelling techniques and designs of dynamic experiment, the task of selecting the dynamic modelling technique and experiments best suited for a given engineering problem, is still a challenge. A multitude of studies have been reported evaluating the performance of different modelling techniques and dynamic experiments; in such studies either the excitation signals are preselected for the identifications purpose [6, 12] or the modelling technique is selected in advance to analyse the effect of different excitation signals on identification process [3, 13]. However, no study has systematically considered the effect of the choice of different excitation signals on different modelling techniques for a given modelling task. The research presented in this paper aims to address this issue, with a study to investigate a co-modelling strategy for selecting a suitable dynamic experiment and a modelling technique, for a system modelling task. The modelling techniques considered in this work are artificial neural networks and neuro-fuzzy models developed using a heuristic construction algorithm.

The paper outlines the review of existing literature in Sect. 2. Section 3 defines the engine case study along with the evaluation criteria. Section 4 presents the proposed methodology. The findings of the research are presented in Sect. 5, and the paper concludes with the discussion of results and future work.

2 Review of Literature

The research in dynamic modelling can be grouped into two main categories:

- design of dynamic experiments;
- Identification of dynamic models.

For design of dynamic experiments, the popular choices are pseudo random binary signals (PRBS), amplitude modulated pseudo random binary signals (APRBS) and varying frequency sinusoidal signals (chirps). The PRBS sequence only alternates between the minimum and maximum value, this leads to poor coverage of input space [1], thus, makes the signal not suitable for the nonlinear system identification as no information regarding the system behaviour is gathered other than at maximum and minimum points. This drawback is addressed by the APRBS type signals which vary in their amplitudes leading to better data coverage over a wide frequency range. However, the harsh nature of the step changes of these types of signals are not suitable to all systems and could be problematic for safe engine operation.

In this regard, the continuous nature (slow varying dynamic) of the chirp signals make them less problematic with regards to safe engine operation as they do not include step disturbances [6, 12]. However, the disadvantage of the chirp signal is the scarce coverage of the centre of the input space [10], thus, they require a long measurement time in order to cover the whole input space which increases with the number of relevant inputs.

The dynamic modelling techniques define the mathematical relationship between input and output without trying to adopt the physical system structure and are extensively used in the automotive industry to build models from measurements. Atkinson and Mott [14] used a hybrid modelling technique by combining an equation based and Neural Network (NN) data-driven method to design dynamic models for engine calibration and then verify the results over transient tests and legislative drive cycles. Röpke et al. [3] have developed non-linear dynamic models for engine calibration using a parametric Volterra series analysis comparing APRBS and chirp type excitation signals for model training. Gühmann and Riedel [12] compared 10 different dynamic modelling techniques for modelling emissions. The models were fitted to measurements obtained from an engine test bed using chirp excitation signals. It was established the Volterra series and NN models performed best. Tan and Saif [15] identified nonlinear dynamic models of the manifold pressure and mass flow processes with recurrent networks. Fang et al. [16] used NN and polynomial models to model torque and lambda and compared the effectiveness and efficiency of these two models for fuel economy optimisation. In [17] and [18] NN were used for the identification of the air fuel ratio (AFR) of a gasoline engine and transient behaviour was emphasised. Another modelling approach which has proven to be a powerful tool is the local model network. There are various studies reported for this type of modelling, such as in [19–22], and the advantage of this modelling approach lies in adapting the complexity of a problem in highly efficient manner. Hafner et al. [23] described the application of a local order linear model tree (Lolimot) for engine control design purposes. Here, the engine and dynamic emission models were developed using APRBS type excitation signals. Hametner and Nebel [8] developed an engine model using a local model network for transient calibration purposes using APRBS type excitation signals and validated its performance on data similar to a legislative drive cycle. Sequenz and Isermann [24] compared four model structures, grid map, Lolimot, kernel method and adaptive polynomial, for implementation on engine control units. The comparison was done in respect to model accuracy, computation time and required memory.

3   Methodology

The exploration of co-modelling strategy in this work is conducted on multi-physic engine simulation (MPES) platform developed by Korsunovs [25]. The principle of the approach behind the framework is to replace engine testing as the basis for mapping and calibration experiments (Model-Based Calibration –(MBC) approach) with a virtual engine simulation framework, coupling airpath simulation modelling (based on Gamma Technologies simulation software GT-Suite) with combustion chemistry solver (SRM). The experimentally validated GT-Suite engine model of a 2.0 L Euro 6c Diesel engine was available. The model was initially developed as a One-dimensional (1D) fluid dynamics model representing the air path of the Diesel engine and then converted into a fast running model (FRM) capable of running in real-time to act as a virtual engine for collecting data to train dynamic models. The probability density function (PDF) based stochastic rector model (SRM), developed by Korsunovs et al. [26], was used to represent the combustion process model and is used to predict the emissions. In addition to this, New European Drive Cycle (NEDC) data measured on a transient engine dynamometer test facility was also available.

3.1   Simulation Case Study

The task of developing the co-modelling framework was approached by partitioning the operational domain of the available drive cycle data into smaller sections, zones, based on engine speed. The segmentation of the drive cycle data is illustrated in Fig. 1. The rationale for this is that the by decomposing modelling problem into zones, compliance to constraints for dynamic experiments can be taken into account more easily [8] and global models at zones can have better accuracy relative to global models generated over the full range, as experiments can be planned to suit the needs of a particular zone [21].

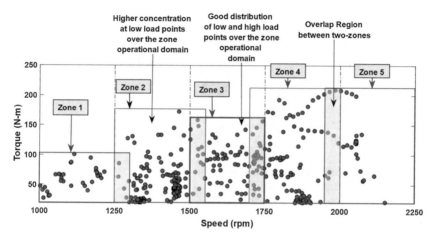

Fig. 1. Operational domain partition of the drive cycle based on engine speed.

For the purpose of this study, zone 3 (illustrated in Fig. 1) was selected. The reasons for specifically choosing zone 3 are as follows:

– An accurate injector model for the case study was available.
– Additionally, there is a good distribution of both low and high loads across the operating range of this zone. The operating range of the selected zone lies between engine speeds of 1500–1750 rpm and torque of 20–160.6 Nm.

3.2   Model Inputs and Outputs

The dynamic metamodel for the GT airpath model was needed to provide a fast mean value estimate for the inputs to the combustion model. Three key control variables were selected for the identification of the dynamic air path; engine speed, engine load and Mass Air Flow (MAF). The desired engine speed and load are the quantities required to simulate the GT-suite engine model and MAF was selected as an input variable because it controls the EGR valve position in a closed loop, which regulates the amount of exhaust gas entering the engine cylinder.

The excitation range of the engine speed and engine load was defined by the operational limit of the case study, zone 3. For MAF, the limit was set to be ±10% of MAF set position (from calibration maps in ECU), to account for the variation between transient and steady-state modes of operation. In addition to excitation range, excitation frequency range was defined based on the frequency analysis of the NEDC drive cycle data and is tabulated in Table 1.

In the study done by Korsunovs et al. [26], it was established that three air path states, inlet pressure (P_inl)/inlet temperature (T_inl)/EGR mass fraction (EGR_mf), have a significant effect on the NOx prediction while the other external parameters have little or no significant effect. Thus, these three quantities were selected as response to be modelled.

Table 1. Input parameters for dynamic air path model.

Inputs	Excitation method	Excitation range	Frequency range
Engine speed	Direct control through Simulink harness for 1D model	1500–1750 rpm	0.003–0.1 Hz
Engine load	Control through the transformation of torque setpoint to fuel injection quantity via maps	20–160.6 Nm	0.01–0.1 Hz
Mass air flow	Control through mass air flow set point in ECU. This is because EGR is in closed loop control depending on the MAF demand	±10%	0.001–0.06 Hz

3.3   Model Evaluation

The quality of each model was evaluated using fit statistics detailed by coefficient of determination (R^2) and root mean square error (RMSE) according to

$$R^2 = 1 - \frac{\sum (\hat{y} - \bar{y})^2}{\sum (y - \bar{y})^2} \tag{1}$$

$$\text{RMSE} = \sqrt{\frac{1}{n} \cdot \sum_{i=1}^{n} (y_i - \hat{y}_i)^2} \tag{2}$$

These statistical criterions provide information about the quality of a model if applied to prediction of training data whereas they give measure of predictive performance if applied to validation data (new set of data not used in training). If there is a significant difference between the model prediction for training and the validation data, this indicates that the model is over-fitted.

4 Development of Diesel Engine Dynamic Air Path

The objective of the co-modelling strategy was to develop dynamic metamodel of air path based on MPES platform. For this purpose, nonlinear Multiple Input Single Output (MISO) models of the selected response quantities, EGR_mf/ P_inl/ T_inl, were developed. The co-modelling strategy, illustrated in Fig. 2 shows, the process of the development of dynamic air path model. The modelling steps are as follows:

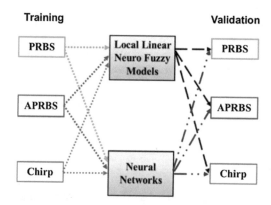

Fig. 2. Illustration of the co-modelling strategy.

- Identify dynamic air path metamodel by data obtained from dynamic tests in virtual diesel engine air path (GT-Suite) model. In this stage, excitations signals are implemented to virtual Diesel engine air path of MPES platform, and the system responses of interest are captured. The inputs signal along with the outputs are used for training the dynamic models.
- Develop dynamic models of Diesel engine air path using a combination of LLNF (Local Linear Neuro-Fuzzy) and NN models with three excitations signals. Validate model performance on three separate validation signals (as shown in Fig. 2).
- Compare the developed combinations of models based on the fit statistics and engineering analysis.

4.1 Model Structure

In this paper, a dynamic local linear neuro-fuzzy model (Lolimot) and recurrent NN (NARX network) was chosen as the dynamic model format. Since, models are used for simulating the air path output rather than predicting the output k-step ahead, a parallel model structure is selected which simulates the current output by input and previously simulated output.

The input and output delays should be optimised to give the best trade off model between model accuracy and complexity. Since, multiple dimension optimisation can be time consuming, in this study they are selected by trial-and-error tests on few settings. For, Lolimot models the optimal number of local linear models is deter-mined by the construction algorithm based on the improvement in modelling accuracy and number of parameters. For NN models, the number of neurons is selected by comparing mean training time and mean RMSE of 10 iteration of a few settings based on trial-and-error tests. For PRBS, APRBS, and Chirp based NN models the number of neurons in hidden layer was determined as 8, 28, and 23 respectively. For the output layer, the number of neurons was selected to be 1 and the network was constructed with a single hidden layer with tan-sigmoid activation function and a single output layer with a linear activation function. So, the relational form of EGR_mf for NN model is given by

$$\hat{y}_{EGR}(k) = f(u(k-1), u(k-2), u(k-3), \hat{y}_{EGR}(k-1), \hat{y}_{EGR}(k-2)) \quad (3)$$

Where u is the input vector and $\hat{y}_{EGR}$ is the simulated output.
The relational form of EGR_mf for Lolimot model is given by

$$\hat{y}_{EGR}(k) = f(u_1(k-1), u_1(k-2), u_2(k-1), u_2(k-2), u_2(k-3),$$
$$u_3(k-1), u_3(k-2), u_3(k-3), y(k-1), y(k-2)) \quad (4)$$

4.2   Excitation Signals

The excitation signals were generated based on the configuration listed in Table 1. Three training and three validation signals were created for three types of excitation

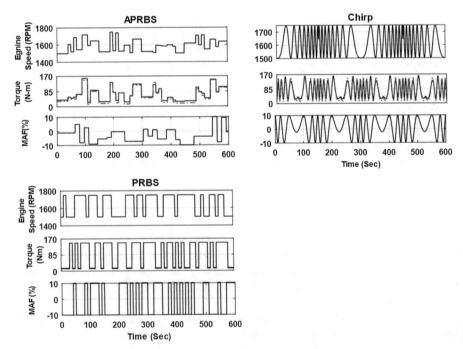

Fig. 3. Identification signals of three types for different input channels (dash-red: original signal & solid: scaled signal).

signals (PRBS/APRBS/Chirps). The identification signals used for training the model are illustrated in Fig. 3, and engine torque signal has been continuously scaled as a function of engine speed to avoid operating points that are not achievable in practice. The MAF did not require scaling, implemented as a percentage factor of the set position in the virtual ECU and scaling is already an inherent part of the engine strategy.

5   Results

It can be observed from the Fig. 4, that models of EGR_mf developed using PRBS type excitation for both NN and Lolimot model structures shows signs of overfitting, as validation RMSE is significantly higher than training RMSE. Thus, the model based on this type of signal was ruled out as a suitable choice.

For Loimot models, APRBS type of excitation signal was selected as the best signal. The reason being models trained using this type of signal performed slightly better than the model trained on chirp signals for both training (<1% EGR_mf) and validation dataset (<2.2% EGR_mf: averaged over three validation signals), as can be observed in Table 2. Also, APRBS based Lolimot model of EGR_mf requires a reduced number of local models (21) compared to chirp based (26) which leads to a reduction in the number of parameters required to identify the system behaviour.

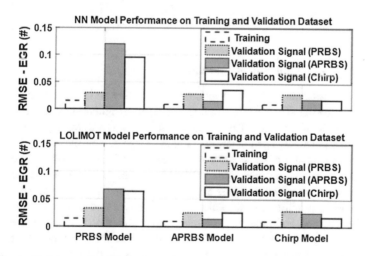

Fig. 4. Training and validation RMSE for NN and Lolimot models of EGR_mf.

Noteworthy, the model trained on a specific signal performs better on a validation signal of same type. This is underpinned by signal properties, such as chirp signals, which are slow varying dynamic signals with less significant step changes, not being able to predict the sudden step changes associated with the APRBS signals.

For NN models, chirp type excitation signals were selected as the best signal based on statistical data presented in Table 3 and Fig. 4, where it can be observed that the chirp signals based model performs slightly better (training error <1% EGR_mf and average validation error <2.2% EGR_mf) than the APRBS based model. Additionally, the model developed using chirp signals requires reduced number of neurons and shorter training time compared to APRBS based model.

Table 2. Evaluation of Lolimot models for three types of excitation signals.

Signal type	Training error - RMSE	Validation error-RMSE			No. Local Linear Models (LLM)	Training time (s)
		PRBS	APRBS	Chirp		
PRBS	0.0145	0.0331	0.0674	0.0637	17	191
APRBS	0.0096	0.0254	0.014	0.0255	21	350
Chirp	0.0097	0.0286	0.0246	0.0166	26	296

Table 3. Statistical evaluation of NN models for three types of excitation signals.

Signal type	Training error- RMSE	Validation error-RMSE			No. of neurons	Training time (s)
		PRBS	APRBS	Chirp		
PRBS	0.0151	0.0298	0.1205	0.0954	8	66.9
APRBS	0.009	0.0281	0.0150	0.0353	28	421.3
Chirp	0.009	0.028	0.0210	0.0150	23	401.2

5.1 Selection of Signal Model Combination

The correlation between EGR _mf from the virtual engine and the selected Lolimot and NN models is illustrated in top left corner of Fig. 5, both of them show a good match and follow the trends in the data. This is corroborated by the training $R^2 > 99\%$, from Table 4, indicating a very accurate model. Figure 5 also depicts the correlation between the selected models and three separate validation data measured from virtual engine.

The Lolimot model, trained with APRBS type signals, captures the trends in all three validation signals and provides a good match over all the points. There is some discrepancy in determining the absolute values for PRBS and chirp type signals which has been highlighted and labelled (encircled and numbered) in Fig. 5 but overall the model is quite accurate with $R^2 \sim 96\%$, average across three validation datasets.

Similar level of accuracy is observed for the NN model trained with chirp type signals providing a good match across all the validation signals. Although, overall model is quite accurate with R2 $\sim$ 95% (average across three validation datasets), this model struggles to capture the step changes associated with PRBS and APRBS type validation signals. This trend is highlighted in Fig. 5 (enclosed with box) and is more prominent in PRBS type validation signals. This effect could be underpinned by the fact that chirp signal is of sinusoidal nature with slow varying dynamics while PRBS and APRBS have sudden changes due to their step nature. Also, the selected NN model requires slightly longer time to train than the Lolimot model. Based on Figs. 4, 5, and Tables 2, 3 and 4, the Lolimot model trained on APRBS signal was selected as the best model for developing a metamodel of air path. Although, the Lolimot model was selected, the NN model is an equally practical choice.

Table 4. Comparison of selected Lolimot and NN model of EGR mass fraction.

Fit statistics	Lolimot-APRBS	NN-Chirp
R^2 – Training	99.24	99.12
R^2 – Validation -PRBS	97.5	95.03
R^2 – Validation-APRBS	98.01	91.57
R^2 – Validation -Chirp	95.34	98.56
Training time (sec)	350	401.2

Residual analysis on the chosen combination, Lolimot-APRBS, was carried out to identify that model prediction does not have any bias or trends associated which would violate the constant variance assumption and is illustrated in Fig. 6. The residuals vs time plot (top left) confirms the degree of randomisation, as there is no negative serial correlation or other discernible trends present in the error terms. Also, observation of residuals vs fitted value plot (bottom-left) shows the residuals are randomly scattered, and the normal probability plot (bottom-right) suggest the distribution of the residuals is approximately linear. Based on the three plots presented, the constant variance assumption for the selected model across the observations is a valid assumption.

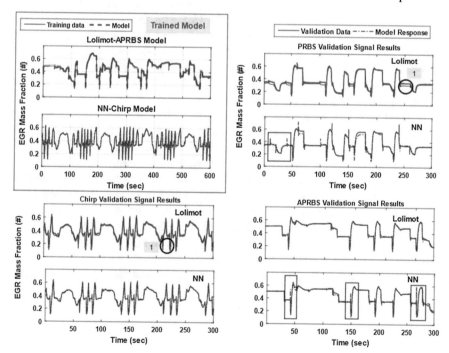

Fig. 5. Response of trained models on validation signals.

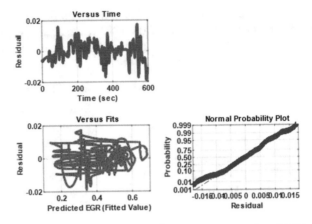

Fig. 6. Residual analysis of APRBS based Lolimot model of EGR_mf.

6   Conclusion

In the co-modelling strategy, it was observed that the model trained on a specific signal performs better on validation signal of same type. This is underpinned by signal properties, such as chirp signals, which are slow varying dynamic signals with less significant step changes, not being able to predict the step changes associated with the APRBS signals. Additionally, the chirp signals have sparse coverage in the centre which is not the case for APRBS type signals. However, the continuous nature (slow varying dynamic) of the chirp signals make them less problematic with regards to safe engine operation when developing global experiments for whole engine operating envelope as step-change disturbances are avoided. On the other hand, APRBS type of signals cover a broader frequency range (both high and low frequency components) and cover a wide range of amplitude so providing the best data coverage. If APRBS is implemented in a similar fashion as in this work, global-zone modelling approach, which allows safe engine operation by designing dynamic experiments with less harsh step changes due to local limits and easier compliance of constraints, the APRBS type excitation signal will be a superior choice for identification purposes.

This paper allows evaluation of a combination of modelling techniques and experiment designs, but further work is needed to address additional factors such as what is the optimal length of the identification signals, incorporation of improved methods of order determination, and to evaluate optimal dynamic experiment methodology in order to create realistic excitation sequences based on expected/multiple drive cycles. This may enhance the identification process, by emulating the characteristics of the drive cycles and compressing them into a sequence which will allow model training on realistic scenarios.

References

1. Nelles, O.: Nonlinear System Identification: From Classical Approaches to Neural Networks and Fuzzy Models, 1st edn. Springer, Heidelberg (2001)
2. Knaak, M., Schoop, U., Barzantny, B.: Dynamic modelling and optimization: the natural extension to classical DoE. In: Röpke, K. (ed.) Design of Experiments (DoE) in Engine Development III, pp. 10–21. Expert Verlag, Berlin (2007)
3. Röpke, K., Baumann, W., Köhler, B.-U., Schaum, S., Lange, R., Knaak, M.: Engine calibration using nonlinear dynamic modeling. In: Lecture Notes in Control and Information Sciences, vol. 418, pp. 165–182 (2012)
4. Sequenz, H.: Emission Modelling and Model-Based Optimisation of the Engine Control, vol. 8, no. 1222. VDI Verlag, Düsseldorf (2013)
5. Baumann, W., Klug, K., Kohler, B.U., Röpke, K.: Modelling of transient diesel engine emissions. In: Röpke, K. (ed.) Design of Experiments in Engine Development, vol. 4, pp. 41–53. Expert Verlag, Berlin (2009)
6. Burke, R.D., Baumann, W., Akehurst, S., Brace, C.J.: Dynamic modelling of diesel engine emissions using the parametric Volterra series. Proc. Inst. Mech. Eng. Part D J. Automob. Eng. **228**(2), 164–179 (2013)
7. Fang, K., Li, Z., Ostrowski, K., Shenton, A.T., Dowell, P.G., Sykes, R.M.: Optimal-behavior-based dynamic calibration of the automotive diesel engine. IEEE Trans. Control Syst. Technol. **24**(3), 979–991 (2016)
8. Hametner, C., Nebel, M.: Operating regime based dynamic engine modelling. Control Eng. Pract. **20**(4), 397–407 (2012)
9. Baumann, W., Schaum, S., Knaak, M., Röpke, K.: Excitation signals for nonlinear dynamic modeling of combustion engines. In: Proceedings of the 17th World Congress, The International Federation of Automatic Control, pp. 1066–1067 (2008)
10. Heinz, T.O., Nelles, O.: Iterative excitation signal design for nonlinear dynamic black-box models. Procedia Comput. Sci. **112**, 1054–1061 (2017)
11. Cheng, C.M., Peng, Z.K., Zhang, W.M., Meng, G.: Volterra-series-based nonlinear system modeling and its engineering applications: a state-of-the-art review. Mech. Syst. Signal Process. **87**, 340–364 (2017)
12. Guhmann, C., Riedel, J.M.: Comparison of identification methods for nonlinear dynamic systems. In: Röpke, K. (ed.) Design of Experiments (DoE) in Engine Development, pp. 41–53. Expert Verlag (2011)
13. Tietze, N.: Model-based calibration of engine control units using gaussian process regression. Ph.D. thesis, Technischen Universität Darmstadt (2015)
14. Atkinson, C., Mott, G.: Dynamic Model-Based Calibration Optimization: An Introduction and Application to Diesel Engines, vol. 2005, no. 724 (2005)
15. Tan, Y., Saif, M.: Nonlinear Dynamic modelling of automotive engines using neural networks. In: Proceedings of the 1997 IEEE International Conference on Control Applications, pp. 407–416 (1997)
16. Fang, K., Li, Z., Shenton, A., Fuente, D., Gao, B.: Black box dynamic modeling of a gasoline engine for constrained model-based fuel economy optimization. In: SAE 2015 World Congress & Exhibition (2015)
17. Arsie, I., Marotta, M.M., Pianese, C., Sorrentino, M.: Experimental validation of a recurrent neural network for air-fuel ratio dynamic simulation in S.I. I.C. engines, no. 47063, pp. 127–136 (2004)

18. Hou, Z., Sen, Q., Wu, Y.: Air fuel ratio identification of gasoline engine during transient conditions based on Elman neural networks. In: Sixth International Conference on Intelligent Systems Design and Applications, vol. 1, pp. 32–36 (2006)
19. Murray-Smith, R.: A local model network approach to nonlinear modelling. Ph.D. thesis, University of Strathclyde (1994)
20. Murray-Smith, R., Johansen, T.A.: Local learning in local model networks. In: 1995 Fourth International Conference on Artificial Neural Networks, pp. 40–46 (1995)
21. Johansen, T.A., Foss, B.A.: Operating regime based process modeling and identification. Comput. Chem. Eng. 21(2), 159–176 (1997)
22. Nelles, O., Fink, A., Isermann, R.: Local linear model trees (LOLIMOT) toolbox for nonlinear system identification. IFAC Proc. 33(15), 845–850 (2000)
23. Hafner, M., Schüler, M., Nelles, O., Isermann, R.: Fast neural networks for diesel engine control design. Control Eng. Pract. 8(11), 1211–1221 (2000)
24. Sequenz, H., Isermann, R.: Emission model structures for an implementation on engine control units. IFAC Proc. 44(1), 11851–11856 (2011)
25. Korsunovs, A.: Multi-physics engine simulation framework. Technical report, University of Bradford (2017)
26. Korsunovs, A., Campean, F., Pant, G., Garcia-Afonso, O., Tunc, E.: Evaluation of zero-dimensional stochastic reactor modelling for a Diesel engine application. Int. J. Engine Res. (2019)

Queries on Synthetic Images
for Large Multivariate Engineering Data
Base Searches

Natasha Micic[(✉)], Ci Lei[(✉)], Daniel Neagu[(✉)], and Felician Campean[(✉)]

Faculty of Engineering and Informatics, University of Bradford, Bradford, UK
{n.micic,C.Lei1,D.Neagu,f.campean}@bradford.ac.uk

Abstract. Engineers are often interested in searching through historical engine testing data sets to explore systems behaviours across experiments or simply to validate the integrity of data by reference to past experience. Engine tests usually consist of design of experiments (DoE) tests carried out at different operating conditions on an engine. While DoE tests record a time or measurement field so could fall under the multivariate time series domain, however, the engineering experiments have no important connection with order. Using this knowledge we have developed a method of visualising the engineering data as a set of normalised histograms to manufacture synthetic images for every test. From these images and their pixel rows we propose a list base indexing method to efficiently index the engine test images and a fast algorithm to search both the univariate and multivariate cases. Querying the data for similar behaviours thus becomes much simpler for the engineer, compared to manually searching files. Our data is of a heterogeneous form which can often mean the absence of columns where they are present in other data sets. It can even sometimes mean the comparison between different data types. This data inconsistency is something that from our research is rarely reported or attempted to traverse, but is consistently occurring in real multivariate data case studies. In order to overcome this realistic problem our method will allow for comparison of univariate data with differing quality scores.

Keywords: List based indexing · List based searching ·
Normalised histograms · Big data · Multivariate data ·
Image manufacturing · Synthetic images · Fast searching ·
Distance measures · Quality scores

1   Introduction

The automotive engineering industry is one that has a multitude of data producing processes from vehicle design and manufacturing to field vehicle health monitoring. Data outputs from engine testing are large in size and are extremely heterogeneous by nature. The engine tests are requested by engineers and then measurements for the experiments are taken and tracked in ETD (Engine Test

© Springer Nature Switzerland AG 2020
Z. Ju et al. (Eds.): UKCI 2019, AISC 1043, pp. 517–528, 2020.
https://doi.org/10.1007/978-3-030-29933-0_43

Data) files and are used in analysis by engineers who are building models of engine behaviour under different set variable circumstances. The data is large and vast and often not revisited past the first intended use, despite its value given by the high cost of the experiments.

To illustrate the volume of data, if we consider 1000 engine calibration multivariate files, each of which can have over 1000 recorded variables, having any where between 5 to 5000 records. Other facts of the data composition are that not all files have the same set of variables recorded and that the data is not always of the best quality. Searching these files rapidly becomes a difficult task for an engineer. Some immediate questions that can be asked include: how can distances between variables be found when variable existence may differ? how can distances between variables in a univariate case be found when variable length may differ? Engine calibration tests [1] are of especial interest to engineers as they take a lot of time and effort to run and are often unknowingly repeated by separate engineers. The data produced in these engineering tests are in fact multivariate time series. Design of Experiment methodology [2] creates ordering of data that is of little importance. In time series analysis methodologies there is high emphasis on data point requiring order another variable that we must control for (such as DTW and other sliding window associated with patter recognition in time series). The findings of this paper show that because of the nature of the data we may be take it instead into a simplified representational domain that will allow us to make quicker and more valid distance measures between sets of multivariate test files.

There are three classes of multivariate data; (1) existing just one table, (2) multiple data tables (same number of columns and rows in each), (3) multiple tables of data each with same variable but difference number of records. Engine test data have a consistently different number of records and variables measured from one test to another. This type of data is atypical of data prominent in machine learning and data analytics benchmark data sets. It is also prone to data quality issues. During our literature review we have found no method of (a) searching data bases, (b) calculating distances and (c) representing more simplistically data with characteristics such as the ETD. Therefore this paper addresses the following issues:

- How do we simplify and visualise large multiple multivariate data sets that are heterogeneous and have data inconsistencies?
- What distance metric should be used to accurately portray the distance between ETD sets?
- How can we index the data for querying?
- What are appropriate queries for an engineer?

The paper is organized as follows. Section 2 covers some related work Sect. 3 defines some notation and gives a more detailed definition of the search and visualisation criteria. Section 4 shows the transformation of data to image. Section 5 defines the algorithm used for indexing and searching the images. Then Sect. 6 shows the algorithm used to search a large engineering data set. And finally a conclusion and description of future work is Sect. 7.

2 Related Work

Data output from engine tests are multivariate and numerous. With such characteristics comes the need for sophisticated visualization techniques summarising whole or sub sections of test. A good summary of the most fundamental multivariate visualisation techniques are given in the paper [3]. Some of which are; iconographic displays [4], parallel coordinates [5], Andrew plots and scatter plot matrix [6]. A lot of these methods suffer from the clutter issue when the number of variables become increasingly large. The data that is used most often with these techniques are one table data.

Multidimensional scaling (MDS) is a key technique in data visualization literature [9]. This method is used in conjunction with data composed of one data table where each data object is a row. The authors in [8] however use of MDS to visualise the distances between icons where the objects are essentially arrays generated from files. This method has motivated the synthetic image generation described in Sect. 4 of this paper. The authors make no mention of dealing with erroneous/poor quality data, this however is a challenge in Engine test source data. On top of this using MDS to visualise images can be a useful way of observing the closeness between files, however due to the large number of variables and files in engineering test to visualise all resulting images using MDS would become confusion and may not yield useful results due to the consistent errors and noise in the data set.

LANSAT images is another domain of image manufacturing and classification. In LANSAT imaging each section of the image can be classified as a land type. From here many image differencing methods can be applied to find the change in land over time [11]. Each pixel of the synthetic images we wish to produce will have some meaning and classification in regards to the original data and this is where these topics share similarities. Although we do not use any of the image differencing/change detection methods (e.g. image regression [10], image rationing, change vector analysis [12], etc.) in our context it can be noted that any applicable distance metric could be substituted into the algorithms seen in this paper. Regarding applicable distance metric between data of differing lengths we can look as several examples in literature of feature extraction on data to then classify and analyse that data more simply. In [7] they show the clustering of univariate time series form several domains, by using diverse features from each of them. The features are reduced to the least redundant and from here observations on patterns will commence. This method uses a clustering algorithms to find similar univariate data however does not extend to the multivariate case.

Data sets from different human activity recordings are used in [13–15] with the intention of classifying the type of activities recorded, learning from labeled historical data. There appear to be no quality issues with the data used for classification but the data structure is closest to that of the engine test data we have found in literature i.e. multivariate class 3. The most appropriate option for ETD closeness discovery seems to be the algorithm from the paper [16]. It uses fast data base querying techniques to find the k-nearest neighbours to a input object. We adapt their method to work instead on summary vectors reduces the run time of the algorithm, using a basic Euclidean distance as the distance metric. From the research for this paper there has been no literature that used data like ours in context of similarity search and data visualisation. The method we have developed for querying the data may be applicable for data in multivariate class 3, where we remove assumption of errors and variable dissimilarity.

3   Background

In this section we describe the general notation used throughout the paper as well as a formal problem definitions.

3.1   Notation

See Table 1.

Table 1. Table describing the notation used throughout the paper

Notation	Definitions
$\mathbf{D} = \{\mathbf{d_1}, ..., \mathbf{d_i}, ..., \mathbf{d_N}\}$	Set of all N/A imputed data sets, where we add non-existent columns with a vector filled with N/A values. Each data set is stored in a separate file in memory, and N represents the total number of imputed data sets in $\mathbf{D}$
$\mathbf{V} = \{v_1, ..., v_j, ..., v_M\}$	Set of all possible variables where each v_j represents a variable name. M is the total number of variables that exist in the union of variables over all data sets in $\mathbf{D}$
$UD(\mathbf{d_i}, v_j)$	Returns the data found for variable v_j in data set $\mathbf{d_i}$
$GMin(v_j)$	Returns a singular global minimum value of a variable over all the data sets in $\mathbf{D}$
$GMax(v_j)$	Returns singular global maximum value of a variable over all the data sets in $\mathbf{D}$
$NU(\mathbf{d_i}, v_j)$	Returns all numeric elements for the list of data retrieved from data set $\mathbf{d_i}$ variable v_j
$ST(\mathbf{d_i}, v_j)$	Returns all String elements for the list of data retrieved from data set $\mathbf{d_i}$ variable v_j
$EM(\mathbf{d_i}, v_j)$	Returns all empty cell elements for the list of data retrieved from data set $\mathbf{d_i}$ variable v_j
$NA(\mathbf{d_i}, v_j)$	Returns all imputed NA cell elements for the list of data retrieved from data set $\mathbf{d_i}$ variable v_j
$L(\mathbf{d_i}, v_j)$	Gives the length of a the list of data retrieved from data set $\mathbf{d_i}$ variable v_j
$QS(\mathbf{d_i}, v_j)$	Returns a quality score 'a', 'b', 'c' or 'd' depending on the following criteria: $$QS(\mathbf{d_{ij}}) = \begin{cases} \text{'}a\text{'}, & \text{if } L_{ij} = |\mathbf{NU_{ij}}| \\ \text{'}b\text{'}, & \text{if } |\mathbf{NU_{ij}}| \geq 1 \text{ and } (L_{ij} - |\mathbf{NU_{ij}}|) \neq 0 \\ \text{'}c\text{'}, & \text{if } L_{ij} = |ST(\mathbf{d_{ij}})| + |EM(\mathbf{d_{ij}})| \\ \text{'}d\text{'}, & \text{if } L_{ij} = |\mathbf{NA_{ij}}| \end{cases} \quad (1)$$
$\mathbf{NH_{i,j}} = Hist(\mathbf{d_i}, v_j)$	Will return a normalised histogram for data set $\mathbf{d_i}$ variable v_j
$dist(\mathbf{NH_{aj}}, \mathbf{NH_{bj}})$	Denotes the Euclidean distance between two calculated normalised histogram vectors of the same variable from different data sets a and b

3.2   Problem Definition

The first issue faced is of representing the data in a way that will allows for comparisons to be made between data sets of different sets of variables, different records and different quality ratings.

Definition 1. *(Building comparable images from data): Given a variable v_j of files $\mathbf{d_a}$ and $\mathbf{d_b}$, (i.e. $UD(\mathbf{d_a}, v_j)$ and $UD(\mathbf{d_b}, v_j)$ respectively), can we produce a simplified representation, $\mathbf{NH_{aj}}$ and $\mathbf{NH_{bj}}$, so that $|\mathbf{NH_{aj}}| = |\mathbf{NH_{aj}}|$. We also require that the representation $\mathbf{NH_{ij}}$ of a data file variable can make a discernible difference between quality score types (Eq. 1). And finally we require that when the Euclidean distance between two vectors is taken, $dist(H(\mathbf{d_a}, v_j), H(\mathbf{d_b}, v_j))$, it will account for and quality score combination of variables as well as an appropriate representation of "closeness" depending on quality score.*

A solution is required for the ϵ-Nearest Neighbour ($\epsilon - NN$) search on the summarised univariate data, $H(\mathbf{d_i}, v_j)$.

Definition 2. *($\epsilon - NN$ search for univariate case): Given a query, consisting of a representative vector, $\mathbf{Q.Hist}$, for a variable of interest, var, and a threshold ϵ, find all data files where the same variable being queried, satisfies $dist(\mathbf{NH_{a,var}}, \mathbf{Q.Hist}) < \epsilon$.*

A definition of the solution to the multivariate problem is required. Find the multidimensional $\epsilon - NN$ of the summary vectors of multiple variables in each data set.

Definition 3. *($\epsilon - NN$ search for multivariate case): Given a query consisting of multiple representative vectors, $\mathbf{Q.Hist} = \{\mathbf{Q\,Hist_1}, \mathbf{Q.Hist_2}, ...\}$, whose corresponding variables are a subset of $\mathbf{V}$ (represented by $\mathbf{Q.vars} = \{\mathbf{var_1}, \mathbf{var_2}, ...\}$), and contains user defined thresholds for each variable, $\epsilon = \{\epsilon_1, \epsilon_2, ...\}$, find the files for which all of $\epsilon - NN$ are satisfied for all the variables in the query.*

4   Image Manufacturing

Discussed in this section is the solution to the problem of defining the vector $\mathbf{NH_{ij}}$ that summarises a variables measurements meeting all the criteria defined in Definition 1.

4.1   Histogram Manufacturing

A very common way of representing data in a summarized fashion is by using histograms. The detail depends of the number of bins used in the histogram, B. As long as B is the same for histograms to be compared, the Euclidean distance can be used to make the comparison. For ease of visual comparison a normalised histogram is created for every recorded data variable. The histograms are normalised by taking global minimum and maximum for every variable type. The global minimum of a variable v_j is defined as:

$$GMin(v_j) = Min\{Min\{UD(\mathbf{d_1}, v_j)\}, ..., Min\{UD(\mathbf{d_N}, v_j)\}\} \quad (2)$$

The global maximum is:

$$GMax(v_j) = Max\{Max\{UD(\mathbf{d_1}, v_j)\}, ..., Max\{UD(\mathbf{d_N}, v_j)\}\} \quad (3)$$

To define a procedure, $Hist(\mathbf{d_i}, v_j)$, that produces normalised histograms for a data set $\mathbf{d_i}$ and variable v_j, firstly the production of the normalised data is required:

$$NUD(\mathbf{d_i}, v_j) = \left(\frac{UD(\mathbf{d_i}, v_j) - GMin(v_j)}{GMax(v_j) - GMin(v_j)}\right) \quad (4)$$

Choosing the number of bins to be B we create a histogram, $\mathbf{H_{i,j}} = \{h_1, ..., h_B\}$, such that:

$$L(\mathbf{d_i}, v_j) = \sum_{l=1}^{B} h_l \quad (5)$$

Where h_l for $l = \{1, ..., B\}$ are the frequency of values from $\mathbf{d_i}$ variable v_j that occur in that bin. Or more formally, h_l is defined as:

$$h_l = \frac{1}{B}(l-1) \leq NUD(\mathbf{d_i}, v_j) < \frac{1}{B}(l), \text{ for } l = \{1, ..., B\} \quad (6)$$

Next a normalised histogram vector, NH_{ij}, is calculated from the normalised data vector, H_{ij}, using

$$NH_{ij} = \{\frac{h_1}{L(\mathbf{d_i}, v_j)}, ..., \frac{h_B}{L(\mathbf{d_i}, v_j)}\} \quad (7)$$

So that the sum over $NH_{i,j}$ is 1. There are also three other data type variable measurements to consider. Above the process from data to histogram for fully numeric data $UD(\mathbf{d_i}, v_j)$ is discussed. We still have to consider the case where the data type is partially numeric, full string values or non-existent, i.e. the 4 quality score types. In every case we want the output to be a vector of length B. So we define a data to image transformation function for a singular variable v_i in data set $\mathbf{d_i}$ as:

$$Hist(\mathbf{d_i}, v_j) = \begin{cases} NH(UD_{ij}) & \text{if } QS(UD_{ij}) = \text{'a'} \\ NH(NU_{ij}) & \text{if } QS(UD_{ij}) = \text{'b'} \\ [(-1)_{\times B}] & \text{if } QS(UD_{ij}) = \text{'c'} \\ [(-2)_{\times B}] & \text{if } QS(UD_{ij}) = \text{'d'} \end{cases} \quad (8)$$

4.2   Full Image Comparison Challenges

Combining histogram vectors on top or each other to create an array of vectors (where each row represents a separate normalised histogram of a specific variable) allows for a heat map representation of the summary data to be produced. These heatmaps are synthetic images which are used as a basis for the data visualisation process. The histograms will function as the basis of the data searching algorithm in the following section. The two synthetic images in Fig. 1 represent two different data files but the same subset of variables. The two images show that in the two data sets not all variables share the same quality score type. In this heat map representation the orange coloured rows represent variable that have a quality score 'c'. The red coloured rows represent variables with a quality score 'd'. The variable subset images that we have showcased here are just a fraction of the total variables that exist in the union over all the files (for the data used in Sect. 6 with there are over 5000). With such a large number of variables, producing an image using the full union of variables would produce and unreadable output, sharing the same clutter issue with many other multivariate visualisation methods. If instead selected are variables of interest to the user then we will get more reasonable and clear images manufactured like those in Fig. 1. Some of the classic methods of analysing and clustering data sets, found in literature, require the development of some form of pair-wise distance matrix. Next generally manifold learning methods can be applied, with the distance matrix as an input, to construct coordinates for the data in a lower dimensional space (from which conclusion of closeness can me drawn using clustering or similar methods). The difficulty is that, with over 5000 variables, a pairwise difference matrix between the fully generated synthetic images (when there are thousands of files to be transformed) is large and takes a long time to compile for clustering. On top of this with experimentation done with these methods using the whole image (the full union of potential variables $\mathbf{V}$), the information is so high that it results in very meaningless clusters. This may be in part due to the distance metric used for experimentation that gives the same weighting for every variables. To overcome this we instead take a more data base querying approach in order to search the images just through variables and threshold of interest to the user. The method we have formulated is described in detail in Sect. 5.

(a) The data transformed image for data file $\mathbf{d_a}$.   (b) The data transformed image for data file $\mathbf{d_b}$.

Fig. 1. Both images a and b demonstrate examples of images that can be generated by histogram vector stacking. Both images are representing the histogram vectors for the same subset of variables, in the same ordering, $\mathbf{v_s}$, where $\mathbf{v_s} \subset \mathbf{V}$ of two separate data files, $\mathbf{d_a}$ and $\mathbf{d_b}$

5   Image Search

This section defines the list based indexing method as well as query searching for the histogram vectors over the same variable. Then a definition is given for the algorithms using the list based indexing method and query searching for histogram vectors over multiple variables of interest.

5.1   Univariate Histogram Search

In Algorithm 1 we use a list based indexing method in order to store references to the data in a way that allows for quicker searching. In order to produce an index of the generated histogram data, $\mathbf{Hist}_{i,var}$, for one variable, var over all data sets, $i = 1, ..., N$, we firstly initialise r reference histograms, $R_1, ..., R_r$, which are taken at random from the set of all histograms in the data of the variable of interest (we only select from random the histograms from 'a' or 'b' class quality type histograms). After this the distance between all the histograms in data for variable var and the reference histogram, R_k, is computed and stored as a sorted list. This list is one reference index. This process is repeated for all reference histograms so that we have in total r reference indices produced for var over data set $\mathbf{Hist}_{i,var}$ (where $i = 1, ..., N$).

Algorithm 1. LBS Make Index (One Variable Case)	**Algorithm 2.** LBS Query Index (One Variable Case)
1: Input: $\mathbf{D}, var$	1: Input: $\mathbf{D}, \mathbf{Q}, \mathbf{Index_1}, ..., \mathbf{Index_r}, \mathbf{R_1}, ..., \mathbf{R_r}, \epsilon$
2: Output: $Index_1, ..., Index_r$	2: Output: $\epsilon - \mathbf{NN}$
3: Init: $R_1, ..., R_r$	3: begin
4: begin	4: for $k = 1, ... , r$ do
5: for $k = 1, ..., r$ do	5: $DIST_Q^k = Dist(R_k.Hist, Q.Hist)$
6: for $i = 1, ..., N$ do	6: $CAND_Q^K = SearchList(DIST_Q^k, \epsilon)$
7: $DIST_i^k = Dist(R_k.Hist, \mathbf{NH}_{i,var})$	7: end for
8: end for	8: $CAND = \{\cap_{k=1}^r CAND_Q^K\}$
9: $Index_k = \{\cup_{i=1}^N \{fname_i, DIST_i^k\}, R_k\}$	9: for Cand in $CAND$ do
10: end for	10: if $Dist(Cand.Hist, Q.Hist) \leq \epsilon$ then
11: for $k = 1, ..., R$ do	11: $\epsilon - \mathbf{NN} = \epsilon - \mathbf{NN} \cup \{Cand\}$
12: $Index_k = \{Sorted\{Index_k\}, R_k\}$	12: end if
13: end for	13: end for
14: return $Index_1, ..., Index_r$	14: return $\epsilon - \mathbf{NN}$

During the searching algorithm on the list based indexing method the input is the query histogram data as well as the indices and their reference histograms produced from Algorithm 1 for a the variable var. The output of the search algorithm is a list of file names that contain histograms for the var that are within the desired threshold distance away from the query histogram, $Q.Hist$ (i.e $\epsilon - NN = \{fname_i$ for i in $1, ..., N$ if $Dist(\mathbf{NH}_{i,var}, Q.Hist) \leq \epsilon\}$). The search method allows this to be satisfied without calculating the distance $Dist(\mathbf{Hist}_{i,var}, Q.Hist)$ for every value of i. Instead the search algorithms firstly, for each reference histograms, searches the sorted list of distances from $\mathbf{R_k}$ to find all the files that lie in the threshold $[Dist(\mathbf{R_k}.Hist, Q.Hist) \pm \epsilon]$. These are the candidate histograms and their associated file names. The union of the candidate are taken from all of the reference searches is taken and stored in the algorithm as the variable $CAND$. Each of the candidate are then checked for their distance to the query histogram, if the candidate is within the required range then the candidate is accepted as a element in the $\epsilon - NN$ output. With this method the index is only required to be calculated once and then can be used for a multitude of histogram queries for that variable. As new data files are produced and processed they can be added to the existing indices by inserting them into the sorted index lists for the distance to each of the reference histograms.

5.2 Multivariate Histogram Search

The multivariate indexing method uses the univariate indexing method, in Algorithm 3. For all the variables of interest, $var_1, ..., var_d$, we produce the indices, **Indices**$_1, ..., $ **d** (where d is the total number of variables of interest), using the randomly chosen reference histograms $R_1^f, ..., R_r^f$ (for $f = 1, ..., d$).

Algorithm 4. LBS Query Index (Multiple Variable Case)

1: Input: **D, Q, Indices, Refs**, ϵ
2: $d = L(\mathbf{Q}.\text{vars})$
3: for $f = 1, ..., d$ do
4:     $CAND_f$ =
        $QueryUni(\mathbf{D}, \mathbf{Q}.\text{Hist}_f, \text{Indices}_f, R^{(f)})$
5: end for
6: $CAND = \{\cap_1^{d-1} Join(CAND_f, CAND_{f+1})\}$
7: for $Cand$ in $CAND$ do
8:     if $Dist(Cand.Hist(var_1), Q.Hist(var_1)) \leq$
       ϵ_1
       and $Dist(Cand.Hist(var_2), Q.Hist(var_2)) \leq$
       ϵ_2
       and ... then
9:         $\epsilon - \text{NN} = \epsilon - \text{NN} \cup \{Cand\}$
10:    end if
11: end for
12: Output: $\epsilon - \text{NN}$

Algorithm 3. LBS Make Index (Multiple Variable Case)

1: Input: **D**, $var_1, ..., var_d$
2: Output: **Indices**$_1, ..., $ **Indices**$_d$
3: **begin**
4: **for** $f = 1, ..., d$ **do**
5:     select $R_1^f, ..., R_r^f$ for var_f
6:     **Indicies**$_f = MakeIndexUni(\mathbf{D}, var_f)$
7: **end for**
8: **return** **Indices**$_1, ..., $ **Indices**$_d$

After the production of the indices for each of the variables, querying on these can be done. The query input is the histogram data for all files in the index, the query consists of a set of query histograms **Q.Hists**$_f$ for variables in $Q.vars_f$ where $f = 1, ..., d$ and d represents the number of variables in the query. The output of the query is a set of file histograms and associated file names that, for all f, a candidate file must satisfy $Dist(\mathbf{NH_{i,var_f}}, \mathbf{Q.Hist}(var_f)) \leq \epsilon_f$. If any variable histogram from a file does not fall inside the threshold then that file will not be added to the $\epsilon - NN$ output. The candidate set that is checked for the distance (just as in the Algorithm 2 but in this case for all variables in query), is created by using the univariate query method for on all of the variables for the corresponding indices, and taking the union of files over the output. We then take the union over all of these lists for each returned candidate list per variable, returning the final candidate list which we check then output as the file query result, $\epsilon - NN$. The query can contain a subset of variables that have

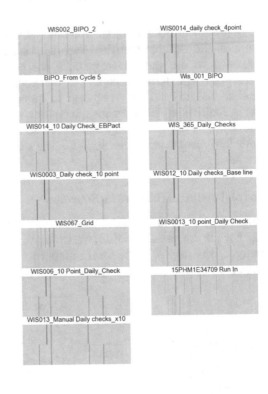

(a) A query on the index for two variable visualised.

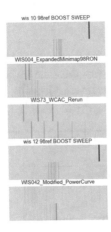

(b) The histograms for the 5 random reference points.

(c) The resulting files and their associated histograms from a query

Fig. 2. The data used for the query input is shown in figure a. The reference histograms taken from real engine calibration test data in the population are shown in figure a. The results from the query are seen in figure c.

been used to produce the indices. Only the variables of interest in the query will be searched in the indices. If additional variables are required from the query, an indices for that variable can be added then left it storing for future use. Here it can be queried again on it's own or in combination with other variables.

6   Experimentation with Engineering Data

As a small validation of the method in this section indices are generated for a subset of real engineering data. We look at the list a files names returned by the query index algorithm in the multivariate case and use the associated synthetic images generated for each of the files over some desired variables to visualise the histograms used for indexing.

From the real data set firstly we can see that none of the query results of reference images contain and variables of quality class lower than 'b'. The first step is to run the indexing matrix for the two variables of interest, and in this case variables D_SPEED and $TORQUE$ were used to generate and index (**Indices$_{D\_SPEED}$, Indices$_{TORQUE}$**) using the sample data provided. Once generated we can query these variables as many times as we like on these same indices for any combination of two histogram vectors. In this case we use the histogram data visualised in Fig. 2a as the query set of histogram vectors, using $\epsilon_{D\_SPEED} = 0.5$ and $\epsilon_{TORQUE} = 0.5$ as the threshold input to the $\epsilon - NN$ algorithm. The query data is taken from a file named 'WIS0365_Dailychecks' and the list of file names output from this search are:

'WIS002_BIPO_2','WIS0014_daily check_4point','BIPO_From Cycle 5',
'Wis_001_BIPO','WIS013_Manual Daily checks_x10',
'WIS014_10 Daily Check_EBPact','WIS_365_Daily_Checks',
'WIS0003_Daily check_10 point','WIS012_10 Daily checks_Base line',
'WIS067_Grid','WIS0013_10 point_Daily Check',
'WIS006_10 Point_Daily_Check','15PHM1E34709 Run In'

Many of the file names in the search output contain similarities with the input query file name. With the visualisation of the synthetic images we can also verify that the images whose file name share these similarities look nearly identical in their synthetic image when compared with the query image.

7   Conclusion and Future Work

From the work presented in this paper we have developed a unique and novel method for visualising large engineering data sets (referred to as synthetic images) in a uniform manner. The uniformity of the output allows the visualisation of all data as images with identical dimensions. We put forward a method for summarising engine test data as normalised histograms even in cases where the data is string only information or even missing/non-existant data. This is beneficial as it allows for appropriate comparison of images representing data regardless of their quality issues/quality scores. On top of this one of the main contributions of this paper is the development of nearest neighbor algorithms to search the synthetic image data base quickly and efficiently using a list based indexing method.

In the future the algorithms developed for searching will be tested of other benchmark data sets to see it's applicability and usefulness in other domains. On top of this we wish so make the generation of queries more accessible to engineers by considering what other ways (apart from selecting a preexisting data set) might there be for meaningfully querying the synthetic data. And we with to explore more meaningful and tailored distance metric which can take into account weightings for more or less important variables.

Acknowledgment. We would like to thank Bradford Soroptimists funding the first author. The research presented in this paper is part of the Intelligent Personalised Powertrain Healthcare research project, in collaboration with Jaguar Land Rover.

References

1. Castagné, M., Bentolila, Y., Chaudoye, F., Hallé, A., Nicolas, F., Sinoquet, D.: Comparison of engine calibration methods based on design of experiments (DoE). Oil Gas Sci. Technol. Rev. l'IFP **63**(4), 563–582 (2008)
2. Antony, J.: Design of Experiments for Engineers and Scientists. Elsevier, Amsterdam (2014)
3. Basalaj, W.: Proximity visualisation of abstract data. University of Cambridge, Computer Laboratory (2000)
4. Pickett, R.M., Grinstein, G.G.: Iconographic displays for visualizing multidimensional data. In: Proceedings of the 1988 IEEE Conference on Systems, Man, and Cybernetics, vol. 514, p. 519, August 1988
5. Inselberg, A.: Parallel Coordinates, pp. 2018–2024. Springer, Boston (2009)
6. Carr, D.B., Littlefield, R.J., Nicholson, W.L., Littlefield, J.S.: Scatterplot matrix techniques for large N. J. Am. Stat. Assoc. **82**(398), 424–436 (1987)
7. Fulcher, B.D., Little, M.A., Jones, N.S.: Highly comparative time-series analysis: the empirical structure of time series and their methods. J. R. Soc. Interface **10**(83), 20130048 (2013)
8. Keogh, E., Wei, L., Xi, X., Lonardi, S., Shieh, J. and Sirowy, S.: Intelligent icons: integrating lite-weight data mining and visualization into GUI operating systems. In: Sixth International Conference on Data Mining (ICDM 2006), pp. 912–916. IEEE, December 2006
9. Cox, T.F., Cox, M.A.: Multidimensional Scaling. Chapman and Hall/CRC, Boca Raton (2000)
10. Singh, A.: Review article digital change detection techniques using remotely-sensed data. Int. J. Remote Sens. **10**(6), 989–1003 (1989)
11. Lu, D., Mausel, P., Brondizio, E., Moran, E.: Change detection techniques. Int. J. Remote Sens. **25**(12), 2365–2401 (2004)
12. Malila, W.A.: Change vector analysis: an approach for detecting forest changes with Landsat. In: LARS symposia, p. 385, January 1980
13. Palumbo, F., Gallicchio, C., Pucci, R., Micheli, A.: Human activity recognition using multisensor data fusion based on reservoir computing. J. Ambient. Intell. Smart Environ. **8**(2), 87–107 (2016)
14. Bruno, B., Mastrogiovanni, F., Sgorbissa, A., Vernazza, T., Zaccaria, R.:. Human motion modelling and recognition: a computational approach. In: 2012 IEEE International Conference on Automation Science and Engineering (CASE), pp. 156–161. IEEE, August 2012
15. De-La-Hoz-Franco, E., Ariza-Colpas, P., Quero, J.M., Espinilla, M.: Sensor-based datasets for human activity recognition-a systematic review of literature. IEEE Access **6**, 59192–59210 (2018)
16. Bhaduri, K., Zhu, Q., Oza, N.C., Srivastava, A.N.: Fast and flexible multivariate time series subsequence search. In: 2010 IEEE International Conference on Data Mining, pp. 48–57. IEEE, December 2010

Signal Categorisation for Dendritic Cell Algorithm Using GA with Partial Shuffle Mutation

Noe Elisa[1], Longzhi Yang[1]([✉]), and Fei Chao[2]

[1] Department of Computer and Information Sciences, Northumbria University,
Newcastle upon Tyne NE1 8ST, UK
{noe.nnko,longzhi.yang}@northumbria.ac.uk
[2] Cognitive Science Department, School of Information Science and Engineering,
Xiamen University, Xiamen, China
fchao@xmu.edu.cn

Abstract. Dendritic Cell Algorithm (DCA) is a bio-inspired system which was specifically developed for anomaly detection problems. In its preprocessing phase, the conventional DC requires domain or expert knowledge to manually categorise the input features for a given dataset into three signal categories termed as safe signal, pathogenic associated molecular pattern and danger signal. The manual preprocessing phase often over-fits the data to the algorithm, which is undesirable. The principal component analysis (PCA) and fuzzy-rough set theory (FRST) based-DCA techniques have been proposed to overcome the aforementioned limitation by automatically categorising the input features to their convenient signal categories. However, the PCA destroys the underlying meaning behind the initial features presented in the input dataset and generates poor classification performance, whilst FRST-DCA is only practical for very simple datasets. Therefore, this study investigates the employment of Genetic Algorithm based on Partial Shuffle Mutation to automatically categorise the input features into the three signal categories. The experimental results of the proposed approach on eleven benchmark datasets have revealed its superiority over other versions of DCA in terms of accuracy, sensitivity and specificity.

Keywords: Dendritic Cell Algorithm · Signal categorisation ·
Genetic Algorithm · Partial shuffle mutation ·
Features-to-Signal mapping

1 Introduction

The DCA is a two-class classification algorithm within the field of Artificial Immune Systems which is developed from the behavioural model of natural

This work has been supported by the Commonwealth Scholarship Commission (CSC-TZCS-2017-717) and Northumbria University, UK.

Z. Ju et al. (Eds.): UKCI 2019, AISC 1043, pp. 529–540, 2020.
https://doi.org/10.1007/978-3-030-29933-0_44

dendritic cells (DCs) and three common immunological signals known as safe signal (SS), pathogenic associated molecular pattern (PAMP) and danger signal (DS). The DCA goes through four stages including preprocessing and initialisation, context detection, context assignment, and labelling in order to process the input signals and classify the data samples. Conventionally, in its pre-processing phase, just after feature selection, each selected feature is mapped into one of the signal categories of either SS, PAMP or DS, usually referred to as signal categorisation.

The initial studies on the DCA used expert knowledge of the problem domain to manually pre-determine the mapping between selected features and the three signal categories [1]. Manual preprocessing phase has been criticised as it may over-fits the data to the algorithm; it is thus application dependent and requires a deep understanding of the problem domain [2]. The principal component analysis (PCA) was also applied to DCA for automatic feature categorisation [2] to overcome the limitations of the manual method. However, the PCA does not produce satisfactory classification performance and destroys the underlying meaning behind the initial features presented in the input dataset by generating a new set of features via dimensionality reduction. Another automatic signal categorisation approach is based on fuzzy-rough set theory (FRST), termed as FRST-DCA, which was proposed in [3] to overcome the shortcomings of the PCA approach. However, the FRST-DCA is an expensive solution to signal categorisation task due to information loss during the discretisation process; also, it is only practically applicable to simple datasets thanks to its computational complexity [4].

This study utilises Genetic Algorithm (GA) based on partial shuffle mutation (GA-PSM) [5] to automatically map the input features into three signals of the DCA for more accurate classification performance. Suppose that m features have been selected, the GA initialises a pre-defined number of solutions in a population each being a permutation of m numbers ranging from 1 to m. Therefore, each value only appears once in each individual, as it represents an index of a feature within the dataset. In order to map each feature to its convenient signal category, the maximum number of features in SS is set as m multiplied by 0.8 and then times the percentage of the normal samples in the dataset; the maximum number of features in PAMP is determined as m multiplied by 0.8 and then times the percentage of anomaly samples; and all the rest of features are mapped to DS. To perform GA-PSM operation on an individual after crossover, the GA selects a number of values (indexes) with a probability of β and permute them to generate a new arrangement of features within an individual. As a result, the new arrangement generates a new feature-to-signal mapping. The same process repeats over a number of iterations until the GA converges. The proposed GA-PSM technique was evaluated using eleven benchmark datasets archived in the UCI machine learning repository [6]; and the experimental results have shown the superiority of the GA-PSM technique over the PCA and FRST-DCA based methods.

The structure of this paper is as follows: Sect. 2 provides the background of biological DCs. Section 3 discusses the DCA. Section 4 details the proposed

GA-PSM system. Section 5 describes the experimentation procedure and reports the experimental results. Section 6 concludes this study and specify the possible future research direction.

2   Biological Dendritic Cells

In the natural immune system, DCs coordinate antigens (e.g., bacteria and virus) presentation from the tissues [7]. The DCs express co-stimulatory molecules (csm) on their cell surface which limit the amount of antigens they sample while in the tissue. The following three signals are responsible for amplifying the DCs in the tissue and cause maturation.

- **PAMP** is proteins produced by virus or bacteria which can be easily detected by DCs and activate immune response. The presence of PAMP indicates an abnormal situation.
- **DS** is released from the disrupted host tissue or stressed cells in the tissue.
- **SS** is produced by a natural programmed cell death in the tissue. The presence of SS indicates normal cell behavior in the tissue.

DCs usually exist in three states which determine the properties of the collected antigens [7] as follows:

1. **Immature DCs (iDCs)** are found in tissues in their pure state. The iDCs collect signals and antigens in the tissue which could be a 'normal' molecule or something foreign. The relative proportions of these signals differentiate an iDC from a mature or semi-mature state.
2. **Mature DCs** are resulted when iDCs are exposed to a greater quantity of either PAMP or DS than SS.
3. **Semi-mature DCs (smDCs)** expose to more SS than PAMP and DS.

3   Dendritic Cell Algorithm

An inspiration from the behaviour of biological DCs leads to the development of the DCA. Firstly, feature selection process is applied with the DCA so as to select the most significant features from the dataset [1]. Subsequently, after feature selection DCA performs four stages of data processing, namely signal categorisation, context detection, context assignment and labelling as discussed below.

3.1   Signal Categorisation

The selected features are categorised into SS, PAMP or DS based on their properties, which was derived from the biological DC model. Signal categorisation can either be performed manually using expert knowledge of the problem domain [1] or automatically achieved using PCA [2] and fuzzy-rough set theory (FRST-DCA) [3].

Briefly, the manual method is performed by trying different permutation of features in a brute force manner. The PCA method creates a new set of features with attributes ranked in terms of their variances. Then, the PCA maps the new ranked attributes into the three signal categories in the order of SS, PAMP and DS. The FRST-DCA employs the core rough set theory concepts and QuickReduct algorithm to map the features to their specific signal categories [3]. The first and second attributes to form the reduct are used to form the SS and PAMP signals respectively while the rest of attributes in the reduct are combined to form the DS signal.

After signal categorisation process, the DCA initialises a population of artificial DCs (often 100) in a sampling pool which are responsible for data items sampling and signal processing. Eventually, a pre-defined number of DCs (often 10) are selected for further data and signal processing in the mature pool and the DCA goes through the following three phases in order to classify each data item sampled by the DCs. Note that, each selected DC from the pool is assigned a migration threshold in order to determine the lifespan spent while sampling data items and signals from the dataset.

3.2   Context Detection

In this phase, DCs use a set of pre-defined weights and the signal values to compute csm, $smDC$ and mDC by employing a weighted summation function as expressed as follows:

$$Contex[csm, smDC, mDC] = \sum_{d=1}^{m} \frac{\sum_{i,j=1,1}^{3}(C_j * W_{i,j})}{\sum_{i,j=1,1}^{3} W_{i,j}}, \tag{1}$$

where $C_j(j = 1, 2, 3)$, are PAMP, DS and SS signal values respectively which are generated by aggregating the assigned attributes. The weights $W_{i,j}(j = 1, 2, 3)$ correspond to csm, mDC and $smDC$ context respectively which are pre-defined or can be derived empirically from the dataset.

Note that, DCs accumulate the context values of csm, $smDC$ and mDC of all the data items they sample overtime. As soon as cumulative csm exceeds the assigned migration threshold, the DCs cease to sample any more data items and then move to the context assignment phase. In this process, a single data instance may be sampled by multiple DCs. The csm's migration threshold is determined from the characteristic behaviour of the dataset and the amount of data instances the DCs can collect.

3.3   Context Assignment

The cumulative context values of $smDC$ and mDC obtained from the detection phase are used to perform context assessment. If the DC has a greater cumulative mDC than $smDC$, it is assigned a binary value of 1; and 0 otherwise. Then, the DC assigns this value to all the data items it has sampled. This information is then used in the classification phase to compute the number of anomalous data items presented in the dataset, that is those with a binary value of 1 are potentially anomalous; otherwise normal.

3.4   Labeling

All sampled data items are analysed by deriving its Mature Context Antigen Value (MCAV), which is used to assess the degree of anomaly of a given data item. Firstly, the anomaly threshold is derived from the training dataset by dividing the total number of anomaly class's samples by the total number of samples presented in the dataset. Then, the MCAV value is calculated as the division of the number of times a data item is presented in the mDC context by the total number of presentations in DCs. Data items with MCAV greater than the anomaly threshold are classified into the anomalous class while the others are classified into the normal class.

4   The Proposed GA-PSM Approach

The proposed GA-PSM based system is illustrated in Fig. 1. Firstly, given a dataset, feature selection process is applied to select the most significant features. Then, the GA-PSM takes place to categorise the selected features into SS, PAMP or DS, each with s, p and m being the maximum number of features respectively. From here, the three signals are fed into the DCA by going through the signal processing phases as described in Sect. 3 to classify the data instances presented in the dataset. Note that the accuracy of the classification results are used during the training process as the fitness function of the GA, as demonstrated in Fig. 1. The key component of the proposed system, that is signal categorisation using GAPSM, is detailed below.

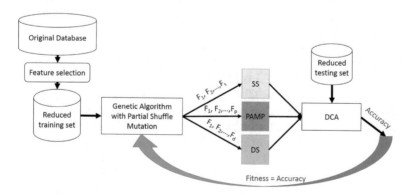

Fig. 1. The proposed GA-PSM based system

The GA-PSM system employs partial shuffle mutation to map the input features to their optimal signal categories of the DCA. GA has been extensively used for locating a near optimal solution to optimisation problems with large parameter search space, such as weight optimisation for the DCA [8,9], training of neural network [10], and rule base optimisation in fuzzy inference systems [11–13].

In order to generate the optimal signal categories for the DCA, the GA-PSM based system firstly divides the m selected features from a particular dataset into three groups containing s features for SS, p features for PAMP, and d features for DS as calculated below:

$$s = 0.8 * percentage\_of\_normal\_samples * m \tag{2}$$

$$p = 0.8 * percentage\_of\_anomaly\_samples * m \tag{3}$$

$$d = the\_rest\_of\_the\_features. \tag{4}$$

Then, it goes through the following steps:

(1) *Individual representation:* An individual (I) within a population ($\mathbb{P}$) is designated as a possible solution that comprises of all m indexes of the selected features, where the first s features belong to SS, followed by p features for PAMP and finally the last d features for DS; which is given by $I = \{F_1^1, F_2^2, .., F_s^e, ..., F_1^f, F_2^g, .., F_p^i, ..., F_1^j, F_2^k, .., F_d^m\}$.

(2) *Population initialisation:* In this study, the initial population $\mathbb{P} = \{I_1, I_2, ..., I_N\}$ is formed by initialising N individuals each containing m unique random numbers ranging from 1 to m, where each number represents an index of a feature within the database.

(3) *Objective function:* The objective function is defined as the classification accuracy of the DCA. The quality of an individual within a population in each iteration is determined by using an objective function.

(4) *Selection:* In this study, the fitness proportionate selection method is employed for selecting a number of individuals who reproduce during crossover and partial shuffle mutation so as to evolve better individuals for the next iterations. Using this technique, the probability of an individual to become a parent is proportional to its fitness.

(5) *Crossover:* A single point crossover genetic operation is applied with a probability of α to increase the exploitation of search space.

(6) *Partial shuffle mutation (PSM):* The PSM was proposed by [5] for fast solving the Travelling Salesman Problem. As its name suggests, in this study, the PSM shuffles the index of features within an individual in the population based on randomly selected indexes of the features with a probability p_m as detailed in Algorithm 1. This process is equivalent to the trying of a different combination or rearrangement of index of features. Suppose that the dataset is of 10 features, 55% of anomaly data samples and 45% of normal samples; then, $s = 0.8 * 0.45 * 10 = 3$, $p = 0.8 * 0.55 * 10 = 4$ and $d = 3$); the PSM process for this example is illustrated in Fig. 2. When the GA-PSM is applied the order of indexes of features within an individual is changed, so as the order and indexes in each signal category as shown in Fig. 2.

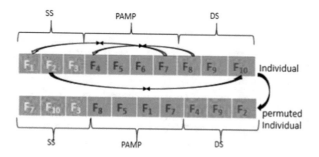

Fig. 2. The Partial Shuffle Mutation

After applying the PSM genetic operator to the individuals, the newly generated individuals and some of the best individual in the current iteration $\mathbb{P}$ jointly form the next generation of the population $\mathbb{P}'$. Note that, the performance of each individual is evaluated by the classification accuracy of the DCA.

(7) *Iteration and termination:* The GA terminates when the pre-defined maximum number of iterations is reached or the classification accuracy of the DCA exceeds the pre-defined threshold of the optimum accuracy. When the GA terminates, the optimal solution of the categorised features is taken from the fittest individual in the current population. From this, the optimal features-to-signals mapping is applied to the DCA for classification tasks.

Algorithm 1. Signal Categorisation using Partial Shuffle Mutation

---

1: **input:** an individual I and mutation probability β
2: **output:** a permuted individual I'
3: $i = 1$;
4: **while** $i \leq m$ **do**
5:     Choose p_m a random number between 0 and 1
6:     **if** $p_m < \beta$ **then**
7:         Choose j a random number between 1 and m;
8:         Permute (F_i, F_j);
9:     **end if**
10:     $i = i + 1$
11: **end while**
12: Return a permuted individual I'

---

5 Experimentation

The datasets for evaluating the performance of the proposed technique were taken from the UCI machine learning repository [6]. The properties of these datasets are detailed in Table 1.

Table 1. Benchmark Datasets

Dataset	#Samples	#Features
Mammographic Mass (MM)	961	6
Pima Indians Diabetes (PID)	768	8
Blood Transfusion Service Center (BTSC)	748	5
Wisconsin Breast Cancer (WBC)	699	9
Ionosphere (IONO)	351	34
Liver Disorders (LD)	345	7
Haberman's Survival (HS)	306	4
Statlog (Heart) (STAT)	270	13
Sonor (SN)	208	61

5.1   Experimental Setup

This work adopted the Information Gain approach to select the most important features from the datasets [14]. Then, each feature was normalised using the min-max normalisation technique. Ultimately, the GA-PSM based system optimally categorises the selected features into SS, PAMP and DS. The sampling pool was initialised with 100 DCs whilst the size of the mature pool was set to 10 DCs. The weights proposed in he original DCA [1] was used in this experiment. The migration threshold was initialised by a Gaussian distribution with the mean of 7.5 and standard deviation of 1 to ensure that the DCs survive over multiple iterations. The population size was set as 50; the mutation rate of the experimentation was set as 0.1, crossover rate as 0.95; and the maximum number of iterations was set as 300.

5.2   Measurement Metrics

The performance of the proposed GA-PSM technique was measured in terms of sensitivity, specificity and accuracy, which can be calculated by:

$$Sensitivity = \frac{TP}{TP + FN} \tag{5}$$

$$Specificity = \frac{TN}{TN + FP} \tag{6}$$

$$Accuracy = \frac{TP + TN}{TP + TN + FP + FN}, \tag{7}$$

where TP, FP, TN, and FN refer respectively to, true positive, false positive, true negative and false negative.

Moreover, since the datasets used in the experimentation have uneven class distribution, the GA-PSM techniques was further validated using precision, recall and F1-score. Note that, high accuracy shows that the model is doing better only when the datasets are symmetric. F1-score is efficient than accuracy when there is imbalanced dataset is used, which is the case in this work. The precision, recall and F1-score are defined as follows:

$$Precision = \frac{TP}{TP + FP} \tag{8}$$

$$Recall = \frac{TP}{TP + FN} \tag{9}$$

$$F1 - score = \frac{2 * Recall * Precision}{Recall + Precision}. \tag{10}$$

5.3   Results and Analysis

Comparison with FRST-DCA and PCA Approaches: Table 2 presents the testing results on sensitivity, specificity and accuracy using the GA-PSM, FRST-DCA and PCA techniques. The results for the GA-PSM technique are presented in the first column and the other two columns present the results for the FRST-DCA and PCA respectively. The best performing sensitivity, specificity and accuracy among the GA-PSM, FRST-DCA and PCA for each dataset are marked in bold.

Table 2. Comparison of GA-PSM with FRST-DCA and PCA approaches

Dataset	Sensitivity(%)			Specificity(%)			Accuracy(%)		
	GA-PSM	FRST	PCA	GA-PSM	FRST	PCA	GA-PSM	FRST	PCA
MM	**100.0**	99.10	88.25	97.18	**99.03**	93.25	98.44	**99.06**	91.28
PID	94.37	**99.40**	94.66	**100.0**	99.25	94.95	97.92	**99.34**	94.80
BTSC	98.88	**99.47**	92.59	**100.0**	98.87	85.52	**99.73**	99.33	87.67
WBC	**100.0**	99.13	99.46	**98.35**	98.33	99.33	**99.41**	98.85	99.39
IONO	**100.0**	97.77	94.44	**100.0**	95.23	95.55	**100.0**	96.86	95.15
LD	**99.50**	89.50	93.82	**100.0**	90.34	90.10	**99.71**	89.85	91.82
HS	**100.0**	88.88	81.60	89.29	94.22	70.90	91.18	**92.81**	79.40
STAT	**98.04**	90.00	82.50	**100.0**	91.33	84.55	**98.89**	90.74	83.75
SN	**100.0**	97.29	93.82	**97.98**	97.93	90.10	**99.03**	97.59	91.82

It can be noticed that, the GA-PSM technique has generated overall better performances on sensitivity, specificity and accuracy compared to FRST-DCA and PCA except for three datasets where the FRST-DCA performed the best on sensitivity for PID and BTSC datasets, accuracy for MM and PID datasets and specificity for MM dataset. Apart from the Haberman's Survival (HS) dataset where GA-PSM's performance was slightly worse, for the rest of datasets, the accuracy results ranged from 97% to 100%, sensitivity range from almost 95% to 100% while specificity range from 97% to 100%. This results demonstrate that, the GA-PSM technique is able to well map the features to their appropriate signal category.

The GA-PSM experiments results on precision, recall and F1-score for all datasets are shown in Fig. 3. Except for Haberman (HS) dataset where GA-PSM generated low precision, the rest of datasets have generated best results with their precision ranging from 98% to 100%. F1-score performance for all datasets range from 98% to 100%, which indicates that, for uneven class distribution datasets, the proposed GA-PSM is notably the best choice for signal categorisation.

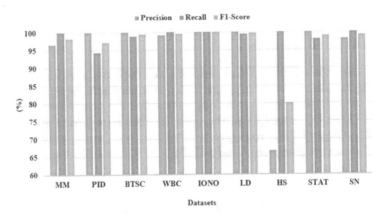

Fig. 3. Precision, Recall and F1-Score

Application of the Proposed Technique to Intrusion Detection Datasets: The GA-PSM approach was also applied to two benchmarked intrusion detection datasets which are more complex compared to the datasets presented in Table 1. Testing sets are available with these original dataset provision, hence, the training set was used for training and testing set for validation. The properties of the two datasets are briefed below.

KDD99 Dataset contains 494,021 records of training set (i.e.; 97,278 normal and 396,743 anomalous) and 311,029 records of testing set (i.e.; 60,593 normal and 250,436 anomalous) [15]; this dataset is of 41 features.

UNSW_NB15 Dataset contains 175,341 records of training set (i.e.; 56,000 normal and 119,341 anomalous) and 82,332 records of testing set (37,000 normal and 45,332 anomalous) [16]; this data set is of 49 attributes.

The GA-PSM is only compared with the PCA since FRST-DCA is not practically applicable to these datasets due to the large size of the data set. Table 3 presents the sensitivity, specificity and accuracy performances obtained on these datasets.

From the results, it is clear that the GA-PSM technique is applicable to intrusion detection datasets with larger number of samples with good classification performances. Except for sensitivity where the PCA has shown superiority on UNSW_NB15 dataset, the GA-PSM has outperformed the PCA on the

Table 3. Sensitivity, Specificity and Accuracy on intrusion detection datasets

Dataset	Sensitivity (%)		Specificity (%)		Accuracy (%)	
	GA-PSM	PCA	GA-PSM	PCA	GA-PSM	PCA
KDD99	**95.05**	88.25	**94.20**	93.25	**94.91**	91.28
UNSW_NB15	89.21	**92.59**	**97.56**	85.52	**91.24**	89.67

KDD99 dataset on all three metrics. The precision, recall and F1-score for the two datasets are shown in Fig. 4.

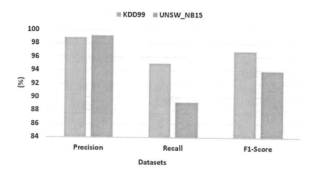

Fig. 4. Precision, Recall and F1-Score for IDS datasets

Both datasets have produced better F1-score performance than accuracy, indicating that, the GA-PSM is notably better choice for feature-to-signal mapping in intrusion detection datasets which is often of uneven class distribution.

6   Conclusions

This study proposed a GA-PSM approach based on the partial shuffle mutation genetic operator of the GA to automatically categorise input features into three signal categories of DCA. The GA-PSM was evaluated using eleven two-class datasets. The results from the experiments reveal that the GA-PSM is applicable to datasets with larger number of samples and performs better than FRST-DCA and PCA techniques in terms of sensitivity, specificity, accuracy and F-1 Score. The possible future work is to investigate the signal categorisation process of DCA using other search and optimisation techniques such as particle swam, ant colony and simulated annealing. Also, the proposed approach will be validated using more challenging datasets.

References

1. Greensmith, J., Aickelin, U., Cayzer, S.: Introducing dendritic cells as a novel immune-inspired algorithm for anomaly detection. In: International Conference on Artificial Immune Systems, pp. 153–167. Springer (2005)
2. Gu, F.: Theoretical and empirical extensions of the dendritic cell algorithm. Ph.D. thesis, University of Nottingham (2011)
3. Chelly, Z., Elouedi, Z.: Hybridization schemes of the fuzzy dendritic cell immune binary classifier based on different fuzzy clustering techniques. New Gen. Comput. **33**(1), 1–31 (2015)
4. Chelly, Z., Elouedi, Z.: A survey of the dendritic cell algorithm. Knowl. Inf. Syst. **48**(3), 505–535 (2016)
5. Abdoun, O., Tajani, C., Abouchabaka, J.: Analyzing the performance of mutation operators to solve the traveling salesman problem. Int. J. Emerg. Sci **2**(1), 61–77 (2012)
6. Dua, D., Graff, C.: UCI machine learning repository (1998)
7. Banchereau, J., Steinman, R.M.: Dendritic cells and the control of immunity. Nature **392**(6673), 245 (1998)
8. Elisa, N., Yang, L., Naik, N.: Dendritic cell algorithm with optimised parameters using genetic algorithm. In: 2018 IEEE Congress on Evolutionary Computation (CEC), pp. 1–8. IEEE (2018)
9. Nnko, N., Yang, L., Fu, X., Naik, N.: Dendritic cell algorithm enhancement using fuzzy inference system for network intrusion detection, pp. 1–8 (2019)
10. Montana, D.J., Davis, L.: Training feedforward neural networks using genetic algorithms. In: IJCAI, vol. 89, pp. 762–767 (1989)
11. Li, J., Yang, L., Yanpeng, Q., Sexton, G.: An extended takagi-sugeno-kang inference system (tsk+) with fuzzy interpolation and its rule base generation. Soft Comput. **22**(10), 3155–3170 (2018)
12. Naik, N., Diao, R., Shen, Q.: Dynamic fuzzy rule interpolation and its application to intrusion detection. IEEE Trans. Fuzzy Syst. **26**(4), 1878–1892 (2018)
13. Elisa, N., Li, J., Zuo, Z., Yang, L.: Dendritic cell algorithm with fuzzy inference system for input signal generation. In: UK Workshop on Computational Intelligence, pp. 203–214. Springer (2018)
14. Witten, I.H., Frank, E., Hall, M.A., Pal, C.J.: Data mining: practical machine learning tools and techniques. Morgan Kaufmann (2016)
15. KDD Cup 1999 Data. http://kdd.ics.uci.edu/databases/kddcup99/kddcup99.html/. Accessed 16 Dec 2018
16. Moustafa, N., Slay, J.: UNSW-NB15: a comprehensive data set for network intrusion detection systems (UNSW-NB15 network data set). In: Military Communications and Information Systems Conference (MilCIS), 2015, pp. 1–6. IEEE (2015)

Video Tampering Detection Algorithm Based on Spatial Constraints and Stable Feature

Han Pu[1,2,3(✉)], Tianqiang Huang[1,2,3(✉)], Gongde Guo[1,2(✉)],
Bin Weng[1,2,3(✉)], and Lijun You[4(✉)]

[1] Mathematics and Informatics, Fujian Normal University,
Fuzhou 350007, China
1619650550@qq.com, {fjhtq,GGD}@fjnu.edu.cn,
wb_371@163.com
[2] Fujian Provincial Engineering Research Center of Big Data Analysis
and Application, Fuzhou 350007, China
[3] Digital Fujian Big Data Security Technology Institute, Fuzhou 350007, China
[4] Fujian Institute of Meteorological Sciences, Fuzhou 350001, China
ylj16003@163.com

Abstract. Most traditional video passive forensics methods only utilize the similarity between adjacent frames. They usually suffer from high false detection rate for the videos with severe motion. To overcome this issue, a novel coarse-to-fine video tampering detection method that combines spatial constraints with stable feature is proposed. In the coarse detection phase, both the low-motion region and the high-texture region are extracted by using spatial constraint criteria. The above two regions are merged to obtain the regions with rich quantitative correlation, which are then used for extracting video optimal similarity features. The luminance gradient component of the optical flow is computed and considered as relatively stable feature. Then, the suspected tampered points are found by combining the above two features. In the fine detection phase, the precise tampering points are located. The similarity of the gradient structure based on the characteristics of the human visual system is utilized to further reduce the false detections. This method is tested on three public video data sets. The experimental results show that compared with the existing works, this method not only has lower false detection rate and higher accuracy for the videos with severe motion, but also has high robustness to regular attacks, such as additive noise, blur and filtering.

Keywords: Spatial constraints · The quantitative correlation rich regions · Stable feature · Videos with severe motion

1 Introduction

The tampering detection of digital video becomes more and more important in judicial forensics and news media. Many digital video forensics methods have been proposed in recent years [1]. However, most of the existing methods only considered the videos with smooth motion. For the videos with severe motion that could be found almost

© Springer Nature Switzerland AG 2020
Z. Ju et al. (Eds.): UKCI 2019, AISC 1043, pp. 541–553, 2020.
https://doi.org/10.1007/978-3-030-29933-0_45

everywhere, these methods usually resulted in high false detection rate [2]. Therefore, the forensics of videos with severe motion is an important open research issue.

Many methods have been proposed for the frame tampering detection of videos. Zhao et al. [3] performed coarse detection for inter-frame tampering by comparing the similarity of HSV (hue, saturation, value) color histograms between adjacent frames of video, and run fine detection to further confirm the tampering points by combining SURF feature extraction with FLANN. Sowmya KN et al. [4] generated a 128-bit information number for any given video as the only fingerprint by STTFR (spatial temporal triad feature relationship) technique. It is based on the assumption that the fingerprint will be destroyed after the tampering in the time domain. Zhang et al [5] utilized the MSSIM (Mean Structure Similarity) quotient between adjacent frames of tampering video. They used Chebyshev inequality to find tampered points. Liu et al. [6] used Zernike's opposite chromaticity space principle to extract frame features for coarse detection, then combined with coarseness roughness for fine detection and positioning. Liu et al. [7] proposed a frame deletion detection scheme specially designed for H.264 encoded video, which proves that the average residual of the P frame sequence would show periodicity in the time domain when video frame deletion tampered. Although many methods have been developed, the following challenges remain:

(1) High computational complexity. Most of the existing methods detect tampering between video frames based on the principle of maximal correlation. While the correlation values are based on multiple similarity calculations or unstable image features, it requires a large amount of computation power.
(2) Application limitations. Many existing methods can only work on videos with specific encoding formats, and cannot detect any length of tampered frames. The application of these detection algorithms is limited.
(3) Lack of robustness. Many detection algorithms are vulnerable to regular attacks, such as Gaussian blur, median filtering, and Gaussian white noise, etc.

To tackle the above issues, this paper proposes a coarse-to-fine detection algorithm based on spatial constraints and stable feature. The proposed method finds a balance among detection efficiency, robustness, and applicability. The main contribution is a method with the following advantages:

(1) Less computational complexity. Based on the fact that the severe motion region of videos is the main factor affecting the accuracy of detection for severe motion videos, the spatial constraint of videos is firstly carried out by the adaptive threshold optimal partition method, and then the luminance component of the optical flow that is relatively consistent in the videos with severe motion is extracted. Since the feature extracted in the quantitative correlation rich regions of the video changes relatively consistent, the algorithm can be applied to videos with different levels of motion. Also, the high complexity of video optical flow calculation is solved using the detected video spatial constraint.
(2) Wider application. It is not limited by the clarity, type and tampering length of videos. Even the tampering of only a few frames could be detected.

(3) More robust. Thanks to the fusion of spatial constraint and stable features, a series of experiments prove that this method is robust to the attacks of filtering, Gaussian noise and Gaussian blur.

This paper is organized as follows: Sect. 2 briefly describes the video space constraints, and the robustly improved optical flow; Sect. 3 introduces the coarse-to-fine detection algorithm in this paper; the experimental results and analysis are presented in Sect. 4; Sect. 5 concludes the paper.

2 Related Work

2.1 Video Spatial Constraints

Most existing video tampering detection algorithms are based on the principle of maximal correlation between adjacent frames. However, tampering is not the only factor accounting for the outliers, other factors may also produce spikes in the detecting phase and thus cause the false detections. For the severe motion video, due to the dramatic change of correlation between adjacent frames, the existing detection algorithms usually resulted in a large number of false detections. Fortunately, the video could be spatially constrained to obtain a spatial region that is favorable for tampering detection. This would improve the accuracy of the detection. The idea is to constrain the video frame in the spatial domain by combining relevant information on the video space-time domain. By defining certain constraints, a part of the area that compromises the tampering detection is filtered out, and the spatial region that is favorable for tamper detection is reserved. The proposed algorithm mainly used adaptive threshold optimal region division method to spatially constrain the detected video [8]. The main process is shown in Fig. 1.

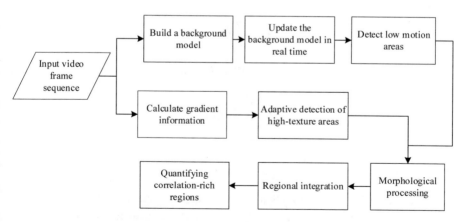

Fig. 1. The flow chart of video spatial constraints

Detecting Low-Motion Region. First, the difference image is obtained by calculating the absolute difference between the current frame P_i and the background model B_i [9], and then the two-dimensional mask corresponding to the difference image is obtained by using the adaptive threshold Th_m to determine the low motion region. As shown in Eq. (1), where Th_m is set according to reference [10].

$$LM_i = \begin{cases} 0 & |P_i - B_i| < Th_m \\ 255 & |P_i - B_i| > Th_m \end{cases} \qquad (1)$$

Detecting High-Textured Region. The gradient structure information of the video frame is calculated, and the gradient image is filtered by the adaptive threshold Th_t to obtain a corresponding binary mask, which determines the high-texture areas of the video [11], as shown in Eqs. (2) and (3), respectively.

$$\|\Delta P_i\| = \sqrt{P_{ix}^2 + P_{iy}^2} \qquad (2)$$

$$TM_i = \begin{cases} 255 & \|\Delta P_i\| > Th_t \\ 0 & \|\Delta P_i\| \leq Th_t \end{cases} \qquad (3)$$

Where P_{ix} is the partial derivative of the frame P_i in the x direction, P_{iy} is the partial derivative of the frame P_i in the y direction, and $\|\Delta P_i\|$ is the gradient structure information of the frame P_i.

Quantifying Correlation-Rich Regions. The low-motion and high-texture regions obtained by the spatial constraints are merged, and the intersection of the two regions is taken to obtain the region MR_i of the video. Perform a morphological operation on the binary mask MR_i, finally merge the MR_i and the frame F_i to obtain the video spatial constrain areas $QCRI_i$, which is also the correlation rich regions. As shown in Eqs. (4), (5) and (6), respectively.

$$MR_i = \prod_{i=1}^{n} LM_i \cap TM_i \qquad (4)$$

$$MR_i = (MR_i \bullet SE) \circ SE \qquad (5)$$

$$QCRI_i = \prod_{i=1}^{n} F_i \cap MR_i \qquad (6)$$

2.2   Robust Optical Flow

Optical Flow (OF) is the distribution of apparent movement velocities of brightness pattern in videos, which can give important information about the image spatial arrangement and change rate of objects [12]. Among them, the Lucas-Kanade Optical Flow (LK), proposed by Lucas and Kanade, is a local least square calculation to compute OF sparsely for each blob [13]. The high computational complexity of traditional OF method, and it is vulnerable to video with large displacement of pixel motion. To tackle the above issues, the specific steps of the improved LK algorithm are as Fig. 2. Firstly, it is necessary to spatially constrain the video frames to quantify the correlation rich regions $QCRI_i$. Then utilize the Shi-Tomasi corner detection method to detect the corner points in the regions. Secondly, the OF values at the corner points are calculated by the LKT Tracker. Finally, the calculated OF values are subjected to pre-backward filtering processing to remove the unmatched optical flow points [14].

The reason why the improved LK algorithm has better performance is based on the following three points. Firstly, calculating the sparse optical flow on the low-motion regions, it can satisfy the small motion assumption of OF algorithm. So the improved LK algorithm can also be applied to video with large motion variation. Secondly, since the gradient of the optical flow in the low-texture regions is relatively small, the optical flow in the low-texture regions can be neglected. So the improved LK algorithm only needs to calculate on the high-texture regions, which reduce computational complexity greatly. Thirdly, LKT Tracker and pre-backward filtering can remove the unreliable matching results. The above points can improve the robustness and speed of the proposed algorithm. In summary, the improved LK algorithm is effective and efficient.

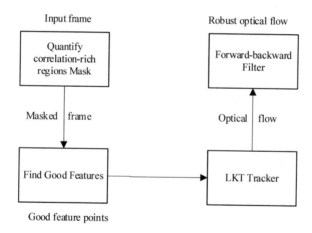

Fig. 2. The extraction process of the robust optical flow

3   The Method

We propose a novel approach to detect frame deletion forgeries in consideration of the detection efficiency, robustness, and applicability. It is a coarse-to-fine detection method based on spatial constraints and stable feature, which can greatly improve the efficiency, robustness and accuracy of the detection. First, in order to reduce false detection, the video could be spatially constrained to obtain a quantitative correlation rich region $QCRI_i$, which is advantageous for the detection method based on the continuity and regularity in video and the similarity between adjacent frames. Then calculating the optical flow luminance information to obtain a spatial constraint correlation value, coarse detection analyzes the consistency of the correlation values to find suspected tampering points. Finally, due to it cannot guarantee that all of the tampering points extracted by coarse detection are juggled points, the precise tampering points are located. Including extract k-nearest frames around the outlier and calculate the gradient structure similarity feature GSSIM based on validation check. The whole detection process is shown in Fig. 3.

3.1 Coarse Detection

Aiming at the problem that most of the existing methods usually resulted in high false detection rate for the video with severe motion, this algorithm extract the feature which perform relatively stable, and the stability is reflected that the feature does not change rapidly. Through research, the luminance gradient direction component of the optical flow E_b can be consistent in video with severe motion. Based on the above characteristic, the algorithm intends to extract values of E_b between the adjacent spatially constrained video frames to detect video tampering. Combining the improved LK

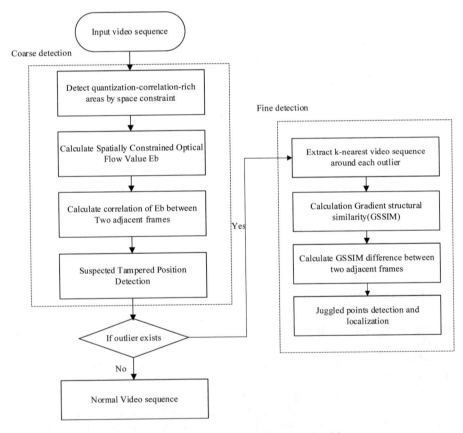

Fig. 3. The flow chart of the detection algorithm

optical flow method described in Sect. 2.2, the calculation of E_b can be expressed as Eq. (7):

$$E_b = - \frac{E_t}{\sqrt{(E_{x^2} + E_{y^2})}} \tag{7}$$

Where E_x, E_y, E_t are the partial derivatives of the optical flow in the x, y direction and time dimension respectively, and the values can be calculated by the following formula:

$$E_x = \frac{\partial E}{\partial x}, \; E_y = \frac{\partial E}{\partial y}, \; E_t = \frac{\partial E}{\partial t} \tag{8}$$

Assuming that N corner points are detected for a spatial constraint region, the following formula (9) portrays the correlation of E_b between adjacent frames.

$$cor(i, i+1) = \frac{\sum_{n=1}^{N} (E_{bi}(n) - \overline{E_{bi}})(E_{b(i+1)}(n) - \overline{E_{b(i+1)}})}{\sqrt{\sum_{n=1}^{N} (E_{bi}(n) - \overline{E_{bi}})^2 \cdot \sum_{n=1}^{N} (E_{b(i+1)}(n) - \overline{E_{b(i+1)}})^2}} \tag{9}$$

For the frame sequence of length n, the correlation of the E_b between adjacent frames is calculated according to Eq. (9), and the spatial constraint correlation value $cor_{(i,i+1)}$ between $n - 1$ adjacent frames can be obtained, a sequence of correlation coefficients cor is generated as the Eq. (10).

$$cor = \{cor_{(1,2)}, cor_{(2,3)}, \ldots cor_{(i,i+1)}, \ldots cor_{(n-1,n)}\} \tag{10}$$

Where $cor_{(i,i+1)}$ represents the correlation of E_b between the frame i and the frame $i + 1$.

In order to further eliminate the impact of video motion changes on the detection results [15], we replace $cor_{(i,i+1)}$ with the quotient of $cor_{(i,i+1)}$, and the definition of $cor_{(i,i+1)}$ quotient is as shown in Eq. (11). Finally, the abnormal point is detected by Chebyshev inequality [16].

$$\Delta cor_{(i,i+1)} = \begin{cases} \frac{cor_{(i,i+1)}}{cor_{(i+1,i+2)}} & cor_{(i,i+1)} \geq cor_{(i+1,i+2)} \\ \frac{cor_{(i+1,i+2)}}{cor_{(i,i+1)}} & cor_{(i,i+1)} < cor_{(i+1,i+2)} \end{cases} \tag{11}$$

3.2 Fine Detection

Coarse detection helps to locate suspected outliers which based on the E_b correlation value, it just calculates the optical flow values on the corner points. This process will reduce multiple calculations, but may lead to false detections, so it needs fine detection based on more detailed features to identify whether the suspected outliers are caused by frame deletion forgery.

In this section, fine detection is proposed, which intends to extract all k-neighbor frames around each outlier and calculate the gradient structure similarity feature MGSSIM of those frames. The feature MGSSIM mainly combines the local information extracted from the original image and the corresponding gradient image [17, 18]. By comparing the brightness, contrast, and structural information of adjacent frames to fully characterize the similarity of k-nearest frames, the tampering point can be accurately located. The steps of fine detection are as follows:

Assume that point v is the outlier, then extract k-nearest frame sequence, the MGSSIM of each frame is calculated, and a sequence of correlation coefficients g is formed, where g_i represents the MGSSIM difference between the frame i and the frame $i + 1$. The calculation of g_i is as shown in the formula (12) and (13), where $i \in [v - k, v + k)$.

$$g_i = |MGSSIM_{i+1} - MGSSIM_i| \tag{12}$$

$$g = \{g_{v-k}, g_{v-k+1}, \cdots g_v, g_{v+1}, \cdots g_{v+k-1}\} \tag{13}$$

For the correlation coefficient sequence g obtained by fine detection, the larger the value, the larger the difference between the frame i and frame $i + 1$, and the more likely it is the tampering point. If the value satisfies the formula (14), (15). The frame can be positioned as a tampering point.

$$g_v = \max(g_{v-k}, g_{v-k+1}, \cdots, g_{v+k-1}) \tag{14}$$

$$g_v \geq \beta((\sum_{i=1}^{2k} g_i - \max(g_{v-k}, g_{v-k+1}, \cdots, g_{v+k-1}) \\ - (\min(g_{v-k}, g_{v-k+1}, \cdots, g_{v+k-1}))/(2k - 4) \tag{15}$$

4   Experimental Results and Analysis

4.1   Experimental Data and Evaluation Standards

In order to evaluate the detection effect of the algorithm, we test experiments on three public data sets, namely SULFA (Surrey University Library for Forensic Analysis), CDNET (A video database for testing change detection algorithms) and South China Institute of Technology video tampering detection database VFDD (Video Forgery Detection Database) Version 1.0. A total of about 400 videos.

In order to evaluate the performance of the algorithm, the precision rate R and recall rate P are used to analyze the experimental results. The calculation formula is as follows:

$$R = \frac{N_c}{N_c + N_f} \tag{16}$$

$$P = \frac{N_c}{N_c + N_m} \tag{17}$$

Among them, N_c is the number of detected correct points, N_f is the number of detected false points, and N_m is the number of tampering points for missed detection.

4.2 Evaluate the Effectiveness and Robustness of the Algorithm

We have done a lot of experiments to evaluate our proposed methods. The experimental data and evaluation indicators have been introduced earlier, and our method is compared with the existing two classical algorithms in terms of precision, robustness, and applicability.

Evaluate the Effectiveness and Robustness of Coarse Detection. In order to evaluate the effectiveness and robustness of the algorithm, some common attacks were simulated as secondary forgery after deletion forgery, including additive Gaussian blur, Gaussian white noise and different filtering. For the frame deletion tampering of the video, the detection effect can be judged by observing whether the tampering position has a sharp peak in the detection result and whether the abnormal point is positioned.

Through the experiments on the videos with different types and tampering numbers, the experimental results are shown in the Fig. 4. By observing the experimental results, it can be found that the coarse detection algorithm can accurately locate the tampering points for videos with common attacks, the accuracy and robustness including anti-noise, anti-filtering, anti-blurred and anti-motion of the coarse detection are evaluated.

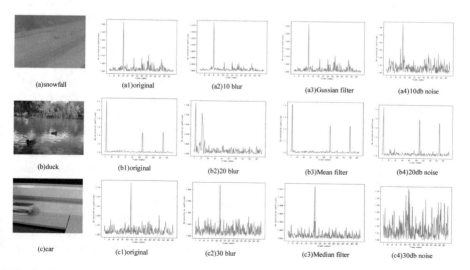

Fig. 4. Evaluate the effectiveness and robustness of coarse detection (The X axis is the number of frames, and the Y axis is the correlation coefficient of E_b)

Verify the Effectiveness and Robustness of Fine Detection. Through the analysis of MGSSIM difference among 4 adjacent frames of the tampering points, the experimental results are shown in the Fig. 5. By observing the experimental results, it can be found that the fine detection algorithm can accurately locate the tampering points for videos with regular attacks, so the fine detection has high accuracy and robustness.

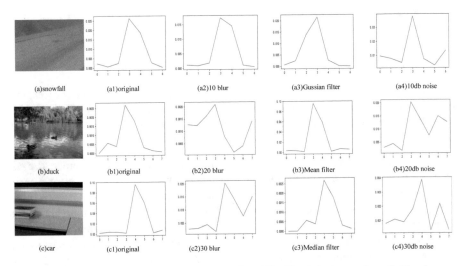

(a)snowfall   (a1)original   (a2)10 blur   (a3)Gussian filter   (a4)10db noise

(b)duck   (b1)original   (b2)20 blur   (b3)Mean filter   (b4)20db noise

(c)car   (c1)original   (c2)30 blur   (c3)Median filter   (c4)30db noise

Fig. 5. The degree of MGSSIM difference to outliers (The X axis is the number of video frames, the Y axis is the MGSSIM difference.)

Verify the Validity of Spatial Constraints. In order to verify whether the spatial constraints can improve the accuracy of detection, the coarse detection results of original video(OriFrame) and spatially constrained video(Contraint-Area) under the same conditions are as shown in Fig. 6. The cyan line depicts the *Eb* correlation between adjacent frames of the Contraint-Area. and the lake blue line depicts the correlation of *Eb* between the OriFrame. By observing the Fig. 6, it is obvious that the fluctuation between adjacent frames of the OriFrame is relatively large, with multiple sharp peaks appearing, it will cause a large number of false detections. However, the correlation fluctuation of Contraint-Area is relatively gentle and consistent, which avoids false detection and improves the accuracy of detection.

Performance Analysis of the Algorithm. The experimental results are shown in Tables 1 and 2. Table 1 shows the tampering detection results on the video datasets processed by different regular attacks. It can be seen that the algorithm has anti-noise, anti-filtering, anti-blurred characteristics and high robustness. Table 2 is the performance comparison between the proposed algorithm and the existing classical algorithms. It shows that the performance of the proposed method is superior to the comparative literature [6, 7], where the detection performance of the night video in the

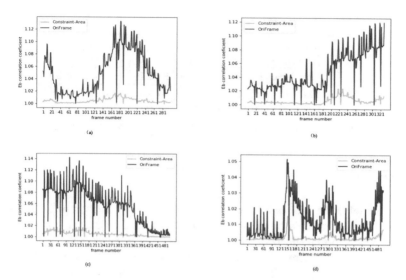

Fig. 6. Verify the validity of spatial constraints. (The X axis is the number of frames, and the Y axis is the correlation coefficient of *Eb*)

CDNET library is relatively unsatisfactory, but it is also better than the comparison algorithm. Maybe because the light of nighttime video is dark, and the extracted frame features represent the frame image content poorly. Therefore, the detection performance is slightly reduced in the night video.

Table 1. Test results of the robustness of the proposed method

Attacks		Precision rate P	Recall Rate R
Under the condition of no attack		99.65%	99.65%
Gaussian blur	10	94.31%	92.10%
	20	95.82%	92.35%
	30	96.78%	95.12%
Gaussian noise	10	96.36%	91.42%
	20	97.35%	92.71%
	30	98.72%	93.58%
5 * 5 filter	Mean filter	97.21%	94.52%
	Median filter	96.32%	94.30%
	Gaussian filter	97.00%	94.21%

Table 2. Comparison results with the existing classical paper

Dataset	Proposed algorithm		Literature [6] algorithm		Literature [7] algorithm	
	Precision	Recall	Precision	Recall	Precision	Recall
SULFA	99.95%	99.45%	95.43%	94.64%	94.43%	93.64%
CDNET	94.62%	92.11%	90.41%	89.62%	92.41%	91.62%
HUANAN	99.71%	96.25%	94.12%	93.11%	96.12%	93.21%

5 Conclusions

In order to solve the problem of high computational complexity, application limitations, and low robustness of existing video tampering detection algorithms. In this paper, a coarse-to-fine video tampering detection algorithm combining spatial constraints with stable feature is proposed. The effectiveness and robustness of the algorithm are evaluated by a series of experiments on different types of video databases. Experiments show that the proposed algorithm performs well on videos with severe motion, and the method has the characteristics of anti-noise, anti-filtering, anti-blurred and anti-motion. At the same time, it has high accuracy rate, low error detection rate, great practicability and strong robustness. The disadvantage is that the tampering detection performance of night video is not very friendly, which is also the work that needs to be studied in the future.

Acknowledgments. This work was supported by grants from the National Key Program for Developing Basic Science (Grant Nos. 2018YFC1505805), and SX201803.

References

1. Chen, W., Yang, G., Chen, R.: Digital video passive forensics for its authenticity and source. J. Commun. **32**(6), 177–183 (2011)
2. Singh, R.D., Aggarwal, N.: Video content authentication techniques: a comprehensive survey. Multimedia Syst. **24**, 211–240 (2018)
3. Zhao, D.N., Wang, R.K., Lu, Z.M.: Inter-frame passive-blind forgery detection for video shot based on similarity analysis. Multimedia Tools Appl. **1**, 1–20 (2018)
4. Sowmya, K.N., Chennamma, H.R., Lalitha, R.: Video authentication using spatial temporal relationship for tampering detection. J. Inf. Secur. Appl. **41**, 159–169 (2018)
5. Zhen-Zhen, Z., Jian-Jun, H., Zhao-Hong, L.I., et al.: Video-frame insertion and deletion detection based on consistency of quotients of MSSIM. J. Beijing Univ. Posts Telecommun. **8**, 84–88 (2015)
6. Liu, Y., Huang, T.: Exposing video inter-frame forgery by Zernike opponent chromaticity moments and coarseness analysis. Multimedia Syst. **23**(2), 223–238 (2017)
7. Liu, H., Li, S., Bian, S.: Detecting frame deletion in H.264 video. In: Proceedings of 10th International Conference, ISPEC, Fuzhou, China, pp. 262–270 (2017)

8. Aghamaleki, J.A., Behrad, A.: Inter-frame video forgery detection and localization using intrinsic effects of double compression on quantization errors of video coding. Signal Process. Image Commun. **47**, 289–302 (2016)

9. Li-xia, X.-Q.U.E., Yan-li, L.U.O., Zuo-cheng, W.A.N.G.: Detection algorithm of adaptive moving objects based on frame difference method. Appl. Res. Comput. **28**(4), 1551–1552 (2011)

10. Jian-jun, Z.U.O., You-fu, W.U.: A real-time adaptive threshold segmentation method based on histogram. J. Bijie Univ. **32**(4), 53–56 (2014)

11. Wang, X., Bi, X., Ma, J., et al.: Nonlinear filtering algorithm using probability statistic and main texture direction analysis based on radon transforms. J. Image Graph. **13**(5), 858–864 (2018)

12. Horn, B.K.P., Schunck, B.G.: Determining optical flow. Artif. Intel. **17**(1–3), 185–203 (1981)

13. Lucas, B.D., Kanade, T.: An iterative image registration technique with an application to stereo vision. In: International Joint Conference on Artificial Intelligence (1981)

14. Tan, H., Zhai, Y., Liu, Y., et al.: Fast anomaly detection in traffic surveillance video based on robust sparse optical flow. In: IEEE International Conference on Acoustics, Speech and Signal Processing (ICASSP 2016). IEEE (2016)

15. Huang, T.-Q., Chen, Z.-W., Su, L.-C., et al.: Digital video forgeries detection based on content continuity. J. Nanjing Univ. (Nat. Sci.) **47**(5), 493–503 (2011)

16. He, Z.: The data flow anomaly detection analysis based on LipChebyshev method. Computer System Application **18**(10), 61–64 (2009)

17. Jianjun, S.U.N., Yan, Z.H.A.O., Shigang, W.A.N.G.: Improvement of SIFT feature matching algorithm based on image gradient information enhancement. J. Jilin Univ. (Sci. Ed.) **56**(1), 82–88 (2018)

18. Nercessian, S., Agaian, S.S., Panetta, K.A.: An image similarity measure using enhanced human visual system characteristics. In: SPIE Defense, Security, and Sensing, pp. 806310–806310-9 (2011)

Author Index

© Springer Nature Switzerland AG 2020
Z. Ju et al. (Eds.): UKCI 2019, AISC 1043, pp. 555–556, 2020.
https://doi.org/10.1007/978-3-030-29933-0

Printed in the United States
By Bookmasters